TRANSONIC FLOW PROBLEMS IN TURBOMACHINERY

A PROJECT SQUID WORKSHOP

TRANSONIC FLOW PROBLEMS IN TURBOMACHINERY

EDITED BY

T. C. ADAMSON, JR.

Department of Aerospace Engineering
The University of Michigan

M. F. PLATZER

Department of Aeronautics
Naval Postgraduate School
Monterey, California

HEMISPHERE PUBLISHING CORPORATION

Washington London

Library of Congress Cataloging in Publication Data

Project Squid Workshop on Transonic Flows in Turbomachines, Naval Postgraduate School, 1976.
Transonic flow problems in turbomachinery.

At head of title: A Project Squid workshop.
"Sponsored by the Office of Naval Research (Project SQUID), the Air Force Office of Scientific Research, and the Naval Air Systems Command."
"Contract N0014-75-C, NR-098-38."
Includes index.
1. Turbomachines—Aerodynamics—Congresses. 2. Aerodynamics, Transonic--Congresses. I. Adamson, Thomas Charles. II. Platzer, M. F. III. Project Squid. IV. United States. Air Force. Office of Scientific Research. V. United States. Naval Air Systems Command. VI. Title.
TJ267.P66 1976 621.4'06 77-22185
ISBN 0-89116-069-8

Proceedings of the Project SQUID WORKSHOP ON TRANSONIC FLOWS IN TURBOMACHINES, held at the Naval Postgraduate School, Monterey, California, February 11-12, 1976, and sponsored by the Office of Naval Research (Project SQUID), the Naval Air Systems Command, and the Air Force Office of Scientific Research.

1 2 3 4 5 6 7 8 9 0 D O D O 7 8 3 2 1 0 9 8 7

This work relates to the Department of the Navy Contract No. N00014-75-C-1143 issued by the Office of Naval Research under contract authority No. NR 098-038. However, the content does not necessarily reflect the position or policy of the Department of the Navy or the Government, and no official endorsement should be inferred. The United States Government has a royalty-free, nonexclusive and irrevocable license throughout the world for Government purposes to publish, translate, reproduce, deliver, perform, dispose of, and to authorize others to do so, all or any portion of this work.

CONTENTS

PREFACE

The problems associated with transonic flows in turbomachinery have been discussed with varying intensity for many years. Recently, interest in this subject has greatly increased because of the growing need to improve engine performance characteristics while minimizing development risks. Engine efficiency, noise generation, and aeroelastic stability are additional aspects that have acquired new and obvious importance in recent years because of the worsening energy situation, the increasingly strict noise regulations, and the occurrence of new flutter problems in high-speed fans and compressors. Progress on solution of these problems clearly depends on a detailed understanding and accurate prediction of the complex transonic flows that of necessity are encountered in modern high-performance engines. Therefore, it appeared that a workshop specializing in the problems found in internal transonic flows would be helpful not only in informing the various workers in this field about the latest work being done, but also in providing a number of opinions on the subject of what important work remains to be done.

This workshop complemented the IUTAM Symposium Transsonicum, which was run so ably by Professor K. Oswatitsch in 1975 in Göttingen. The Symposium Transsonicum provided a grand, large-scale review of work completed in the general field of transonic flows. Here, in a more modest effort, only those problems peculiar to turbomachines were considered in a workshop atmosphere designed to encourage everyone to discuss ideas and work in progress. This desire also led to the format used for the meeting, in which lead-off speakers set the stage for each session, followed by several speakers who discussed work in progress.

The purpose of the workshop, then, was first to assess current advances in transonic flow analyses and experiments, and their applicability in predicting flow in transonic turbomachines, and second, to ascertain what work should be done in the future to improve these techniques. To this end, we tried to form a mix of external and internal aerodynamicists, analytical and computational experts, and experimentalists and theoreticians. The attendance list shows attendees from as far as Eurpoe and from as near as the Naval Postgraduate School.

After much discussion, it was decided, finally, not to include a special session on unsteady flows. This is because any serious discussion of this subject area should include the interaction of the flow field and the structure, i.e., aeroelastic effects, and it appeared that this kind of discussion would be too broad for just one session. Indeed, this subject could be and often is the subject of a whole workshop. Therefore, we decided to limit unsteady flow considerations to the assessment of the effects of unsteady flow on performance.

The first two sessions were concerned with analysis. In the first, the formulation of transonic flow problems to various orders of approximation was considered, including governing equations and boundary conditions, the validity of the steady flow assumption, the validity of simplified flow models including two-dimensional formulations and linearized governing equations, and the use of asymptotic methods. In the second, computational methods of solution were discussed. Here, the problems considered included comparison of various methods of computation, the special problems that exist in turbomachine calculations as opposed to single airfoil studies, increasing the efficiency of given methods of computation, and the inclusion of shock fitting methods.

The third session on viscous effects in transonic flows included studies of shock wave–boundary layer interactions, the general inclusion of boundary layer effects, and separation. The solutions presented employed both numerical and analytical methods.

The fourth session, on experiment, covered reviews of data and experimental techniques presently available and under active development at the present time. Some promising new experimental methods were discussed, with emphasis being given to the use of the laser as a diagnostic tool.

The fifth and final session consisted of a general discussion in which each of the session chairmen presented a brief review of his session.

The format of these proceedings follows, with a few minor exceptions, that of the meeting. At the sponsors' request, a brief statement of conclusions was formulated.

The workshop was held under the sponsorship of the SQUID Project (Office of Naval Research), Naval Air Systems Command, and the Air Force Office of Scientific Research. It took place at the Naval Postgraduate School, Monterey, California, February 11–12, 1976. The workshop was recorded, the discussions were transcribed and edited, and the discussions and the papers submitted by the speakers now appear in the form of these proceedings.

Special mention should be made of the support and encouragement given by Mr. J. Patton, Power Branch, ONR, Dr. H. J. Mueller,

NAV-AIR, and Lt. Col. R. C. Smith, AFOSR. Their interest in this project is evident in the welcoming remarks made by each at the meeting and reproduced here (pp. xvii–xix); without their continuing support, this workshop would not have been possible. Special thanks are also due to Dr. S. N. B. Murthy, SQUID Project director, were contributed his support and ideas concerning the organization of both the workshop and these proceedings.

Admiral I. Linder and the staff of the Department of Aeronautics of the Naval Postgraduate School are to be congratulated for having handled 67 visitors with gracious efficiency; Mr. John C. King and his staff performed very well indeed the difficult task of recording the workshop. The transcription of the tapes and the retyping of all the papers was done at the Department of Aerospace Engineering, the University of Michigan. Our gratitude goes to Ms. Sharon Bauerle, who performed the Herculean task of transcribing the tapes and retyping the papers, and to Ms. Margaret Fillion, who aided both in the meticulous work of making corrections and in the original organizational work for the workshop. Finally, we gratefully acknowledge the work done by Professor M. Sichel and Professor A. F. Messiter, both of the University of Michigan, in editing the discussions and proofreading the manuscripts.

T. C. Adamson, Jr., and M. F. Platzer
Editors and Workshop Cochairmen

THIRTY YEARS OF RESEARCH

The primary objective of the workshop on Transonic Flows in Turbomachines was to establish the current status of design procedures in transonic turbomachinery, to determine the extent of agreement in explaining some of the phenomenology associated with transonic blading and nozzles, and to determine the directions in which further research may be expected to yield important and useful results. A secondary goal was to bring theoreticians and practical design groups together from industry, universities and government, in an effort to use their combined talents. There is a close relation of transonic flow problems in turbomachines with the disciplines covered in other SQUID workshops, such as measurement techniques and aerodynamic-structural integration (blade flutter), etc.

There has been a resurgence of interest in recent years in basic problems relating to air-breathing engines due to increasing demands for higher performance, for smaller and lighter-weight power plants, and for operation over wider ranges of operating conditions. This interest is exemplified in the current need to develop deeper fundamental understanding of the physical phenomena involved in all aspects of engine design and development.

To satisfy this need, the Power Program of the Office of Naval Research, through Project SQUID, initiated a series of workshops to consider selected topics from the standpoint of (1) a critical evaluation of current efforts, (2) determining the extent of agreement in explaining various phenomena associated with the subject, and (3) discussion of possible new approaches to solution of problem areas. These workshops have been held:

- Research in Gas Dynamics of Jet Engines, ONR/Chicago, December 4–5, 1969, Project SQUID Report
- Fluid Dynamics of Unsteady 3-D Separated Flows, Georgia Tech., June 10–11, 1971, AD736248
- Laser Doppler Velocimetry Flow Measurements, Purdue University, March 9–10, 1972, AD753243
- Aeroelasticity in Turbomachines, Detroit Diesel Allison, June 1–2, 1972, AD749680

- Laser Raman Diagnostics, G.E. Research & Development Center, May 10-11, 1973, GE-2-PU, Project SQUID Report
- Laser Doppler Velocimetry Flow Measurements, II, Purdue University, March 1974, PU-R1-75, Project SQUID Report
- Turbulent Mixing: Non-Reactive and Reactive Flows, Purdue University, May 20-21, 1974, Plenum Press, New York
- Unsteady Flow in Jet Engines, United Aircraft Research Laboratory, (UARL) now (UTRC), July 11-12, 1974, UARL-3-PU, Project SQUID Report
- Measurement Techniques in Combustors, Purdue University, May 22-23, 1975; R. Goulard, ed. 1976. *Combustion measurements: Modern techniques and instrumentation.* Washington, D.C.: Hemisphere Publishing Corp.
- Transonic Flows in Turbomachines, Naval Postgraduate School, February 11-13, 1976 (*this volume*)
- Turbulence in Internal Flows, Airlie House, June 1976

At the time of this workshop, the Office of Naval Research was actively planning its thirtieth anniversary for 1976. The year 1976 also marks the thirtieth anniversary of Project SQUID. This volume is part of a set of scientific publications released by ONR in recognition of, and on the occasion of, its thirtieth anniversary.

James R. Patton
Power Program
Office of Naval Research
U.S. Department of the Navy

Anniversary Theme:

Exploring new horizons to protect our heritage

WELCOMING REMARKS

James R. Patton
Power Program, Office of Naval Research

We appreciate the very cordial welcome of this workshop group to the Naval Postgraduate School by Admiral Linder. The Postgraduate School is certainly an appropriate place to hold a workshop on the subject of "Transonic Flows in Turbomachinery"; not only is there pertinent research by capable investigators under way here, but in addition, the fine facilities and beautiful environment should be conducive to accomplishing our purpose.

On behalf of the Office of Naval Research, it's a great pleasure to welcome you here today and especially to see the large number of people who have turned out for this workshop. We are particularly happy to join with the U.S. Air Force Office of Scientific Research and the Naval Air Systems Command in sponsoring the workshop. The year 1976 happens to be the thirtieth anniversary of the Office of Naval Research. In recognition of this anniversary, under Project SQUID, we have planned this workshop plus another to be held in June on the subject of "Turbulence in Internal Flows." These two workshops are part of a series of workshops we have held starting in 1969, all on basic subjects relating to air breathing engines for aircraft and missile applications.

It is not necessary to explain to this group the importance of the subject of this workshop. Suffice it to say that, from our standpoint in the Department of Defense, operational requirements have been dictating the need for greater performance in the applications of air-breathing and turbine-type power plants. This usually requires using higher temperatures and extracting more work per unit of blade surface in the engine. One result is to use the kind of flows that will be talked about at this workshop. From the research standpoint, we consider that developing a detailed, accurate understanding of the internal flows and interactions within modern jet engines is a vital and integral part of an air-breathing propulsion system development program just as much as is the development of the component parts of the engine. To this end, through workshops such as this, we seek to pinpoint specific areas of concern, learn about the current state of knowledge, take an unbiased look at the problem areas, and, we hope, develop realistic approaches to understanding and solving the problems. Our current research program, and I'm sure this holds for the other services as well, has been strongly influenced by holding workshops such as this.

On behalf of the sponsors I wish to thank the two organizers, Dr. Tom Adamson, of the University of Michigan, and Dr. Max Platzer, of the Naval Postgraduate School, for arranging and planning the workshop. We look forward to the deliberations of this outstanding group of participants and wish to thank you all for contributing your knowledge and time, essential to the success of the workshop.

Herbert J. Mueller
Research Administrator, Naval Air Systems Command

Gentlemen, the Naval Air Systems Command has the responsibility for Naval Aviation Technology and Systems. Consequently, the command has considerable interest in sponsoring symposia in aerodynamics—not only in order to provide a forum for technical interchange but primarily as a basis for developing the command's research programs. There is no doubt that the navy needs a strong research program in this field! In the future we will see the development of a new, in many regards unconventional, navy. Yet, in many instances, basic knowledge and fundamental technology is not yet available to the extent that would guarantee the timely and cost-efficient design of those advanced propulsion systems and platforms the navy will want to have. Such workshops are therefore a significant first step in the homework we have to do in order to ultimately achieve our ambitious goals.

As a cosponsor of this workshop it is my pleasant duty to thank Dr. Platzer and Dr. Adamson for assuming the chairmanship. I would also like to thank Admiral Linder on behalf of the Naval Air Systems Command for hosting this event. And, last but not least, I welcome all of you who are here to share with us their expertise and experience. I sincerely wish you a very successful, rewarding and enjoyable meeting.

Robert C. Smith, Lt. Col., USAF
Program Manager, Directorate of Aerospace Sciences,
Air Force Office of Scientific Research

I think all of you are familiar with what AFOSR's job is: we are the basic research managers for the Air Force. We are spending approximately $400,000 this year on basic research programs concerned with the fluid mechanics of turbomachinery. I am the program manager for this area. We hope, depending on what happens to the FY77–78 budgets, there will be additional funds for this area.

Rather than belabor this point, I would like to use my time to ask you to do AFOSR a favor. As you probably are all aware, Tony Ferri died in December unexpectedly. The management of AFOSR has decided to set up an R&D award, which this year will honor Tony Ferri. It will include a lecture series similar to the von Karman lecture series established by the AIAA. The favor I need from you is this: do you have any suggestion as to who should be considered for this award and/or a lecture topic that might be appropriate. If you have any suggestions you may either give them to me during the meeting or else forward them to my office in care of Milt Rogers. Milt is handling the details. Our address is AFOSR/NA, Building 410, Bolling AFB, Washington, D.C. 20332. Any suggestions you have would be sincerely appreciated.

Now back to this meeting. I hope to learn a great deal here. The research area to be discussed, Transonic Flow in Compressors, is one in which we must develop new predictive methods so we can design better engines. This requires an increase in our understanding of what is happening in the flow field. I think that the representatives from the commercial companies here will agree that we still don't understand everything that is going on inside the turbojet. I hope this meeting will clarify some of the unknowns.

TRANSONIC FLOW PROBLEMS IN TURBOMACHINERY

SESSION I

Chairman:
WILLIAM D. McNALLY
NASA Lewis Research Center

ANALYSIS: BASIC FORMULATION FOR TRANSONIC FLOW PROBLEMS IN ROTORS

INTRODUCTORY REMARKS FOR SESSION ON BASIC FORMULATION FOR TRANSONIC FLOW PROBLEMS IN ROTORS

WILLIAM D. McNALLY
NASA Lewis Research Center

In the past several years, significant advances have been made in developing methods for solving problems in fluid dynamics. Equally significant advances have been made in extending these techniques to applications in turbomachinery. Many good two-dimensional solutions now exist for blade-to-blade and hub-shroud flows through turbomachinery blade rows. More recently, three-dimensional, transonic, viscous, and unsteady effects have begun to be incorporated, bringing with them subsequent increases in computer run time. Several years ago, full three-dimensional, numerical, inviscid flow solutions were generated for the first time in transonic blade rows. Most recently, viscous three-dimensional solutions have been attempted.

In this Session, and again in Sessions II and III, we will discuss some of the formulations of equations and boundary conditions for transonic turbomachinery problems, as well as solution algorithms for these formulations. An orderly presentation and resultant clear understanding of this material will be facilitated by referring to the outline on Figure 1. As the figure indicates, no single problem is being considered, but a tremendous variety of problems depending upon a series of key choices made by the engineer.

First, the exact problem to be solved must be chosen. Several decisions must be made at this point concerning type of turbomachine, number of blade rows, blade row geometry parameters, spatial dimensions in the solution, velocity level, and viscosity and time effects. Second, after a well-defined problem is established, the appropriate governing flow equations and boundary conditions for that problem must be determined. Often the derivation of boundary condition equations and of equations governing treatment of shock discontinuities is more difficult than the derivation of basic equations for the through-flow. Third, several very important decisions must be made concerning the algorithm for solution of the equations. If the algorithm must be numeric, fine details of the numerics will determine the eventual stability and accuracy of the solution, as well as the proper modelling of true physics in key regions such as upstream and downstream boundaries, blade row edges, and shock discontinuities. Finally, the solution procedure on a computer must be established in such a way that computer storage and especially run time are kept to reasonable limits.

Anyone who has worked through the above process for a realistic turbomachinery problem is aware of two very important facts. First, depending upon whether or not the true physics of the problem are modelled correctly in equations, and upon how these equations are subsequently solved, it is possible to generate impressive looking results which are utter nonsense. The utmost care must be exercised in the formulation and solution of any problem to ensure that the engineer receives a true solution to the problem at hand.

Secondly, there are some substantial tradeoffs to be considered in making many of the decisions inherent in Fig. 1. In many of these decisions, simplifications to the modelling and analysis of the problem generally lead to decreased computation time and storage, but at the same time to decreased accuracy and loss of true physics. As always, it requires engineering artistry to be able to simplify a problem to where it is practical for design, yet maintain enough of the problem so that answers are meaningful.

As Fig. 1 indicates, many approaches can be taken to the general problem of transonic flow in turbomachinery. Care must be exercised in developing and applying any solution so that the two extremes of inaccurate results and overkill of the problem are avoided. These considerations should be kept in mind in evaluating the presentations which follow.

1. DEFINITION OF EXACT PROBLEM FOR SOLUTION

   Geometry - Compressor or turbine, axial or centrifugal;
   Single blade row, camber, twist, lean, thickness
   Full stage, or multistage

   Dimension - 2-D or 3-D
   Radius change and streamsheet convergence

   Velocity - Subsonic
   Transonic (mixed flow), choked flow
   Supersonic

   Time - Steady
   Unsteady

2. APPROPRIATE FORM OF FLOW EQUATIONS

3. ALGORITHM TO CALCULATE SOLUTION

   Analytic

   Numeric - First, second, third order
   Stability and accuracy

4. BOUNDARY CONDITION TECHNIQUE

   Proper physics

   Solution algorithm - First, second, third order
   Stability and accuracy

5. TREATMENT OF SHOCK DISCONTINUITIES

   Shock capturing

   Mesh fitting

   Shock fitting

6. COMPUTER STORAGE AND TIME

Figure 1. Analytic Solution of Problems in Turbomachinery

BASIC FORMULATION FOR TRANSONIC FLOW PROBLEMS IN ROTORS

KLAUS OSWATITSCH

Deutsche Versuchs- und Forschungsanstalt für
Luft- und Raumfahrt and Technical University of Vienna

ABSTRACT

The basic equations for transonic flow in stators and rotors are derived. Herewith vortex free flow conditions in the entrance is assumed. The specific properties of steady transonic flow in turbomachines are considered. The characteristic parameters and similarities are discussed. Special attention is given to the two dimensional approximation. The mutual influence of stators and rotors is considered.

The different sources of loss: shock waves, boundary layers and separations are compared. The displacement by boundary layers and secondary flow is estimated.

INTRODUCTION

In this lecture I would like to treat some particular properties of the steady transonic flow in turbomachines. I assume that the flow far upstream is vortex-free with respect to the inertial system and steady with reference to the rotor-fixed system. This means that the particles have constant entropy and the entropy changes from particle to particle only smoothly.

BASIC EQUATIONS

To begin with I assume cylindrical coordinates, x, r, β (Fig. 1) with u, W_1, W_2 as velocity components in the x, r, β - directions. Then one has the following relations

$$u = \phi_x; \qquad W_1 = \phi_r; \qquad W_2 = \frac{1}{r}\phi_\beta$$

$$-\phi_t = h + \frac{1}{2}(u^2 + W_1^2 + W_2^2) + \text{Const} \tag{1}$$

where ϕ is the velocity potential, t the time and h the specific enthalpy.

Changing now to the rotor-fixed system with ω as angular velocity of the rotor and primed quantities referring to the new system (Fig. 2) one obtains the following relations

$$r' = r; \qquad \beta' = \beta - \omega t; \qquad t' = t;$$

$$W_1' = W_1; \qquad W_2' = W_2 - \omega r \tag{2}$$

For the derivatives one finds

$$\frac{\partial}{\partial \beta'} = \frac{\partial}{\partial \beta}; \qquad \frac{\partial}{\partial t'} = \frac{\partial}{\partial t} + \omega\frac{\partial}{\partial \beta} \tag{3}$$

From the fact that the velocity potential must be steady in the rotor-fixed system it follows

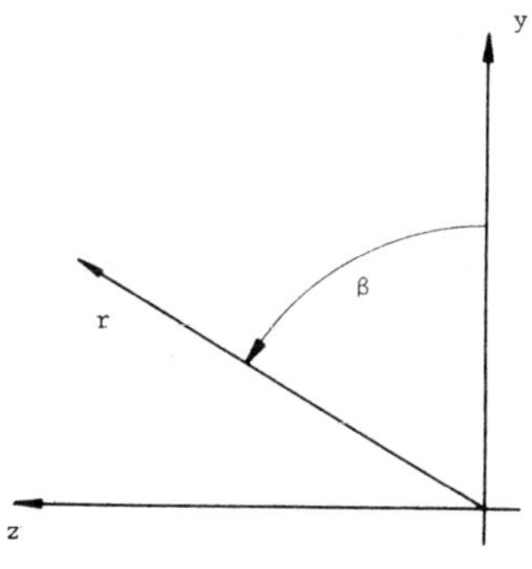

Figure 1. Cylindrical Coordinates in a Plane x = Const. of the Inertial System

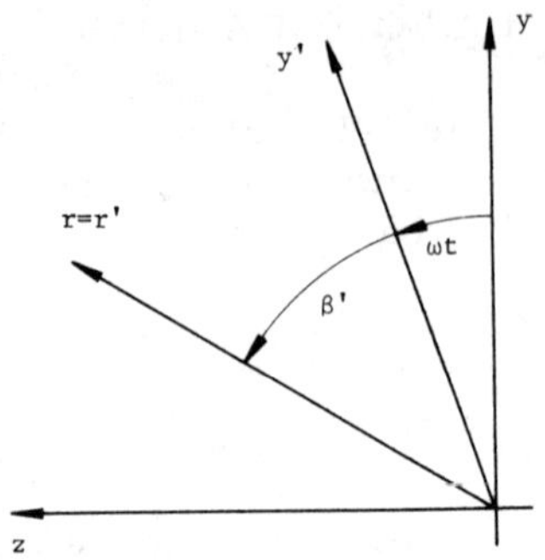

Figure 2. Relation to the Rotor-Fixed System

$$\frac{\partial \phi}{\partial t'} = 0 = - [h + \frac{1}{2}(u^2 + W_1^2 + W_2^2)] + \omega W_2 + \text{Const.}$$

and hence

$$h' + \frac{1}{2}[u'^2 + W_1'^2 + W_2'^2 - \omega^2 r'^2] = \text{Const.} \tag{4}$$

which is the energy equation in the rotor-fixed system.

For the remainder of the paper rotor-fixed coordinates are used and therefore, for simplicity, the primes are dropped. One finally obtains the energy equation in the following form

$$h + \frac{1}{2}(W^2 - \omega^2 r^2) = \text{Const;} \quad W^2 = u^2 + W_1^2 + W_2^2 = u^2 + v^2 + w^2 \tag{5}$$

where W is the magnitude of the velocity vector and v, w are the velocity components in the cartesian y, z directions (Fig. 1).

Then the following relations exist for the cylindrical system

$$u = \phi_x; \quad W_1 = \phi_r; \quad W_2 = \frac{1}{r}\phi_\beta - \omega r \tag{6}$$

and for the cartesian system

$$u = \phi_x; \quad v = \phi_y + \omega z; \quad w = \phi_z - \omega y \tag{7}$$

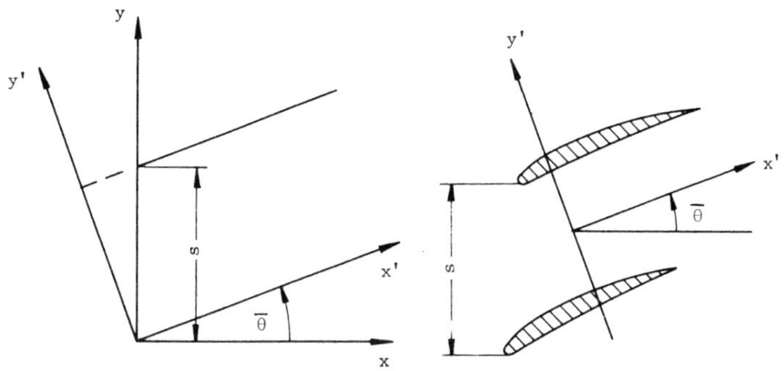

Figure 3. Natural Coordinate System

A coordinate system which is most suitable to describe the flow through blade cascades of axial machines is now introduced where x' and y' are the directions parallel and perpendicular to the local flow (Fig. 3). r_o is the hub radius, s the cascade pitch, $\bar{\theta}$ is the angle between the rotor axis and the x'-axis and $z' = z - r_o$. Then the following relations between this new system and the rotor-fixed cartesian coordinate system hold

$$
\begin{aligned}
x' &= x \cos \bar{\theta} + y \sin \bar{\theta} \\
y' &= - x \sin \bar{\theta} + y \cos \bar{\theta} \\
z' &= z - r_o
\end{aligned}
\qquad (8)
$$

$$
\begin{aligned}
u' &= u \cos \bar{\theta} + v \sin \bar{\theta} \\
v' &= - u \sin \bar{\theta} + v \cos \bar{\theta} \\
w' &= w
\end{aligned}
\qquad (9)
$$

Using Eqs. (7), (8), (9) and after partial differentiation, Eqs. (10) are obtained

$$
\begin{aligned}
u' &= W + \ldots = \phi_{x'} + \omega(r_o + z') \sin \bar{\theta} \\
v' &= \phi_{y'} + \omega(r_o + z') \cos \bar{\theta}
\end{aligned}
\qquad (10)
$$

$$w = w' = \phi_{z'} - \omega y = \phi_{z'} - \omega(x' \sin\bar{\theta} + y' \cos\bar{\theta})$$

Note that the u' velocity component is approximately the magnitude of the velocity.

Because of the independence of the continuity equation from the special coordinate system used it immediately follows that

$$\frac{\partial \rho u'}{\partial x'} + \frac{\partial \rho v'}{\partial y'} + \frac{\partial \rho w'}{\partial z'} = 0$$

and because of the orientation of the x', y', z'-system, Eq. (11) is obtained after retaining first order small disturbance quantities only

$$\frac{u'}{\rho}\frac{\partial \rho'}{\partial x'} + \frac{\partial u'}{\partial x'} + \frac{\partial v'}{\partial y'} + \frac{\partial w'}{\partial z'} = 0 \tag{11}$$

With $\partial/\partial x'$ being the derivative in the flow direction, in which one has isentropic changes of state only, the first law of thermodynamics gives

$$\frac{\partial h}{\partial x'} = \frac{1}{\rho}\frac{\partial p}{\partial x'} = \left(\frac{\partial p}{\partial \rho}\right)_s \frac{1}{\rho}\frac{\partial s}{\partial x'} = a^2 \frac{1}{\rho}\frac{\partial \rho}{\partial x'}$$

where here s is the specific entropy, ρ the density, p the pressure and a the velocity of sound.

The energy equation (5) assumes the following form

$$h + \frac{1}{2}u'^2 - \frac{1}{2}\omega^2\left[(x' \sin\bar{\theta} + y' \cos\bar{\theta})^2 + (r_o + z')^2\right] = \text{Const.}$$

and hence for the density derivative one has

$$\frac{a^2}{\rho}\frac{\partial \rho}{\partial x'} = \frac{\partial h'}{\partial x'} = - u'\frac{\partial u'}{\partial x'} + \omega^2 (x' \sin\bar{\theta} + y' \cos\bar{\theta}) \sin\bar{\theta}$$

Together with Eq. (11) finally the basic gasdynamic equation in small disturbance form is obtained

$$\left(1 - \frac{u'^2}{a^2}\right)\frac{\partial u'}{\partial x'} + \frac{\partial v'}{\partial y'} + \frac{\partial w'}{\partial z'} + \frac{\omega^2 u'}{a^2}(x' \sin\bar{\theta} + y' \cos\bar{\theta}) \sin\bar{\theta} = 0 \tag{12}$$

Defining a^* as the critical velocity of sound on $r = r_o$ for a perfect gas of constant specific heats the following form of the energy equation results

$$\frac{2}{\gamma-1}\left(\frac{a^2}{a^{*2}}-1\right)+\left(\frac{u'^2}{a^{*2}}-1\right)=\frac{\omega^2 r_o^2}{a^{*2}}\left(\frac{r^2}{r_o^2}-1\right) \qquad (13)$$

and for small disturbances one gets the special form

$$\frac{2}{\gamma-1}\left(\frac{a}{a^{*}}-1\right)+\left(\frac{u'}{a^{*}}-1\right)=\frac{\omega^2 r_o^2}{a^{*2}}\left(\frac{r}{r_o}-1\right)=\frac{\omega^2 r_o^2}{a^{*2}}\,\frac{z'}{r_o} \qquad (14)$$

where both equations contain the characteristic parameter $\omega r_o/a^*$.

LATERAL INFLUENCE

Transonic flow is characterized by nearly constant mass flow density which therefore causes very strong lateral influences. This effect is shown very clearly in older experiments by J. Ackeret and N. Rott [1], Fig. 4, who measured and computed the

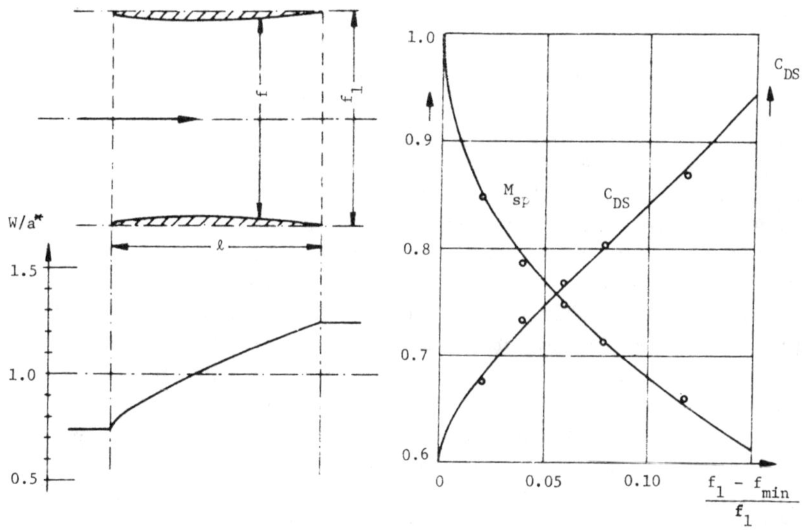

Figure 4. Mach Number M_{sp} and Drag Coefficient C_{DS} for Choked Flow through an Unstaggered Cascade as a Function of Area Contraction Ratio $(f_1 - f_{min})/f_1$ according to [1].

flow between two profiles. The gap-to-chord ratio in these experiments was $f_1/\ell = 2/3, 1, 4/3, 2$ and 4, respectively. In all cases accelerated flow was studied. The choking Mach number in the entrance M_{sp} and the maximum drag coefficient C_{DS} was calculated with simple one-dimensional theory. The right-hand side of Fig. 4 shows the choking Mach number and the drag coefficient as a function of the contraction ratio. This example illustrates a wide applicability of the one-dimensional flow approximation.

In many important cases the cross-flow terms with v' and w' in Eq. (12) dominate in comparison with the axial term of the gasdynamic equation. This represents a situation similar to R. T. Jones' low aspect ratio wing theory [2]. The gasdynamic equation then degenerates to Eq. (15) where ϕ is the velocity potential in Eq. (10).

$$\phi_{y'y'} + \phi_{z'z'} + \frac{\omega^2}{a^*} y' \cos\bar{\theta} \sin\bar{\theta} = 0 \tag{15}$$

The boundary condition can be given, e.g., on the suction and p pressure side of neighboring blades as well as on the hub and casing wall

$$\begin{aligned} y' = 0 \text{ and } y' = s\cos\bar{\theta}: \quad v' &= \phi_{y'} + \omega(r_o + z')\cos\bar{\theta} \\ z' = 0 \text{ and } z' = b: \quad w' &= \phi_{z'} - \omega y' \cos\bar{\theta} \end{aligned} \tag{16}$$

where b is the blade span, Fig. 5.

It should be noted that this boundary value problem leaves an additive function of x undetermined resulting in a flow problem similar to the flow past low aspect ratio wings of a given thickness distribution at zero angle of attack treated by K. Oswatitsch [3]. In the present case this function can be found from continuity considerations because the total mass flow in each cross-section has to be constant. If only force and moment calculations are of interest this function makes no contribution to the pressure difference between pressure and suction side of the blade. This may be the reason for the applicability of linearized theory to the calculation of momentum in transonic flow, see e.g., McCune [4]. If, on the other hand, choking and similar effects are to be considered the additive function of x plays an important role. These latter considerations are all restricted to the flow between the blades, i.e., the passage flow which can be regarded to be bounded by the

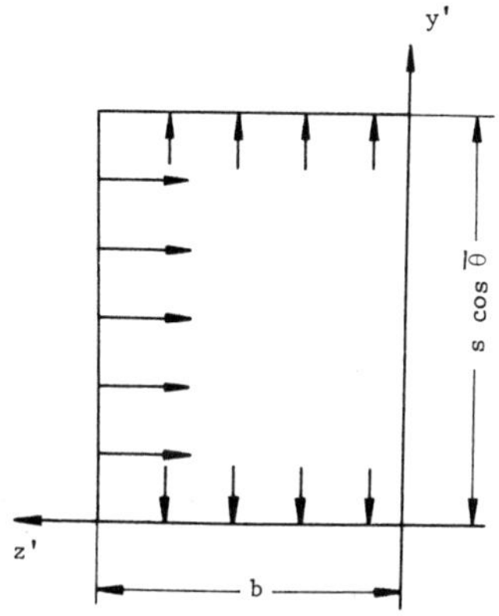

Figure 5. Simplified Boundary Condition in the Cross-Section x = Const.

potential lines extending from the blade leading and trailing edge to the neighboring blade, Fig. 6. For the passage flow good

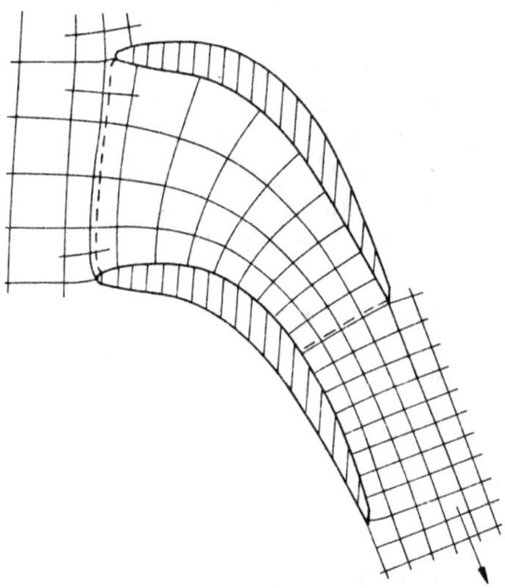

Figure 6. Boundaries of the Passage Flow Region (-----) in the Potential - Streamline Net

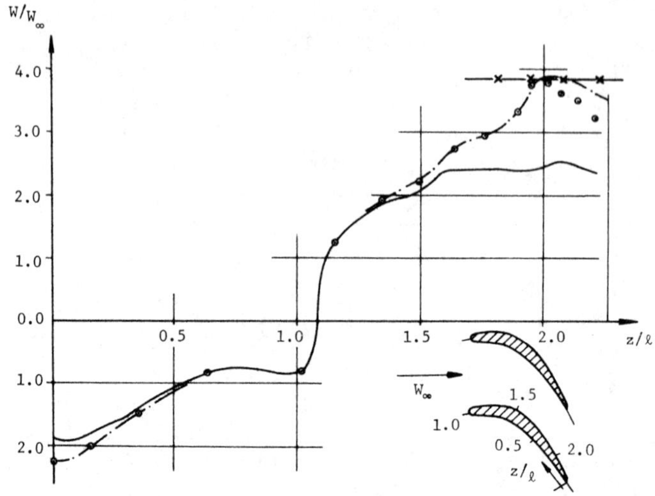

Figure 7. Comparison of Approximation [5] (ooooo) with Exact Theory [6] (—·—·—·—) at M = 0.24.
-*-*-*- Sonic; ———— Incompressible.

approximation theories can be developed, as e.g., Oswatitsch and Ryhming [5], Fig. 7. There is very good agreement between the approximation and the exact theory by Uchida [6] in the blade passage. In the entrance region and in the present example most noticeably in the exit region the results are unsatisfactory. This is mainly due to the significant lateral influence in transonic flow mentioned earlier. Therefore, the question arises whether theories and experiments which do not take into account the mutual blade row interference are meaningful.

Transition from subsonic to supersonic flow, in general, takes place in the blade passage, Fig. 8 [7], and can generally be computed much more easily than for the single airfoil in unbounded flow. It is essential to note, however, that the supersonic exit flow is quite different in character from the subsonic case due to the supersonic wave structure.

SIMILARITY LAWS

The similarity laws valid for single profiles and wings are

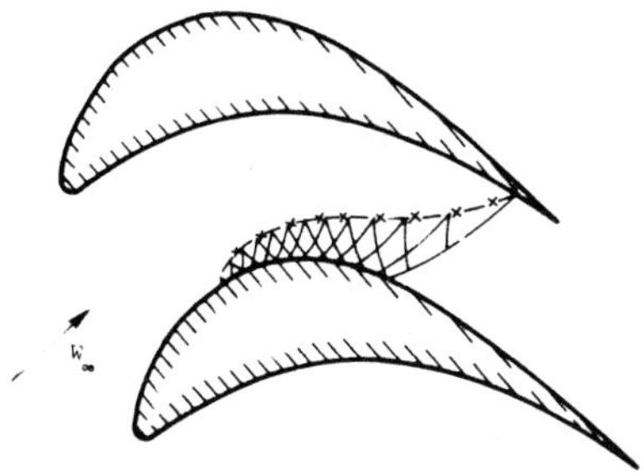

Figure 8. Acceleration to Supersonic Flow in a Cascade

equally applicable to transonic cascades and three-dimensional flows in turbomachines. Disregarding rotational effects the following reduced quantities should be introduced. Defining τ as the flow deflection angle and γ as the ratio of specific heats the following reduced variables have to be used, Fig. 9.

$$\bar{y} = [(\gamma+1)\tau]^{1/3} y'$$

$$\bar{z} = [(\gamma+1)\tau]^{1/3} z'$$

$$\left(\frac{u'}{a^*} - 1\right)(\gamma+1)^{1/3}\tau^{-2/3} = \bar{u} \tag{17}$$

$$\frac{v'}{a^*}\tau^{-1} = \bar{v}$$

$$\frac{w'}{a^*}\tau^{-1} = \bar{w}$$

Only flows with the same reduced variables can be compared. For the stagger angle Λ the following relation holds

$$\mathrm{tg}\,\Lambda \sim \sqrt{M^2-1} = (\gamma+1)^{1/2}\left(\frac{u'}{a^*} - 1\right)^{1/2} = [(\gamma+1)\tau]^{1/3}\bar{u}^{1/2} \tag{18}$$

showing that the blade staggering has to change with the Mach line

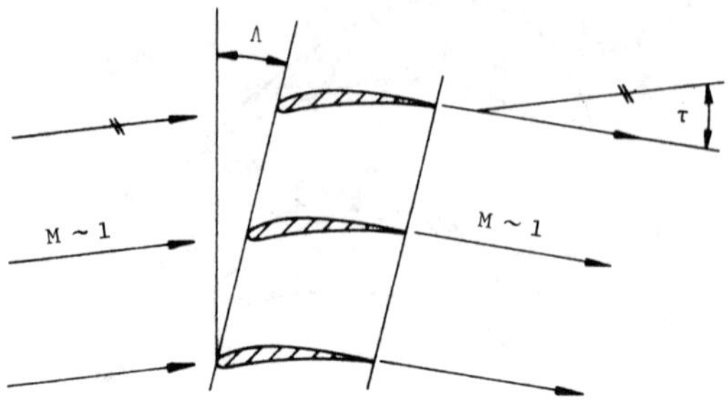

Figure 9. Small Flow Turning in Transonic Cascade

inclination and blade spacing has to vary according to Eqs. (17).

TWO-DIMENSIONAL AND QUASI-TWO-DIMENSIONAL FLOW

It can easily be shown that two-dimensional flow can also be transformed into certain three-dimensional flows. A simple example shall be given here. If we assume the boundary condition on the hub and on the casing to be given by

$$z' = 0:\ w' = 0; \qquad z' = b:\ w' = w_o \tag{19}$$

where w_o is constant, then the following relation for

$$w' = \frac{w_o z'}{b}$$

satisfies the boundary condition. The gasdynamic equation for small disturbances and the irrotationality condition are now given by

$$(1 - M^2)\frac{\partial u'}{\partial x'} + \frac{\partial v'}{\partial y'} + \frac{w_o}{b} = 0; \qquad \frac{\partial u'}{\partial y'} - \frac{\partial v'}{\partial x'} = 0 \tag{20}$$

Introducing now the transformation, Eq. (21)

$$v' + \frac{w_o}{b} y' = \bar{v} \tag{21}$$

reduces the inhomogeneous system Eqs. (20) to the well known two-dimensional Eqs. (22) for u' and $\bar{v}$.

$$(1-M^2)\frac{\partial u'}{\partial x'} + \frac{\partial \bar{v}}{\partial y'} = 0; \qquad \frac{\partial u'}{\partial y'} - \frac{\partial \bar{v}}{\partial x'} = 0 \tag{22}$$

It should be noted that there are more general transformations which will be reported at the forthcoming GAMM-meeting by H. Sobieczky [8]. Also, it is important to realize that useful approximation theories can be developed for high aspect ratio blades whose shape changes only smoothly from hub to tip. This is similar to Prandtl's lifting line theory and Keune's approximation for thick high-aspect-ratio wings [9].

SOME REMARKS ON LOSSES

It appears that the shock losses are sometimes overestimated. In this regard it is important to keep in mind that a shock with a pressure ratio of $\hat{p}/p = 2$ has only a stagnation pressure loss of about 3%. In contrast to this, diffusion and mixing losses can be much more important. Since the isentropic flow density is nearly constant an area increase of the available flow duct must be compensated mainly by stagnation pressure decrease.

Also, secondary flows in cascades have to be considered. The displacement thickness on the suction side of the blades can be a multiple of the two-dimensional displacement thickness. More fundamental investigations are needed on this problem.

REFERENCES

1. Ackeret, J. and Rott, N., "Ueber die Stroemung von Gasen durch ungestaffelte Profilgitter," Schweiz. Bauzeitung, Vol. 67, pp. 40-41 and pp. 58-61 (1949).

2. Jones, R. T., "Properties of Low-Aspect Ratio Pointed Wings at Speeds Below and Above the Speed of Sound," NACA Report 835 (1946).

3. Oswatitsch, K., "Die theoretischen Arbeiten ueber schallnahe Stroemung am Flugtechnischen Institut der Koeniglich-Technischen Hochschule Stockholm," Proc. 8th Int. Congress Theor. and Appl. Mechanics, Istanbul (1952).

4. Okurounmu, O. and McCune, J. E., Lifting Surface Theory of Axial Compressor Blade Rows, Journ. of Aerospace Sciences, Oct. (1974), pp. 1363 bis 1380.

5. Oswatitsch, K. and Ryhming, I., "Ueber den Kompressibilitaetseinfluss bei ebenen Schaufelgittern starker Umlenkung, DVL-Bericht 28, Deutsche Versuchsanstalt fuer Luftfahrt, (1957).

6. Uchida, S., "Calculation of Compressible Cascade Flow by the Method of Flux Analysis," Journ. Aeron. Sci., 21, pp. 237-250 (1954).

7. Sobieczky, H., Application of generalized potentials on plane transonic flow, Proceedings Symposium Transsonicum II, Springer Berlin (1976).

8. Sobieczky, H., Eine Verallgemeinerung der Gleichung schallnaher Strömung, GAMM-Tagung (Society for Applied Mathematics and Mechanics) Graz 5.-9.4.(1976).

9. Keune, F., "An Approximate Method of Calculating the Subsonic Velocity Distribution on High-Aspect-Ratio Swept Wings of Small Thickness at Zero Lift, KTH Aero TN 26, (1953)or DVL Bericht 48 (1958).

DISCUSSION OF THE OSWATITSCH PAPER

MC CUNE: You were using a slide to indicate three-dimensional effects were very important in the transonic regime? They were very successful in predicting the drag in this particular case up to a certain critical Mach number?

OSWATITSCH: Yes. It is so. I think it can be that the internal flow gives the dominating term for the forces. And then one has not to trouble too much perhaps with the external flow if one only is interested in the forces. This internal flow, even if it is three-dimensional, may be calculated quite simply.

SICHEL: You had one picture where you showed a sonic line that was going across the cascade blade passage - I think you had those fairly curved blades with your characteristics drawn on them. In the development of the transonic flow, wouldn't you expect at first that the sonic line would bend around and form sort of a bow on one surface? I wonder if you could talk about that.

OSWATITSCH: I haven't talked about that because I think it's very disagreeable! The change from the local supersonic region to the sonic line which goes from one side to the other side is, I think - I am not sure at the moment, but I think - that it is a real unsteady change. You can only get it by unsteady solutions in between. Unsteady solutions in internal flow are also quite problematical. We have done research - with Dr. Kluwick who is now in Blacksburg - we have done the change you have mentioned in the very simple case of a one-dimensional Laval nozzle and then he has studied the change from the symmetrical to the nonsymmetrical - from the accelerated solution to the other one. This is already quite a little paper, and the change from one to the other solution takes a logarithmically infinite time so it is an asymptotic solution, which is a very good asymptotic solution, but in any case it shows that even with very simple examples you have trouble. So the change you mentioned should be considered - but people like Landahl should do it. The main wave is an upstream going wave, and this upstream going wave is very disagreeable at the sonic line because it is turning around absolutely. In the subsonic flow, it goes ahead and in the supersonic flow, it does not go ahead; therefore, small disturbance wave theory doesn't fit, but it may be that there are some better ways to find. I haven't seen them found yet.

THREE-DIMENSIONAL INVISCID FLOW THROUGH A HIGHLY-LOADED TRANSONIC COMPRESSOR ROTOR

JAMES E. McCUNE

Massachusetts Institute of Technology

ABSTRACT

A newly-developed approach to the theoretical description of three-dimensional inviscid flow in axial compressor rotors or fans is applied to the case of transonic operation at pressure ratios of practical interest. Results are compared with recently acquired data for a transonic ducted fan operating in the M. I. T. Blowdown Facility. In the theory, nonlinear inviscid effects are retained to any desired accuracy in the pitch-wise averaged flow (axisymmetric through flow), while blade-to-blade variations are computed by linearizing around this mean flow. In this latter step, no quasi-two-dimensional (cascade) assumption is necessary. The comparison between experiment and theory is sufficiently encouraging to suggest that the improved theory, appropriately refined, may become of practical use in compressor design. Further, the results have interesting implications in the study of compressor noise.

I. INTRODUCTION

Earlier theoretical attempts [1-8] to describe three-dimensional inviscid flow in axial compressor rotors or fans were carried out within the realm of strictly linear theory, analogous to classical thin airfoil theory. While these original theories strongly suggest the presence of several interesting three-dimensional effects, with possible important implications both for rotor design

[1, 6, 8] and compressor acoustics [9, 10], this could not be verified in practical applications until now, because of the limitations inherent to the basic linearization procedure. Even improved, higher-order perturbation theories [11], so long as they are based on an expansion around the undisturbed inlet flow, do not seem to offer immediate application to practical engine design or performance studies.

A recent fundamental advance in the analytic theory, however, now enables one to consider three-dimensional effects in blade rows which induce flow turning angles and/or pressure ratios of truly practical interest. This new theory, strictly speaking, is still "linearized," but it is now linearized about the circumferential mean flow (axisymmetric through-flow), and nonlinear terms are kept in the latter to any desired accuracy. This allows the treatment of arbitrary mean turning and back pressures, and, of course, also includes the satisfaction of the (nonlinear, mean-flow) requirement of radial equilibrium in the annular case. This "hybrid" or "quasi-linear" theory is outlined in some detail, for the incompressible case, in Refs. [12] and [13]. In the first of these the new approach is introduced, with the help of Beltrami-flow concepts, in a three-dimensional rectilinear geometry with the blades represented by a cascade of ducted lifting lines. In the second, the corresponding theory is developed in the annular case. A first approach to the compressible (higher Mach number) problem is presented in Ref. [14] for rectilinear geometry.

In the present paper our purpose is to apply this hybrid theory to a highly-loaded ducted fan in transonic operation, and to compare the results with recent data acquired in the M. I. T. transonic rotor blow-down facility [15, 16].

This paper is constructed as follows. In the next section (II) the well known basic equations are summarized and then specialized to the case of Beltrami flow (isentropic, inviscid, uniform rothalpy, isolated blade row). The corresponding mean-flow equations are discussed in the light of Beltrami flow concepts and certain points of special use to us in the present application are noted and compared with classical results [17, 18]. Because of our intention to perturb around the mean flow, it is of practical importance to describe the properties of the latter as efficiently as possible; we attempt to do so in this section.

In Section III the perturbation equations are derived, emphasizing the linearization about the mean flow as suggested above.

These equations are a generalization, to the compressible, annular case, of those obtained earlier in Refs. [12-14]. The basic small parameter which allows such an expansion about the mean flow actually turns out to be $2\pi/B$ (B=number of blades in a given row) or, roughly, the inverse of the solidity. This is, of course, a measure of the blade "discreteness," and the reader may well note that the smallness of this parameter is really only marginal for a fan (B typically about 20). However, one of the important three-dimensional effects of blade discreteness, as we shall see, is that of the vorticity trailing behind the loaded blades [12]. This becomes weak, and eventually vanishes, for rotor designs approaching "constant work." Such designs are typical for many fans. We take advantage of this fact in the present study. Nevertheless, as we shall discover, the theory works remarkably well (perhaps better than could be expected, a priori) even when applied to a fan which actually departs substantially from constant work [19].

Section IV is devoted to a brief discussion of practical methods of treating both the mean flow and blade-to-blade variations computationally in the transonic regime; typical results are shown for design parameters of interest in this range. In Section V direct application of the theory [19] to the experiment mentioned above [15, 16] is shown and discussed. Here, for purposes of obtaining results relatively quickly we resort to the familiar technique of applying a basically incompressible (low Mach number) theory to the experimental situation by simply adjusting the channel height (hub-tip ratio) to obtain the proper (average) axial velocity ratio. More detailed comparison with the full transonic theory is still in progress.

Section VI of this paper provides a summary and some conclusions, as well as a few suggestions for extensions to more general situations. Some interesting implications regarding compressor noise spectra are also discussed.

II. BASIC EQUATIONS AND BELTRAMI FLOW

Consider the flow in an annulus, using the right-handed cylindrical coordinates (r, θ, x) in a duct-fixed system. Let the absolute velocity, $\underline{V} = \underline{V}(\underline{x}, t) = (u, v, w) = u\hat{e}_r + v\hat{e}_\theta + w\hat{e}_x$, represent the fluid flow measured in this system. Using the concept discussed by Marble [17] of describing the presence of the blade row (or rows) by a distributed body force $\underline{F}$ (force per unit mass, really an acceleration), acting on the fluid, the momentum equations can be written in Lamb's form

$$\frac{\partial \underline{V}}{\partial t} + \underline{\Omega} \times \underline{V} = - \nabla H + T \nabla s + \underline{F} \qquad (2.1)$$

where $\underline{\Omega} \equiv \text{curl } \underline{V}$, $H = h + (1/2)q^2$ is the total specific enthalpy in duct-fixed (absolute) coordinates, $q^2 \equiv u^2 + v^2 + w^2$, s is the specific fluid entropy, T the fluid temperature, and $\underline{F} = \underline{F}(r, \theta, x, t)$. Actually, of course, $\underline{F}$ could if we chose contain specific viscous stresses other than those occurring solely due to contact with the blading, so (2.1) is actually general enough to include the Navier-Stokes equations with proper interpretation of $\underline{F}$. In our present work, however, we shall treat $\underline{F}$ to be zero away from the blading, so we will here be neglecting, for example, possible important viscous interactions in the wakes. Treatment of this latter problem is especially interesting in rotating machinery, but is beyond the scope of this article. In writing (2.1) we have of course used the standard thermodynamic equation

$$T ds = dh - \frac{1}{\rho} dp.$$

Analogous use of a volumetric specific power input to the fluid, due (at least) to the blading, leads to the familiar energy equation

$$\frac{DH}{Dt} - \frac{1}{\rho}\frac{\partial p}{\partial t} = P \qquad (2.2)$$

where D/Dt is the usual convective derivative in absolute coordinates. As with $\underline{F}$ in (2.1), P in (2.2) could, if we choose, include viscous dissipation, net heat transfer in the wakes and at the blading, etc., as well as any actual power input associated with an ordered movement of $\underline{F}$ through the fluid. From (2.1) and (2.2)

$$T \frac{Ds}{Dt} = P - \underline{F} \cdot \underline{V} \qquad (2.3)$$

which simply states that the rate of specific entropy increase of a fluid particle moving through the machine is proportional to the difference between the actual net power input and the rate of work done by the body forces $\underline{F}$ (in particular, by the blading). To illustrate, suppose $\underline{F}$ is due only to a single isolated blade row, rotating at the speed ω (clockwise looking from the inlet, i.e. $\underline{\omega} = \omega \hat{e}_x$), so that $\underline{F} = 0$ outside the blade row and otherwise $\underline{F} = \underline{F}(r, \theta - \omega t, x)$. Suppose also that $P = \omega r F_\theta$ so that there is no net cooling at the blades, etc. Then, in that case,

$$T \frac{Ds}{Dt} = - \underline{F} \cdot \underline{W} \tag{2.4}$$

where $\underline{W} \equiv \underline{V} - \omega r \hat{e}_\theta$ is the "relative velocity" -- the velocity field viewed by an observer riding on the (isolated) rotor. Note that, as Marble [17] pointed out, $\underline{F} \cdot \underline{W} = 0$ if $\underline{F}$ represents entirely the reaction on the fluid of ideal pressure forces on the blades. Indeed, this picture then assigns the cause of entropy increase solely to the drag on the blades.

A. Isolated Rotors

We now specialize to the case of an isolated rotor, i.e., to situations for which rotor-stator or interstage interaction is negligible. All quantities are then steady in the system fixed in the rotor (the relative system), each quantity $g = g(r, \theta - \omega t, x)$, and

$$\frac{\partial}{\partial t} \rightarrow - \omega \frac{\partial}{\partial \theta}$$

$$\frac{D}{Dt} \rightarrow \underline{W} \cdot \nabla \tag{2.5}$$

Then, correspondingly, neglecting blade cooling, viscous dissipation and heat transfer in wakes, as well as viscous stresses in the wakes, etc., $P = \omega r F_\theta$ and (2.3) or (2.4) becomes

$$T \, \underline{W} \cdot \nabla s = - \underline{F} \cdot \underline{W} \tag{2.4a}$$

while (2.2) takes the form

$$\underline{W} \cdot \nabla H = - \frac{\omega}{\rho} \frac{\partial p}{\partial \theta} + \omega r F_\theta \tag{2.2a}$$

Simple manipulation of (2.1) yields

$$\underline{\Omega} \times \underline{W} = - \nabla I + T \nabla s + \underline{F} \tag{2.6}$$

where $I \equiv H - \omega r v$ (the rothalpy) and we recall that H is the absolute specific enthalpy. Dotting (2.6) with $\underline{W}$ and using (2.4a) gives the important and familiar relation

$$\underline{W} \cdot \nabla I = 0 \tag{2.7}$$

This latter result will be recognized as a differential form of Euler's turbine equation; in particular, for the special solution I = (spatial constant) we have the usual statement

$$\Delta H = \omega \Delta(rv) \tag{2.7a}$$

where the changes defined by the "Δ" are to be taken at corresponding points along a streamline in the steady (relative) system definable for an isolated rotor.

One can show trivially that (2.7), with (2.2a), is consistent with the θ-component of momentum in the isolated rotor case. It is important to note that (2.7) is applicable throughout the flow, even at the blading where $\underline{F} \neq 0$, provided only that $\partial/\partial t = -\omega(\partial/\partial\theta)$ and $P = \omega r F_\theta$. By contrast, even in the regions where $\underline{F}$ or $P = 0$,

$$\underline{W} \cdot \nabla H = -\omega \frac{1}{\rho} \frac{p}{\partial\theta} = \omega T \frac{\partial s}{\partial\theta} - \omega \frac{\partial h}{\partial\theta}$$

which vanishes only in the mean (average over θ). Thus, in regions where H is increasing along $\underline{W}$, so is ωrv in like amount. More important, if

$$(\bar{\ }) \equiv \frac{1}{2\pi} \int_0^{2\pi} d\theta\,(\ ),$$

then in the mean $\overline{\underline{W} \cdot \nabla H} \cong \underline{\bar{V}} \cdot \nabla\bar{H} = 0$ in regions where $\underline{\bar{F}} = 0$, and $\underline{\bar{V}} \cdot \overline{rv} = 0$ there as well, i.e., $\bar{H} = \bar{H}(\psi)$ upstream and downstream of the blade row, and similarly for $\overline{rv}$.* It is very useful in applying (2.7) to recall that $\bar{I} = \bar{H} - \omega\overline{rv} = \bar{I}(\psi)$ right through the blade row, even though $\underline{\bar{V}} \cdot \nabla\bar{H}$ and $\underline{\bar{V}} \cdot (\overline{rv})$ do not individually vanish in the blade passages. Here, ψ is defined by $(\bar{\rho}\bar{u}, 0, \bar{\rho}\bar{w}) = \nabla\theta \times \nabla\psi$.

*In making the above remark we have introduced the approximation inherent to our expansion about the mean. In fact, $\overline{AB} = \bar{A}\bar{B} + \overline{(\tilde{A}\tilde{B})}$, and in the following we shall generally neglect quadratic terms in the fluctuations $\tilde{A}$, $\tilde{B}$, etc., i.e. $\overline{AB} \cong \bar{A}\bar{B}$. Exceptions occur when one of the quantities, A or B, is singular [cf. Eqs. (2.14)-(2.20) and Section III].

B. Mean Flow in Nearly Loss-Free Rotor Fields

Before focussing on our main subject—three-dimensional Beltrami flows through rotors—we note some useful properties of the mean (axisymmetric) flow in the important case for which $\nabla I \neq 0$, even approximately, because of inlet conditions. An important example is that of incoming "wheel flow," ($v \cong Kr$), for instance, set up by inlet guide vanes. In that case, upstream of the rotor and downstream of the I. G. V., $\bar{I} \sim$ ct. $+ \omega K r^2$, $|\nabla \bar{I}| \sim r$, etc. Nevertheless, to a good approximation in the upstream flow, $\overline{\nabla H}$ and $\bar{T}\, \nabla \bar{s} \cong 0$ (mean inlet total temperature and total pressure nearly uniform). Moreover, $P \cong \omega r F_\theta$ in the absence of significant net heat loss or input, and (2.7) holds approximately. Then, in the mean, $\bar{I} = \bar{I}(\psi)$ everywhere, and $\overline{\nabla I} \equiv \nabla \bar{I} = \bar{I}'(\psi)\, \nabla \psi$ where $\psi(r, x)$ describes the mean flow to be determined.

Now, for nearly loss-free flows, we must have, outside of the blading, $\underline{\bar{F}} \cong 0$, and in the rotor passages $\underline{\bar{F}} \cdot \underline{\bar{W}} \cong 0$ (body forces dominated by pressure forces [17], so $\bar{s} \cong \bar{s}(\psi)$, according to (2.4a), throughout the flow field. But, from the stated inlet conditions, this requires $\bar{s}'(\psi) \cong 0$ and hence $\overline{T \nabla s} \cong 0$ everywhere. Finally,* $\overline{\underline{W} \times \underline{\Omega}} \cong \underline{\bar{W}} \times \underline{\bar{\Omega}}$, so, outside the blading, (2.6) reduces to

$$- \underline{\bar{W}} \times \underline{\bar{\Omega}} \cong - \bar{I}'(\psi)\, \nabla \psi \tag{2.8}$$

Equation (2.8) applies upstream and downstream of the rotor. We discuss it here to illustrate how techniques useful to (mean) Beltrami flow, described below, can be generalized immediately to other important cases.

The treatment given here (in this example) will be limited to incompressible flow for the sake of brevity. In that case $\nabla \alpha \times \nabla \psi = \underline{\bar{W}}$, $\alpha \equiv \theta - f(r, x)$ without loss of generality, and

$$\underline{\bar{W}} = \left(\frac{1}{r}\frac{\partial \psi}{\partial x},\ r\underline{\bar{V}} \cdot \nabla f,\ -\frac{1}{r}\frac{\partial \psi}{\partial r}\right).$$

The dot product of $\underline{\bar{\Omega}}$ with (2.8) tells us that $\underline{\bar{\Omega}} \cdot \nabla \psi = 0$; i.e.,

---

*Some care must be exercised here in interpreting this statement when, for example, $\underline{\Omega}$ is singular—i.e., concentrated in "vortex sheets" as in idealized inviscid flow. No really fundamental problem arises, however, so we reserve discussion of this point for Sections II. C and III.

$\underline{\bar{\Omega}}$ lies in the surfaces ψ = const. This offers a useful conceptual picture of the mean vorticity vector. (Of course $\underline{\bar{W}}$ also lies in such surfaces, so mean Beltrami flow can be regarded as the special case of the present situation in which $\underline{\bar{\Omega}}$ and $\underline{\bar{W}}$ are mutually aligned.) Formally, $\underline{\bar{\Omega}} \times \underline{\bar{W}} = \underline{\bar{\Omega}} \times (\nabla\alpha \times \nabla\psi)$, or

$$\underline{\bar{\Omega}} \times \underline{\bar{W}} = (\underline{\bar{\Omega}} \cdot \nabla\psi)\nabla\alpha - (\underline{\bar{\Omega}} \cdot \nabla\alpha)\nabla\psi$$

$$= -(\underline{\bar{\Omega}} \cdot \nabla\alpha)\nabla\psi,$$

in view of $\underline{\bar{\Omega}}$ lying in stream surfaces. Clearly, then, from (2.8),

$$\underline{\bar{\Omega}} \cdot \nabla\alpha = + I'(\psi) \tag{2.9}$$

and $I'(\psi)$ is known from the inlet conditions to the rotor (or outlet conditions from the I. G. V. or upstream stator). Recalling that $\underline{\bar{\Omega}}$ is the curl of the mean absolute velocity, $\underline{\bar{V}}$, we have

$$\underline{\bar{\Omega}} = [-\frac{1}{r}\frac{\partial}{\partial x}(\overline{rv}), \frac{1}{r}\frac{\partial^2\psi}{\partial x^2} + \frac{\partial}{\partial r}(\frac{1}{r}\frac{\partial\psi}{\partial r}), \frac{1}{r}\frac{\partial}{\partial r}(\overline{rv})]$$

$$= \frac{1}{r}[-g'(\psi)\frac{\partial\psi}{\partial x}, \Delta^*\psi(r,x), g'(\psi)\frac{\partial\psi}{\partial r}]$$

and

$$\underline{\bar{\Omega}} \cdot \nabla\alpha = \frac{1}{r^2}\Delta^*\psi + g'(\psi)\,\underline{\bar{V}} \cdot \nabla f = + I'(\psi) \tag{2.10}$$

where $g(\psi) \equiv \overline{rv}(\psi)$ (see section II. A) and, in the cylindrical coordinates of interest here,

$$\Delta^*\psi \equiv \frac{\partial^2\psi}{\partial x^2} + \frac{\partial^2\psi}{\partial r^2} - \frac{1}{r}\frac{\partial\psi}{\partial r} = \nabla^2\psi - \frac{2}{r}\frac{\partial\psi}{\partial r}.$$

Returning to our earlier definitions of $\underline{\bar{W}}$, g and α, we see that

$$\underline{\bar{V}} \cdot \nabla f = \frac{\bar{v}}{r} - \omega = [g(\psi)/r^2 - \omega]$$

and hence

$$\Delta^* = -g'(\psi)[g(\psi) - \omega r^2] + r^2 I'(\psi) \tag{2.11}$$

which is the (highly nonlinear) equation for $\psi(r, x)$ determining the mean flow—upstream and downstream of the rotor—for the type of flow under discussion here. The reader will recognize (2.11) as a special case of the mean flow equation given by Marble [17], applicable under the approximations stated above.

Solution is "direct" for the design problem; given the blade loading and inlet conditions, the turning angles are specified, and hence $g(\psi) = \overline{rv}$ is known upstream and downstream of the blading. Chen [19] has developed a practical and quick iterative method for solving nonlinear equations of the type (2.11) for turbomachinery, based on the idea that $\psi = \psi(r)$ to lowest order in $1/B$ (or $2\pi r_T g'/B$) (see Section V).

The off-design problem is more tedious, involving as usual an integral equation (in this case, for $f \equiv \theta - \alpha$) but solution can again be obtained by iterative techniques.

Matching of the "upstream" and "downstream" problems can be accomplished simply, as in the following sections, using an actuator disc approximation for the mean flow (and lifting lines to represent the discrete blades in the full three-dimensional problem). In principle, the analysis can be carried out more correctly by solving the rotor-passage equations where $\underline{F} \neq 0$. In practice, however, judicious use of rotor or cascade data is preferable at this stage of our knowledge, especially as regards internal flows in transonic rotor or fan passages, in the corresponding matching of upstream and downstream flows.

The special case of (2.11), obtained by putting $I'(\psi) = 0$, corresponds to the mean-flow Beltrami equation ($\overline{\underline{W}} \times \overline{\underline{\Omega}} = 0$) in the compressible case [cf. Eq. (2.22)]. This situation obtains, for example, for loss-free, purely axial inlet flow, in which case g(upstream) = 0 and g(downstream) = $B\Gamma/2\pi$, where $\Gamma(\psi)$ is the circulation per blade. It also applies for free-vortex inlet flow with uniform (inlet) total absolute enthalpy, as well as for several variations with incoming swirl such that the vorticity is aligned with $\overline{\underline{W}}$.

In a very useful sense, the measured misalignment of $\underline{\Omega}$ and $\underline{W}$ [15, 16] can be used, with careful interpretation, as an experimental indication of the departure of real flows from the idealizations employed in theoretical treatments such as the one to be outlined here.

In fact, for transonic fans in practical use, the inlet conditions are often such that Beltrami flow should be expected (in detail, not just in the mean) except where viscous stresses and heat transfer play a key role—for example, in the wakes. Recognizing this fact, we have developed such a theory for a transonic ducted fan, with the intention [13, 19] of comparing the (idealized) three-dimensional theory with the actual data available from the M. I. T. Blowdown Facility [15, 16].

The basic elements of this treatment, following the ideas given in the Introduction, are outlined in Sections II. C and III. For further details, the reader may refer to the developments given in Refs. [12], [13] and [19].

C. Transonic Beltrami Flow

Finally, in this section, we specialize our treatment to ideal (inviscid, adiabatic) three-dimensional flow through an isolated axial rotor with uniform inlet conditions. However, we include compressibility in order to study ducted fans in transonic operation. Under these assumptions the flow is isentropic. We then obtain Beltrami flow—in detail, not just in the mean—outside of the blading [see Eq. (2.6)]. The fact that this situation may pertain under less restrictive assumptions we reserve for later discussion. We now have

$$\operatorname{div}\,(\rho\,\underline{W}) = 0 \tag{2.12}$$

and, outside the blading,

$$\underline{\Omega} \times \underline{W} = 0 \tag{2.13}$$

where we recall that $\underline{\Omega} \equiv \operatorname{curl}\,\underline{V}$, $\underline{W} \equiv \underline{V} - \omega r\hat{e}_\theta$.

The general solution of (2.13) can be written conveniently in the form

$$\underline{\Omega} = \lambda\,\underline{W} = \frac{\lambda}{\rho}\,(\rho\,\underline{W}) \tag{2.14}$$

where λ is a scalar function of $\underline{x} = (r, \theta, x)$. Equation (2.12) is automatically satisfied if $\rho\,\underline{W}$ is put equal to $\nabla\alpha \times \nabla\beta$, once again we can take $\alpha = \theta - f(r, x)$ without loss of generality, and taking the divergence of (2.14) shows that $\lambda/\rho = Q(\alpha, \beta)$ [12, 13]. The special case $\lambda = Q = 0$, of course, returns us to potential flow, and this type of flow actually pertains, under the assumptions stated, upstream of the fan. Downstream of the fan, however, $\lambda \neq 0$ and $\underline{\Omega} \neq 0$. Nevertheless, the vorticity present in this region is not due, here, to viscosity or any other departure of the flow from "ideal" conditions; it is simply the free vorticity trailing downstream of the blades, which appears whenever the spanwise distribution of the blade loading (circulation) is nonuniform. In other words, in this simplified picture, $\underline{\Omega}$ is the trailing vorticity of classical wing theory. Indeed, we shall be able to relate λ, in the following, to $d\Gamma/dr$.

In exact Beltrami flow, this trailing absolute vorticity is precisely aligned with the relative velocity vector, $\underline{W}(r, \theta, x)$. (Contrary to remarks made in Ref. [12], this does not mean that $\underline{\Omega}$ is "convected with" the downstream flow; in fact, $\underline{W} \cdot \nabla \underline{\Omega} \neq 0$ in general [20, 21].)

Progress can be made in developing a practical theory from these general concepts if we adopt the expansion (linearization) about the mean flow ($\overline{\underline{\Omega}}$, $\overline{\underline{W}}$, $\bar{\rho}$, ...) suggested in the Introduction—and already used in Section II. A. Before doing this, however, it is useful first to employ some physical ideas consistent with the idealized three-dimensional flow we ultimately wish to describe. First, we note that in the downstream flow we should expect the actual vorticity, as opposed to its mean, $\overline{\underline{\Omega}}$, to be zero between the blading (since we have assumed uniform inlet conditions), and to remain zero between the corresponding wakes. The vorticity should be highly concentrated (singular) in the wakes represented here by trailing vortex sheets.* Further, this trailing vorticity, averaged pitchwise, should equal $\overline{\underline{\Omega}}$, and, if we define our perturbations correctly, our mean (reference) flow should also be a Beltrami flow with

$$\overline{\underline{\Omega}} \times \overline{\underline{W}} = 0$$

$$\mathrm{div}\,(\bar{\rho}\,\overline{\underline{W}}) = 0 \qquad (2.15)$$

In analogy with the above, $\overline{\underline{\Omega}} = \hat{\lambda}\overline{\underline{W}}$, $\bar{\rho}\overline{\underline{W}} = \nabla\alpha_o \times \nabla\psi$, $\alpha_o \equiv \theta - f_o(r, x)$, and

$$\hat{Q} \equiv \hat{\lambda}/\bar{\rho} = \hat{Q}(\alpha_o, \psi) = \hat{Q}(\psi).$$

---

*Of course, it is essential here that we know where the wakes are located; their location will emerge as one of the results of the mean-flow calculation which is the main subject of this section.

(The last equality follows immediately from the definition of α_o and the independence from θ of all mean-flow quantities.) Now we have already observed that the full flow can be described by $\rho \underline{W} = \nabla \alpha \times \nabla \beta$, so with the perturbations about the mean we also expect the variations $\delta \alpha \equiv \alpha - \alpha_o$, $\delta\beta \equiv \beta - \psi$, such that, formally,

$$(\rho \underline{W}) = \bar{\rho} \underline{\tilde{W}} + \tilde{\rho} \underline{\tilde{W}} + \ldots = \nabla(\delta \alpha) \times \nabla\psi + \nabla\alpha \times \nabla(\delta\beta) \qquad (2.16)$$

and [cf. Eq. (2.14) and below]

$$Q(\alpha, \beta) = Q_o(\alpha_o, \psi) + \frac{\partial Q_o}{\partial \alpha_o} \delta\alpha + \frac{\partial Q_o}{\partial \psi} \delta\beta + \ldots \qquad (2.17)$$

Because of the physical interpretation, discussed in the following, of the meaning of the surfaces α = const. and β = const., we require $\delta\beta = 0$. Then, in view of the fact that $\delta\alpha = f_o - f$ is by definition independent of θ, we must conclude, to be consistent, that

$$\delta\alpha = 0; \quad f_o(r, x) = f(r, x); \quad \alpha_o = \alpha, \qquad (2.18)$$

in order to satisfy the requirement $\overline{(\widetilde{\rho \underline{W}})} = 0.$*

This remarkable feature of the present treatment, although formal, is extremely useful, since it, together with the consequent form of (2.17)—viz., $Q(\alpha, \beta) = (\partial Q_o/\partial\psi)\delta\beta$—will enable us (see Section III) to express the perturbation flow field in terms of the single scalar function, $\delta\beta(r, \theta, x)$, and quantities such as $\bar{\rho}$, ψ and α, which are determined by the mean flow. Apparently, the fact that this is possible results physically from the pitchwise periodicity inherent to turbomachinery flows.

To summarize, the mean flow is characterized by α, ψ and $\bar{I}(\psi)$ (= ct., here), while the perturbation flow is characterized by

$$\bar{\rho} \underline{\tilde{W}} + \tilde{\rho} \underline{\bar{W}} \simeq \nabla\alpha \times \nabla(\delta\beta) \qquad (2.19)$$

and analogous expressions for $\tilde{\Omega}$, $\tilde{\rho}$, etc., developed in III. Algebraically, the theoretical formulas developed throughout this report will by and large be written in as compact a form as possible. We do this to focus here primarily on comparison between theory

---

*Actually, to complete this proof one needs to note that although $\nabla\alpha \equiv \hat{e}_\theta/r - \nabla f(r, x)$ depends on θ through $\hat{e}_\theta$, nevertheless $\overline{\nabla\alpha \times \nabla(\delta\beta)} = 0$ if $\overline{(\delta\beta)} = 0$.

and experiment and, generally, to avoid the lengthy algebraic equations which seem so inevitable. Nevertheless, it is important that these general results be written out in more familiar terms for various cases of particular practical interest, and this is being done in a companion report currently in preparation.

In the remainder of the present section we concentrate our attention on the mean compressible Beltrami flow equations. However, a treatment as formal as this one requires physical interpretation as we go along—both to make it believable and to make it useful. For example, we will find below that the surfaces, α = constant, are actually the surfaces in which the trailing vorticity lies, while of course the stream surfaces for the mean flow, ψ = constant, are already familiar, conceptually, even if they are by no means trivial to calculate [17, 19]. By analogy, the surfaces, $\beta = \psi + \delta\beta$ = const., are the stream surfaces of the detailed three-dimensional flow, so $\delta\beta$ represents the "ripples" of these surfaces, between the blades and/or their wakes, relative to the axisymmetric through-flow ψ-surfaces. For this reason, $\overline{\delta\beta} = 0$.

Our choice of writing the flow-field vector $\rho \underline{W}$ in the form $\nabla\alpha \times \nabla\beta$ also has the classical interpretation that the streamlines associated with $\rho \underline{W}$ lie on the intersections of the surfaces α = const., β = const., which is consistent with the above description. However, it is important to note that this latter statement also implies that the quantities α and β have physical significance only in terms of the set of surfaces α = const., β = const. This is why our putting $\alpha = \theta - f(r, x)$ is in no way restrictive, since the statements $\theta - f(r,x) = \alpha_1, \alpha_2, \alpha_3, \ldots$ (= constants) provide a perfectly general way of specifying a set of three-dimensional surfaces in the space (r, θ, x).

Choosing α in this particular way can best be described as an "educated guess;" indeed, a fortuitous one. We will verify below that $f(r, x)$, and hence α, is completely determined by the mean flow (design) problem. Thus, our proof above that $\overline{\delta\alpha} = 0$, to second order in the perturbations, in fact, implies that the trailing vortex sheets actually do contain the mean flow streamlines, neglecting only terms of second order. This is a result which this author, at least, was unable to anticipate in advance; as already mentioned, it apparently arises from the periodicity constraint in turbomachine flows.

To derive the downstream compressible Beltrami equations for the mean flow, we return to Eqs. (2.15), take advantage of the fact that $\hat{Q} = \hat{Q}(\psi)$, and write

$$\underline{\bar{\Omega}} = \hat{Q}(\psi), (\nabla\alpha \times \nabla\psi) = \hat{Q}(\psi)\, \bar{\rho}\underline{\bar{W}} \tag{2.20}$$

Looking at the r- and x- components of $\underline{\bar{\Omega}}$, noting that $\overline{rv}^{\,d} = \frac{B\Gamma}{2\pi}(\psi)$ in this case, and comparing components in (2.20), we conclude that

$$Q(\psi) = -\frac{B\Gamma'}{2\pi}(\psi) \tag{2.21}$$

where $(\ )' = d/d\psi$. Then the θ-component of (2.20) yields

$$\frac{\partial \bar{u}}{\partial x} - \frac{\partial \bar{w}}{\partial r} = -\frac{B\Gamma'}{2\pi}\bar{\rho}\left(\frac{B\Gamma}{2\pi r} - \omega r\right)$$

or

$$\frac{\partial}{\partial x}\left(\frac{1}{\bar{\rho}}\frac{\partial\psi}{\partial x}\right) + r\frac{\partial}{\partial r}\left(\frac{1}{\bar{\rho} r}\frac{\partial\psi}{\partial r}\right) = -\frac{B\Gamma'}{2\pi}\bar{\rho}\left(\frac{B\Gamma}{2\pi} - \omega r^2\right) \tag{2.22}$$

This, again, is Marble's nonlinear equation for $\psi(r, x)$ [17], specalized to the idealized situation (pure Beltrami flow) discussed here. Combining it with the statement

$$\bar{I} = \text{ct.} = \bar{h}_{-\infty} + (1/2)\bar{w}^2_{-\infty} = \left(\bar{h} + q^2/2 - \frac{\omega B\Gamma}{2\pi}\right),$$

and using the constancy of entropy applicable to the present situation allows the appropriate derivatives of $\bar{\rho}$ to be determined in (2.22) in terms of local Mach numbers in the (downstream) mean flow. Equation (2.22) can then be cast in the form

$$\begin{aligned}(1 - \bar{M}_x^2)\frac{\partial^2\psi}{\partial x^2} + (1 - \bar{M}_r^2)\left[\frac{\partial^2\psi}{\partial r^2} - \frac{1}{r}\frac{\partial\psi}{\partial r}\right] - \bar{M}_\theta^2\frac{1}{r}\frac{\partial\psi}{\partial r}\\ - 2\bar{M}_x\bar{M}_r\frac{\partial^2\psi}{\partial x\partial r} + \bar{M}_x\bar{M}_r\frac{1}{r}\frac{\partial\psi}{\partial x} = -\frac{B\Gamma'}{2\pi}\bar{\rho}^2\left[\frac{B\Gamma}{2\pi} - \omega r^2\right]\end{aligned} \tag{2.23}$$

The homogeneous (potential flow, $\Gamma' = 0$) version of this result is in agreement [14] with the corresponding equation given by Shapiro [22]. A convenient form of the companion downstream "energy" equation, to be used with (2.23), can be written, for a perfect gas, in the form

$$\frac{T_t^{-\infty}}{\overline{T}} = [1 + \frac{\gamma-1}{2}\overline{M}^2 - (\gamma-1)\frac{\omega B\Gamma(\psi)}{2\pi\bar{a}^2}] = (1 + \frac{\gamma-1}{2}\overline{M}^2$$

$$- (\gamma-1)\overline{M}_\omega \overline{M}_\theta) \tag{2.24}$$

where

$$\overline{M}^2 = \overline{M}_r^2 + \overline{M}_\theta^2 + \overline{M}_x^2 = \frac{\bar{u}^2 + \bar{v}^2 + \bar{w}^2}{\bar{a}^2},$$

$\bar{a}^2 = \gamma R\overline{T}$, $\overline{T}$ is the mean static temperature, γ is c_p/c_v, and $T_t^{-\infty}$ is the inlet total temperature. Also,

$$\overline{M}_\theta = \frac{B\Gamma}{2\pi r\bar{a}}$$

downstream of the fan, and $\overline{M}_\omega \equiv \omega r/\bar{a}$.

One of the practical design purposes of solving (2.23) with (2.24), given the loading Γ, is to determine the axial rate of relaxation to the radial equilibrium state downstream of the rotor, which has a bearing on interstage spacing. A relatively simple rule for determining that relaxation rate has been given by Marble [17], and this is frequently used for preliminary design.

For our present aim, however, namely that of comparing theory with experiment, a somewhat different point of view in the use of the mean flow equations is helpful. Among other things, we wish to keep in mind that the mean-flow solution provides the "reference" quantities for our three-dimensional perturbation theory. In addition, we can directly compare our solutions for the mean flow problem with pitchwise-averaged results from the experiment (see Ref. [19] and Section V of this paper).

To complete this section, we note briefly that f(r, x), and hence $\alpha = \theta - f$, is determined from the θ-component of $\bar{\rho}\underline{W} = \nabla\alpha \times \nabla\psi$, namely

$$\bar{\rho}\,(\bar{v} - \omega r) = \frac{\partial\alpha}{\partial x}\frac{\partial\psi}{\partial r} - \frac{\partial\alpha}{\partial r}\frac{\partial\psi}{\partial x}$$

$$= \bar{\rho}\,(\bar{u}\frac{\partial f}{\partial r} + \bar{w}\frac{\partial f}{\partial x})$$

and hence

$$\underline{\overline{V}} \cdot \nabla f = \underline{\overline{W}} \cdot \nabla f = \left(\frac{B\Gamma(\psi)}{2\pi r} - \omega r\right) \tag{2.25}$$

Thus, for the design problem (given Γ, ω), f is directly computable by integration of the given data along the mean-flow stream-surfaces. Fortunately, the latter are obtainable from the ψ-equation given by Marble [17], (specializations being represented by Eqs. (2.23), (2.22) and (2.11)), which is itself independent of f.

The quantity f(r, x), which to an excellent approximation determines where the wake surfaces are located (see also the data discussed in Section V), has some interesting features, some of which are described in Refs. [12] and [13]. It is to be expected that the location of the wakes should be relatively independent of compressibility, and, in fact, $\bar{\rho}$ has cancelled out in (2.25). More interesting, however, is the fact that the (inviscid) wakes quickly "wrap-up" on themselves [12, 13]. This has the physical implication that the inviscid three-dimensional flow (see Section III) rapidly reconverts itself into an essentially axisymmetric one—after an axial distance downstream of the blading roughly comparable to the span ($r_t - r_h$). The induced velocities at the blades associated with the (inviscid) trailing vortex sheets must therefore be computable "locally," i.e., from the structure of the wakes near the blading. A rigorous proof of this admittedly intuitive notion is lacking at the moment, but it is verified for a special case in Ref. [12].

One striking implication of these observations is that any observed pitchwise structure of the flow sufficiently far downstream of an isolated rotor [15] is likely due to viscous effects in rotating flows such as those described by Kerrebrock [21], and these should not show substantial interferences with the (essentially inviscid) trailing vorticity arising from spanwise variation of the blade loading. In this way one can perhaps understand how the observed flow changes, as it moves downstream, from one with a pitchwise structure associated with the blade number, B, to one with a pitchwise structure not (directly) related to B [15, 16, 19, 21].

III. PERTURBED THREE-DIMENSIONAL TRANSONIC BELTRAMI FLOW

The three-dimensional perturbation equations for transonic flow can be derived by returning to Eqs. (2.14), (2.17) [with δa=0], and (2. 19), and applying the physical ideas discussed above (2. 15). If $\underline{\Omega}$ is singular (compressed into idealized vortex sheets) then so is $\overline{(\lambda/\rho)} = Q(\alpha, \beta)$. This is why we have refrained, so far, from taking the mean of $\underline{\Omega} = Q(\rho \underline{W})$ in (2. 14) and relating it immediately to the quantities in (2. 15) or (2. 20). (Because Q is singular, care must be exercised as already mentioned, in applying the simple expansion rule given in the footnote below the start of Section II. B.). Despite its singular nature, however, $Q(\alpha, \beta)$ can be expanded in a Taylor series, in the sense of generalized functions, as in (2. 17) (see Lighthill, Ref. [23]).

If we now expand (2. 14) with the help of (2. 17), we find in the downstream flow

$$\underline{\Omega} = \overline{\underline{\Omega}} + \tilde{\underline{\Omega}} = Q_o(\alpha, \psi)\bar{\rho}\, \overline{\underline{W}} + Q_o'(\alpha, \psi)\delta\beta\bar{\rho}\, \overline{\underline{W}} + (Q_o(\alpha, \psi)\widetilde{(\rho \underline{W})} + \ldots$$

$$= (Q_o(\alpha, \psi) + Q_o'\delta\beta)(\nabla\alpha \times \nabla\psi) + Q_o(\alpha, \psi)(\nabla\alpha \times \nabla(\delta\beta)) \qquad (3.\ 1)$$

where $(\ )' = d/d\psi$. Now we see that our physical description of the flow (see above Eq. (2.15)) is satisfied by assuming simply that

$$Q_o(\alpha, \psi) = F(\psi)\, \delta_p(\alpha) \qquad (3.2)$$

where $\delta_p(\alpha)$ is the periodic delta function

$$\delta_p(\alpha) = \sum^{\infty} \delta(\alpha - \frac{\pi}{B} - \frac{2\pi n}{B}), \quad n=0, \pm 1, \pm 2, \ldots .$$

With the trailing edge of the blades located at $x = 0$,

$$\theta = \pm\pi/B, \pm\frac{3\pi}{B},$$

etc. (This choice requires $f(r, x) = 0$ at $x = 0$.), use of (3.2) in (3.1) indeed makes $\underline{\Omega}$ in the downstream region zero between the wakes and singular at the wakes, $\alpha = \alpha_n = (2n+1)\pi/B$. We shall use the properties

$$\int_{-\epsilon+\alpha_n}^{\epsilon+\alpha_n} d\theta\ \delta(\alpha - \alpha_n) = 1, \text{ and } \overline{\delta_p(\alpha)} = B/2\pi.$$

Hence, invoking periodicity, we readily find that, at each fixed (r, x):

$$\underline{\overline{\Omega}} = \frac{B}{2\pi}[F + F'(\psi)\delta\beta_B](\nabla\alpha \times \nabla\psi) + \frac{BF}{2\pi}(\psi)(\nabla\alpha \times \nabla(\delta\beta))_B \qquad (3.3)$$

where the subscript $(\ )_B$ means "evaluated at the wakes." For example,

$$\delta\beta_B = \delta\beta(r, x, \theta = f + \pi/B) = \delta\beta(r, x, \theta = f + \frac{3\pi}{B}) = \ldots .$$

Now, to maintain the necessary generality, suppose $\delta\beta = \tilde{\beta}(\alpha, \psi) + \hat{\beta}(r, \theta, x)$, whence

$$\delta\beta_B = \tilde{\beta}(\frac{\pi}{B}, \psi) + \hat{\beta}(r, f + \pi/B, x).$$

The part, $\tilde{\beta}$, represents a portion of $\delta\beta$ which is constant along the streamlines lying in the wakes (α = const., ψ = const.), and hence a portion of $(\widetilde{\rho \underline{W}})$ proportional to $\overline{\rho \underline{W}}$. The part, $\hat{\beta}$, on the other hand, represents a portion of the disturbance field which can decay (or propagate) along such streamlines and we know from experience that such a portion of the disturbance field must be present (Namba [1], Falcao [2], Okurounmu [24], and McCune [3]). Okurounmu [24] also noted that such a portion vanishes on the wakes unless propagating modes (modes above "cut-off") occur. Suppressing details, all this means is that the surviving terms in (3.3) contain only $\tilde{\beta}(\pi/B, \psi)$ (= $\tilde{\beta}_B(\psi)$, for short), so that (3.3) collapses* to

$$\begin{aligned} \underline{\overline{\Omega}} &= \frac{B}{2\pi}[F(\psi) + \frac{d}{d\psi}(F\tilde{\beta}_B)](\nabla\alpha \times \nabla\psi) \\ &= \frac{B}{2\pi}[F(\psi) + \frac{d}{d\psi}(F\tilde{\beta}_B(\psi))]\,\overline{\rho \underline{W}} \end{aligned} \qquad (3.4)$$

Comparison of this with (2.20) and (2.21) shows that

$$[F(\psi) + d(F\tilde{\beta}_B)/d\psi] = \Gamma'(\psi).$$

On using the classical operational property of delta functions [23], viz. $\delta_p(\alpha)(\delta\beta) = (\delta\beta)_B\delta_p(\alpha)$, we find, after some manipulation [12,19], that

*The remaining constraint [25] on the perturbations is actually that $F'(\psi)\,\overline{\rho \underline{W}}\,\hat{\beta}_B(r, x) + F(\psi)(\nabla\alpha \times \nabla\hat{\beta})_B = 0$. This can be satisfied for decaying or propagating perturbation modes with appropriate θ-symmetry, a special case being that observed in Ref. [24] and used in Ref. [4].

$$\underline{\tilde{\Omega}} = -\Gamma'(\psi)(\delta_p(\alpha) - \frac{B}{2\pi})(\nabla\alpha \times \nabla\psi)$$

$$\equiv \nabla\Gamma(\psi) \times \nabla S(\alpha) \tag{3.5}$$

where $S(\alpha)$ is the "sawtooth" function [1, 3-6, 24, 13] such that $S'(\alpha) = \delta_p(\alpha) - B/2\pi$. (Since

$$(1/2\pi)\int_0^{2\pi} \delta_p(\alpha) = \frac{B}{2\pi}, \quad \overline{S'(\alpha)} = 0.)$$

This periodic sawtooth function [24] (see also Reissner [26] passes through the origin, $\alpha = 0$, with unit slope, $dS/d\alpha = 1$, has negative jumps of magnitude $2\pi/B$ at $\alpha = \pi/B$, $3\pi/B$, ..., etc., and strictly speaking is to be understood in the operational sense of a "generalized" function. (As made clear by Lighthill [23], however, such functions can be generally regarded, pragmatically, as useful tools in engineering analysis. They arise most often in analyses of physical problems which have been strongly idealized, as in the present (assumed) case of inviscid, isentropic flow.)

The sawtooth function also has a useful Fourier expansion [23, 24, 25, 4-6, 19]

$$S(\alpha) = \frac{B}{2\pi}\sum_{n\neq 0} \frac{(-1)^n}{inB} e^{inB\alpha} \tag{3.6}$$

with sin α symmetry. The convergence of this series is again to be understood in the sense explained in Ref. [23]. For our purposes, we may simply proceed to regard the quantities in (3.5) and (3.6) as ordinary, well-behaved functions and/or their Fourier representations.

Now, let us assume that $\underline{\tilde{V}} = (\tilde{u}, \tilde{v}, \tilde{w}) = \underline{A} + \nabla\phi$. Comparing this with Eq. (3.5), and recalling that $\underline{\Omega} \equiv \text{curl}\ \underline{V}$ and $\overline{\text{curl}\ \underline{V}} = \text{curl}\ \overline{\underline{V}}$, so that $\underline{\tilde{\Omega}} \equiv \text{curl}\ \underline{\tilde{V}}$, we obtain

$$\underline{A} = -S(\alpha)\nabla\Gamma(\psi) \tag{3.7}$$

(plus any gradient), and hence, with (3.5) thus automatically satisfied,

$$\nabla^2\phi = -\text{div}\ \underline{A} + \text{div}\ \underline{\tilde{V}} = +S\nabla^2\Gamma(\psi) + \nabla S \cdot \nabla\Gamma + \text{div}\ \underline{\tilde{V}} \tag{3.8}$$

Special solutions for the perturbation flow in the incompressible limit of (3.8)—div $\underline{\tilde{V}} \cong 0$ —(appropriate for low Mach numbers) have been discussed in detail in Refs. [12] and [13], and also applied in a special way (described in Section V) to the M.I.T. transonic ducted-fan experiment [19]. Compressibility effects in (3.8) turn out to be of importance primarily in determining the existence and the aerodynamic effects of propagating (acoustic) modes arising in the flow fields of transonic ducted fans [9, 27, 28, 3]. The nature of these modes in the present, "quasi-linear," treatment is still under study, (see also, Section VI); as one might expect, some striking differences from earlier, strictly linear, analyses [1, 28] are emerging.

By contrast, insofar as one is primarily interested in wake effects, the incompressible version of (3.8) largely suffices, provided adjustments are made for compressibility in the mean flow. Solution techniques in both approaches are discussed briefly in Section IV, with some useful simplifications (primarily in the boundary conditions and the driving term involving $\Gamma(\psi)$) applied. Before proceeding, however, we complete Eq. (3.8) by determining div $\underline{\tilde{V}}$.

To begin, it is useful to note that, since I = ct. here, the perturbation flow satisfies

$$\tilde{h} + \underline{\bar{V}} \cdot \underline{\tilde{V}} - \omega r \tilde{v} = 0$$

or

$$\tilde{h} + \underline{\bar{W}} \cdot \underline{\tilde{V}} = 0$$

For a perfect gas this yields, using notation already introduced in Section II.C.:

$$\frac{\tilde{T}}{\bar{T}} + (\gamma - 1)\frac{\underline{\bar{W}} \cdot \underline{\tilde{V}}}{q^2}\bar{M}^2 = 0 \tag{3.9}$$

Further, to the same approximation, the flow is also isentropic, so

$$\frac{\tilde{T}}{\bar{T}} + (\gamma - 1)\frac{\tilde{\rho}}{\bar{\rho}} = \frac{\gamma - 1}{\gamma}\frac{\tilde{p}}{\bar{p}} \tag{3.10}$$

and Eq. (3.9) can be used to relate $\tilde{\underline{V}}$ to the fluctuations of any of these thermodynamic quantities.

Evaluation of div $\tilde{\underline{V}}$ in terms of $\nabla\phi$, $\underline{A} = -S\nabla\Gamma$, and the mean flow quantities now follows straightforwardly, but involves some tedious manipulations. One reasonably efficient procedure begins with noting that div $(\overline{\rho \underline{W}}) = 0$, hence div $(\widetilde{\rho \underline{W}}) = 0$, and hence

$$0 \cong \text{div}\,(\bar{\rho}\,\tilde{\underline{V}}) + \text{div}\,\left(\frac{\tilde{\rho}}{\bar{\rho}}\,\overline{\rho \underline{W}}\right).$$

Thus,

$$\text{div}\,\tilde{\underline{V}} = -\overline{\underline{W}} \cdot \nabla\left(\frac{\tilde{\rho}}{\bar{\rho}}\right) - \tilde{\underline{V}} \cdot \nabla \ln \bar{\rho}$$

and, using (3.9) and (3.10), we obtain, with (3.7),

$$\text{div}\,\tilde{\underline{V}} = \overline{\underline{W}} \cdot \nabla\left(\frac{W}{\bar{a}^2} \cdot (\nabla\phi - S\nabla\Gamma)\right) - (\gamma-1)^{-1}\,(\nabla \ln \bar{T}) \cdot (\nabla\phi - S\nabla\Gamma) \qquad (3.11)$$

which, as required, tends to zero with low mean-flow Mach numbers (see Eq. (2.24)). Further useful simplifications of (3.11) can be achieved using (Eq. (2.24). For the moment, however, we content ourselves with completing (3.8) by inserting (3.11):

$$\nabla^2\phi - \left[\frac{\overline{\underline{W}}}{\bar{a}} \cdot \nabla\left(\frac{\overline{\underline{W}}}{\bar{a}} \cdot \nabla\phi\right) - \frac{\nabla\bar{a} \cdot \nabla\phi}{(\gamma-1)\bar{a}}\right] = S\nabla^2\Gamma + \nabla S \cdot \nabla\Gamma$$
$$- \left[\frac{\overline{\underline{W}}}{\bar{a}} \cdot \nabla\left(\frac{\overline{\underline{W}}}{\bar{a}} - S\nabla\Gamma\right) - \frac{\nabla\bar{a} \cdot S\nabla\Gamma}{(\gamma-1)\bar{a}}\right] \qquad (3.12)$$

which is our perturbation equation for ϕ, given $\Gamma(\psi)$ and the mean-flow solution.

It is perhaps surprising that such a dreadful-looking equation affords relatively simple solutions [13, 19] and, in fact, useful ones (Section V).

As already suggested, certain simplifications can be obtained using Eq. (2.24). Moreover the corresponding perturbation equation for "strictly-linear" theory [1-8] ($\overline{\underline{W}} = (o, \omega r, U)$, $\bar{a} = a_{-\infty}$, $\alpha = \omega x/U$, etc.) can be recovered relatively easily from (3.12).

IV. PRACTICAL SOLUTIONS IN THREE-DIMENSIONAL, TRANSONIC, FLOW THROUGH DUCTED FANS

Useful application of (3.12), as well as the governing mean-flow equations (Section II.C), begins with the making of reasonable physical approximations [17, 19]. Many possibilities suggest themselves, but for the moment let us consider a ducted fan operating at nearly "constant work," $\Gamma(\psi) \cong$ ct. In incompressible flow with uniform, zero-swirl inlet conditions [12, 13], a fan or rotor operating at exactly constant work in a straight annular duct* induces no mean streamsurface deflection; i.e., $\bar{u} \sim \partial\psi/\partial x$ is zero throughout the flow, and $\psi = \psi(r)$. Radial equilibrium, in the mean, is established immediately downstream of the fan; indeed, in this very special case, it applies everywhere, except, perhaps, in the blade passages.

A fan operating at constant work in compressible flow under the same conditions does induce some mean stream-surface deflection, because of the coupling between $\bar{\rho}$, $\bar{p}$, $\bar{v}(\sim K/r$, here) and mean-flow continuity requirements. In this case, therefore, $\partial\psi/\partial x \neq 0$ near the blading (generally) but, in practice, it is still quite small, even at inlet axial Mach numbers as high as 0.6-0.7, and radial equilibrium, with $\bar{w} = \bar{w}(r)$, $\bar{\rho} = \bar{\rho}(r)$, $\bar{T} = \bar{T}(r)$, $d\bar{h}/dr = \bar{v}^2/r$, etc., is attained rapidly behind the blading [17, 29].

A useful approximate picture of the mean flow [19, 29, 17], therefore, is to treat $\bar{u}/\bar{w} \sim \bar{u}/\bar{v} \sim \bar{u}/\omega r$ as small. This approximation is useful for many purposes, such as in the driving term in (3.12) or (3.8). Also, in the coefficients of the perturbations $\partial\phi/\partial x$, $\partial^2\phi/\partial x^2$, $(\partial/\partial r)(r\ \partial\phi/\partial r)$, etc., in (3.12) we can treat certain key mean-flow quantities as functions of r only.** To

---

*For the moment, we overlook radial hub and/or shroud convergence. The extent to which this approximation is useful will be discussed subsequently.

**Formally, the smallness of $\partial\psi/\partial x$ can be related to $(B)^{-1}$, $(r_T - r_h)/r_T$, $\Gamma'\bar{\rho}r_T$, or any combination of these. In applying the approximation $\nabla\Gamma \cong \Gamma'(\partial\psi/\partial r)\hat{e}_r$ in the dot product with $S'(\alpha)\nabla\alpha$, care must be exercised, because $\partial f/\partial r$, although initially small, can increase without bound [12, 13].

illustrate, consider $\nabla\Gamma = \Gamma'(\psi)\,\nabla\psi$, which occurs in (3.12). In the above approximation

$$\nabla\psi = \frac{\partial\psi}{\partial r}\hat{e}_r + \text{(higher order terms)} = -\bar{\rho}\bar{w}r\hat{e}_r + \ldots,$$

and $\nabla\Gamma \cong -\bar{\rho}\bar{w}r\Gamma'\hat{e}_r$. Similarly, $\underline{\overline{W}} \cong (0, (B\Gamma(r)/2\pi r) - \omega r, \bar{w}(r))$ can be used approximately in the coefficients of the derivatives of ϕ.

In the mean flow (cf. Eqs. (2.23), (2.24)) this picture is very useful in obtaining results for the mean flow quantities by iterative techniques [19, 29], even when departure from constant work is substantial. The method is most effective when allowance for axial relaxation to asymptotic conditions, based essentially on Marble's analysis [17], is included. Alternatively, WKB methods can be used [13]. We shall see (Section V, Ref. [19]) that the results, compared with actual experiment [15, 16], are encouraging.

The treatment used in Ref. [19] is essentially incompressible [i.e., (3.8) is used with div $\underline{\tilde{V}} = 0$] —with due adjustment for changes of state (in the mean) upstream and downstream of the fan. These results, then, are focused on wake effects.

As a first step in a more detailed analysis, including the effects of acoustic modes above cut-off, we have used the ideas described above to obtain useful approximations in both the mean-flow and perturbation equations. In particular, with the blades represented by B lifting lines, and with the circulation given in the form [29]

$$\frac{\Gamma(r)}{\bar{\Gamma}} = 1 - \epsilon \cos\left[\left(\frac{r - r_h}{r_T - r_h}\right)\pi\right] \tag{4.1}$$

we compute the flow angles at the blades. Here,

$$\bar{\Gamma} \equiv \frac{2}{r_T^2 - r_h^2}\int_{r_h}^{r_T} r\,dr\,\Gamma(r).$$

In Figs. 1-3, comparisons are made of the computed deflection angles at the blades vs. spanwise station (with "δ" representing the difference between the inlet air angle and the mid-chord air angle at each r/r_T) for descriptions of differing degrees of

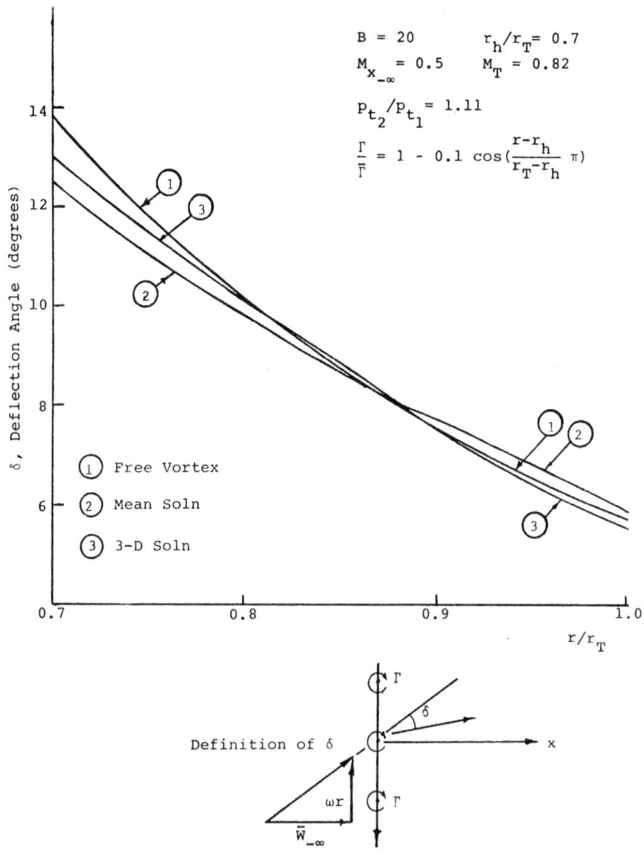

Figure 1. Computed Deflection Angles at the Blades vs. Spanwise Station for Various Flow Descriptions for a Lightly Loaded Subsonic Fan.

"completeness". For example, in Fig. 1, we have a lightly-loaded subsonic fan (B = 20, h = 0.7, $\overline{M}_x^2 + \overline{M}_\omega^2 = (0.782)^2$, $p_t^\infty/\bar{p}_t^{-\infty} = 1.11$, $\epsilon = 0.1$), and compare predictions of the air angles at the blades according to the differing descriptions of the flow. Curve (1) shows the predictions that would be obtained ignoring ϵ($\epsilon = 0 \rightarrow$ pure free vortex flow). Curve (2) gives the results for the "correct" mean flow (ignoring $\underset{\sim}{V}$, but including $\epsilon = 0.1$). Finally, curve (3) illustrates the air angles including—in addition—the three-dimensional induced-velocity effects. Even for such a lightly-loaded fan these three-dimensional changes are roughly comparable (and, incidentally, partially compensating) to those associated with adjusting the mean flow for departures from constant work ($\epsilon \neq 0$).

Curve (2) corresponds to the results obtained by methods in current use. The 3-D corrections, computed here, have not been available previously, to our knowledge, except in strictly linear analysis [1-8].

Figures 2 and 3 show corresponding results for much more highly loaded transonic fans or rotors (B=40, h=0.8), with $\epsilon = 0.2$, so that the design differs more significantly from constant-work. The pressure ratios and relative Mach numbers are (1.5, 1.1) and

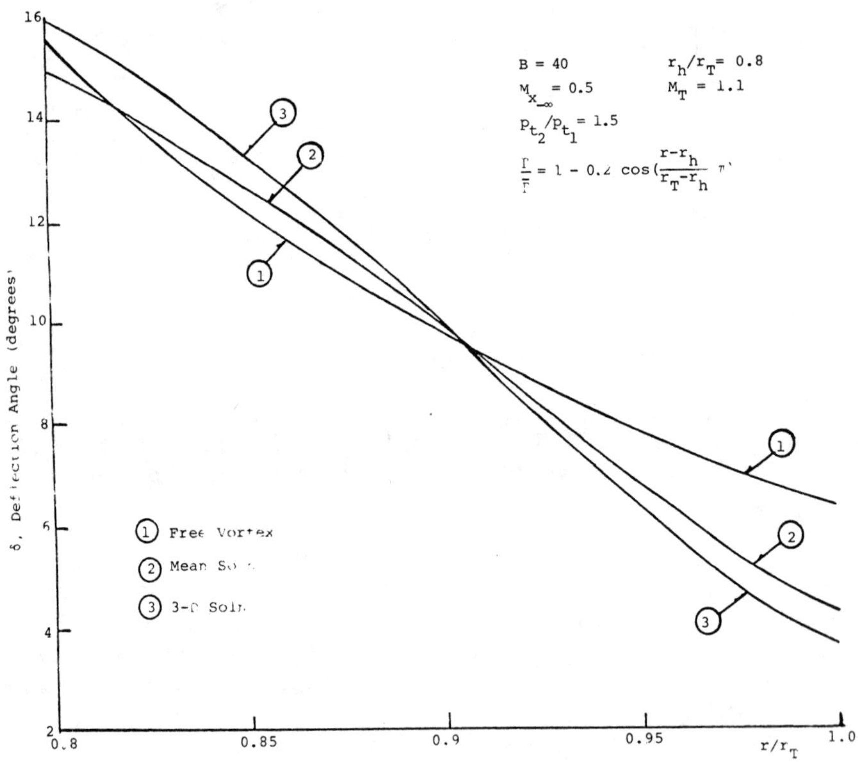

Figure 2. Computed Deflection Angles at the Blades vs. Spanwise Station for Various Flow Descriptions for a Highly Loaded Transonic Fan or Rotor.

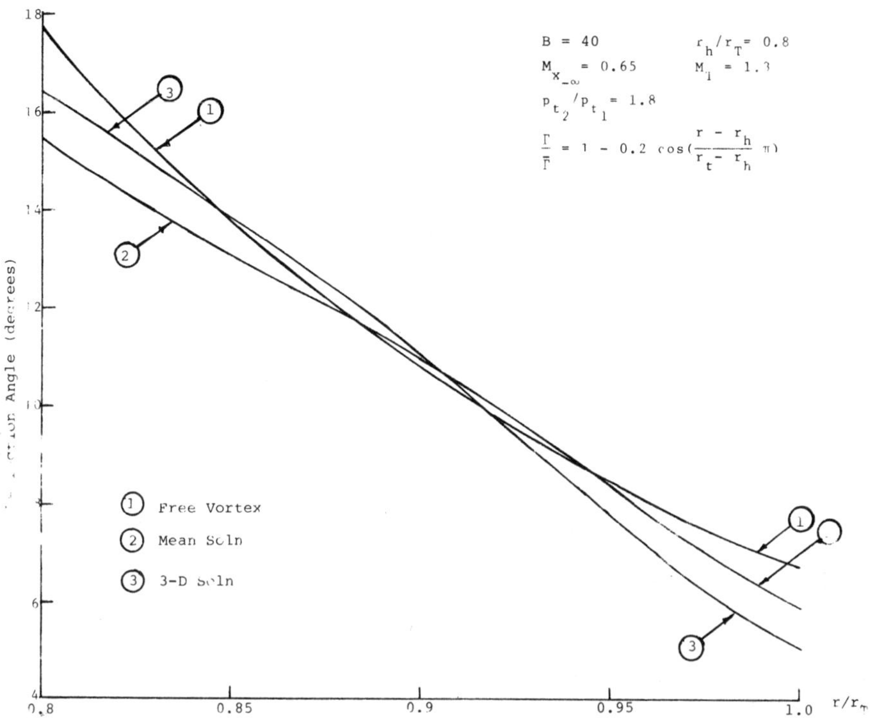

Figure 3. Computed Deflection Angles at the Blades vs. Spanwise Station for Various Flow Descriptions for a Highly Loaded Transonic Fan or Rotor.

(1.8, 1.3), respectively;* these pressure ratios are well out of the range of earlier perturbation theories.

In Figures 2 and 3 we again see that three-dimensional induced velocity effects are certainly comparable with, and in some spanwise regions more important than, adjustments made due to non-constant Γ in the mean flow.

---

*The relative Mach number, $\bar{M}_r$, is given by $\bar{M}_r^2 = \bar{M}_x^2 + \bar{M}_\omega^2\,(r_T)$. The combination (1.5, 1.1) means $p_t^{\infty}/p_t^{-\infty} = 1.5$, $\bar{M}_r = 1.1$, and so on.

These results indicate that, if the present three-dimensional theory has in it some additional elements of the actual physical processes occurring in real turbomachinery flows not normally included in standard analyses, then these effects are likely to be important in turbomachinery design (and, by implication, in off-design studies as well).

Before we can be sure, however, that the effects predicted here are important to include, we must be certain that they are real. This will be the subject of Section V.

The overall theory [29] leading to the results shown in Figs. 1-3 also has important implications regarding the noise spectra to be expected upstream of transonic rotors. A few comments on this point will be included in Section VI.

V. THEORY AND EXPERIMENT

The ducted fan experiment in the M. I. T. blow-down facility was intended to operate at, or near, constant work at the transonic design point. The design methods used correspond essentially to those presently being employed in the most careful and complete procedures available at the present time. For a variety of reasons, including almost certainly the effects of boundary layer-shock wave interaction in the transonic portions of the rotor passage, the actual experimental rotor does not operate, even approximately, with constant spanwise circulation. It does, however, produce very nearly the expected overall compression ratio and operates, at design, with relatively low losses.

Interestingly, the work distribution (or, alternatively, the span-wise circulation distribution) that actually develops [15, 16, 19] is qualitatively similar to those observed in other related transonic rotor experiments. For the M. I. T. experiment [15] the actual observed distribution of $\Gamma(r)$—deduced from the pitchwise-averaged downstream tangential velocity—is as shown in Fig. 4. The fact that $\Gamma(r)$ departs from its spanwise average most drastically outboard of the radius at which the relative velocity is sonic is suggestive of some of the processes which very likely cause the observed departure from design.

If, for any reason—intended or otherwise—$\Gamma(r) \neq$ ct., the three-dimensional effects associated with the trailing vorticity described in Section IV come into play. The theories introduced

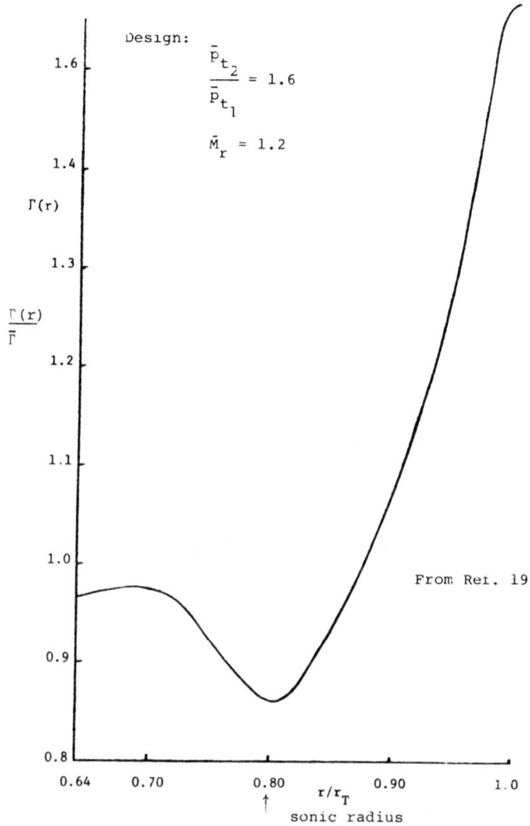

Figure 4. Γ(r) Distribution for the M. I. T. Transonic Fan Experiment.

in Refs. [12] and [13] were worked out for this case, originally, under the assumption (discussed in the previous section) that $(r/\bar{\Gamma})(d\Gamma/dr)$ was small; also the incompressible limit was taken. In the comparison between theory and experiment to be made here we can use the incompressible theory to a good approximation for both the mean flow (cf. Eq. (2.22) with $\bar{\rho} \rightarrow 1$) and for the perturbation flow (cf. Eq. (3.8) with div $\tilde{V} \cong 0$) insofar as we are not concerned primarily with acoustic effects. For the mean flow the reason for this lies in the radial convergence of the hub in the experiment, which is so arranged as to keep $\bar{\rho}$ very nearly constant through the rotor passage [15, 19]. For the perturbation flow, as pointed out by Reissner [26], we can expect div $\underline{\tilde{V}}$ to be relatively unimportant as regards wake effects.

Then the mean downstream flow would be described, in the present notation (compare Ref. [19]), by

$$\frac{\partial^2 \psi}{\partial x^2} + r \frac{\partial}{\partial r}\left(\frac{1}{r}\frac{\partial \psi}{\partial r}\right) \cong -\frac{B}{2\pi}\left(\frac{1}{\bar{w}r}\frac{d\Gamma}{dr}\right)\left[\frac{B\Gamma}{2\pi} - \omega r^2\right] \tag{5.1}$$

if $\partial\psi/\partial x$ were small enough, or $\nabla\Gamma \cong \bar{w}r(d\Gamma/d\psi)\hat{e}_r$ as suggested in Section IV. However, it is obvious from Fig. 4 that $d\Gamma/dr$ is not sufficiently small, in the actual experiment, for this approximation to apply without modification. In Ref. [19] an iteration technique is described, whereby Eq. (2.22) is solved by successive approximation, beginning with Eq. (5.1). The design data ($\Gamma(r)$ at the rotor), for which the theory was devised, is replaced here by the measured Γ distribution from the experiment, as illustrated in Fig. 4. The iteration procedure [19] converges rapidly; four iterations suffice to bring the theoretical error within the uncertainty limits of the experimental data.

On this basis, our present theory predicts the mean flow results shown by the solid lines in Fig. 5, giving the predicted mean downstream radial and axial velocity components associated with the blade loading shown in Fig. 4. In Fig. 5, the experimental data for the corresponding quantities are also shown. The agreement between theory and experiment, particularly with regard to the axial velocity, seems to us to be quite encouraging, especially since the mean-flow analysis was carried out within the actuator-disc approximation. The disagreement between the data and theory for the mean radial velocity outboard of the sonic radius might well be expected in view of the presence of shock wave-boundary layer interaction effects in the sonic portions of the rotor blade passages. Note, however, that $\bar{u}$ (or $\bar{M}_r$) is close to zero throughout, as expected.

The three-dimensional perturbation flow associated with the wakes is computed on the basis of Eq. (3.8)—with appropriate boundary conditions at the hub and shroud, and with div $\underline{\tilde{V}} \cong 0$. For this reason any possible propagating acoustic modes and/or decaying (elliptic) potential modes are ignored in the present comparison. Their effects are treated in some detail in Ref. [29], but they are of minor importance with respect to the comparisons we

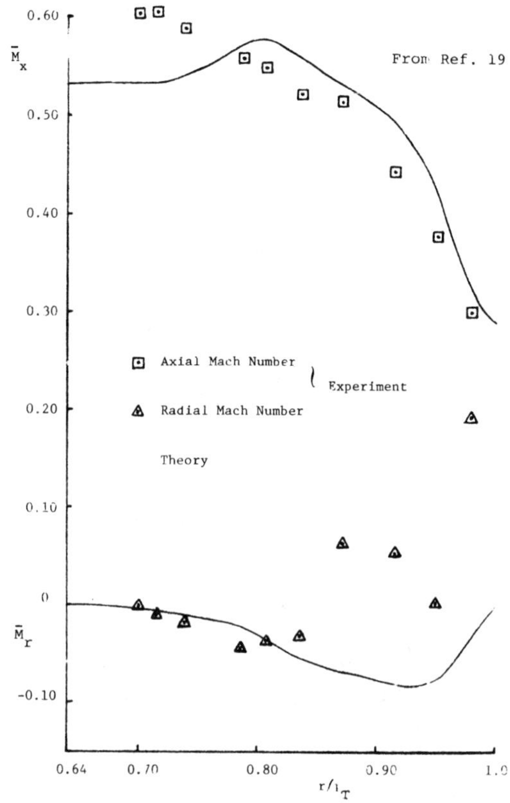

Figure 5. Mean Axial and Radial Mach Numbers; Theory vs. Experiment.

wish to make here.

With regard to the pitchwise perturbations we concentrate our attention here on the effects of the perturbations driven by the forcing term in (3.8), a characteristic feature of which is the sawtooth function, $S(\alpha)$, already described. We refer to these inhomogeneous solutions of the perturbation equation as the "wake functions." If we have correctly predicted and interpreted the location of the α-surfaces (see Section II. C. and also a more complete discussion in Refs. [12], [13] and [19]), we should be able to identify the phases of the downstream pitchwise disturbances at each (r, x) with experimental data. Moreover, to the extent that the Beltrami-flow (nearly loss-free flow) concept is applicable, we should see evidence of the sawtooth structure, especially in the radial and axial velocity components.

In Figs. 6-9, comparison between the ideal three-dimensional predictions and experimental observations are shown for all three velocity components at various spanwise stations. The comparison is made at a downstream station with axial position equal to 0.1 chord lengths. As one would expect, the actual (measured) vorticity is somewhat diffused as compared to the idealized picture of Sections III and IV but, otherwise, the three dimensional theory appears to be remarkably successful in locating the wakes. [One might remark here that earlier, strictly linear, theories [1-8] would exhibit considerable error in this respect.] At the same time, the sawtooth-like structure of the pitchwise variation for the appropriate velocity components is quite evident (cf. Figs. 6a, 6b, for example).

As regards the pitchwise variations, both qualitative and quantitative agreement between the present theory and actual experiment are best at the outboard stations (e.g., $r/r_T = 0.95$) where $d\Gamma/dr$ is large and the trailing vorticity dominates other types of vorticity, such as that generated by losses [20], or by vorticity introduced at the inlet [25]. Whereas overall agreement is good in the comparison shown in Fig. 6, it is certainly poor, at least quantitatively, in Fig. 9 which is for a station at which $d\Gamma/dr \cong 0$.

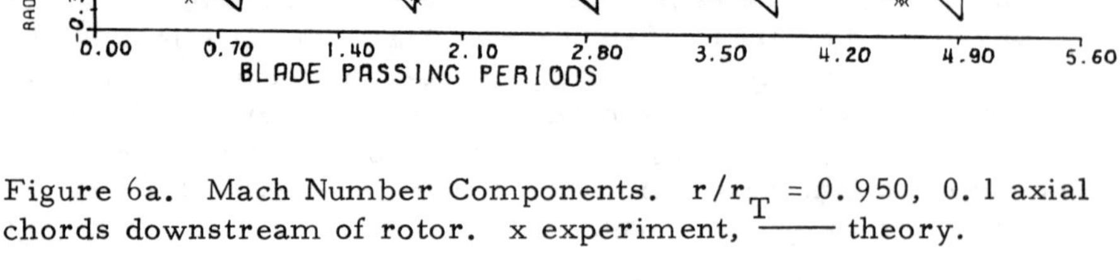

Figure 6a. Mach Number Components. $r/r_T = 0.950$, 0.1 axial chords downstream of rotor. x experiment, —— theory.

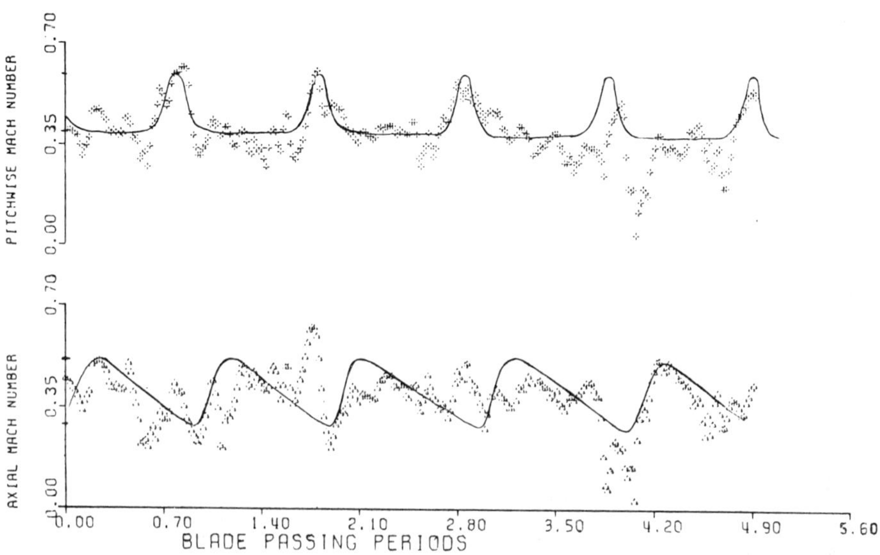

Figure 6b. Mach Number Components. $r/r_T = 0.950$, 0.1 axial chords downstream of rotor. +, Δ experiment, —— theory.

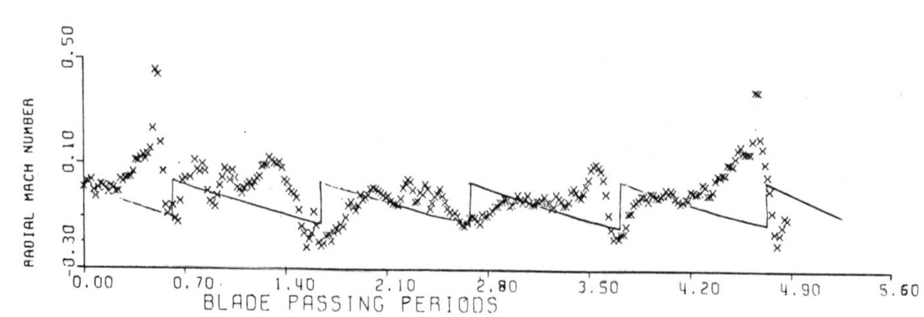

Figure 7a. Mach Number Components. $r/r_T = 0.845$, 0.1 axial chords downstream of rotor. x experiment, —— theory.

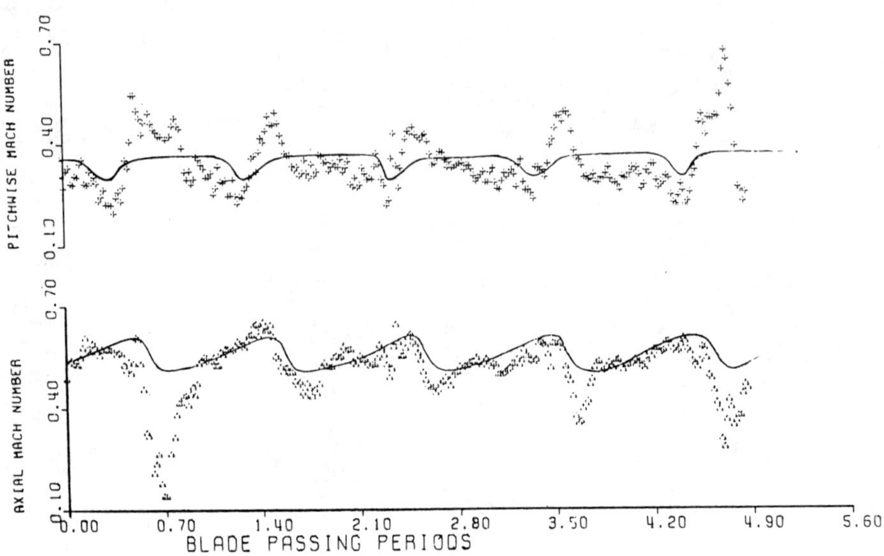

Figure 7b. Mach Number Components. r/r_T = 0.845, 0.1 axial chords downstream of rotor. +, Δ experiment, —— theory.

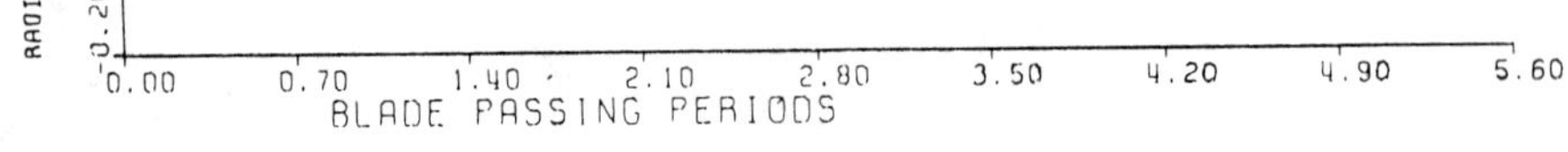

Figure 8a. Mach Number Components. r/r_T = 0.785, 0.1 axial chords downstream of rotor. x experiment, —— theory.

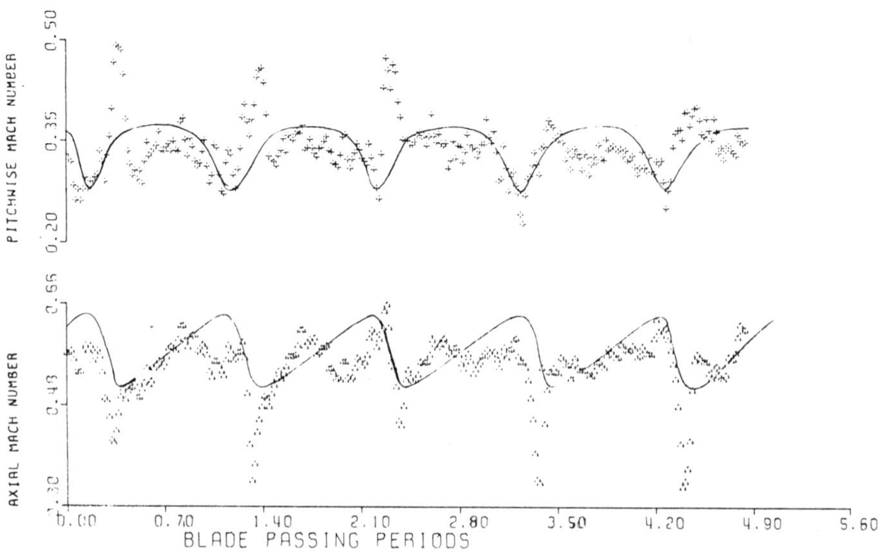

Figure 8b. Mach Number Components. $r/r_T = 0.785$, 0.1 axial chords downstream of rotor. +, Δ experiment, —— theory.

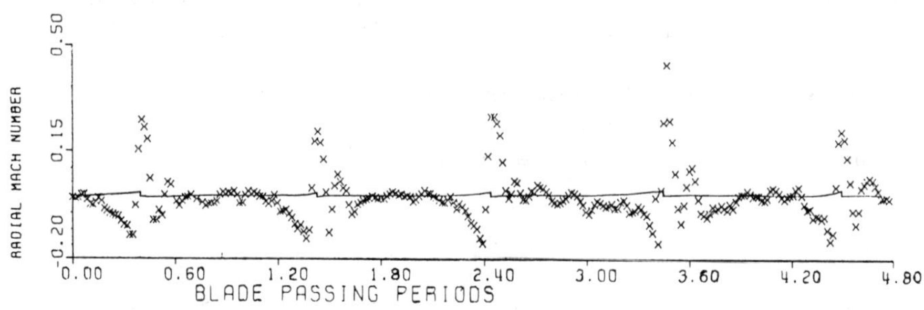

Figure 9a. Mach Number Components. $r/r_T = 0.700$, 0.1 axial chords downstream of rotor. x experiment, —— theory.

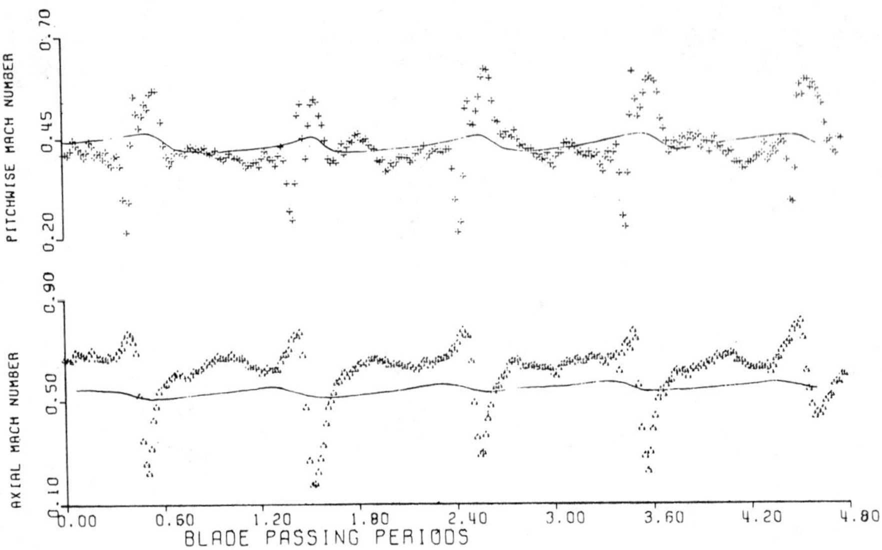

9b. Mach Number Components. $r/r_T = 0.700$, 0.1 axial chords downstream of rotor. +,Δ experiment, —— theory.

Based on work carried out since completion of Ref. [19], we now realize that this is as it should be. Both vorticity introduced at the inlet (cf. Section II. B) and vorticity associated with losses at the blades and in the viscous wakes will change the magnitude of the pitchwise variations of the flow field without substantially altering the characteristic structure exhibited in the figures.

Generally, we find comparison between experiment and our analytic studies sufficiently encouraging to justify attempts to refine and extend the present analysis.

VI. CONCLUDING REMARKS

It has been shown in Ref. [13], and further verified in Ref. [19], that for rotors operating near constant work the "wake functions" generated by the trailing vorticity can be described to a good approximation by a portion of $\tilde{\underline{V}}$ of the form $\tilde{V}_p = \nabla\phi_p$, where

$$\phi_p \cong \sum_{n=1}^{\infty} R_n(r) e^{inB\alpha} \tag{6.1}$$

and where $\alpha = \theta - f(r, x)$ is the quantity discussed in Section II. The radial eigenfunctions in turn satisfy approximately (cf. Eqs. (2.61) and (2.62) of Ref. [19]):

$$\frac{d^2 R_n}{dr^2} + \frac{1}{r}\frac{dR_n}{dr} - n^2 B^2 R_n \left[\left(\frac{\omega}{\bar{w}_{-\infty}} - \frac{B\Gamma}{2\pi r^2 \bar{w}_{-\infty}} \right)^2 + \frac{1}{r^2} \right] = G_n(r) \tag{6.2}$$

and $G_n(r)$ is determined in terms of $d\Gamma/dr$ and mean-flow quantities such as α, ψ, $\overline{V}$, etc. The quantity in the brackets in (6.2) can readily be related to the relative outlet air angle associated with rotor blading which induces "free-vortex" downstream swirl.

In the case of a stator ($\omega = 0$), or very lightly loaded rotors ($\omega r \gg B\Gamma/2\pi r$), the $R_n(r)$ can be reduced to Bessel functions of various types, but outside these limits they are not reducible to cylinder functions of any kind. The study of their general properties is underway [29].

The point that we wish to make here is simply that the presence in the downstream flow of perturbations of this type, which do not match in a simple "natural" way to the upstream eigenmodes, has interesting and important implications with regard to the upstream acoustic field—aerodynamic noise—associated with ducted fans or rotors in transonic operation.

A simple way to see this is to recognize that one way to accomplish detailed three-dimensional matching of the upstream and downstream flows is to expand the $R_n(r)$ in (6.2) in terms of the natural upstream radial eigenfunctions which are, in fact, cylinder functions like those described in Refs. [1-8]. Physically, this has the implication that the presence of the downstream wake perturbations will excite upstream acoustic modes that would not have been present in the strictly linear analysis appropriate to very light loading (cf. Refs. [9] and [10]). Thus, nonlinear (or quasi-linear) effects associated with highly-loaded rotors result in feeding acoustic energy into regions of the upstream noise spectra of practical ducted fans which would not otherwise be excited. A study of this effect is included in Ref. [29].

In conclusion, we find that a three-dimensional perturbation theory for the flow through highly-loaded ducted rotors is not only feasible but offers the possibility of helpful comparison and interpretation of existing practical experiments. Appropriately refined, the new theory may eventually help in attaining improved and more efficient design techniques.

REFERENCES

[1] Namba, M., "Lifting Surface Theory for a Rotating Subsonic or Transonic Blade Row," Aero. Res. Counc. R&M No. 3740.

[2] Falcão, A. F., de O., "Three-dimensional Flow Analysis in Axial Turbomachines," Ph. D. Thesis, Engineering Dept., Cambridge University (1970).

[3] McCune, J. E., "A Three-dimensional Theory of Axial Compressor Blade Rows - Subsonic, Supersonic and Transonic," Ph. D. Thesis, Cornell University (1958). (See also: US AFOSR TN-58-72; J. Aero. Space Sci., 25, 544, and 616, 1958.)

[4] Okurounmu, O., and McCune, J. E., "Transonic Lifting Surface Theory for Axial Flow Compressors," United Aircraft Res. Labs., E. Hartford, Ct., Report No. K 213580-1 (1971).

[5] Okurounmu, O. and McCune, J. E., "Lifting Surface Theory of Axial Compressor Blade Rows: Part I - Subsonic Compressor," AIAA J. 12, 1363 (1974).

[6] Okurounmu, O., and McCune, J. E., "Lifting Surface Theory of Axial Compressor Blade Rows: Part II - Transonic Compressor," AIAA J., 12, 1372 (1974).

[7] McCune, J. E. and Okurounmu, O., "Three-dimensional Flow in Transonic Axial Compressor Blade Rows," Proc. Int. Symp. Fluid Mech., Acoustics and Design of Turbomachinery, Pa. State Univ., (ed. B. Lakshminarayana, W. R. Britisch & W. S. Gearhart)(1970). [See also NASA Special Publication SP-304-Pt-1 (N75-11174), p. 155.]

[8] McCune, J. E. and Dharwadkar, S. P., "Lifting-Line Theory for Subsonic Axial Compressor Rotors," M. I. T. Gas Turbine Lab. Report No. 110, July (1972).

[9] Lordi, J. A., "Noise Generation by a Rotating Blade Row in an Infinite Annulus," Conf. AIAA 4th Fluid and Plasma 71-617 (1971).

[10] McCune, J. E. and Kerrebrock, J. L., "Noise from Aircraft Turbomachinery," Annual Review of Fluid Mechanics, 5, 13, 281-300 (1973).

[11] Rae, W. J., "Finite-Difference Calculations of Three-dimensional Transonic Flow through a Compressor Blade Row using the Small Disturbance Nonlinear Potential Equation," Workshop of Transonic Flow Problems in Turbomachinery, Monterey, Feb. (1976).

[12] McCune, J. E., and Hawthorne, W. R., "The Effects of Trailing Vorticity on the Flow through Highly-Loaded Cascades," J. Fluid Mechanics, in press (1976).

[13] Cheng, W. K., "A Three-dimensional Theory for the Velocity Induced by a Heavily Loaded Annular Cascade of Blades," S.M. Thesis, Aero & Astro, M.I.T., June (1975).

[14] Morton, K. B., Jr., "Three-dimensional, Compressible Flow through a Highly-Loaded, Rectilinear Cascade," S. M. Thesis, Aero & Astro, M.I.T., September (1974).

[15] Thompkins, W. T., Jr. and Kerrebrock, J. L., "Exit Flow from a Transonic Compressor Rotor," Proc. AGARD PEP 46th Meeting on Unsteady Phenomena in Turbomachinery, Monterey, Sept. 22-26, 1975; see also M.I.T. Gas Turbine Lab. Report No. 123, Sept. (1975).

[16] Kerrebrock, J. L., "The M.I.T. Blowdown Compressor Facility," M.I.T. Gas Turbine Lab. Report No. 108, May (1972).

[17] Marble, F. E., "Three-dimensional Flow in Turbomachines," High Speed Aerodynamics and Jet Propulsion, Vol. X, C, 83, Princeton University Press.

[18] Horlock, J. H., Axial Flow Compressors, (1966 orig. ed.) Krieger Publ. Co., Huntington, New York, Reprint (1973).

[19] Chen, L. T. and McCune, J. E., "Comparison of Three-dimensional Quasi-Linear Large Swirl Theory with Measured Outflow from a High-Work Compressor Rotor," M. I. T. Gas Turbine Lab. Report No. 128, Sept. (1975).

[20] Tan, C. S., Hawthorne, W. R. and McCune, J. E., "Vorticity Modeling of Real Flow Wakes in Turbomachinery and their Aerodynamic Effects," to be published as M. I. T. Gas Turbine Report (1976).

[21] Kerrebrock, J. L., "Small Disturbances in Turbomachine Annuli with Swirl," M. I. T. Gas Turbine Lab. Report No. 125, Oct. (1975). (Submitted for publication in AIAA Journ.)

[22] Shapiro, A. H., The Dynamics and Thermodynamics of Compressible Fluid Flow, vol. 1, Ronald Press, New York, (1953).

[23] Lighthill, M. J., An Introduction to Fourier Analysis and Generalized Functions, Cambridge University Press (1959).

[24] Okurounmu, O., "Wave Drag in Transonic Axial Compressors," Sc. D. Thesis, M. E., M. I. T., Feb. (1968).

[25] Adebayo, A. O., "Three-dimensional Beltrami Flows in Turbomachinery with Strong Swirl," to be published as M. I. T. Gas Turbine Lab. Report (1976).

[26] Reissner, H., "On the Vortex Theory of the Screw Propeller," J. Aero. Sci., 5, 1 (1937).

[27] Hawkins, D. D., "Multiple Tone Generation by Transonic Compressors," Symp. Aerodynamic Noise, Loughborough University of Technology, England (1970). [See also C. L. Morfey, J. Sound Vib., 14, 37 (1971).

[28] Tyler, J. M. and Sofrin, T. G., "Axial Flow Compressor Noise Studies," SAE Aeronautics Meeting Paper 345D (1962); SAE Trans. 70-309 (1961).

[29] Cheng, W. K., "Uniform-Inlet Three-dimensional Transonic Beltrami Flow through a Ducted Fan," to be published as M.I.T. Gas Turbine Lab. Report (1976).

DISCUSSION OF MCCUNE PAPER

CHENG: From your model with the lifting line, I assume that you have a near field around your blade which includes also a part of the shed vorticity. And then you have a far part which inclues your mean velocity calculation. Now, I just want to make sure that in your near field calculation that you do account for the local wake vorticity.

MCCUNE: Oh, yes.

CHENG: Then, in that case, the type of the approximation that you mentioned–namely, you said you neglect certain types of vorticity changes due to changes in circulation–that has to do with the far wake, am I correct?

MCCUNE: Well, it depends on what you're comparing it to. Here I am simply trying to compare it to a strict mean flow theory, actuator disc theory to be exact, which does not account for the near wake.

CHENG: So then there is one related question, namely, can we say anything about the characteristic wavelengths in your wake, due to oscillation?

MCCUNE: I'm very glad you brought up the question. The characteristic length before these wakes wrap up on themselves and disappear is about one span ... one to two spans. After that, other structures associated with viscous flows and those predicted by Kerrebrock and his students become more important.

CALCULATION OF 3-DIMENSIONAL CHOKING MASS FLOW IN TURBOMACHINERY WITH 2-DIMENSIONAL FLOW MODELS

THEODORE KATSANIS

NASA Lewis Research Center

INTRODUCTION

There are several 2-D and quasi-3-D flow models available. Many of these can be used to estimate choking mass flow. However, it is often difficult to get a sufficiently accurate estimate with 3-D flow. The problem lies with getting accurate velocity components normal to a given passage cross section and over the entire cross section simultaneously. The most accurate calculation can be made when the cross section considered is approximately normal to the flow. It turns out that this gives a sufficiently accurate inviscid flow model. The largest uncertainty is due to viscous effects; e.g., boundary layer displacement thickness and flow separation.

CHANNEL CROSS-SECTIONAL FLOW ANALYSIS

An approximate flow solution can be obtained on a cross-sectional flow surface within a guided channel. If the cross-sectional surface is chosen to be orthogonal to the wall of the passage (Figure 1), then, with a calculated velocity variation, a maximum or choking mass flow can be accurately calculated. The velocity variation over the cross-sectional surface can be calculated for a guided passage for either stationary or rotating blades by using momentum equations to obtain a velocity gradient solution.

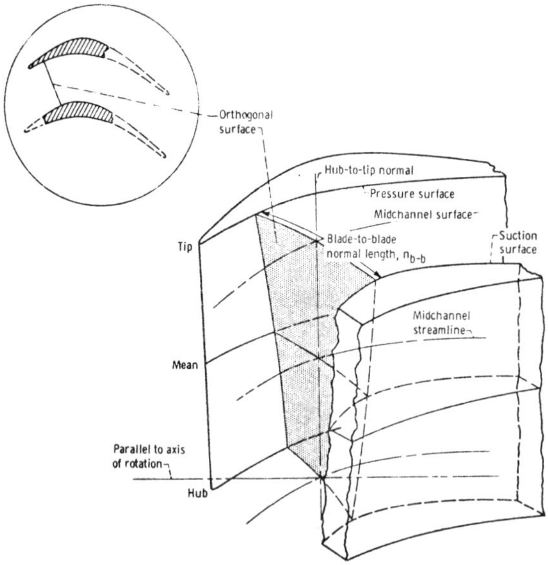

Figure 1. Pair of typical turbine blades with three-dimensional orthogonal surfaces across flow passage.

To illustrate the principle involved, consider a 2-D passage with curved walls at the throat as shown in Figure 2. The maximum possible choking mass flow is determined by the cross sectional area at the throat. However, since the walls are curved, there is a velocity variation so that the choking mass flow is determined by calculating the maximum mass flow that can flow through the throat with a velocity variation that will satisfy the momentum equation normal to the flow. This momentum equation, used to calculate the velocity variation normal to the flow, is called a velocity gradient equation.

The mass flow across the throat is calculated by:

$$w = \int \rho \, W dA \tag{1}$$

where

w	mass flow
ρ	density
W	flow velocity relative to the blade passage
dA	area element normal to the flow.

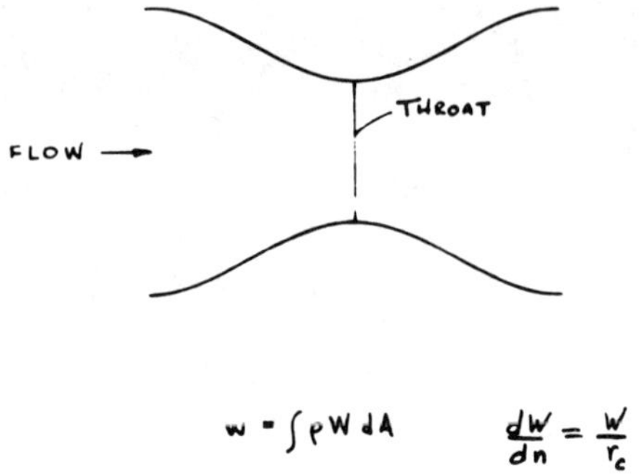

Figure 2. 2-D Nozzle

Figure 3 shows how the integrand, ρW, varies as a function of W/W_{cr} for $\gamma = 1.4$. Of course the maximum occurs when $W = W_{cr}$ (Mach No. is 1). Also, notice that there is a relatively small variation of ρW for W near W_{cr}. This makes possible an accurate

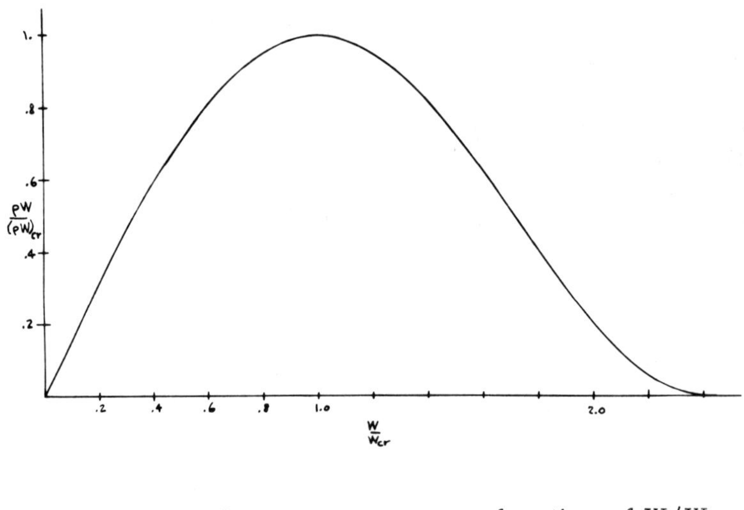

Figure 3. Mass flow parameter as a function of W/W_{cr}.

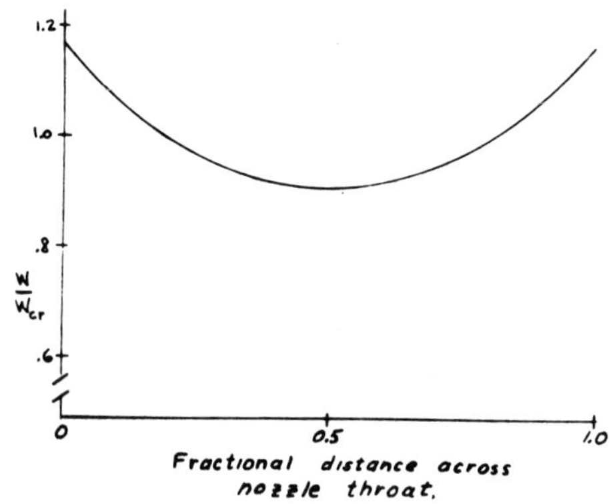

Figure 4. Velocity distribution for nozzle shown in Figure 2.

estimate of the inviscid choking mass flow even with a curved passage. Figure 4 indicates the variation of velocity across the 2-D choked passage of Figure 2, and Figure 5 shows the corresponding variation of ρW. The variation of W across the passage is calculated by the velocity gradient equation for 2-D flow:

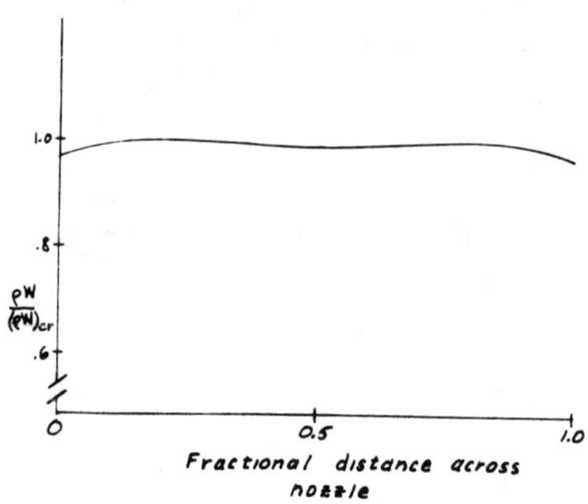

Figure 5. Mass flow distribution for nozzle shown in Figure 2.

$$\frac{dW}{dn} = \frac{W}{r_c} \tag{2}$$

where

n distance normal to streamlines

r_c radius of curvature of the streamline.

In a guided passage, r_c is known at both walks and can easily be estimated across the passage. The exact curvature distribution across the passage is not critical for a choking mass flow calculation, although it may be important in calculating the wall velocity. With the estimated curvature distribution, equation (2) can be solved numerically, with the constant of integration determined by continuity; i.e., equation (1) must be satisfied. To determine the choking mass flow, we find the solution (Figure 4) to equation (2) that will result in the maximum mass flow, w, as calculated by equation (1).

The choking solution for a three-dimensional guided passage in a blade row can be obtained in a very similar manner. The difference is that two momentum equations (analogous to equation (2)) must be satisfied; one in the blade-to-blade direction, and the other in the hub-to-tip direction. Complete details are given in Reference [1].

The accuracy of the calculated choking mass flow is dependent on using a cross-sectional surface which is approximately orthogonal to the passage. The reason for this is that the velocity vectors will be close to orthogonal to the surface so that small errors in the flow angle will have a negligible effect on the solution.

It is usually not evident just where the cross sectional surface should be located to determine the choking mass flow. Therefore, it is usually necessary to check several cross sections to find the cross section which has the minimum choking mass flow.

EXAMPLE CALCULATION OF CHOKING MASS FLOW

It is desired to calculate the choking mass flow for a centrifugal compressor impeller for an automotive application. The design of this impeller is reported in Reference [2]. Briefly, this is a 4-to-1 pressure ratio backswept impeller with approximately a 9 cm inducer tip diameter. Design mass flow is 0.61 kg/sec., and

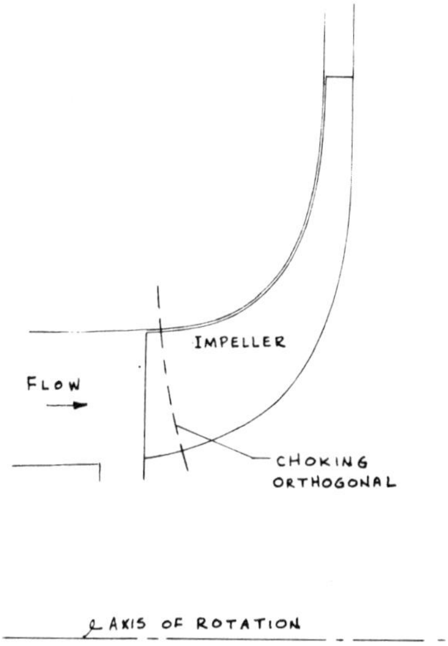

Figure 6. Compressor impeller.

predicted total efficiency is .775.

Figure 6 shows the impeller profile and the predicted choking orthogonal. The location of the choking orthogonal was arrived at by checking several orthogonals in the inducer region to determine the minimum calculated choking mass flow. The estimated total pressure loss at the choking orthogonal varied from 0.7% to 1.1%. Figure 7 shows the calculated blade-to-blade variation of velocity at the hub, mean streamline and tip for the choking orthogonals. Figure 8, shows the corresponding variation of $\rho W/(\rho W)_{cr}$. Note that ρW is very close to $(\rho W)_{cr}$ over the entire cross section even though there is a 73% variation in W over this cross section. $(\rho W)_{cr}$ varies about 16% from hub to tip because of the variation in wheel speed.

Comparison with experimental results was very good; the calculated choking mass flow was 97.7% of the designed value versus an experimental choking mass flow of 99.0% of design. The difference between the calculated and experimental mass flow is due to a number of factors such as fabrication accuracy, and possible small errors in measurements of wheel speed, temperature, density, etc.

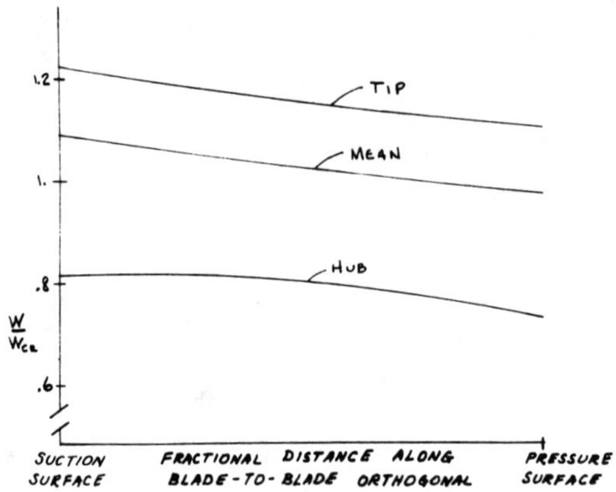

Figure 7. Velocity distribution on choking orthogonal of compressor impeller.

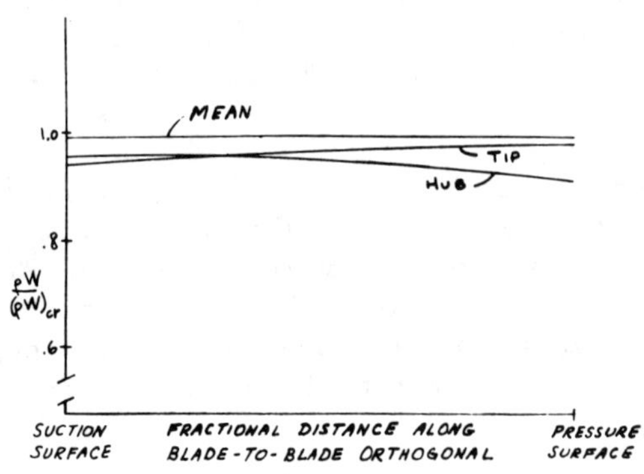

Figure 8. Mass flow distribution on choking orthogonal of compressor impeller.

CONCLUDING REMARKS

Experience has indicated that the choking mass flow for complex three-dimensional flow can be accurately calculated by this method. The success of the method hinges on using a passage cross section which is approximately orthogonal to the flow so that the proper area is used and so that any errors in estimated flow angles have a very small effect on the solution.

A fair amount of work can be required to obtain the necessary geometrical coordinates, but this has been reduced by a recent addition to the MERIDL Program of Reference [3]. This program addition calculates all necessary geometrical data required for this method.

REFERENCES

[1] Katsanis, Theodore: FORTRAN Program for Quasi-Three-Dimensional Calculation of Surface Velocities and Choking Flow for Turbomachine Blade Rows. NASA TN D-6177 (1971).

[2] Galvas, Michael R.: A Compressor Designed for the Energy Research and Development Agency Automotive Gas Turbine Program. NASA TM X-71719 (1975).

[3] Katsanis, Theodore and McNally, William D.: FORTRAN Program for Calculating Velocities and Streamlines on the Hub-Shroud Mid-Channel Flow Surface of an Axial- or Mixed-Flow Turbomachine. I - User's Manual. NASA TN D-7343 (1974).

DISCUSSION OF KATSANIS PAPER

RUNSTADLER: Ted, a couple of questions. Was the test data that you were talking about taken with the impeller alone? In other words, was the vane diffuser in there or was it a vaneless diffuser?

KATSANIS: I guess I couldn't say ... there was a reference on this in a paper by Galvas and in the printed version. I can give you the reference.

RUNSTADLER: The reason I was asking is, you know, usually you design for the choking flow in the vane diffuser to be very close to what you would assume the choking flow in the inducer to be. The question is were you measuring the choking flow?

KATSANIS: I can't answer that right now; I'm not sure.

RUNSTADLER: And, secondly, relative to this technique, can that be extended to take care of what I presume what happens—that is, once you get choking along some radial stream path, like it gets close to the mean path that you showed in that last slide for ρW product, can you account for redistribution of the flow so to speak and hence the inducer attempting to pass higher than the mass flow that you get when you assume that you get a choking flow in just one stream sheet?

KATSANIS: I'm not sure I understand exactly what you're saying. Our surface is an orthogonal surface, it goes from hub to tip.

RUNSTADLER: Right. But you have the tip and the hub at ρW products appreciably below the critical values, I think.

KATSANIS: Yes. This is based on the hub to shroud velocity gradient equation based on the curvatures of the walls and the change in flow angle and so forth. In other words, you would use the complete momentum equation going from hub to shroud to predict what that variation would be.

RUNSTADLER: May I ask the question differently? If the flow were to reach choking conditions along some stream sheet, say the mean stream sheet, which is close to the value I think you have there, would the flow in reality try to redistribute itself if you wanted to open the back throttle more and try to pass more flow?

KATSANIS: Well, what I'm saying is the choking mass flow I'm predicting here is going to be the absolute maximum for the rotor and there'd be nothing you could do downstream that's going to influence that, I don't think. It could be reduced below that because of the choking of the diffuser.

RUNSTADLER: Maybe I misunderstood your figure, then. I thought the tip and the hub values which you showed were indicative of Mach numbers being less than sonic on the stream sheet.

KATSANIS: It was subsonic at the hub and supersonic at the tip.

MC NALLY: I understood you to say that the geometrical input is difficult to obtain but that you have other programs that gave you that.

KATSANIS: Right.

BROWN: I have a question about your assumed variation of radius of curvature with position. How sensitive are your results to that assumption?

KATSANIS: It's not sensitive at all as far as predicting choking mass flow. It could affect the velocity ... that is, you can use this same technique for predicting surface velocities and then it becomes more critical because then you want to know what they are much more accurately; then it's necessary to determine the choking mass flow accurately.

THREE-DIMENSIONAL TRANSONIC SHEAR FLOW IN A CHANNEL*

T. C. ADAMSON, JR.
University of Michigan

ABSTRACT

Steady inviscid transonic shear flow is considered as it passes through a three dimensional channel with a constriction; this flow, in its essential features, is similar to that through a linear rotor cascade, with the blades aligned parallel to the incoming flow. The case considered is that where the difference in velocity across the channel due to the velocity gradient in the incoming flow is of the same order as the change in velocity induced by the flow constriction. Analytical solutions are given in terms of asymptotic expansions about the incoming flow conditions. Results show that choking can occur at a flow constriction even though the flow at the minimum area is mixed, and indicate the range of pressures downstream of the flow constriction for which shock waves appear in the supersonic part of the mixed flow.

*The support of the Office of Naval Research, Project SQUID, under Contract N00014-67-0226-0005, is gratefully acknowledged.

INTRODUCTION

The problem considered is that in which inviscid transonic shear flow in a channel passes through a constriction in the flow. It may be interpreted as the flow through a linear rotor cascade, with the blades aligned parallel to the incoming flow, in which case the parallel walls seen in the top view of the channel are symmetry boundaries, and the flow constrictions are half blades. The velocity gradient arises as a result of the radial variation of the tangential velocity component of the rotor. It is thus an extension of the problem considered by Ackeret and Rott [1], to three dimensions, and similar to a problem solved numerically by Oliver and Sparis [2]. This note is intended as a report of work in progress and thus presents a brief discussion of results obtained to date.

PROBLEM FORMULATION AND SUMMARY OF RESULTS

A sketch illustrating the three dimensional channel considered, the incoming shear flow, and the notation employed, is given in Figure 1. Thus, in the top view, one sees walls which are parallel except for a flow constriction representing half blades, the parallel walls being a distance 2s apart. All lengths are made dimensionless by the half chord of the blades. In the side view, the parallel walls are at $z=0$ and $z=b$, so that the span of the "blades" is b; the dotted lines in the side view represent the leading and trailing edges of the flow constriction, or blades. The incoming flow has a gradient in the U velocity component such that the sonic surface position is at $z=z_{so}$. Thus, if $z_{so} \geq b$, the incoming flow in the channel is all subsonic, while if $z_{so} \leq 0$, it is all supersonic, and for $0 < z_{so} < b$, it is mixed.

Velocity components and flow properties are made dimensionless with respect to their critical values on the sonic surface (z_{so}) in the incoming flow. If T and P refer to the temperature and pressure respectively, the velocity components are as shown in Figure 1,.and the subscript 0 refers to the incoming flow, then

$$U_0 = 1 + \delta m(z - z_{so}) \qquad V_0 = W_0 = 0$$
$$T_0 = P_0 = 1 \tag{1}$$

where for transonic flow, $\delta \ll 1$, and m is an arbitrary constant, $m = O(1)$.

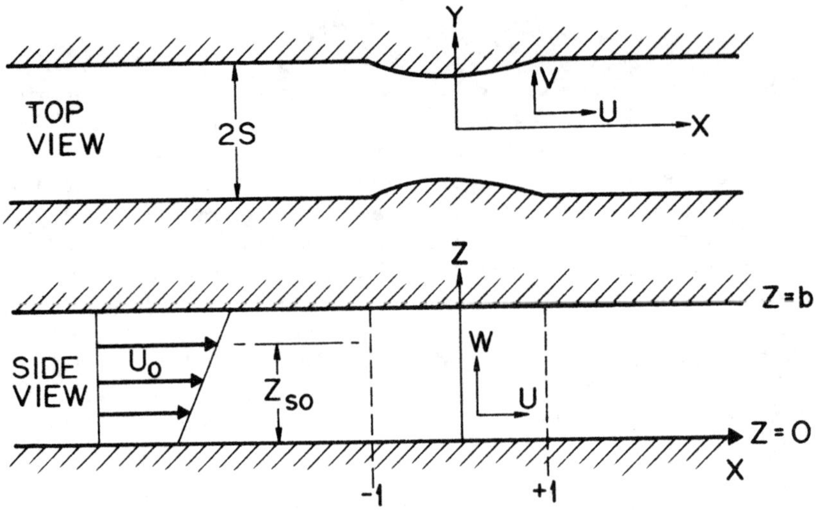

Figure 1. Sketch Showing Problem Considered and Notation; z_{so} is the distance to the sonic surface in the incoming flow.

The equation for the walls in the top view is

$$y_w = \pm (s - \epsilon^2 f(x)) \tag{2}$$

where ϵ is a parameter ordering the blade thickness, with $\epsilon \ll 1$, and f(x) is the blade shape function with f=0 for $|x| > 1$.

The various problems associated with the above boundary and initial conditions are characterized by the relative orders of δ and ϵ. The problem discussed here is that where $\delta = O(\epsilon)$, and in fact since the arbitrary constant m is available for numerical variation in initial conditions, $\delta = \epsilon$ is considered. Physically, this problem is that for which the ΔU which exists across the channel in the incoming flow is of the same order as the ΔU induced by the flow constriction.

Solutions are written in the form of asymptotic expansions about the incoming flow conditions. Because the incoming flow is rotational, velocity perturbations are written in terms of a potential plus additional functions, which, since $\delta = \epsilon \ll 1$, arise first in second order terms in U and third order terms in V and W. Thus,

$$\begin{aligned} U &= U_0 + \epsilon\phi_{1x} + \epsilon^2(\phi_{2x} + u_{a2}) + \epsilon^3(\phi_{3x} + u_{a3}) + \ldots \\ V &= \epsilon^2\phi_{2y} + \epsilon^3(\phi_{3y} + v_{a3}) + \ldots \\ W &= \epsilon^2\phi_{2z} + \epsilon^3(\phi_{3z} + w_{a3}) + \ldots \end{aligned} \tag{3}$$

where the result $\phi_1 = \phi_1(x)$, found in solving the first order equations, has been used in writing V and W.

The governing equations are those for inviscid, steady, rotational flow. For the case considered, the governing equations for each order of approximation are linear, as in comparable two dimensional transonic nozzle flow calculations [3, 4]. However, there are regions in which for certain flow conditions, the solutions may not be uniformly valid so that inner regions with nonlinear governing equations must be employed. A discussion of these regions will appear in a subsequent paper.

Analytical solutions have been found to second order. Here, only those results involving first order solutions are summarized. Thus, the first order solution for U and the corresponding equation for the sonic surface, z_s, found by equating U and the sound speed, a, are as follows:

$$U = 1 + \epsilon\left\{m(z - z_{so}) - \frac{m}{(\gamma+1)}(b - 2z_{so})\left[1 \mp \sqrt{1 - \frac{(\gamma+1)}{2sm^2}\frac{f(x)}{(\frac{b}{2} - z_{so})^2}}\right]\right\} + \ldots \tag{4}$$

$$z_s = \frac{b}{2} \mp \left(\frac{b}{2} - z_{so}\right)\sqrt{1 - \frac{(\gamma+1)}{2sm^2}\frac{f(x)}{(\frac{b}{2} - z_{so})^2}} + \ldots$$

It should be noted that higher order terms for z_s would involve y as well as x; that is, in general, $z_s = z_s(x, y; \epsilon)$.

Calculations for z_s, made for an airfoil shape consisting of two circular arcs,

$$f(x) = 2\left[1 - \left(\frac{x - x_m}{1 \pm x_m}\right)^2\right] \tag{5}$$

(upper sign for $x < x_m$, lower sign for $x > x_m$) are presented in Figure 2, for various initial positions z_{so} and for the parameters noted. In all cases, the flow above the sonic surface is supersonic, and that below subsonic. For $z_{so} = 4.10 (z_{so} > b)$, it is seen that as the flow passes through the flow constriction (between the blades), it remains subsonic except for a supersonic pocket. As z_{so} decreases, there is more and more supersonic flow, with a dip in the sonic surface as the point of maximum blade thickness (minimum cross sectional area) is passed, until at $z_{so} = 3.01$, the sonic surface drops to $z = b/2$. In all the calculations mentioned so far, the upper sign in equation (4) has been used, thus satisfying the boundary conditions that $\phi_{1x} = 0$ and $z_s = z_{so}$ at $x = -1$, where $f(x) = 0$. However, for the case where z_s passes through $b/2$, it is seen from the equation for z_s (equation (4)), that for $x > x_m$, either

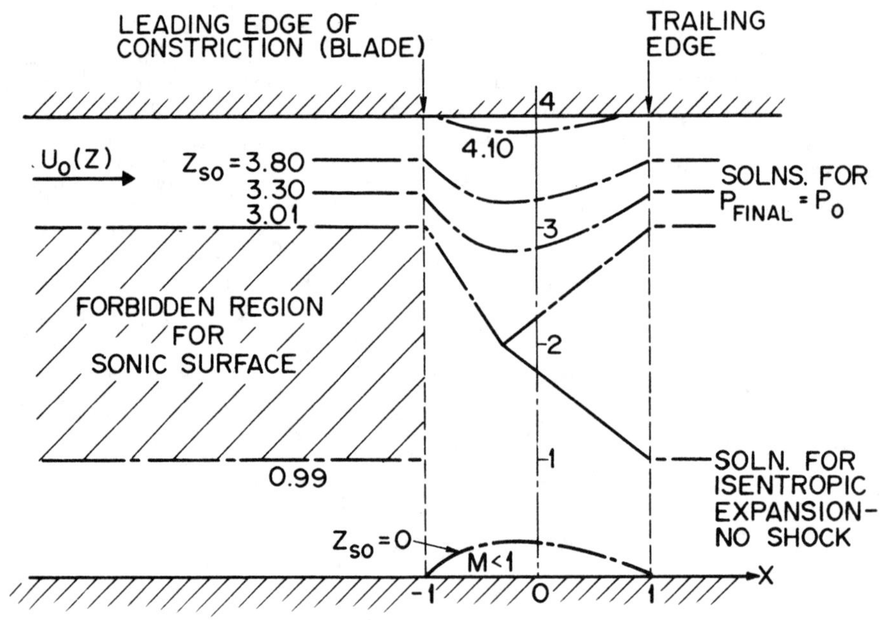

Figure 2. The Sonic Surface, z_s, as a Function of x for a Circular Arc Airfoil (Equation 5) with Maximum Thickness at $x_m = -1/3$. $\epsilon = 0.1$ (therefore a two percent thick airfoil), $b = 4$ (AR = 2), $s = 6$, $U_o(b) - U_o(0) = 0.25$.

the upper or lower sign may be chosen, leading either to the case where z_s returns to z_{so} (upper sign) or decreases to $z_s = b - z_{so}$ (lower sign) as $x \to 1$. There is, of course, a downstream pressure associated with each solution. The solutions corresponding to the upper signs refer to those where the pressure returns to P_o as $x \to 1$. The solution passing through b/2 and involving the lower sign in equations (4), is that for isentropic expansion, with no shock waves. It appears that since the flow has, everywhere, a linear gradient, the interpretation of these solutions is as follows: for those values of z_{so} for which $z_s > b/2$, the average Mach number is everywhere subsonic. Hence, as pressure downstream of the flow constriction drops, signals pass upstream and change the initial conditions (decrease z_{so}) until those conditions are reached at which $z_s = b/2$ at $x = x_m$. Then the average Mach number at the minimum area is unity and the flow, in the average, is choked. There is then one more solution found from equation (4), corresponding to complete isentropic expansion with no shocks, (lower sign, equation (4)). Evidently for back pressures between those corresponding to the solutions with upper and lower signs in equation (4), shock waves appear in the supersonic part of the flow.

For $z_{so} = 0$, where the incoming flow in the channel is completely supersonic, one finds that as the flow passes through the constriction, a subsonic pocket forms (Figure 2). Furthermore, as z_{so} increases, one would find for the solutions for z_s mirror images of those mentioned above for z_{so} slightly less than b; that is, the sonic surface goes through a maximum at $x = x_m$, and decreases to its original value. Finally, z_{so} reaches a value where z_s goes through b/2 and two shockless solutions are again available for $x > x_m$, one corresponding to an isentropic compression. Evidently, flows with shock waves in the supersonic part of the flow will again appear for back pressures in the proper range.

It is clear from the above discussion that there is a region in the incoming flow within which the sonic surface cannot be positioned for a steady flow solution. It is bounded by those values of z_{so} for which $z_s = b/2$ at $x = x_m$; these limiting values for z_{so} are found by setting equal to zero the term within the square root bracket in equations (4). That is, if real solutions are desired

$$\left| z_{so} - \frac{b}{2} \right| \geq \sqrt{\frac{(\gamma+1)t}{2\,sm^2}} \tag{6}$$

where t is the maximum value of f(x), i.e., $f(x_m)$. The values of

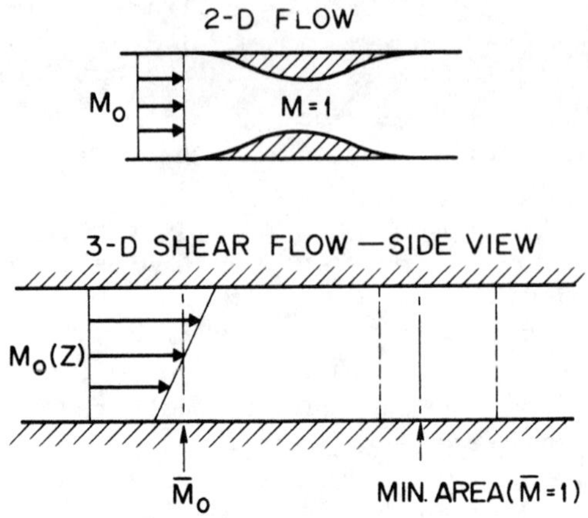

Figure 3. Sketches Used in Explaining the Apparent Choking of a Mixed Flow through a Channel; The Overbar Indicates an Averaged (over z) Quantity.

z_{so} calculated using the quality in equation (6) are those forming the boundary of the so-called forbidden region in Figure (2).

An explanation for the existence of the forbidden region can again be given by considering the average Mach number of the flow and drawing upon experience with flows in two dimensional channels, as illustrated in Figure 3. For a 2-D nozzle flow with a given entrance to throat area ratio, there is a given subsonic M_o, say M_{oa}, and a supersonic M_o, say M_{ob}, for which M=1 at the throat. Now, M_o can be less than M_{oa} (subsonic flow at the throat) or greater than M_{ob} (supersonic flow at the throat), but M_o cannot be between M_{oa} and M_{ob}. In the present 3-D case, there is also, an average subsonic $\overline{M}_o$, say $\overline{M}_{oa}$, and an average supersonic $\overline{M}_o$, say $\overline{M}_{ob}$, for which $\overline{M}$=1 at the minimum area, and $\overline{M}_o$ cannot be between $\overline{M}_{oa}$ and $\overline{M}_{ob}$. Since the value of $\overline{M}_o$ is associated with a value of z_{so}, the limit Mach numbers $\overline{M}_{oa}$ and $\overline{M}_{ob}$ are associated with limiting values of z_{so}.

It should be noted that the above results depend upon the condition that the incoming flow velocity distribution is constrained to that considered (i.e., a linear gradient) no matter what the downstream pressure condition may be; this condition is met, of course, in flows through rotating machinery.

It appears that a choking condition similar to that described here, may occur in turbomachinery, and that there will therefore be a forbidden region for the sonic surface in the incoming flow, for steady flow.

A similar problem has been considered by Oliver and Sparis [2]. Numerical solutions were found for flow through a channel equivalent to the one described here, in which at time t=0, there was zero velocity gradient. As time progressed, the hub speed was kept constant, and the tip speed increased, always with a linear velocity gradient. At the highest tip Mach number, wavelets were found in the Mach number distribution. Based on the present analysis, it appears that these wavelets could have been caused by unsteadiness in the flow resulting from the fact that the mass flow associated with the conditions in question, could not be accommodated.

REFERENCES

[1] Ackeret, J. and Rott, N.: "Ueber die Stromung von Gasen Ungestaffelte Profilgitter," Schweiz Bauzeitung, 67 (1949), pp. 40-41 and 58-61.

[2] Oliver, D. and Sparis, P.: "Computational Studies of Three Dimensional Transonic Shear Flow: Work in Progress," G.T.L. Report No. 101, Massachusetts Institute of Technology, (1970).

[3] Adamson, Jr., T. C., Messiter, A. F., and Richey, G. K.: "On the Matching of Solutions for Unsteady Transonic Nozzle Flows," Archiwum Mechaniki Stosowanej, 26 (1974), pp. 617-628.

[4] Messiter, A. F., and Adamson, Jr., T. C.: "On the Flow Near a Weak Shock Wave Downstream of a Nozzle Throat," J. Fluid Mech., 69 (1975), pp. 97-108.

DISCUSSION OF THE ADAMSON PAPER

MC NALLY: I think Dave Oliver is out, which is too bad, because this is a very interesting question. [i.e., Adamson's proposed explanation of wavelets found in a Mach number distribution by Oliver and Sparis]. Jim McCune is here. Jim, do you remember enough about Oliver and Sparis' solution to answer that question?

MC CUNE: As a matter of fact, that's what we speculated but I don't think we ever came to any conclusion, although Dave may have. It (Adamson's explanation) looks very likely.

ADAMSON: In the Oliver-Sparis paper, there was some speculation about whether the wavelets they found, were similar to the standing waves that Namba found in his calculation; I don't think that that's the case. I don't think that those are real physical waves that Namba found, but I may be mistaken.

SOME FORMULATION CONSIDERATIONS IN 3D TRANSONIC FLOW COMPUTATION

D. S. PARIS, A. A. GANZ, AND J. F. LIUTERMOZA

Pratt and Whitney Aircraft Division

ABSTRACT

Some preliminary results for 3D subsonic and transonic shear flows in non-lifting compressor stators are presented. Certain non-uniqueness conditions arising from "thin" blade boundary conditions are described and some of the many problems inherent in a 3D solution are discussed. Issues of numerical and physical shock propagation are discussed in terms of weak solutions of first order hyperbolic systems.

INTRODUCTION

The aerodynamic design of transonic compressor fans traditionally proceeds in two basic steps. For a specified radial work distribution, axisymmetric stream surfaces are calculated by the well known streamline curvature or matrix through-flow techniques. From this computation, the flow incident to and exiting from the blade row section on each surface is obtained. A cascade of airfoils is then designed, or selected from an experimental "library," which produces the desired turning and flow conditions. This procedure has proven satisfactory in the subsonic regime, but has led to disturbing difficulties for transonic and supersonic relative flow. As a consequence of strong normal and oblique shocks in the rotor passage and upstream (for subsonic axial Mach number), extremely large spanwise static pressure gradients exist and produce a radial redistribution of the flow field which can vary substantially from the design condition. Unique incidence may force a further alteration such that the fan blade effectively always operates at off-design conditions.

The design resolution of this problem has generated strong interest in the 3D computational effort for transonic rotors in progress at MIT [1], [2], under NASA sponsorship. In that analysis, the two-step Lax-Wendroff scheme is applied to 3D inviscid flow. A similar effort is underway at Pratt & Whitney Aircraft Division to develop a computational program whose format would be consistent with the Pratt & Whitney design system and would utilize various display capabilities. Conceptually there exist many similarities between the two analyses. In this paper, the formulation and some preliminary results of subsonic and transonic shear flows in non-lifting stators are presented. Additionally, some observations are made on the non-uniqueness of the "thin blade approximation" for airfoil boundary conditions and some experiences with 3D shock propagation in transient numerical schemes are described.

BASIC FORMULATION

It is assumed that viscous effects do not significantly dominate the aerodynamics near the design condition and that an ex post facto boundary layer calculation will be sufficient for loss estimation. The 3D flow field is assumed steady with respect to the blades and periodic from blade to blade. The hub and casing surfaces are extended up and downstream analytically and then transformed by a set of non-orthogonal mappings to a rectangular computational domain as shown in Figure 1. The compressible inviscid gas dynamic equations are expressed in the rotating coordinate system fixed to the blades as:

$$\begin{aligned} &\nabla \cdot \rho \underline{w} = 0 \\ &(\underline{w} \cdot \nabla)\underline{w} + 2\underline{\omega} \wedge \underline{w} - \omega^2 r + \nabla p/\rho = 0 \qquad (1) \\ &(\underline{w} \cdot \nabla)e + p(\nabla \cdot \underline{w})/\rho = 0 \qquad (p = \rho R T) \end{aligned}$$

and are then rewritten in the transformed coordinate system. The boundary condition $\underline{w} \cdot n = 0$ is applied on the hub and casing and on the mean camber surface of the fan blades in the "thin blade approximation" as is frequently done in compressible potential flow analysis.

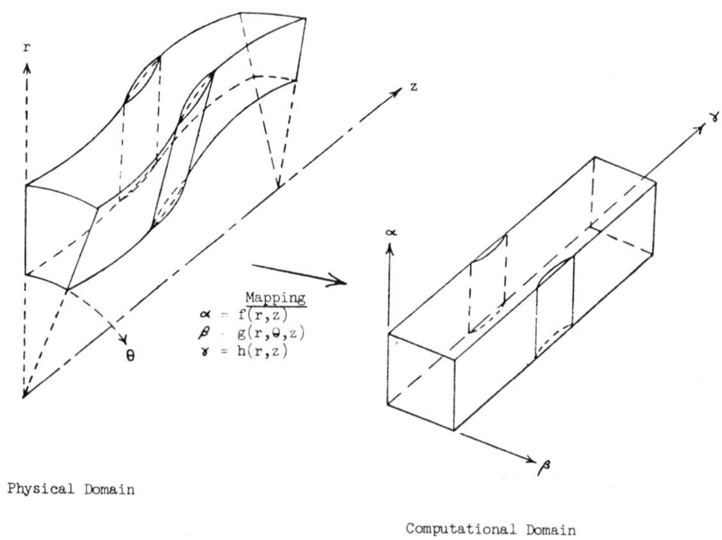

Figure 1. Coordinate Transformations

The upstream flow boundary in the relative coordinate system consists of mixed flow with shocks emerging from the supersonic region. The mass flow is primarily determined by the downstream pressure condition which feeds through the subsonic portion of the flow field and redefines the upstream transonic flow distribution. The geometric position of the sonic surface in the inlet is part of the solution to be determined. Appropriate supersonic and subsonic boundary conditions are utilized on the entrance and exit planes [3].

The equation system is recast in transient form and solved by a 3D extension of the two-step Lax-Wendroff technique [4 - 6]. While the 2D Lax-Wendroff scheme normally employs smoothing in one or two dimensions, the most efficient smoothing for the 3D problem is still an open question. At present, smoothing is used only in the axial flow direction,

$$U_{smooth} = U + c \frac{\Delta t}{\Delta x} \frac{d}{dz} \left(\left| \frac{dw_z}{dz} \right| \frac{dU}{dz} \right) \tag{2}$$

where U is the vector of conserved properties. However, similar smoothing operators have been included in the code for the other two coordinate directions [7]. More efficient techniques are found in the literature for two-dimensional problems [8], but at project inception it was decided to remain within established state-of-the-art insofar as the basic numerics were concerned. For the rotor problem, the Kutta condition is enforced by matching static pressures at the suction and pressure sides of the trailing edge, but no attempt is currently made to explicitly incorporate the trailing vortex sheet. Computational stability is obtained using the standard CFL condition [4],

$$(|w| + a) \frac{\Delta t}{\Delta x} \leq 1. \quad (3)$$

THIN BLADE APPROXIMATION

The "thin blade approximation" requires application of the boundary condition $\underline{w} \cdot \underline{n} = 0$, where $\underline{n}$ is the local blade normal, on the mean camber surface of the airfoil and is frequently selected for analytical and numerical solutions in compressible potential flow and for logical simplicity in 3D flow programs. The condition insures uniqueness when the domain is totally isentropic and a velocity potential exists. It has the effect of introducing mass into the flow field through the forward portion of the "blade" and removing mass through the rearward portion (See Figure 2). A separating streamline actually defines the "blade" (providing it closes). In this sense, the boundary condition does not really constrain the blade to be "thin"; for large mass injection (corresponding to thick blades) the actual blade shape is not *a priori* known and is determined from the final converged solution.

When the stagnation properties through the flow field are variable, as might be caused by non-isentropic processes, the thin blade boundary condition introduces a non-uniqueness or

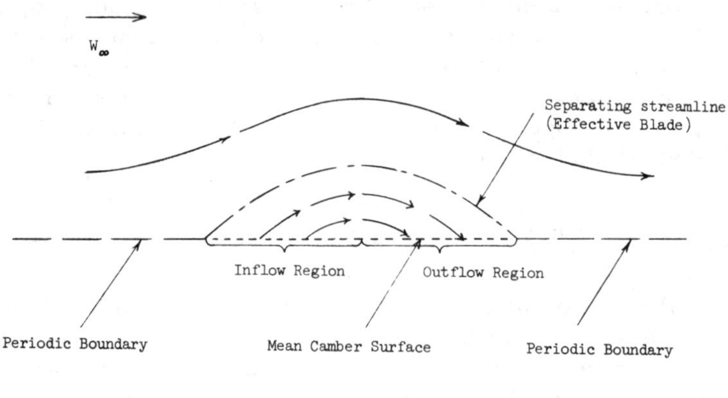

Figure 2. "Thin Blade" Boundary Condition

indeterminacy into the numerical solution. For positive flow angles, the inflow region along the "blade" originates in some reservoir, and stagnation properties or other additional data must be specified as part of the boundary conditions. If left indeterminate, the transient flow field appears to be characterized by a high frequency "pumping" in thermodynamic properties which can trigger instabilities in the calculation and slow or prohibit ultimate convergence. Experience has shown that when the flow redirection implied by the $\underline{w} \cdot \underline{n} = 0$ condition is impulsively introduced on the sides of a uniform flow numerical divergence generally occurs. Further numerical experimentation has shown that good results can be obtained if the normal vector is slowly introduced as a function of time. Difficulties using this procedure are anticipated in shocked flow, however, and the transformations are currently being revised to explicitly incorporate the solid surface.

SHOCK PROPAGATION IN THE GRID

In multi-dimensional numerical computations, a shock may be located or propagated according to several considerations: viscous balance; equation system demands; numerical differencing scheme. For example, in uniform flow with a single normal shock in a constant area duct, the position of the discontinuity is completely arbitrary from inviscid considerations and should remain stationary at its initial axial location throughout the computation. When real viscosity is present, however, the thickening boundary layer and the total friction loss balance alters the effective cross sectional area and causes the shock to propagate forward into the supersonic region. Should a geometric area contraction exist, the shock will stabilize in a lambda or oblique configuration for high Mach numbers, but will propagate upstream for lower Mach numbers. See Figure 3.

The issue of propagation rates for discontinuities of first order hyperbolic systems has received considerable research attention from the point of view of weak solutions (piecewise continuous in this setting). Early formulative work is given in Refs. [9], [10] and recent discussions are presented in Refs. [11] and [12], where smoothing requirements are treated. It is emphasized that piecewise continuous solutions of the steady state equations in general do not satisfy the transient equation; for a given jump magnitude, there exists a unique rate at which the discontinuity must propagate. A study of various proposed differencing techniques is presented in [12]; comments concerning the Lax-Wendroff technique and the

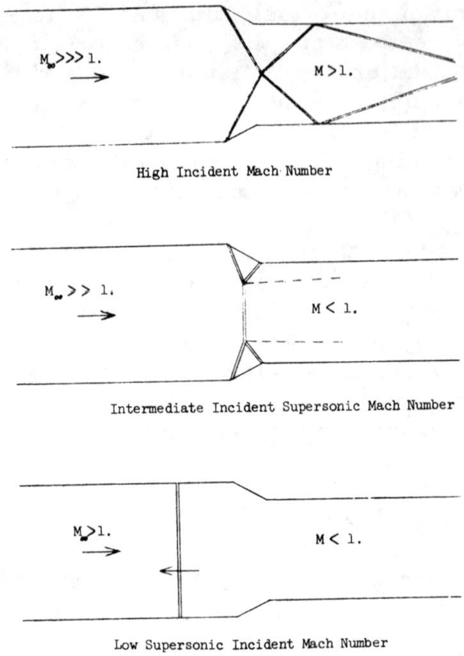

Figure 3. 1-D Normal Shock Behavior

nonphysical evolution of discontinuities were made previously in [13] and [14]. The effect of the difference scheme selected is readily seen through the one-dimensional wave equation. In conservation form, this is

$$\frac{\partial}{\partial t} P(x,t,u) + \frac{\partial}{\partial x} Q(x,t,u) = 0 \qquad P(x,t,u) = u \;;\; Q(x,t,u) = u^2/2 \tag{4}$$

and a weak solution exists in the sense that discontinuities <u>must</u> be connected on the line

$$\frac{dt}{dx} = \frac{P^+ - P^-}{Q^+ - Q^-} \tag{5}$$

as in Figure 4 for the initial conditions

$$U(x,0) = \begin{cases} 2, & x < 0 \\ 1, & x > 0. \end{cases}$$

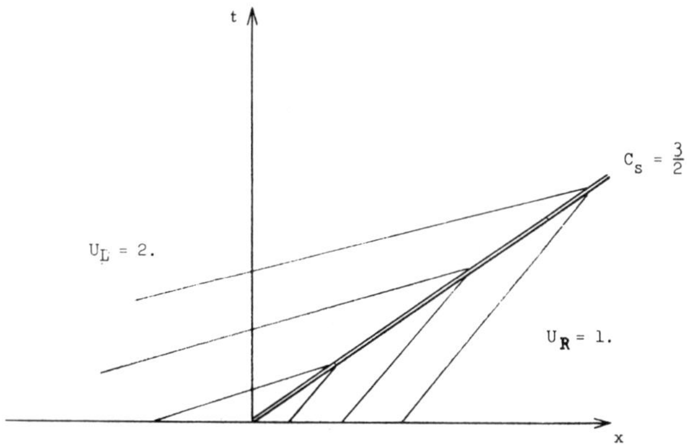

Figure 4. Transient Shock Propagation for Model Equation

The slope of this line gives the propagation rate in the (x, t) plane.

If a second order difference scheme such as the two-step Lax-Wendroff procedure with smoothing operators is used, the equation is effectively altered to

$$\frac{\partial u}{\partial t} + u \frac{\partial u}{\partial x} = S\left(\frac{\partial u}{\partial x}, \frac{\partial^2 u}{\partial x^2}, \mu\right) \tag{6}$$

and the jump conditions $P^+ - P^-$, $Q^+ - Q^-$ can change, depending on how the first derivative term and μ are selected. In general, μ is a function of Δt, Δx and a characteristic velocity; for coarse grid computations (as currently required for 3D calculations) the change in shock velocity due to these additional terms can be substantial.

For the gas dynamic equations, the Rankine-Hugoniot relations satisfy the transient problem when posed as a first order hyperbolic system. In the Lax-Wendroff difference scheme, propagation can occur depending on the shock strength and the manner in which the additional difference terms are selected. The relation between the shock speed and jump conditions for proposed algorithms can be established by the above theory.

The above discussion illustrates that when fluid problems involving shocks are solved numerically, the difference schemes must be carefully devised in order to assure that the correct shock stability and propagation features are obtained.

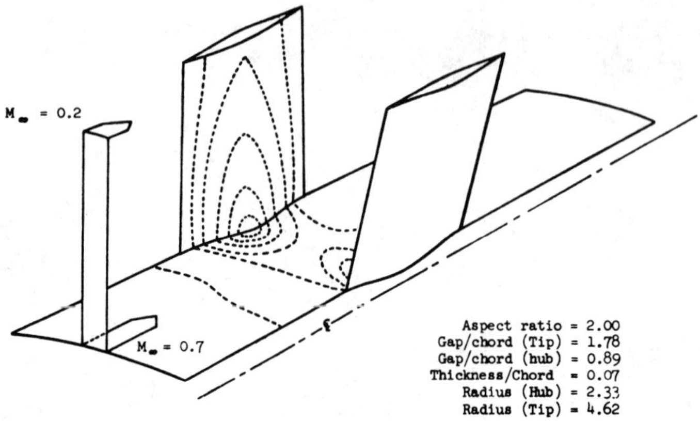

Figure 5. Subsonic Shear Flow in Nonlifting Stator: Static Pressure Contours on Blades and Hub

Although not formulated in the sense of generalized functions, a substantial analytical effort in shock "capturing" has been pursued by G. Moretti and others [15-17]. As will be shown below in our preliminary computations, a normal shock occurring on a supercritical airfoil appears stationary while a strong incident normal shock in the supersonic blade region propagates rapidly forward, all in the same three-dimensional grid!

SUBSONIC SHEAR FLOW

The first application of this program was to subsonic shear flow in a low hub/tip ratio non-lifting stator. The particulars of the geometry are shown in Figure 5. Because of the radial area divergence, a weak spanwise pressure gradient exists in the blade passage when the upstream flow is uniform. For a sheared incident flow, stronger pressure gradients are formed (Fig. 5) and are seen to skew across the blade by roughly 60° to the flow direction. The largest flow acceleration is shown in the root/endwall corner, where the incident Mach number to the blade section is highest. Comparison with 2D computations at the hub gap/chord ratio have demonstrated that substantial 3D pressure relief occurs.

The grid used was $9 \times 9 \times 33$ and required 35 minutes on the

IBM 370/168 to achieve practical pointwise convergence starting from uniform flow at M = .5. Convergence based on integrated mass flow on the exit plane was within $3. \times 10^{-5}$.

TRANSONIC SHEAR FLOW

The primary application of this program is directed at 3D flows with strong static pressure gradients. To maximize this phenomenon for demonstrative purposes, a transonic shear case was constructed by initializing the flow field from "layered" or "stacked" 2D flows in the strip theory sense. The initial 3D flow field was constructed such that the incident flow was sheared with M varying from .5 at the hub to 1.5 at the casing (see Table 1.) and a normal shock stood upstream of the blades in the supersonic portion (see Figure 6).

Table 1

Initial Conditions for 3-D Shear Flow

Radius (ft.)	P_{T_∞} (Psf)	T_{T_∞} (^{o}R)	M_∞
(hub) 2.00	1326	↑	.50
2.25	1464	|	.63
2.50	1649	|	.77
2.75	1891	|	.90
3.00	2386	518.7	1.1
3.25	2710	|	1.2
3.50	3098	|	1.3
3.75	3557	|	1.4
(casing) 4.00	4104	↓	1.5

In order to initialize this 3D shear flow, nine subsonic 2D computations were made with initial conditions corresponding to conditions downstream (as defined by the Rankine-Hugoniot relations) of the normal shock (see Table 2). These nine layers were then stacked and for cases 5-9 conditions upstream were overwritten with the supersonic conditions from Table 1.

For the higher incident Mach numbers, the subsonic 2D flows produced supercritical regions terminating in shocks on the airfoil surfaces. When the cases were stacked, these supersonic

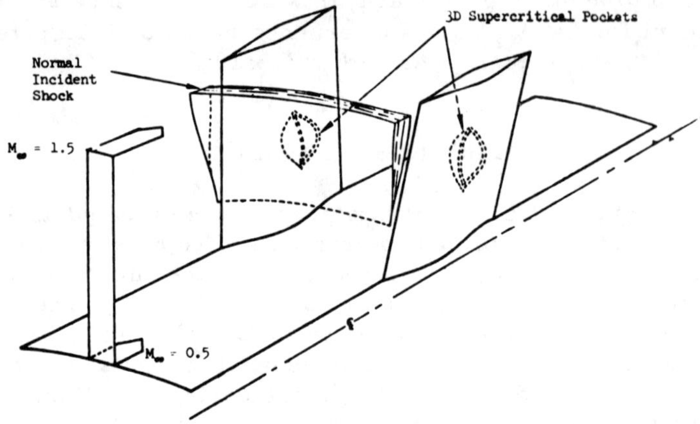

Figure 6. Initial Transonic 3D Flow Field

Table 2

Initial Conditions for 2-D Cases

Radius (ft.) when stacked	P_{t_∞} (Psf)	T_{T_∞} (°R)	M_∞
2.00	1326	↑	.50
2.25	1464		.63
2.50	1649		.77
2.75	1891		.90
3.00	2384	518.7	.91
3.25	2691		.84
3.50	3033		.79
3.75	3408		.74
4.00	3815	↓	.70

regions formed a 3D supersonic bubble terminated by a normal shock over the mid-span region of the blades. The initial 3D flow field contained a maximum static pressure to minimum static pressure ratio of 5.24.

Preliminary transient 3D computations are shown in Figure 7. A very strong pressure wave appears to pass down the blade toward the hub and then to proceed upstream in the subsonic portion of the flow below the sonic cylinder. Concurrently, the normal incident

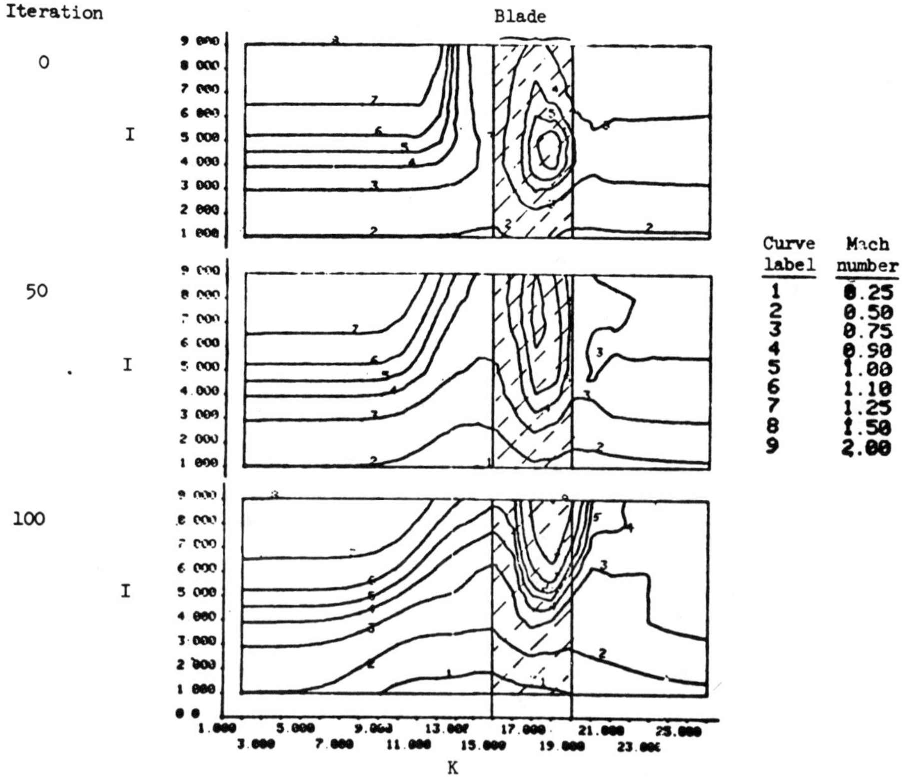

Figure 7. Transient Mach Number Contours on Periodic and Blade Surfaces

shock smears out, propagates upstream and becomes oblique as shown by its rotation through 50 degrees or more. The supersonic pockets on the blades spread into the tip region, and weaken considerably, but little streamwise propagation of the shocks occurs.

It is noteworthy that the supercritical flow shock on the blade remains essentially stationary, while rapid propagation occurs from the incident normal shock. At the present time, it is conjectured that the interior shock is geometrically stabilized, while the exterior normal shock is physically unstable due to the downstream static pressure gradient and would require strong smoothing to prevent its propagation. Under these circumstances, it appears unwise to attempt to stabilize the exterior normal shock.

Current efforts are now aimed at carrying transonic stator calculations further and applying the method to more general geometries. Presently work will begin on subsonic rotor calculations and subsequently on transonic rotor calculations.

CONCLUSIONS

The "thin" blade boundary condition does not necessarily require that the blades be physically thin, but it can introduce instabilities into the Lax-Wendroff solution. Gradual imposition of the boundary condition is sufficient to stabilize non-shocked flow. Great care must be exercised in devising numerical schemes for 3D shocked flow in order to obtain the correct shock propagation behavior.

Results for 3D subsonic flow indicate substantial 3D pressure relief in sheared flow through blade passages. Preliminary results for transonic sheared flows indicate stability of interior shocks and instability of upstream normal shocks accompanied by radial redistribution of flow.

ACKNOWLEDGEMENTS

The authors would like to acknowledge the early efforts of C. Weber and R. Riefer of Pratt & Whitney Aircraft Division on this project. Additionally, fruitful discussions have been held with J. Kerrebrock, D. Oliver and W. T. Thompkins at MIT.

REFERENCES

1. Oliver, D. A. and Sparis, P., "A Computational Study of Three-Dimensional Shear Flow in Turbomachine Cascades," AIAA Paper No. 71-83, January (1971).

2. Oliver, D. A. and Sparis, P., "Computational Aspects of the Prediction of Multidimensional Transonic Flows in Turbomachinery," presented at NASA Conference on Aerodynamic Analyses Requiring Advanced Computers, NASA Langley Research Center, March (1975).

3. Ganz, A. A. and Serra, R. A., "Boundary Conditions and Uniqueness in Internal Gas Dynamics Flows," AIAA Journal, vol. 12, March(1974), pp. 263-264.

4. Richtmyer, R. D. and Morton, K. W., Difference Methods for Initial-Value Problems, Interscience, New York (1967).

5. Lax, P. D. and Wendroff, B., "Systems of Conservation Laws," Communications on Pure and Applied Mathematics, vol. 12, May(1960), pp. 217-237.

6. Lapidus, A., "A Detached Shock Calculation by Second-Order Finite Differences," Journal of Computational Physics, vol. 2 (1967), pp. 154-177.

7. Weber, C. R., "Comments on Stability Analysis of Explicit Finite Difference Operators," Pratt & Whitney Aircraft Div. internal memo to A. A. Ganz, April 11 (1975).

8. MacCormack, R. W. and Paullay, A. J., "Computational Efficiency Achieved by Time Splitting of Finite Difference Operators," AIAA Paper No. 72-154, January(1972).

9. Lax, P. D., "Weak Solutions of Nonlinear Hyperbolic Equations and Their Numerical Computation," Communications on Pure and Applied Mathematics, vol. 7 (1954), pp. 159-193.

10. Courant, R. and Hilbert, D., Methods of Mathematical Physics, Volume II, Partial Differential Equations, Interscience, New York (1962).

11. Yoshihara, H., "A Survey of Computational Methods of 2D and 3D Transonic Flows with Shocks," General Dynamics, Convair Aerospace Division Report GDCA-ERR-1726, December 1972; also in Transonic Aerodynamics, AIAA Professional Study Series (1974).

12. Warming, R. F. and Beam, R. M., "Upwind Second-Order Differencing Schemes and Applications in Unsteady Aerodynamic Flows," Proceedings, AIAA Second Computational Fluid Dynamics Conference, Hartford, Conn., June(1975), pp. 17-28.

13. Moretti, G., "A Critical Analysis of Numerical Techniques: The Piston-Driven Inviscid Flow," Chapter 8 in Numerical Methods in Fluid Dynamics, AGARD Lecture Series No. 48, Ed. J. J. Smolderen, AGARD, NATO, May (1972).

14. Moretti, G., "The Choice of a Time-Dependent Technique in Gas Dynamics," Chapter 11 in Numerical Methods in Fluid Dynamics, AGARD Lecture Series No. 48, Ed. J. J. Smolderen, AGARD, NATO, May (1972).

15. Moretti, G., "On the Matter of Shock Fitting," Proc. of the Fourth International Conf. on Numerical Methods in Fluid Dynamics, Boulder, Colorado, June 1974, Ed. R. D. Richtmyer, Lecture Notes in Physics, vol. 35, Springer-Verlag, Berlin (1975).

16. Moretti, G., "A Circumspect Exploration of a Difficult Feature of Multi-dimensional Imbedded Shocks," Proceedings, AIAA Second Computational Fluid Dynamics Conference, Hartford, Conn., June (1975), pp. 10-16.

17. Salas, M. D., "The Anatomy of Floating Shock Fitting," Proceedings, AIAA Second Computational Fluid Dynamics Conference, Hartford, Conn., June (1975), pp. 47-54.

DISCUSSION OF THE PARIS, et al. PAPER

FARN: You mentioned the jump in P_T and T_T. Is this due to the artificial viscosity you use or have you used any?

GANZ: No. What you're asking here is this question: If you had in fact what you claim to be a thin blade, which say is in here somewhere, what you say in a thin body approximation is that the normal here on the blade is in fact going to be applied on the mean camber surface which means, from a computational point of view, that you're going to say, as a boundary condition, that this is the flow angle here. What you're really doing then is saying that from here to here is a mass inflow region in which you really have to set, say the flow angle, as well as the stagnation conditions. Whereas here, of course, being outflow region you would specify only, say a flow angle, or static pressure. But what this means is that it does draw in the fluid and if you look across the channel in this sense and you compute the mass flow that runs across each of these planes - as you run along the blade you will in fact see an increase in the mass flow and then as you pass the mid-chord of the blade it will drop back down again. By the time you reach the end of the blade you'll hopefully have the same mass flow as when you started.

FARN: Have you compared your result with Peter McNaughten's method results or did he apply his method to your problem?

GANZ: Well, one thing to bear in mind is that his is only a planar calculation. It's basically a finite area type of method whereas here we were interested in pushing as fast as we could to a full 3D flow. What we have compared against is a nozzle calculation where we basically look at an area constriction and we have exactly matched, at least to the number of points that we use in the grid, both the Mach number and the static pressure distribution along the blades. At least from that point of view we do have a reasonable degree of confidence in the calculation.

MC NALLY: Do you use Lax-Wendroff methods or have you used MacCormack's later schemes - and what do you use - do you have any artificial viscosity?

GANZ: Yes. We are very much aware of the MacCormack scheme. But we were concerned at the beginning that what we were developing was a scheme for the design groups which, in a sense, meant that we had to work with the state of the art. It would be very hard to support a project in which we said we were looking at advanced numerical schemes instead of working within the state of the art. So from that point of view we are in fact using the two step Lax-Wendroff procedure. It's much easier to work a prototype first and then come back and say we can accelerate it and make it go faster.

MC NALLY: Did you put in artificial viscosity at any time?

GANZ: Yes. It's the fairly standard sort ... you're multiplying the magnitude of the first derivative by the second derivative of the scalar functions that represent all the terms.

MORETTI: This is not a question; it's a comment. Enforcing the boundary conditions on the camber line, say, was customary when people were trying to get analytical solutions and there is no reason whatsoever to do that when we are trying to get numerical solutions. So you can put the RIGHT condition, and mind you, the RIGHT condition, on the right place. It's funny that in the same place, not the same room but around here, I was saying this thing seven years ago.

GANZ: The only response I might make there is that it is true that you can include the whole blade shape but you are introducing more transformations in the equations which, of course, brings up the increased possibility of bugs and at least at the beginning, our experience has been that it's a lot easier to try to push through with as easy or as straightforward a formulation as possible.

STAHARA: I would be interested in knowing what some of the typical computational times are for some of your converged solutions.

GANZ: You have to be a little bit careful about exactly what is meant by convergence in these problems. You can ask for, say, pointwise convergence or you can ask for, say, mass average convergence. Some of our experiences have been, in tracking the static pressure as a function of iteration number in these flows, is that if this is iteration number or time and this, say, is the static pressure, that, of course, you will go through some kind of an initial transient and after that it may die out and then as the waves which result from the blades being brought in, proceed up and downstream, you may, in fact, get a glitch of this type and then it may again oscillate for awhile and in this way reflects off the entrance boundary and comes back. Once again you get another glitch. Now if one stops the calculation there, one is scared. On the other hand, if you go all the way out until there aren't any glitches anymore in the flow field you may burn out a very large amount of computer time. So, in our experience, what we do is to run through three or four, say, of wave passing times, and then do an extrapolation over the mean result of the last, say 100, iterations. Now in the subsonic shear flow case which I showed, that was for a grid of 9x9x33 and that was about a one-half hour calculation on the IBM-one-370. My understanding is that there are some very long computation times that they found at MIT and I'm sure we'll be hearing more about that this afternoon. On the order of like 50 hours.

MCNALLY: In all these methods, depending on how fine a grid, too, it's very very grid dependent - you get into a finer mesh and your time goes up fantastically and then, of course, you go to a stronger convergence critieria and it goes up, too.

COMPUTATION OF STEADY AND PERIODIC TWO-DIMENSIONAL NONLINEAR TRANSONIC FLOWS IN FAN AND COMPRESSOR STAGES

JOHN ERDOS, EDGAR ALZNER, AND PAUL KALBEN

Advanced Technology Laboratories, Inc.,
and General Applied Science Laboratories, Inc.

INTRODUCTORY REMARKS

We have three computer programs under development for NASA Lewis Research Center which employ "time-marching" finite difference methods to achieve solutions of the compressible Euler Equations on two-dimensional surfaces in transonic fan and compressor stages. These programs use the same basic solution algorithm but differ with respect to statement of the problem and character of the boundary conditions (i. e., steady vs. periodic). The three programs can be briefly described as follows:

(a) M2DATL - Solution of the unsteady three-dimensional axisymmetric inviscid equations of motion on a meridional (hub-to-shroud) surface. The basic equations are integrated over a blade-to-blade passage to obtain a system describing circumferentially averaged values of the dependent variables. Only asymptotically steady solutions are sought, but up to five stages (each consisting of rotor and stator) can be analyzed simultaneously.

(b) B2DATL - Solution of the unsteady two-dimensional inviscid equations of motion on a circumferential (blade-to-blade) stream surface through a single stage (rotor and stator) machine operating in an undistorted inlet flow. An asymptotically periodic solution describing the interaction of rotor and stator at transonic conditions is sought. Steady solutions for a single blade row can also be obtained.

(c) BCDATL - A modification of B2DATL to permit analysis of a transonic rotor operating under circumferentially distorted inlet (or discharge) conditions.

The statement of equations and boundary conditions and outline of the method of solution for Programs M2DATL and B2DATL are given in [1]. The present paper describes the formulation of BCDATL, discusses the problem of modelling inlet flow distortion produced by screens, and reports on some more recent results obtained with B2DATL.

FORMULATION

An axisymmetric surface $S(r, z) = 0$ through a rotor is shown schematically in Figure 1. A curvilinear coordinate system (m, n, θ) is defined on the surface, and flow variations in the n direction are neglected. The following system of equations is thereby produced:

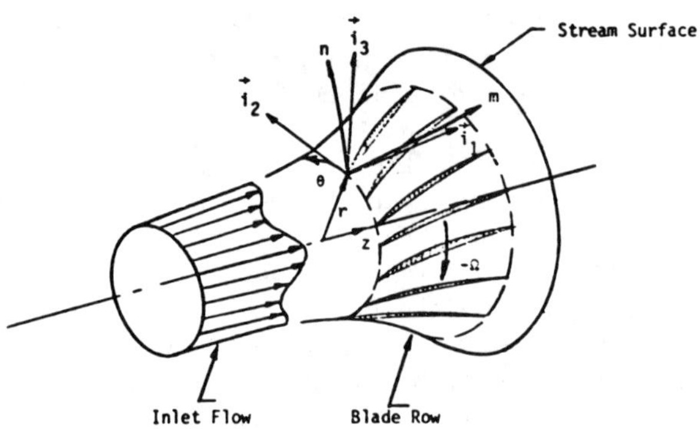

Figure 1. Schematic of Coordinate System.

$$\frac{\partial \rho}{\partial t} + \frac{\partial \rho u}{\partial x} + \frac{\partial \rho v}{\partial y} = - \frac{\rho u}{rb} \frac{drb}{dx} \tag{1}$$

$$\frac{\partial \rho u}{\partial t} + \frac{\partial (\rho u^2 + p)}{\partial x} + \frac{\partial \rho uv}{\partial y} = - \frac{\rho u^2}{rb} \frac{drb}{dx} + \rho (v + \Omega r)^2 \frac{1}{r} \frac{dr}{dx} \tag{2}$$

$$\frac{\partial \rho v}{\partial t} + \frac{\partial \rho uv}{\partial x} + \frac{\partial (\rho v^2 + p)}{\partial y} = - \frac{\rho uv}{rb} \frac{drb}{dx} - \rho u (\frac{v}{r} + 2\Omega) \frac{dr}{dx} \tag{3}$$

$$\frac{\partial \rho E'}{\partial t} + \frac{\partial \rho u H'}{\partial x} + \frac{\partial \rho v H'}{\partial y} = - \frac{\rho u H'}{rb} \frac{drb}{dx} \tag{4}$$

The nomenclature is defined as follows:

t	time
$x = m$	meridional distance
$y = r(\theta, \Omega t)$ (at m=constant)	circumferential distance
$u = V_m$	meridional component of velocity
$v = V_\theta - \Omega r$	relative circumferential velocity component
$H' = C_p T - \Omega r V_\theta$	"relative" total enthalpy "rothalpy")
$E' = H - p/\rho$	"relative" total energy rotor speed
Ω	rotor speed
$b = (r_1 - r_2)$	radii of upper and lower bounds of the streamsheet
$r = (r_1 + r_2)/2$	mean radius of the streamsheet

The computational plane extends axially from an inlet station to a discharge station and laterally over a fraction of the circumference encompassing one blade-to-blade passage, as indicated in Figure 2. Extensions of the mean camber line from the leading edges forward form the periodic boundaries ahead of the blade row, while the blade slipstreams form natural boundaries of the passage downstream of the blades. Note that the computational plane therefore rotates with the rotor blades. The computational

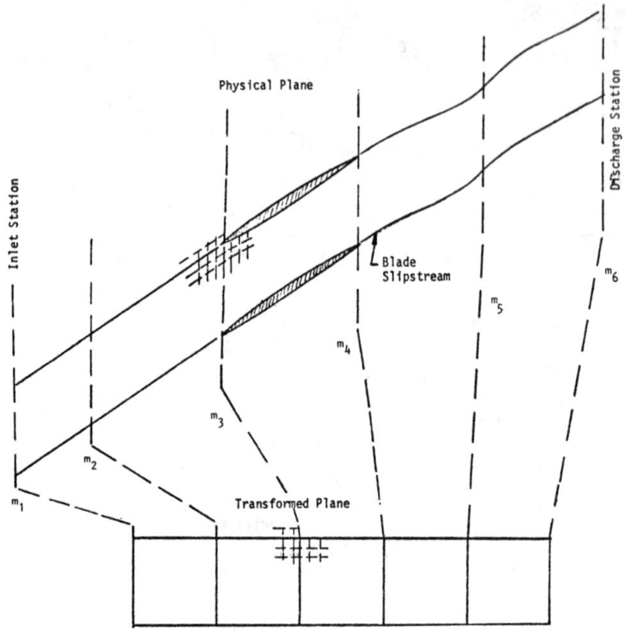

Figure 2. Schematic of Computational Domain.

plane is further divided into 5 domains. The third (central) domain extends from the blade leading edge line, m_3, to the trailing edge line m_4, and the second and fourth domains extend equal distances upstream and downstream, i.e.,

$$m_2 = 2m_3 - m_4 \tag{5}$$

$$m_5 = 2m_4 - m_3 \tag{6}$$

The inlet station, m_1, and discharge station, m_6, may be located an arbitrary distance from the blade row, or may be selected coincident with m_2 and m_5 respectively. The 5 domains thereby described (or 3 if $m_1 = m_2$ and $m_5 = m_6$) are transformed into a corresponding series of square domains with coordinates:

$$\sigma = (m-m_i)/(m_{i+1}-m_i) \tag{7}$$

$$\nu = (\theta-\theta_1)/(\theta_2-\theta_1) \tag{8}$$

where i identifies the domain, and θ_1 and θ_2 refer to the lower and upper periodic boundaries or slipstreams. Each domain is spanned by a rectangular grid network of equal spacing $\Delta\sigma$ by $\Delta\nu$.

The differential equations, Equations (1) through (4), are expressed in finite difference form and solved at all interior grid points by the MacCormack algorithm, [2].

BOUNDARY CONDITIONS

At the inlet and discharge boundaries a reference plane method-of-characteristics procedure is employed, which is described in [1]. To recapitulate briefly, the equations are recast in characteristic form in a plane which is aligned normal to both the stream surface and inlet (or discharge) plane. The influence of the flow gradients in the θ direction is minimized by allowing the reference plane to translate in the θ direction at a velocity v, which also effectively transforms the inlet and discharge boundary point calculations back from a rotating (relative) frame to a fixed (absolute) frame. The circumferential momentum equation is restated as:

$$\frac{\partial V_\theta}{\partial t} = \frac{T}{r}\frac{\partial \Delta S}{\partial \theta} - \frac{1}{r}\frac{\partial H}{\partial \theta} - V_m \zeta \tag{9}$$

when ζ is the vorticity, ΔS is the entropy, and H is the (absolute) total enthalpy. Note that if the vorticity is known, Equation (9) involves no streamwise gradients and thus the MacCormack algorithm used at the interior points can be applied to this equation at the inlet and discharge boundaries.

The energy and momentum equations can also be restated to show that:

$$\frac{D\Delta S}{Dt} = \frac{D}{Dt}\left(\frac{\zeta}{\rho}\right) = 0 \tag{10}$$

Thus the entropy and ratio of vorticity to density are properties of the inlet flow which convect across the inlet station and cannot be altered by outward radiating acoustic waves. Therefore, these two variables should be stated as boundary conditions. A third variable, representing information carried by the inward travelling waves (downstream characteristics) must also be supplied. The compatibility relation on the outward travelling waves* (in

*A subsonic axial velocity is assumed at both the inlet and discharge stations.

the reference plane) and Equation (9), then complete the solution set at the inlet. The specific requirements to model distortion screens in the inlet are discussed below.

Equations (9), (10) and the characteristic compatibility relations [1] also apply at the discharge station. However, it is evident that the entropy and ratio of vorticity to density convect across this station from within the computational domain, and hence must be determined by a Lagrangian method. The only boundary condition to be specified here is one replacing the upstream travelling characteristic,* e.g., the back pressure.

Distortion screens are usually designed to produce a controlled spatial variation of flow properties which simulates in some respect the flow distortions due to inlet design, external disturbances, and/or engine-inlet interactions. For example, the screens developed by Bruce [3], are designed to produce the sinusoidal variation in axial velocity of the form:

$$u = u_{ref}(1 + A \sin N\theta) \tag{11}$$

This design was validated by experimental measurements of the velocity distribution downstream of the screens in the absence of the rotor [3] and [4]. In this particular laboratory arrangement the large distance between the rotor and the distortion screens implies that the influence of rotor generated waves on the screen characteristics should be small. However, it cannot always be neglected and in general data measured downstream of a screen (with or without the rotor operating) must be interpreted as time-averaged wave field. Consequently, the quantities measured should be used to calculate variables which are unaffected by an upstream radiating wave field, namely, the entropy and the ratio of vorticity to density. For example, in this case the measurements can be considered to consist of:

$$\bar{p} = \frac{1}{t}\int_0^t p(\theta, t)dt = f(\theta) \simeq \text{constant} \tag{12}$$

$$\bar{u} = \frac{1}{t}\int_0^t V_m \, dt = f(\theta) \tag{13}$$

---

*A subsonic axial velocity is assumed at both the inlet and discharge stations.

$$\bar{v} = \frac{1}{t}\int_0^t V_\theta \, dt = f(\theta) \ll \bar{u} \tag{14}$$

The vorticity produced by the screen can therefore be approximated by:

$$\zeta = -\frac{1}{r}\frac{\partial \bar{u}}{\partial \theta} \tag{15}$$

The entropy can be determined by integrating Equation (10):

$$\Delta S/C_v = \gamma \log \left[1 - \frac{\gamma-1}{2\gamma} \bar{p}^{-(\gamma-1)/\gamma} k^{-1/\gamma} (\bar{u}^2 - \bar{u}_0^2)\right] \tag{16}$$

in which the equations of state

$$p = k\rho^\gamma \exp(\Delta S/C_v) = \rho RT \tag{17}$$

have been introduced. Therefore, the data yield the inlet boundary conditions:

$$\zeta/\rho = f(\bar{p}, \bar{u}) = f(\theta) \tag{18}$$

$$\Delta S/C_v = f(\bar{p}, \bar{u}) = f(\theta) \tag{19}$$

These conditions must be supplemented by a condition representing the information carried on the downstream travelling waves. If the magnitude of the rotor generated wave field is negligible at the inlet station, then either the design or measured velocity distribution, Equations (11) or (13), or the measured pressure distribution, Equation (12), provide a useful boundary condition. If it is not, then a suitable acoustic far field representation is required which in its simplest form might be adequately approximated by the Riemann Invariant for a one-dimensional flow:

$$u - 2a/(\gamma-1) = f(\theta) \tag{20}$$

where a is the local sound speed.

Enforcement of a periodic condition along the lateral boundaries of the computational domains is achieved through use of a row of exterior grid points at which the solution (at each time step) is equated to the solution at a row of interior grid points at an earlier time. The points along the boundary extending upstream from the blades therefore become interior points which are treated in the same fashion as all other interior points. However, the solution along the slipstream points is double-valued, since jumps in the component of velocity tangent to the slipstream and in the entropy are permitted. Only the pressure and normal component of velocity are necessarily continuous. The reference plane method-of-characteristics technique is invoked to carry out the solution on each side of the slipstreams and to determine their motion. In this case, the reference plane is aligned normal to the slipstream (at each point) and allowed to translate parallel to it at the local tangential velocity of the flow. The first-order compatibility relations provide a direct (non-iterative) solution for the pressure and normal component of velocity along the slipstreams, which is also applied along the blade surfaces by setting the normal velocity component to zero. The details of this method are outlined in [1]. The point to be emphasized in this matter is that application of the conservation equations to a contact surface, such as a blade slipstream, only yields the condition that the pressure and normal component of velocity are continuous and that jumps in the remaining variables are admissible. In contrast to the Rankine-Hugoniot shock relations, the magnitude of the jumps is not determined from the equations, but by the boundary conditions which generate the contact surface. Therefore, the slipstream cannot be "captured" by a finite difference solution of the conservation equations in the same fashion that shock waves are "captured."

A Kutta condition is applied at the trailing edge by requiring the pressure to be continuous; the slipstream angle at this point is iterated to achieve this condition together with zero normal velocity. The leading edge of the blade is assumed to be sharp, and a similar procedure is employed at this point. However, a more careful treatment of the leading edge, and in particular, the bow shock produced by finite leading edge bluntness, is considered warranted.

The nature of the cyclic procedure for specifying the periodic solution along the exterior grid rows is illustrated in Figure 3, which represents the interaction of a 3-bladed stator and a 4-bladed rotor. The stator blades can also represent the lobes of a 3-cycle distortion pattern. It is evident that if the flow field at time t_o is considered to be centered about blade #2, then the same field must be centered about blade #3 at time $t_o + \Delta t$. In particular, the solution in the passage between blades a and b at time t_o should exist in the passage between blades b and c at time $t_o + \Delta t$, and, consequently, the solution on exterior grid row δ at time

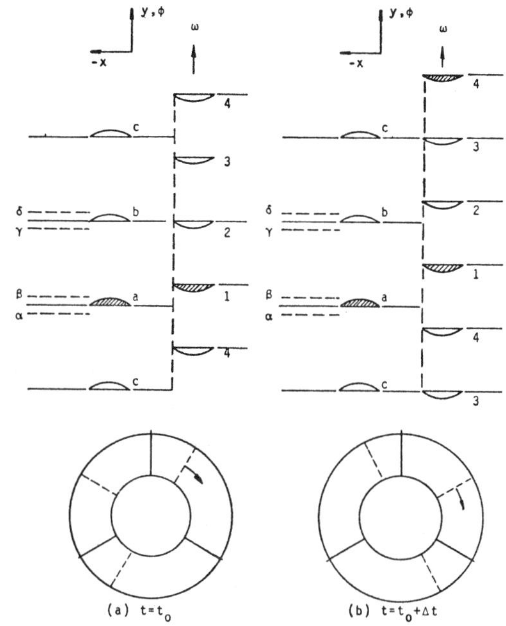

Figure 3. Illustration of Cyclic Algorithm for Stage with Unequal Number of Blades in Stator and Rotor ($N_1 = 3$, $N_2 = 4$).

$t_o + \Delta t$ can be equated to that on the interior row β at time t_o. The solution on exterior row α must be similarly equated to that on interior row γ at a time earlier than t_o. Obviously, this phase lag in specification of periodic boundary conditions is not compatible with an initial value problem, since the required information is not known a priori. However, it can be approximated with improving accuracy as time passes and the desired periodic solution attained asymptotically.

NUMERICAL RESULTS

The basic capability of Programs B2DATL and BCDATL has been examined with respect to several transonic fan rotors for which pressure surveys (from fast response pressure gauges mounted in the shroud) and holographic visualizations are available, [5] and [6]. These examples have been carried out with undistorted inlet conditions, i.e., $\zeta = \Delta S = 0$ and $u - 2a/(\gamma-1) =$ constant, and a uniform discharge condition, i.e., $u + 2a/(\gamma-1) =$ constant. Accordingly, steady solutions were obtained

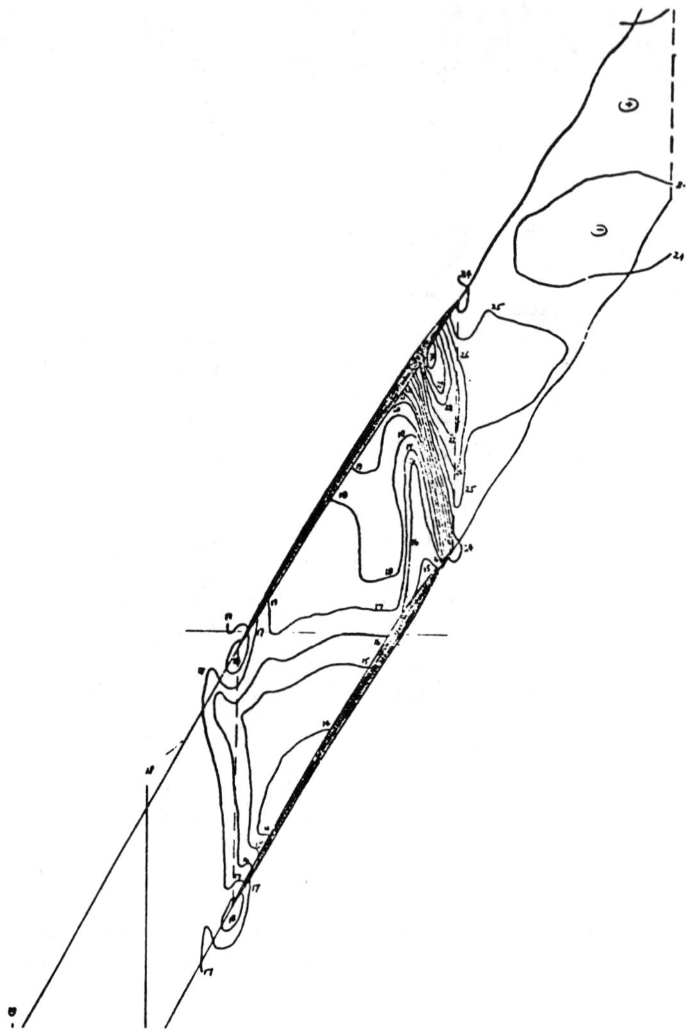

Figure 4. Computed Isobars for Reading 128 of Reference [5]. M_∞ = 1.6 and Pressure Ratio = 1.5.

asymptotically.

Pressure contours at the shroud are shown in Figure 4 for conditions corresponding to Reading 128 from [5]. The design tip speed is 1600 ft/sec., and the (relative) inlet Mach number is approximately 1.62. The weak oblique shock off the leading edge of the upper blade and the stronger oblique shock near the trailing edges implied by the compressions seen in Figure 4 are in general agreement with the shock structure observed from the holographic visualization. However, the computed second shock is slightly downstream of the observed shock, possibly due to the leading edge shock being somewhat weaker than that produced by the true nose bluntness. Additionally, the observations necessarily include the effects of tip vortices and boundary layer interactions that are absent from the present model. Nevertheless, the agreement is considered very encouraging. The mesh spacing included 18 points in the axial direction and 10 in the circumferential direction in each of the 3 domains used in the computation. A steady solution was achieved in 10^3 time steps, which required less than 3 minutes on a CDC 7600 computer system.

The pressure ratio for the above case was about 1.5. In the next case, which corresponds to Reading 126, the pressure ratio was increased to about 1.7. As can be seen in Figure 5, an anticipated forward movement of the second (passage) shock was obtained; however, no flow visualization was obtained at this condition. In the third case, shown in Figure 6, which corresponds to Reading 106 from [5], the tip speed was reduced ten percent while maintaining the pressure ratio of 1.5. A normal shock located in the aft portion of the passage was obtained in this case, as seen in Figure 6; however, the holographic observations indicate that the rotor was "unstarted" at this condition, i.e., this shock stands at the leading edge of the upper blade. The solution was perturbed several times, e.g., by increasing the back pressure, without causing the shock to jump to the unstarted position. This disparity may be attributable to either the three-dimensionality of the actual flow, the aforenoted viscous effects, or a combination thereof.

A rotor-stator interaction case is currently in progress which represents Reading 137 of the 1500 ft/sec. fan stage of [6]. This configuration has 44 blades in the rotor and 46 in the stator; consequently the fundamental period of the circumferential variations is $2\pi/(46-44) = \pi$. Therefore the rotor must complete half a revolution before a spatial repetition of the solution in the considered

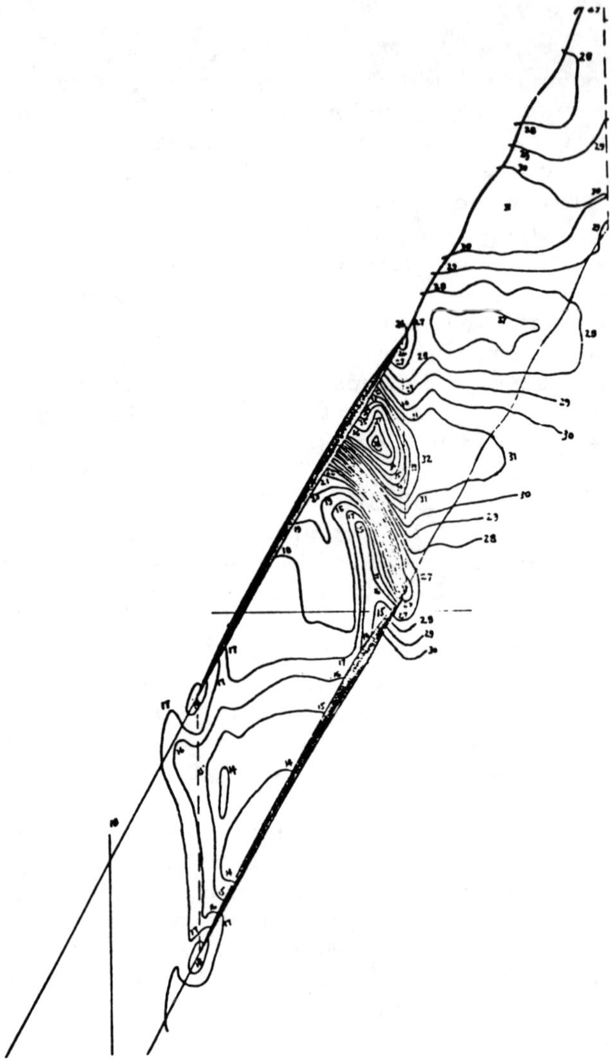

Figure 5. Computed Isobars for Reading 126 of Reference [5]. M_{∞} = 1.6 and Pressure Ratio = 1.7.

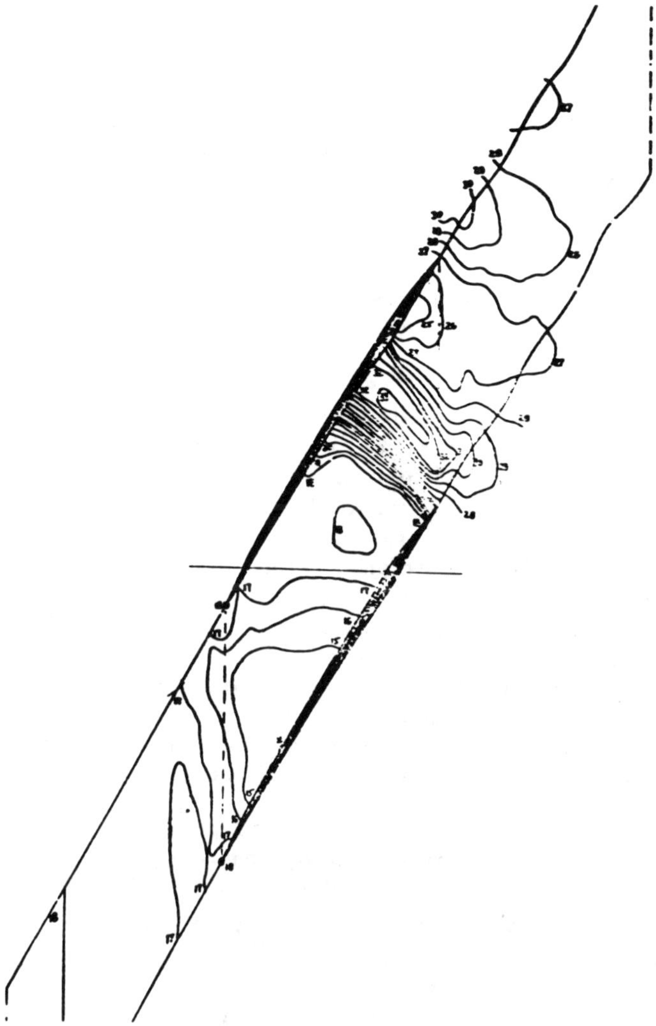

Figure 6. Computed Isobars for Reading 106 of Reference [5]. $M_\infty = 1.4$ and Pressure Ratio = 1.5.

passage can be expected.

The computed pressures at the mid-chord positions of the rotor and stator blades are presented in Figure 7 for a period of 30 rotor blade passings (more than half a revolution). It is evident that the initial transient in the rotor dies within the first 6 rotor passings, i.e., about 1/8 of a revolution, but may be influencing the stator solution (where the flow is entirely subsonic) over at least the first 10 rotor passings, i.e., about 1/4 revolution. Therefore the solution must be continued through at least 3/4 revolution to clearly establish attainment of a periodic solution.

In an example which should prove somewhat more computationally efficient, a case representing the laboratory rotor system of [4] operating in a circumferentially distorted inlet flow is planned. The selected configuration has a 6-bladed rotor and a 6-cycle distortion pattern, as indicated in Figure 8. The blades are uncambered non-lifting airfoils of ten percent thickness. Hence the observed normal force and moment coefficients are perturbations about mean values of zero. The selected configuration is a resonant combination in the context of small disturbance theory; in the context of the present computational model the phase lag in the periodic boundary conditions is zero. Therefore the initial transient should only be associated with the accuracy of the initial data, and not with the periodic boundary data.

CONCLUDING REMARKS

The significant features in the formulation of two computer codes for determination of steady and periodic transonic flows in fan and compressor stages have been outlined. The method employs the familiar "shock capturing" capability of the conservation form of the equations, but the blade slipstreams are treated as moving surfaces of discontinuity. A cyclic procedure has been developed to specify the periodic boundary conditions with a phase lag. Results obtained to date indicate good agreement with data for steady transonic rotors which are "started", but a possible problem exists at "unstarted" conditions which may be a manifestation of significant three-dimensional and/or viscous interaction effects at off-design operation. Computations to demonstrate periodic blade row interactions are still in progress, but results should be reported very soon.

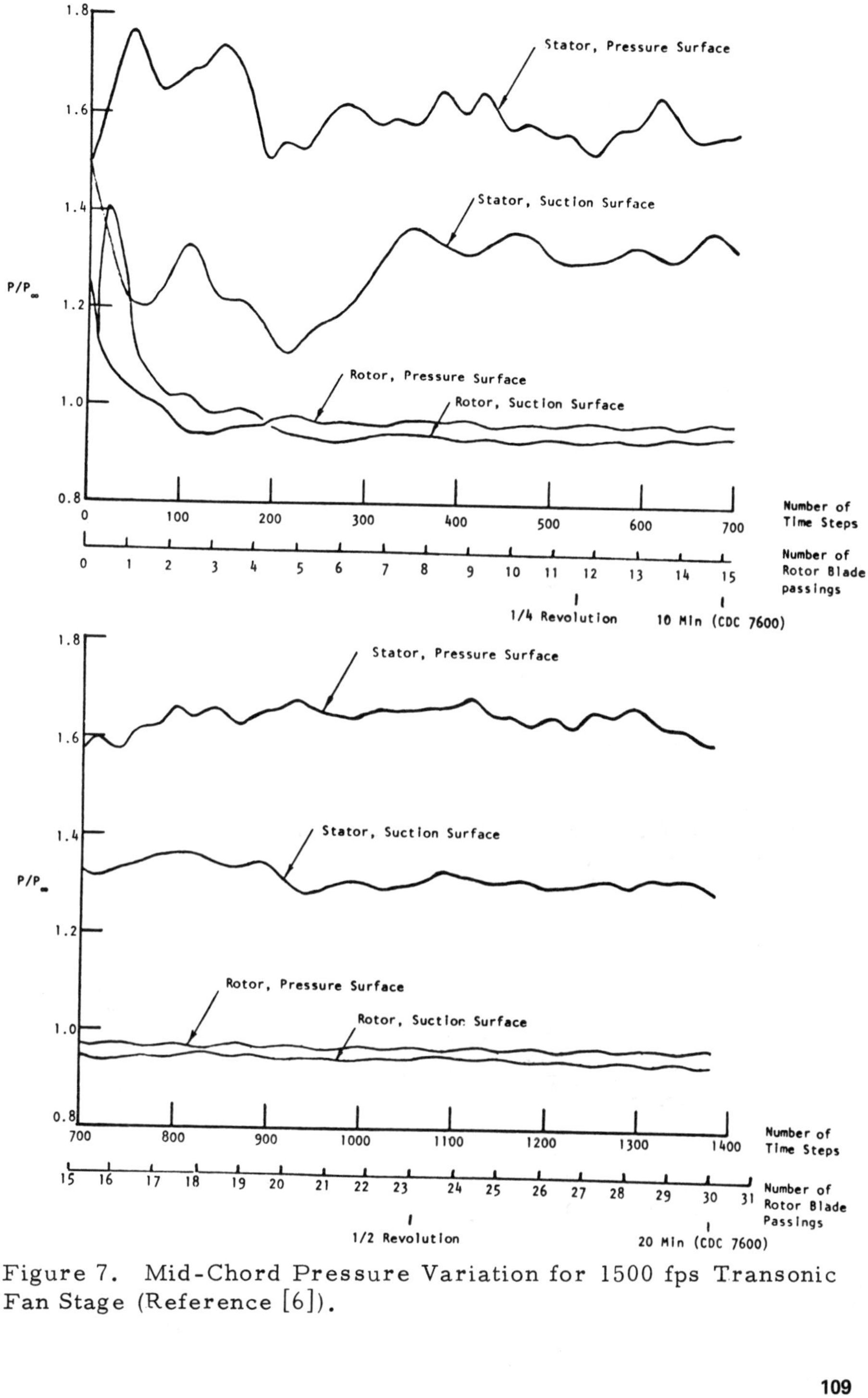

Figure 7. Mid-Chord Pressure Variation for 1500 fps Transonic Fan Stage (Reference [6]).

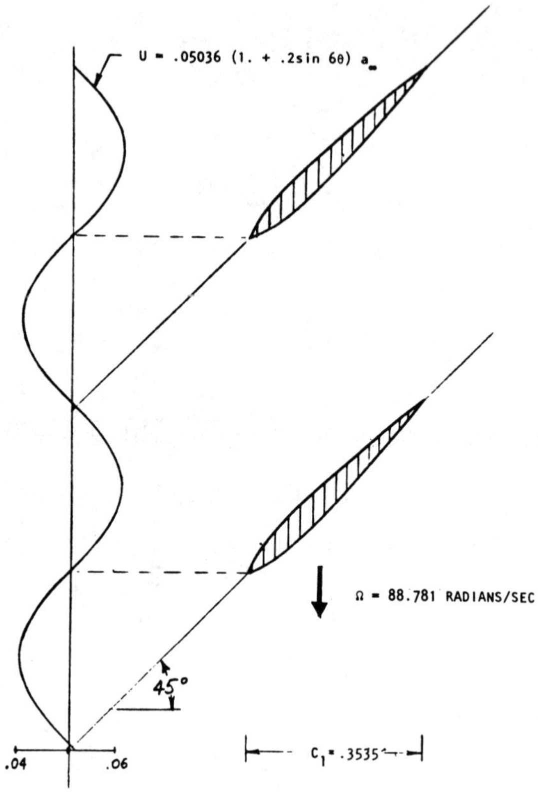

Figure 8. Configuration for Rotor Operating in Circumferentially Distorted Inflow. (Reference [4]).

REFERENCES

[1] Erdos, J., Alzner, E., Kalben, P., McNally, W. and Slutsky, S., "Time Dependent Transonic Flow Solutions for Axial Turbomachinery," in Aerodynamic Analyses Requiring Advanced Computers, NASA SP-347 (1975).

[2] MacCormack, R. W., "Numerical Solution of the Interaction of a Shock Wave with a Laminar Boundary Layer," in Proc. 2nd Int'l. Conf. on Numerical Methods in Fluid Dynamics, Springer-Verlag, New York (1971).

[3] Bruce, E. P., "Design and Evaluation of Screens to Produce Multicycle ± 20% Amplitude Sinusoidal Velocity Profiles," AIAA Paper No. 74-623, July (1974).

[4] Bruce, E. P. and Henderson, R. E., "Axial Flow Rotor Unsteady Response to Circumferential Inflow Distortions," Project SQUID Report PSU-13-PU, Purdue University, September (1975).

[5] Wuerker, R. F., Kobayaski, R. J., Heflinger, L. O. and Ware, T. C., "Application of Holography to Flow Visualization within Rotating Compressor Blade Row," NACA CR-121264, February (1974).

[6] Bilwakesh, K. R., Koch, C. C. and Prince, D. C., "Evaluation of Range and Distortion Tolerance for High Mach Number Transonic Fan Stages," NASA CR-72880, June (1972).

SESSION II

Chairman:

GINO MORETTI
Polytechnic Institute of New York

ANALYSIS: COMPUTATIONAL METHODS

COMPUTATION OF TRANSONIC POTENTIAL FLOWS IN TURBOMACHINERY

EARLL M. MURMAN
Flow Research, Inc.

ABSTRACT

This paper will review the computation of transonic potential flow by finite difference methods and consider their application to turbomachinery problems. It will begin with a brief description of the physical approximations, governing equation, and boundary conditions describing transonic potential flow in rotating machinery. Following this will be a development of the finite difference equations for mixed flow problems with embedded shock waves. The development will include the requirements of consistency, conservative differencing, correct domain of dependence (including Jameson's rotated difference scheme) and calculation of embedded shock waves. The importance of correctly enforcing boundary conditions will be emphasized including choice of coordinate systems and treatment of Dirichlet and Neumann boundary conditions. Finally, various solution procedures for the finite difference equations will be described including relaxation methods, semidirect methods, and accelerated convergence methods. Computed examples will be given for isolated bodies and cascade flows to illustrate the application of these techniques.

A number of special topics will be discussed as time permits. These topics include: hybrid direct-inverse design methods of Tranen and Carlson; the optimization design methods of Hicks, Murman, and Vanderplaats; the treatment of unsteady flows for isolated airfoils by Ballhaus and Lomax; calculation of three-dimensional flows; the effects of viscosity including a simplified shock wave-boundary layer interaction model; and application of finite element methods to transonic flow problems.

INTRODUCTION

The computation of transonic flow provides a challenging problem for both the numerical analyst and the design engineer. The numerical analyst must solve a nonlinear partial differential equation of mixed elliptic-hyperbolic type, for solutions that contain embedded shock waves. Two principal methods are presently available. They are the mixed finite-difference method treated in this paper and the time-dependent finite-difference method covered in the following paper by Yoshihara. Both methods for treating inviscid flow are reasonably well developed and have been applied to a number of practical problems. Currently, research on these numerical techniques is directed towards increasing the computational efficiency and extending the techniques to more complete model equations. The design engineer is faced with incorporating the inviscid solution techniques into computer programs that include the effects of viscosity and flow unsteadiness, and that treat increasingly complex geometries. In addition, the basic analysis methods must be developed into practical design tools that are capable of producing new geometric shapes to meet various design goals. In the field of external aerodynamics, the symbiotic relationship between the numerical analyst and the design engineer has led to a rapid development of transonic computational techniques during the past five years. A parallel development is currently in progress in the field of turbomachinery.

The purpose of this paper is to provide a brief review of the computation of transonic potential flows by finite-difference methods and to consider their application to turbomachinery problems. This paper is not intended to be an exhaustive review of either subject, but to illuminate the current state-of-the-art so as to provide a reference point for new developments. The numerical methods considered here are based on mixed finite-difference equations solved by relaxation and semi-direct solution techniques.

POTENTIAL FLOWS IN TURBOMACHINERY

Before we discuss the solution of potential flows in

turbomachinery, it is useful to consider the situations under which a potential flow exists. A comprehensive discussion of this subject is found in the book by Vavra [16], and the remainder of this section is taken directly from this reference.

Consider a rotor turning at a constant angular velocity, ω, with an oncoming flow that, in an absolute frame of reference, is axisymmetric and steady, has constant total energy, and is of the free vortex type (i. e., $rV_\theta = K_1$ = constant). Only isentropic-flow processes may be considered; thus, viscous effects are neglected, and all shock waves are weak (typically the normal Mach number at the shock is less than 1. 3.). Under these conditions, a potential flow exists such that

$$\nabla \times \vec{W} = -2\vec{\omega} = \text{constant}, \tag{1}$$

where $\vec{W}$ is the velocity vector in the rotating coordinate system. Two consequences of these assumptions are that the elementary rotation of fluid parcels are equal and opposite to the rotation of the rotor and that the energy of each fluid parcel is changed by the same amount.

We may define a reduced velocity field by

$$\vec{V}_A = \vec{W} + \vec{i}_\theta \omega r, \tag{2}$$

and, using (1), we obtain

$$\vec{V}_A = \nabla\phi_A, \tag{3}$$

so that

$$\vec{W} = \nabla\phi_A - \vec{i}_\theta \omega r. \tag{4}$$

Thus, in the rotating frame of reference, the velocity field may be decomposed into a potential component and a solid-body rotation component. Inserting these expressions into the equations of motion yields the following basic potential equations for rotor flows:

$$\nabla^2\phi_A = \frac{\nabla\phi_A}{a^2} \cdot \nabla \left[\frac{V_A^2}{2} - \omega \frac{\partial\phi_A}{\partial\theta}\right] - \frac{\omega}{a^2} \frac{\partial}{\partial\theta} \left[\frac{V_A^2}{2} - \omega \frac{\partial\phi_A}{\partial\theta}\right] \tag{5}$$

$$a^2 = a_0^2 - (\gamma - 1)\left[\frac{V_A^2}{2} + \omega\left(K_1 - \frac{\partial \theta_A}{\partial \theta}\right)\right]. \qquad (6)$$

On the solid boundary of the rotor,

$$\vec{n} \cdot \vec{W} = 0 \Longrightarrow \vec{i}_\theta \cdot \nabla\phi_A = \omega r. \qquad (7)$$

Boundary conditions for the upstream and downstream flow, any vortex wake, and any shock waves must be included. Equation (5), as written, is not in divergence form and, hence, cannot be integrated to obtain the weak solution for the potential-flow shock jump relations. Vavra [16] does not cover these aspects, and they need to be worked out. Note, however, that a potential flow does exist in rotors under certain conditions. Vavra devotes considerable discussion to the points that axisymmetric and two-dimensional flows do not truly exist in rotors and that the full three-dimensional Eq. (5) is required to describe the flow field. However, most researchers have adopted the traditional approach of computing blade-to-blade (2D) and hub-to-tip (axisymmetric) solutions. Exceptions to this include the small-disturbance model of Rae reported elsewhere in this workshop, and the three-dimensional time-dependent solution of Oliver and Sparis [14] for non-potential flows. A direct computation of Eq. (5) (in divergence form), using the methods outlined below, seems warranted.

RELAXATION METHODS FOR SOLVING POTENTIAL FLOW

A direct solution to the potential-flow equation may be obtained by using finite-difference techniques or finite-element methods. Of these, the former are more highly developed. The general requirements are that the discretized equations be consistent with the governing partial differential equation, the shock jump relations, and the boundary conditions. The iterative solution procedure must be stable and computationally efficient. For the transonic equation, there are no general proofs that the numerical solution approaches the solution of the differential equation and shock jumps in the limit of the discretization error going to zero. An *a posteriori* comparison with exact solutions and model problems, together with Lax's Equivalence Theorem for linear equations, gives us reason to believe that the solutions do indeed converge for properly formulated difference equations.

Small-Disturbance Equation

The numerical techniques for solving Eq. (5) can be most simply described by considering first the solution of a model transonic small-disturbance equation and then the potential equation in a steady frame of reference for an airfoil in free air. We can write the simplest small-disturbance equation that illustrates the features of transonic flow as

$$f_x + g_y = 0, \tag{8}$$

where

$$f = K\phi_x - \frac{(\gamma+1)}{2}\phi_x^2, \tag{9a}$$

and

$$g = \phi_y . \tag{9b}$$

Inserting (9) into (8) and differentiating, we obtain

$$[K - (\gamma+1)\phi_x]\phi_{xx} + \phi_{yy} = 0 . \tag{10}$$

Equation (10) is elliptic or hyperbolic, depending upon the sign of

$$V = [K - (\gamma + 1)\phi_x], \tag{11}$$

which is the local velocity relative to the sonic velocity. A more complete discussion of this model equation may be found in Murman and Cole [10].

The boundary conditions for an airfoil with a thickness ratio $\delta \ll 1$ (the usual small-disturbance assumption) and with a shape described by $y = S(x) = \delta\tilde{s}(x)$ for $x = (0,1)$ are written as

$$\phi_y = \frac{d\tilde{s}}{dx} \quad \text{on } y = 0, \quad x = (0,1) \tag{12}$$

$$\phi_y, \phi_x \to 0 \quad \text{as } x^2 + y^2 \to \infty . \tag{13}$$

For numerical applications, a far-field formula for ϕ is derived to facilitate application of Equation (13).

To illustrate the numerical method, we establish the difference grid shown in Figure 1 and define the quantities f and g of Eq. (9) as centered differences midway between mesh points.

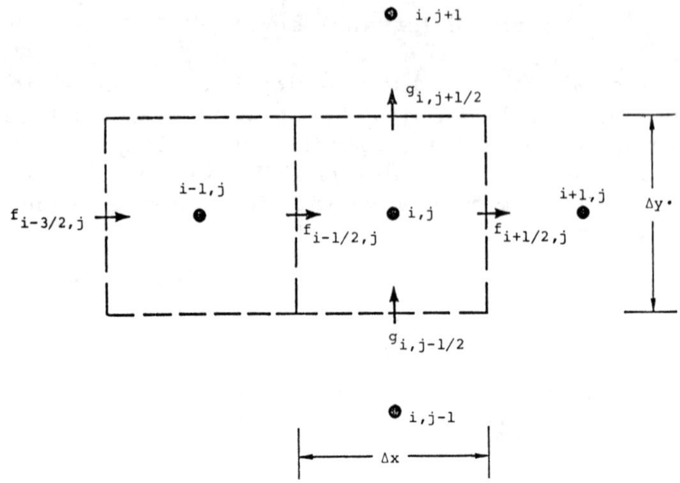

Figure 1. Mesh System Notation

$$f_{i+1/2,j} = K\left(\frac{\phi_{i+1,j} - \phi_{ij}}{\Delta x}\right) - \frac{\gamma+1}{2}\left(\frac{\phi_{i+1,j} - \phi_{ij}}{\Delta x}\right)^2 \tag{14a}$$

$$g_{i,j+1/2} = \frac{\phi_{i,j+1} - \phi_{ij}}{\Delta y} . \tag{14b}$$

We now define a switching function whose value is 0 or 1, depending upon the value of the V, which is computed from a central difference formula, as follows:

$$\mu_{ij} \equiv \begin{Bmatrix} 0 \\ 1 \end{Bmatrix} \text{ if } V_{ij} \begin{Bmatrix} > 0 \\ < 0 \end{Bmatrix} \begin{matrix} \text{"elliptic"} \\ \text{"hyperbolic"} \end{matrix} \tag{15}$$

$$V_{ij} \equiv K - (\gamma + 1) \frac{\phi_{i+1,j} - \phi_{i-1,j}}{2\Delta x} \tag{16}$$

We may then write the difference equation for Eq. (8) as

$$(1-\mu_{ij}) \frac{f_{i+1/2,j} - f_{i-1/2,j}}{\Delta x} + \mu_{i-1,j}\left(\frac{f_{i-1/2,j} - f_{i-3/2,j}}{\Delta x}\right) + \left(\frac{g_{i,j+1/2} - g_{i,j-1/2}}{\Delta y}\right) = 0 \tag{17}$$

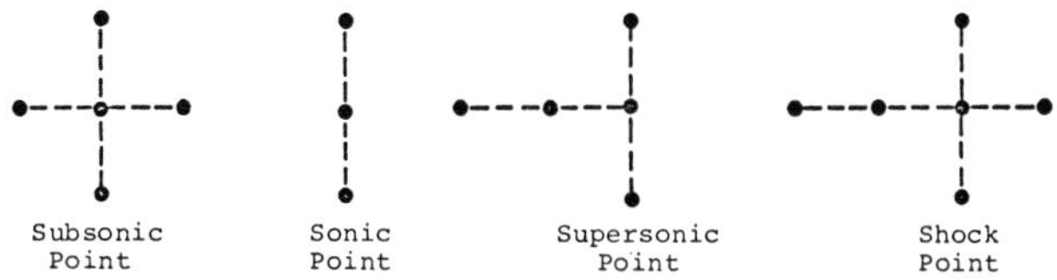

Figure 2. Mesh-Point Patterns for Fully Conservative Difference Equations

By selecting the four combinations of $\mu_{i,j}$ and $\mu_{i-1,j}$, we obtain the subsonic, sonic, supersonic, and shock-point difference equations. The mesh-point arrangement for each of these is shown in Figure 2. Of particular note is that the implicit supersonic difference operator is unconditionally stable since it satisfies the CFL stability requirement that the numerical domain of dependence must always include the mathematical domain of dependence. The latter is described by the characteristic relation

$$\tan\beta = \pm[(\gamma+1)\phi_x - K]^{-1/2} . \tag{18}$$

The characteristics for point i, j always lie behind the row i = constant, as shown in Figure 3. We note that the difference equation (17) can be considered as an integral equation for the flux of the quantities f and g (i.e., the mass) through the box surrounding the mesh point i, j as shown in Figure 1. Consequently, the total flux of f and g into the entire mesh domain is exactly conserved. Murman [11], [12] has shown that these difference equations are consistent

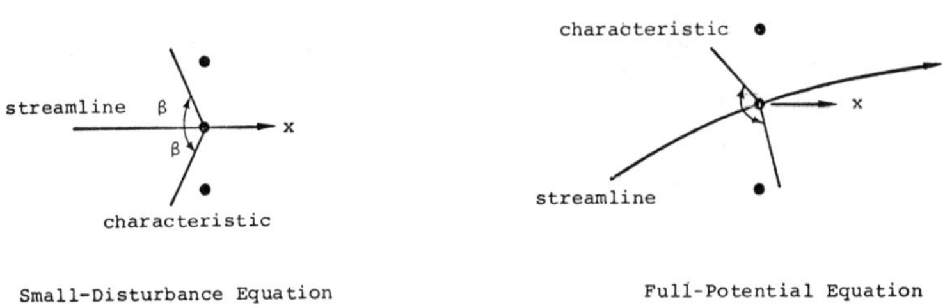

Figure 3. Domains of Dependence Relative to Mesh Points

with both the governing partial-differential equation and the shock jump relations. These difference equations are generally referred to as the fully conservative relaxation equations (FCR).

An earlier version of these difference equations introduced by Murman and Cole [10] may be written as

$$(1-\mu_{ij})\left(\frac{f_{i+1/2,j}-f_{i-1/2,j}}{\Delta x}\right)+\mu_{ij}\left(\frac{f_{i-1/2,j}-f_{i-3/2,j}}{\Delta x}\right)$$
$$+\frac{g_{i,j+1/2}-g_{i,j-1/2}}{\Delta y}=0. \qquad (19)$$

We can see that these equations do not contain the special sonic and shock-point difference equations. Murman [11][12] has shown that these equations do not guarantee that the correct shock jumps are calculated or that the total mass flow of the system is conserved. The error in the shock jumps introduces spurious sources into the flow field. There is also evidence to indicate that the numerical solutions to Eq. (19) obtained by an iterative solution (see below) may depend on the initial guess for the iteration. The difference Eqs. (19) are generally considered not fully conservative relaxation (NCR) equations.

A third difference scheme was introduced by Garabedian and Korn [2] and Jameson [4] for the full-potential equation written in non-divergence form. We can illustrate this by considering Eq. (10) and defining

$$f' = \phi_x . \qquad (20)$$

The difference scheme is then

$$(1-\mu_{ij})V_{ij}\left(\frac{f'_{i+1/2,j}-f'_{i-1/2,j}}{\Delta x}\right)+\mu_{ij}V_{ij}\left(\frac{f'_{i-1/2,j}-f'_{i-3/2,j}}{\Delta x}\right)$$
$$+\left(\frac{g_{i,j+1/2}-g_{i,j-1/2}}{\Delta y}\right)=0. \qquad (21)$$

We see that only the second-order derivatives are switched from centered to upwind and that the first-order derivatives are always centered differenced. This scheme is also non-conservative and is not consistent with the shock jump relations, nor does it conserve total mass flow for solutions with embedded shock waves. It is generally called the Garabedian-Korn-Jameson (GKJ) method.

Full-Potential Equation

The finite-difference equations illustrated above for the small-disturbance equation may be applied to the full-potential equation. Consider a steady two-dimensional flow in an absolute (non-rotating) frame of reference. We may write the governing potential in divergence form as

$$(\rho \Phi_x)_x + (\rho \Phi_y)_y = 0, \tag{22}$$

where

$$\rho = \text{function } (\Phi_x, \Phi_y, \rho_\infty),$$

and in non-divergence or quasi-linear form as

$$(a^2 - \Phi_x^2)\Phi_{xx} - 2\Phi_x \Phi_y \Phi_{xy} + (a^2 - \Phi_y^2)\Phi_{yy} = 0, \tag{23}$$

where

$$a^2 = \text{function } (\Phi_x, \Phi_y, a_\infty).$$

The boundary conditions for the above are that

$$\nabla\Phi \cdot \vec{n} = 0 \quad \text{on the body} \quad y = S(x) \tag{24}$$

$$\nabla\Phi \rightarrow 0 \quad \text{as} \quad x \rightarrow -\infty,$$

where S(x) is the body shape. We must consider several factors when choosing the finite-difference equations for Eqs. (22) and (23). First, the hyperbolic difference equation must satisfy the CFL stability condition that the numerical domain of dependence include the mathematical domain of dependence. Figure 3 shows a typical example where this condition would be violated if a cartesian grid system were used and if only the x derivatives were upwind differenced. Two basic approaches have been used to avoid this. The first approach is to use a coordinate system that is approximately aligned with the streamline directions in the supersonic zones. Examples of this approach include the "body normal" coordinate system of Steger and Lomax [15], the conformal mapping to the circle plane by Garabedian and Korn [2], and the incompressible stream-line potential-function coordinate system of Colehour [1]. For external flows, it is difficult to find a coordinate system that will be suitable for a broad range of Mach numbers and angles of attack. However, for internal flows, finding a coordinate system may not be as difficult since the streamline trajectories are more confined. Each of the above approaches has been used in cascade problems, as discussed in the next section.

The second and more general approach is the rotated difference operator introduced by Jameson [5]. Consider either Eq. (22) or (23). In the smooth regions of flow, these equations are equivalent to

$$\left(1 - \frac{q^2}{a^2}\right)\Phi_{ss} + \Phi_{nn} = 0, \tag{25}$$

where the intrinsic coordinates (s, n) represent the tangential and normal direction to the streamline, q^2 is the velocity magnitude, and

$$\Phi_{ss} = \frac{1}{q^2}\left(\Phi_x^2\Phi_{xx} + 2\Phi_x\Phi_y\Phi_{xy} + \Phi_y^2\Phi_{yy}\right) \tag{26a}$$

$$\Phi_{nn} = \frac{1}{q^2}\left(\Phi_x^2\Phi_{xx} - 2\Phi_x\Phi_y\Phi_{xy} + \Phi_y^2\Phi_{yy}\right). \tag{26b}$$

All the quantities on the right-hand side of Eq. (26) may be computed by finite-difference expressions for the cartesian mesh. For subsonic flows, centered differences are used everywhere. For supersonic flows, upwind differences are used for the x and y derivatives in the Φ_{ss} expression, and centered differences are used for the x and y derivatives in the Φ_{nn} expression. Depending upon the direction of the flow relative to the coordinate system, the upwind direction may lie in any quadrant.

The next consideration in deriving suitable difference operators for the full-potential equation is how to properly switch between the supersonic and subsonic operators to ensure that the shock waves are being correctly computed. The simple shock-point operator of Murman [11] is not directly extendable to the full-potential equation in divergence form, Eq. (22). Abruptly switching the x derivatives destroys the diagonal dominance of the difference equation matrix and leads to divergence. Jameson [6] has noted, however, that the switching operators of Eq. (17) are equivalent to the addition of an artificial viscosity to the basic centered difference expression. Consider the following x derivatives:

$$(1 - \mu_{ij}) \frac{f_{i+1/2,j} - f_{i-1/2,j}}{\Delta x} + \mu_{i-1,j} \frac{f_{i-1/2,j} - f_{i-3/2,j}}{\Delta x},$$

which can be written as

$$f_x - \left(\mu_{ij} f_{x_{ij}} - \mu_{i-1,j} f_{x_{i-1,j}}\right),$$

where the second term may be considered as an approximation to

$$\Delta x (\mu f_x)_x .$$

Thus, the difference Eq. (17) is equivalent to the partial differential equation

$$f_x + g_y = \Delta x (\mu f_x)_x . \tag{27}$$

The right-hand side represents an artificial viscosity, which vanishes in the limit of $\Delta x \to 0$.

Jameson [6] noted that the FCR difference equations use an artificial viscosity in divergence form while the NCR and GKJ methods [Eq. (19) and (21)] add an artificial viscosity in non-divergence form to give

$$f_x + g_y = \Delta x \mu (f_x)_x . \tag{28}$$

The use of divergence expressions or conservative difference operators together with the introduction of an artificial viscosity were first recognized by Lax [7] as the essential requirements for computation of purely hyperbolic problems with embedded shock waves.

The concept of adding an upwind artificial viscosity in conservative form has been used by Jameson [6] to construct a fully conservative scheme for the full-potential equation in divergence form. To illustrate the procedure, consider the non-rotated difference equations for the x derivative of Eq. (22), which may be written as

$$\frac{\partial}{\partial x}(\rho \Phi_x) = \rho \left(1 - \frac{\Phi_x^2}{a^2}\right) \Phi_{xx} - \rho \frac{\Phi_x \Phi_y}{a^2} \Phi_{xy} . \tag{29}$$

We can construct an artificial viscosity term to effectively provide an upwind difference for the Φ_{xx} term by adding to the central difference formula for Eq. (22)

$$- \mu_{ij} P_{ij} + \mu_{i-1,j} P_{i-1,j} , \tag{30}$$

where

$$P_{ij} = \frac{\Phi_{i+1,j} - 2\Phi_{ij} + \Phi_{i-1,j}}{(\Delta x)^2} \tag{31}$$

and

$$\mu_{ij} = \min \left[0, \left(1 - \frac{u^2}{a^2} \right) \right] . \tag{32}$$

The switching function provides a smooth transition of the difference equations across the sonic line. The equivalent artificial viscosity that is added is

$$\Delta x \, (\mu \Phi_{xx})_x . \tag{33}$$

Jameson [6] notes that a second-order term may be added to Eq. (31) to provide a second-order accurate hyperbolic scheme and that suitable artificial viscosities are also available for the rotated schemes.

Iterative Solution Procedures

The system of nonlinear difference equations for one of the above methods must be solved by an iterative algorithm. The original iteration scheme proposed by Murman and Cole [10] was the successive line over-relaxation scheme. Jameson [6] has given a more suitable line over-relaxation algorithm. Consider the small-disturbance difference Eq. (17) written as

$$f_{x_{i,j}} + g_{y_{i,j}} - \mu_{ij} f_{x_{ij}} + \mu_{i-1,j} f_{x_{i-1,j}} = 0 . \tag{34}$$

If we introduce the residual at iteration n as

$$R^n_{ij} = (\Delta x)^2 \left(f^n_{x_{ij}} + g^n_{y_{ij}} - \mu_{ij} f^n_{x_{ij}} + \mu_{i-1,j} f^n_{x_{i-1,j}} \right) \tag{35}$$

and the correction to ϕ as

$$c_{ij} = \phi_{ij}^{n+1} - \phi_{ij}^{n}, \tag{36}$$

a suitable iteration procedure is given by

$$\left(\frac{\Delta x}{\Delta y}\right)^2 (c_{i,j+1} - 2c_{ij} + c_{i,j-1}) + A_{ij}(-\frac{2}{\omega} c_{ij} + c_{i-1,j})$$

$$+ A_{i-1,j}(c_{ij} - 2c_{i-1,j} + c_{i-2,j}) = -R_{ij}^{n}, \tag{37}$$

where

$$A_{ij} = [K - (\gamma+1)\phi_x]_{ij}^{n} \mu_{ij},$$

and ω is a relaxation factor between 0 and 2. If c_{ij} is regarded as a temporal finite-difference operator, Eq. (37) may be recast as a time-dependent differential equation and analyzed to determine the convergence rate of the iterations. If we update the values of c_{ij} in the direction of increasing ij, Eq. (37) becomes a tridiagonal matrix, which may be easily inverted. The iteration scheme given by Eq. (37) is generally convergent, but it may require many iterations when a fine mesh is used. Using several sets of successively refined meshes aids the initial convergence rate.

Martin [9] and Jameson [6] have devised a much faster iterative procedure, which uses direct elliptic solvers. Consider the equation

$$Nc = -\omega R, \tag{38}$$

where N is a difference operator, c is the correction, and R is the residual. The most rapid convergence of the iteration will occur if N is as close to R as possible. However, for computational efficiency, N must also be a rapidly invertible matrix. The relaxation procedure described above does not bring in the full elliptic behavior of the difference equations in the subsonic flow on each iteration. Martin and Jameson proposed choosing N to be an elliptic operator for which a direct matrix inversion may be constructed, for example,

$$\alpha_1(c_{i+1,j} - 2c_{ij} + c_{i-1,j}) + \alpha_2(c_{ij} - c_{i-1,j})$$
$$+ \left(\frac{\Delta x}{\Delta y}\right)^2 (c_{i,j+1} - 2c_{ij} + c_{i,j-1})$$
$$= - \omega R_{ij}^n \; . \qquad (39)$$

An iteration scheme based on Eq. (39) converges very rapidly when the flow is subsonic everywhere, and it provides a computationally efficient algorithm. For cases with embedded supersonic zones, the iteration scheme is divergent (Jameson, [6]). Martin [9] proposed using a "desymmetrized" operator, N, which is stable in the supersonic zone and directly invertible. Jameson [6] has proposed a scheme of using one step of a fast Poisson solver followed by one or more line relaxation calculations to stabilize the solution in the supersonic region. Examples of the accelerated convergence rates of this method are given in the paper by Jameson in these proceedings.

Other methods of improving convergence rates include the multi-grid method reported by South in these proceedings and the acceleration methods of Hafez and Cheng [3].

Remarks

The above discussion is not complete since we do not cover the treatment of boundary conditions, nor do we explain the details of the computational methods. The reader must examine the references for this information. The basic solution methods have been applied to both internal and external flows with subsonic and supersonic free streams. Generally, convergent solutions are obtained.

Two important observations can be made. First, difference equations that are not fully conservative do not yield the correct solutions when there are embedded shock waves. Non-conservative schemes under-predict the shock pressure rise. Figure 4 shows a solution for the external flow past an airfoil that was computed by using the full-potential equation (Eq. 22) with conservative differencing (Jameson [6]). Figure 5 shows the same airfoil calculation using the non-conservative or quasi-linear form of the full-potential equation (Eq. 23) differenced with the GKJ scheme (Jameson [6]). Note both the reduced shock-wave strength and the

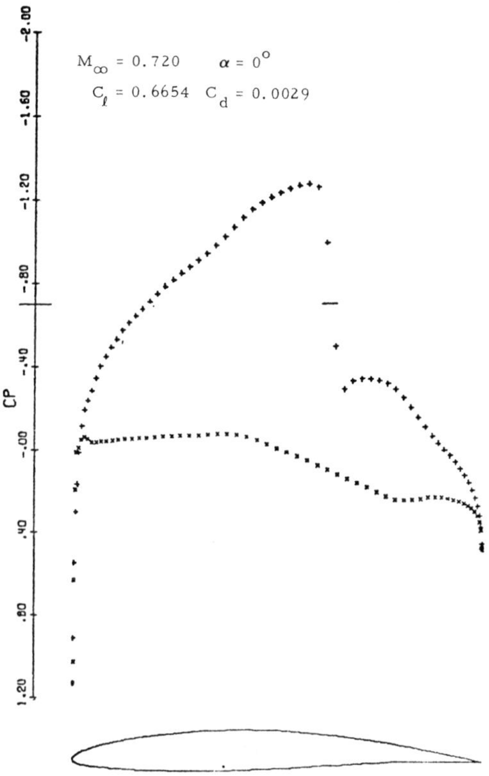

Figure 4. Potential-Flow Solution for Flow Past an NACA 64 A 410 Airfoil. Conservative Difference Equations (Conservation Form; Jameson, [6])

difference in the drag coefficient (see Murman and Cole, [10]). The non-conservative forms are clearly incorrect. If shock waves are explicitly fitted, as proposed by Hafez and Cheng [3], then non-conservative schemes may be useful. However, most researchers and design engineers continue to use the non-conservative schemes since their results agree better with experiments where the shock wave is weakened by viscous effects (not by numerical errors). More research should be directed towards rational and correct modeling of the viscous-inviscid interation. Other papers in these proceedings treat this topic.

The second observation is that non-conservative schemes do not conserve total mass flow (Murman [11]; Murman and Cole [12]).

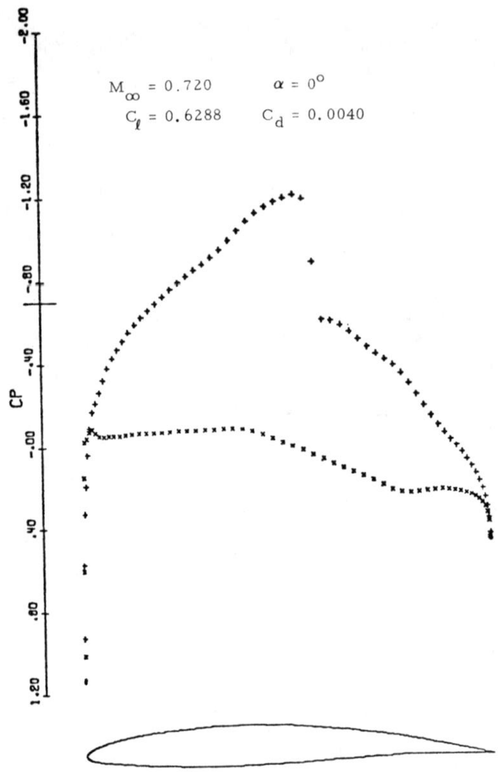

Figure 5. Potential-Flow Solution for Flow Past an NACA 64A410 Airfoil. Non-conservative Difference Equations (Quasilinear Form; Jameson, [6])

For flows with shock waves, the error is substantial. In a recent note, Newman and South [13] give a good example of this observation for flow past a parabolic arc of M = .95. Figures 6 and 7 show the flow patterns computed with non-conservative and conservative methods, and Figure 8 illustrates the computed streamline deflections. Note that for the non-conservative method, the streamlines do not close as $x \to \infty$ because of the spurious sources introduced at the shock wave. For internal flow computations, such as in wind tunnels and turbomachinery, non-conservation of mass may produce completely incorrect results.

We note that the computed examples for turbomachinery mentioned in the next section use non-conservative schemes, and thus, suffer from the above deficiencies. Hopefully, the methods will

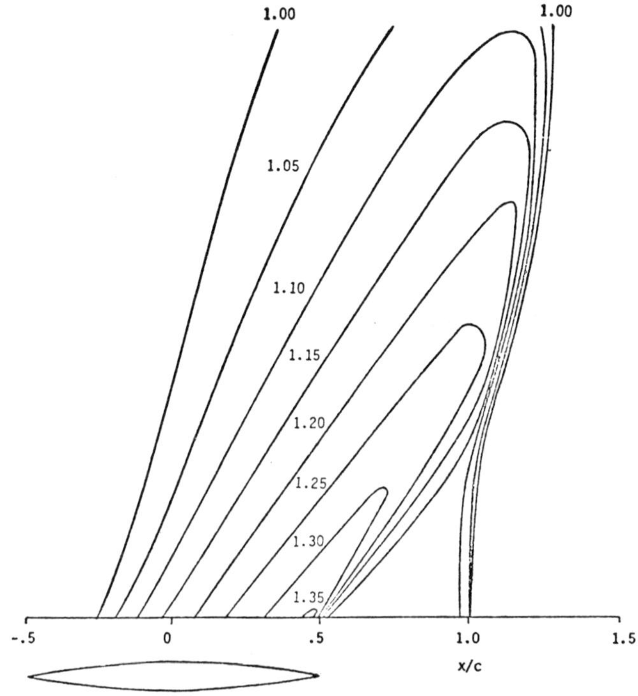

Figure 6. Mach Number Contours in Supersonic Bubble of Strong Supercritical Flow Obtained using Conservative Finite Differencing (Newman and South, [13])

be improved in the future.

COMPUTED RESULTS FOR TURBOMACHINERY

Several studies using the above methods have been reported in the literature or are presented in the proceedings of this workshop. For the latter category, the reader is referred to the papers of Rae and Dodge. In the former category is the paper by Luu and Coulmy [8]. Luu and Coulmy use the full-potential equation in non-conservation or quasi-linear form and consider the two-dimensional cascade problem for compressors and turbines. They use a conformal coordinate system, consisting of the incompressible stream function and the potential function. Allowance for open trailing edges is made to include boundary-layer corrections made at a

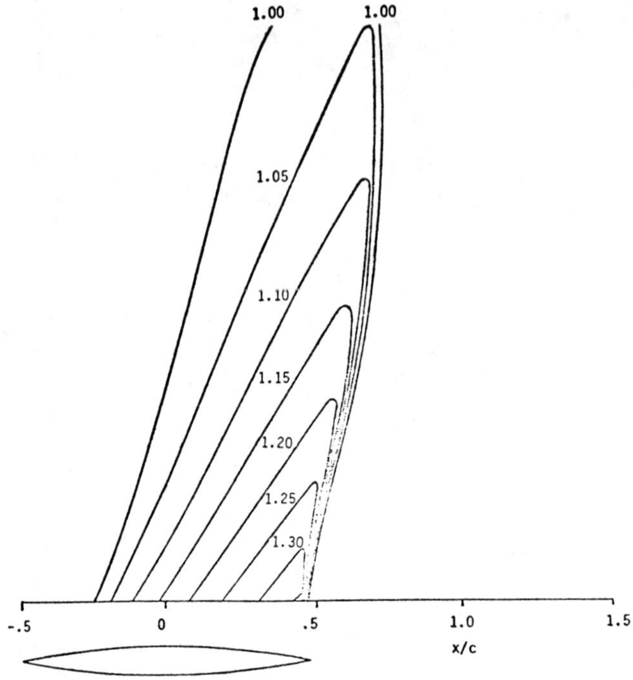

Figure 7. Mach Number Contours in Supersonic Bubble of Strong Supercritical Flow Obtained using Non-Conservative Finite Differencing (Newman and South, [13])

later date. Figure 9 shows one result from Luu and Coulmy's paper for a turbine cascade at Mach number 0.37. A disadvantage of this coordinate system is that mesh resolution at the leading edge is poor. Finally, studies that are underway at Pratt and Whitney by Dr. Ives use the full-potential equation and a mapping technique similar to Garabedian and Korn. These results will be reported soon.

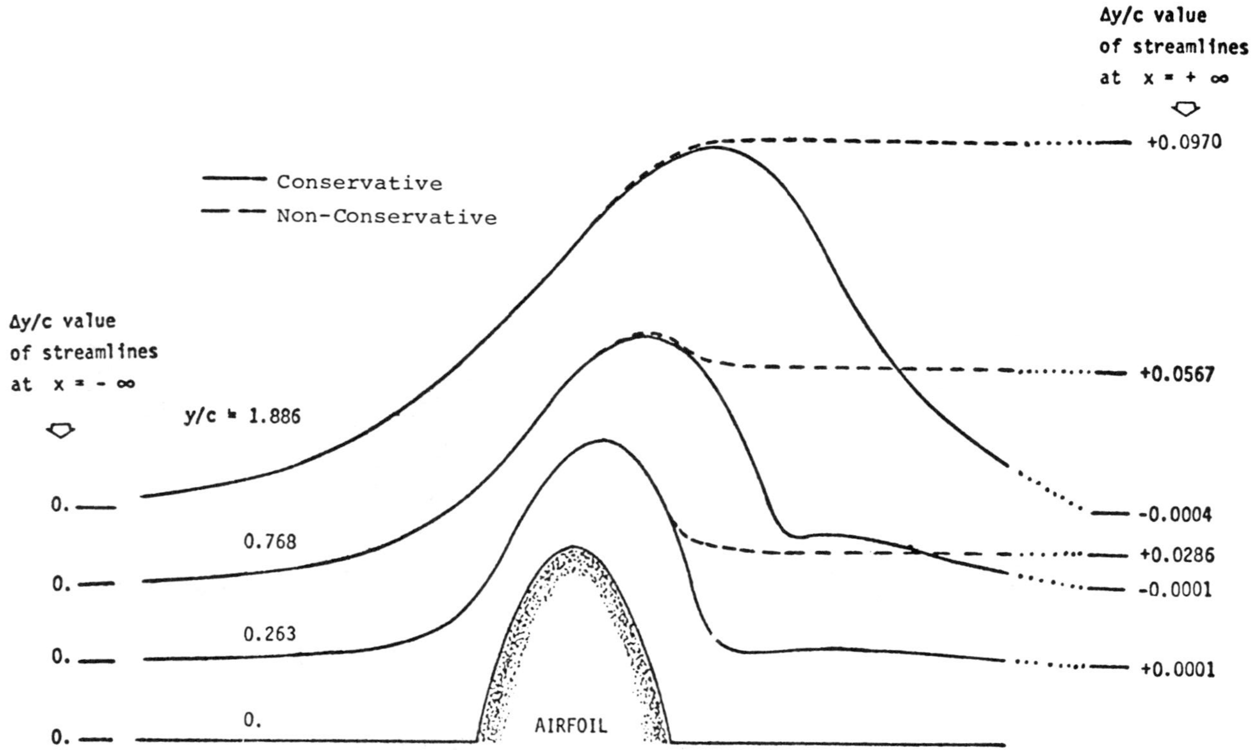

Figure 8. Comparison of Computed Streamline Deflections ($\Delta y/c$) for Strong Supercritical Flow ($M = 0.95$) Past a 10-Percent Parabolic Arc Airfoil at Zero Incidence (Newman and South, [13])

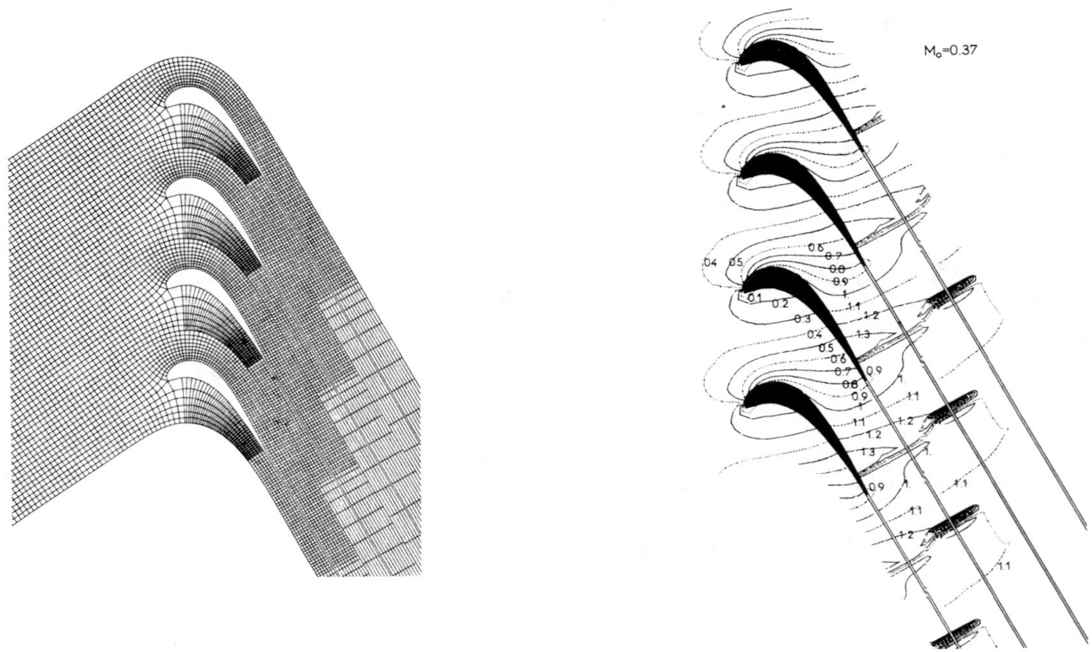

Mesh System

Mach Number Contours

Figure 9. Transonic Flow Through a Cascade (Luu and Coulmy [8])

REFERENCES

[1] Colehour, J. (1973), "Transonic Flow Analysis Using a Streamline Coordinate Transformation Procedures," AIAA Paper No. 73-657.

[2] Garabedian, P. and Korn, D. (1971), "Analysis of Transonic Airfoils," Comm. of Pure & Appl. Math., 24, pp. 841-851.

[3] Hafez, M. M. and Cheng, H. K. (1975), "Convergence Acceleration and Shock Fitting for Transonic Flow Computations," USC Report USCAE-132.

[4] Jameson, A. (1971), "Transonic Flow Calculations for Airfoils and Bodies of Revolution," Report 390-71-1, Grumman Aerospace Corporation.

[5] Jameson, A. (1973), "Numerical Calculation of Three Dimensional Transonic Flow Over a Yawed Wing," Proceedings of AIAA Computational Fluid Dynamics Conference, pp. 18-26.

[6] Jameson, A. (1975), "Transonic Potential Flow Calculations using Conservation Form," Proceedings of AIAA 2nd Computational Fluid Dynamics Conference, pp. 148-161.

[7] Lax, P. D. (1954), "Weak Solution of Non-Linear Hyperbolic Equations and their Numerical Computation," Comm. Pure & Appl. Math., 7, pp. 283-309.

[8] Luu, T. S. and Coulmy, G. (1975), "Relaxation Solution for the Transonic Flow through a Cascade," Proceedings of Transsonicum Symposium II, Springer-Verlag.

[9] Martin, E. (1975), "A Fast Semi-Direct Method for Computing Transonic Aerodynamic Flows," Proceedings AIAA 2nd Computational Fluid Dynamics Conference, pp. 162-174.

[10] Murman, E. M. and Cole, J. D. (1971), "Calculations of Plane Steady Transonic Flows," AIAA Journal, 9:1, pp. 114-121.

[11] Murman, E. M. (1974), "Analysis of Embedded Shock Waves Calculated by Relaxation Methods," AIAA Journal, 12:5, pp. 626-632.

[12] Murman, E. M. and Cole, J. D. (1974), "Inviscid Drag at Transonic Speeds," AIAA Paper No. 74-540.

[13] Newman, P. and South, J. (1976), "Conservative Versus Non-Conservative Differencing: Transonic Streamline Shape Effects," NASA TMX-77827.

[14] Oliver, D. and Sparis, P. (1975), "Computational Aspects of the Prediction of Multidimensional Transonic Flows in Turbomachinery," NASA SP-347, pp. 567-586.

[15] Steger, J. and Lomax, H., (1972), "Numerical Calculation of Transonic Flow and Two-Dimensional Airfoils by Relaxation Procedures," AIAA Journal, 10:1, pp. 49-54.

[16] Vavra, M. H. (1960), Aero Thermodynamics and Flow in Turbomachines, John Wiley & Sons, Inc.

NOMENCLATURE

Symbol	Definition
A	coefficient in iteration equation
a	speed of sound
a_o	stagnation speed of sound
c	correction to ϕ
f	flux quantity in x direction
g	flux quantity in y direction
i	mesh point subscript
$\vec{i}_\theta$	basis vector in θ direction
j	mesh point subscript
K	constant in transonic equation
K_1	constant in potential equation for rotating flow
N	matrix operator
$\vec{n}$	unit normal to body surface
n	normal direction to streamline iteration counter
P	artificial viscosity term
q	velocity magnitude
R	residual
r	radial coordinate
S	shape function for airfoil
s	tangential direction of streamline
$\tilde{s}$	scaled shape function for airfoil
V	velocity relative to sonic velocity

Symbol	Definition
V_A	reduced velocity for absolute rotor flow
V_θ	circumferential velocity
$\vec{W}$	velocity vector in rotating coordinate system
x	cartesian coordinate
y	cartesian coordinate
α_1, α_2	coefficients in iteration equation
β	characteristic angle
γ	ratio of specific heats
Δ	finite difference
δ	body thickness ratio
μ	switching coefficient for difference equation
ρ	density
Φ	velocity potential
ϕ	velocity perturbation potential
ϕ_A	velocity potential in rotating coordinate system
ω	angular velocity of rotor; relaxation factor

DISCUSSION OF THE MURMAN PAPER

DIXON: In that semidirect method do they apply the Poisson solver to the whole flow field or just to the part of the flow field which is subsonic?

MURMAN: They apply it to the whole flow field. They could eliminate the part that is supersonic but then they would have to recompute the direct matrix reduction every time. What happens then when they apply the elliptic solver to the whole flow field in the supersonic region is that the results are kind of bad. I guess the errors grow in the supersonic regions. But then after doing this Poisson step the calculation follows from the relaxation sweep which in the supersonic zone is a marching calculation and all the errors are swept out. From 1 to 5 relaxation steps are used, depending on which of all these various forms of equations are chosen.

MELNIK: It's my understanding that in the use of the streamline coordinates there's still a problem in that when Colehour gave his early papers - the stagnation point did not move. In other words, you do the incompressible flow and you get a coordinate system that has a stagnation point there, then you do the relaxation solution, and you find the compressible flow. You find that the stagnation point did not move in that computation. It's clearly wrong. Do you know if that situation has changed?

MURMAN: I don't know. That wasn't brought up in Luu and Coulmy's paper and I wondered about that, but since I couldn't find an answer, I didn't say anything about it. Maybe Paul Dodge will have some comment on that when he talks, if he's going to cover that coordinate system.

CASPER: In the one-sided differencing thing, as you pointed out, it is important that the domain of dependence of the difference equation overlap that of the differential equation. Now, if center differences were used, this could certainly be the case. If you use center differences without using the switching operators, you get divergence. If you use the switching operators, but do not have the proper domain of dependence, do you also get divergence?

MURMAN: The best way to answer that question is to say that one should do a stability analysis using the von Neumann type stability test to see whether an error introduced at a mesh point would grow or decay as you marched on downstream. You find out that if you use a center difference operator and you solve for the middle mesh point - like if it was an elliptic equation - then it would be divergent. If you use a center difference operator, you speed up the convergence of the screen. If you solve for this point of the flow as supersonic, you can show from the stability test, that it would be divergent. If you solve for this point, the forward point, considering it as a marching problem, you would have enough initial conditions that you start out solving for this point. Then the procedure is stable as long as the characteristics lie within these two points back here. That's the correct interpretation of the difference equation containing the domain of dependence of the differential equation. But that won't happen near Mach 1 - where the characteristics are more or less normal. So this explicit method is not very useful for transonic flow computations, but Paul Dodge will be explaining some computations later on where he has tried to compute the coordinate systems where this always happens. But the implicit method, this method here, is unconditionally stable for the small disturbance equation.

FINITE DIFFERENCE PROCEDURE FOR UNSTEADY TRANSONIC FLOWS: A REVIEW

HIDEO YOSHIHARA

Convair Division of General Dynamics

ABSTRACT

The finite difference procedure of Ref. [1] will be described in detail. This procedure is based upon the Euler equations and utilizes the Lax-Wendroff second order explicit difference scheme. The boundary conditions are fulfilled using a localized unsteady plane wave emitted from the surface. Shock waves acquire a profile due to the numerical viscosity, and they are captured automatically using a fine mesh about the shock. The procedure is illustrated by the example of an airfoil oscillating in pitch at supercritical conditions.

The rationale and the background for the selection of the mesh configuration, the difference scheme, and the plane wave procedure for fulfilling the airfoil boundary condition used in the above procedure are briefly reviewed showing the advantages and disadvantages relative to other well-known finite difference schemes or procedures. Here implicit difference schemes and other explicit schemes as the MacCormack and alternating direction schemes are described as well as other boundary condition and shock capture procedures.

INTRODUCTION

Although titled a "review," the primary attention here will be focused on the procedure of Ref. [1] based upon the Euler equations. To rationalize a posteriori the selection of the numerical ingredients used therein, as the choice of the difference scheme, we shall briefly describe alternatives presently available. This then will serve as the review.

In developing a numerical procedure it is helpful to begin with a physical understanding of the flow of interest, particularly its salient features and the influential flow mechanisms that control them. Such knowledge serves not only to enable the formulation of a viable and tractable flow model, but to guide the construction of the corresponding numerical analogue, the debugging of the resulting computer program, and finally the assessment of the final results.

We shall be concerned with unsteady supercritical (transonic) flows over planar airfoils. In Figure 1 we show schematically the principal features of such a flow where the unsteadiness is created by the motion of the airfoil. The central feature shown here is the strong boundary layer-shock wave interaction whose displacement effects, magnified by the extreme sensitivity of transonic flows, lead to major modifications of the flow. In an unsteady flow the movement and strength of the shock wave are governed by the instantaneous state of flow just upstream and downstream of the

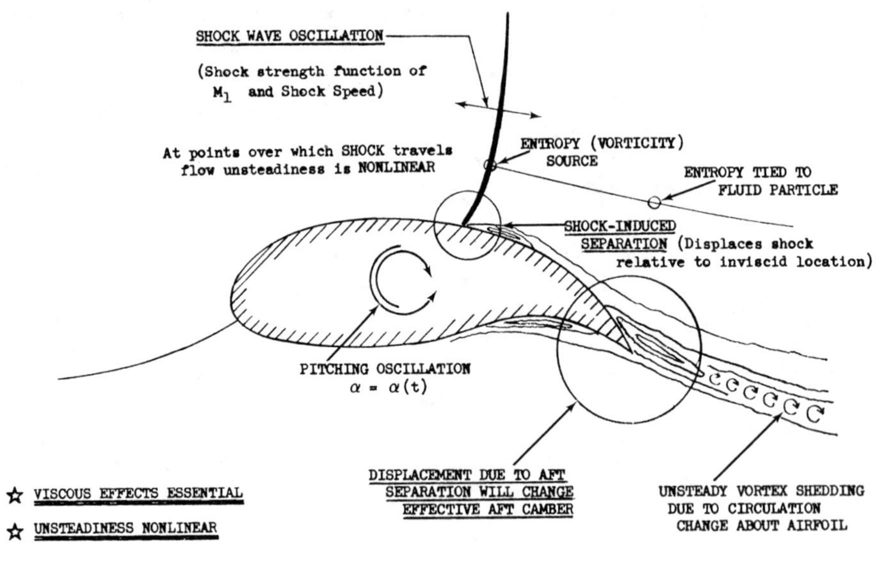

Figure 1. The Anatomy of the Unsteady Supercritical Flow

shock. The conditions here are in turn established by the arrival of piston waves generated at the moving airfoil at prior times retarded by the wave propagation times. Significant lags in the shock positioning and strength relative to the airfoil motion can accordingly arise resulting in shock configurations having no counterpart in the more familiar quasi-steady case.

Boundary layers of practical relevance are turbulent. It is well known that we do not have adequate knowledge of the turbulent transport of sufficient generality to formulate the necessary flow equations. As a tentative necessity we must shelve the real problem and consider instead an inviscid flow. The development of the numerical procedures even with this simplification has proven to be challenging enough with the numerical development plagued by the familiar extreme sensitivities of transonic flows to perturbations. The resulting experience gained with the inviscid model clearly will be relevant in the eventual treatment of the real flow.

The inviscid initial-boundary value problem appropriate for the problem on hand is formulated in Figure 2. Here the unsteady

● BOUNDARY CONDITIONS

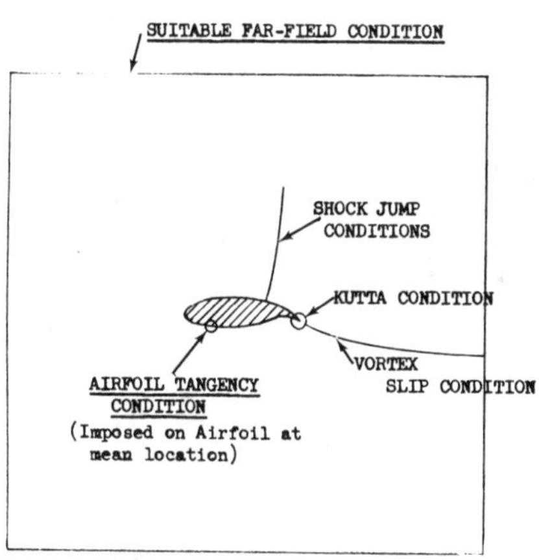

● SUITABLE INITIAL FLOW MUST BE PRESCRIBED

● UNSTEADY EULER EQUATIONS (In proper conservation form)

$$\underbrace{W_t}_{\text{time derivative}} = \underbrace{-F_x - G_y}_{\text{space derivatives}}$$

$$F = F(W) \qquad G = G(W)$$

$$W \equiv \begin{Bmatrix} \rho \\ \rho u \\ \rho v \\ e \end{Bmatrix}$$

$$F \equiv \begin{Bmatrix} \rho u \\ \rho u^2 + p \\ \rho uv \\ u(p+e) \end{Bmatrix}$$

$$G \equiv \begin{Bmatrix} \rho v \\ \rho uv \\ \rho v^2 + p \\ v(p+e) \end{Bmatrix}$$

$$e = (\gamma - 1)^{-1} p + \rho(u^2 + v^2)/2$$

SYSTEM OF EQUATIONS IS HYPERBOLIC

Figure 2. Formulation of the Inviscid Problem (Magnus and Yoshihara)

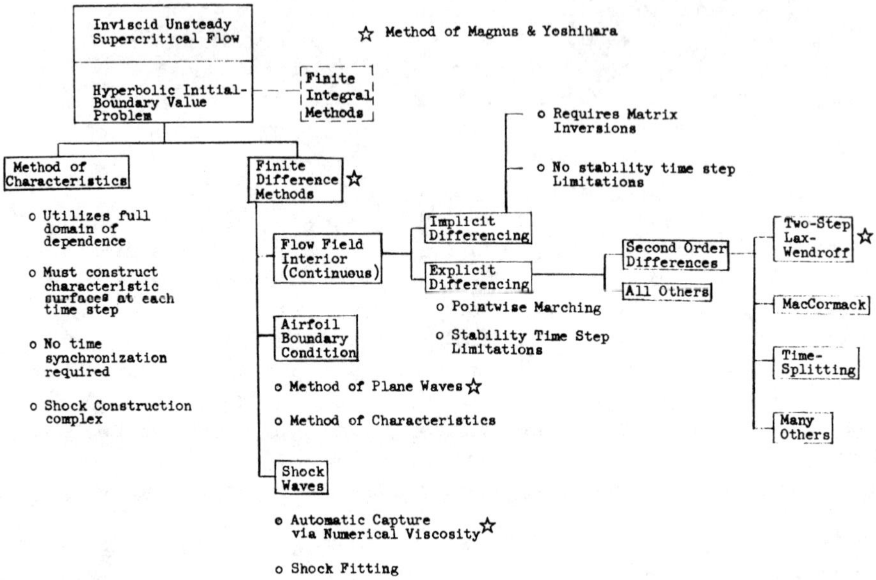

Figure 3. Flow Chart for Finite Difference Procedure

Euler equations represent a fully hyperbolic system, and they are given here in the proper divergence form to enable the correct capture of shocks.

In Figure 3 we next show some alternative routes for the construction of the numerical analogue. In Ref. [1] we have selected the finite difference method using the differential equations. Personal bias shaped by past experience has led us to this choice. In many respects the method of characteristics offers a more elegant and sounder approach by taking advantage of the natural features of the hyperbolic system. Finite integral methods also have their advantages, particularly in the treatment of boundary conditions.

With the selection of the finite difference procedure, we must now select a suitable mesh configuration, a difference scheme to construct the system of difference equations, a scheme to fulfill the boundary conditions, and finally a routine to treat the shock waves. In the following we shall take up each of these items.

THE MESH CONFIGURATION

The selection of the mesh configuration is an important task

GENERAL REQUISITES

- Mesh spacing must be TAILORED to the expected FLOW GRADIENTS

  (For example fine spacing about blunt noses, shocks, and vortex lines)

- o Mesh must be CONFORMAL to airfoil surface (Generally)

PROCEDURE I USE OF MAPPING (Flow domain mapped for example into rectangle with suitable coordinate stretching)

Advantage: Uniform mesh spacing in transformed space

Disadvantage: Flow equations become much more complex

PROCEDURE II HYBRID CONFIGURATION (Cartesian system in general with variable spacing plus embedded curvilinear system conformal to the airfoil)

Advantage: Equations complicated only in the inner embedded domain.

Tailoring of the mesh is direct and controllable

Disadvantage: Overlap of the various mesh subsystems required as well as a data transfer logic between mesh systems.

Figure 4. Selection of the Mesh Configuration

that is frequently not given its due attention. The integrity of the final results and less importantly the computing time rest heavily on the proper choice of the mesh system. In Figure 4 we indicate some of the alternatives available in constructing the mesh system, giving advantages and disadvantages of each; while in Figure 5 we show the mesh system used in Ref. [1]. In

DETAILS OF THE NOSE MESH

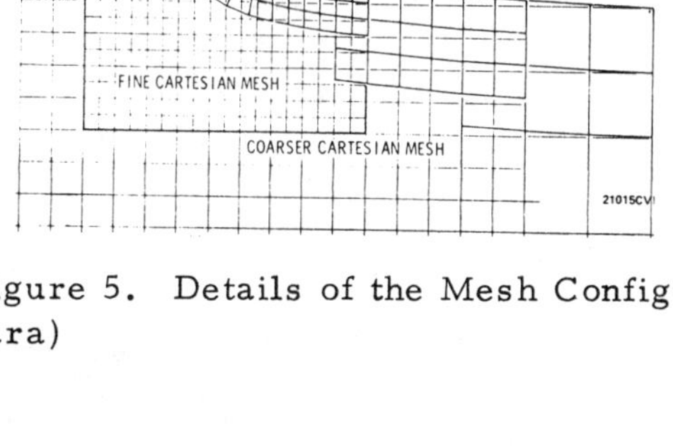

TOTAL NUMBER OF MESH POINTS ~ 6000

- o Mesh sizes range from .0066 chords at nose to .05 chords about the airfoil as a whole.
- o .0125 chord mesh through shock
- o .4 chord mesh in outermost region

Figure 5. Details of the Mesh Configuration (Magnus and Yoshihara)

the latter a hierarchy of cartesian subsystems with varying refinement covers the domain, while a non-cartesian subsystem conformal to the airfoil surface locally embeds the airfoil to facilitate the imposition of the boundary condition. A refined cartesian system also covers the subdomains where the shocks are expected.

THE DIFFERENCE SCHEME - ACCURACY AND STABILITY

Consider next the selection of the difference scheme to be used in an elemental marching step to advance the solution in the flow interior from the initial data to a new time an increment Δt later. In Figure 6 we show two criteria of importance in selecting the difference scheme, the order of accuracy and the stability which establishes the allowable time step. In the definition of the order of accuracy given in Figure 6, the comparison basis is the linearized solution which should suffice. It is important to remember here that, aside from the order of accuracy itself, the proportionality factor K, characteristic of a given difference scheme, is of importance, since the arithmetic magnitude of the error itself is the relevant quantity inasmuch as the limit $\Delta t \rightarrow 0$ is never taken.

Consider the elemental marching step

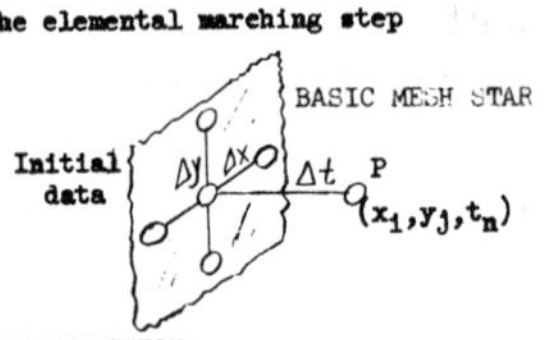

CRITERIA FOR SELECTION

- **Order of ACCURACY** (W = Dependent Variable)

$$\left| W(x_i, y_j, t_n) - W_{ij}^{n} \right| \leq O(\Delta t^{h+1}) = K(\Delta t)^{h+1}$$

EXACT (linearized) solution; FINITE DIFF. solution

h = ORDER OF ACCURACY

NOTE that proportionality factor is important since the limit $\Delta t \rightarrow 0$ is never taken.

THAT IS, SOME SECOND ORDER SCHEMES ARE "MORE" SECOND ORDER THAN OTHERS

- **STABILITY** (Establishes allowable time step)
  - Solve linearized equations
  - Determine conditions to avoid exponentially growing Fourier components

    This leads to CFL limits on the time step (CFL = Courant, Friedrichs, & Lewy)

NOTE: Stability time step limits are very much more STRINGENT than that required to resolve the flow unsteadiness

Figure 6. Selection of the Difference Scheme

The stability of the elemental marching step is determined from the locally linearized difference solution and is defined as the absence of any exponentially growing Fourier component. Absence of such components can be assured by suitably restricting the time step in accordance with the Courant-Friedrichs-Lewy (CFL) condition which can be derived in a straightforward manner from the linear solution.

The CFL condition is a necessary but not a sufficient condition for the stability of the overall difference solution. It is a necessary condition since if the elemental marching step itself is unstable, the overall solution will surely be unstable. On the other hand, stability in the elemental marching step does not insure stability of the overall solution since the latter involves in the limit $\Delta t \rightarrow 0$ an infinite number of elemental time steps, and additionally destabilizing influences can be and frequently are generated at the boundaries in the process of fulfilling the boundary conditions. The general practice is to take a time step less the CFL limit adding diffusive damping (which we shall shortly describe) in local trouble spots until an acceptable stable solution is obtained.

The CFL condition can be interpreted in terms of the domain of dependence of the flow differential equations. Frequently in the case of multidimensional problems the CFL conditions are so complex that one turns instead to the domain of dependence as a guide in determining the allowable time step. In Figure 7 the domain of dependence for a given point P is shown, and it represents that portion of the initial data plane intersected by the characteristic forecone of Pt. P. That is, only the initial data within the domain of dependence can influence Pt. P. As a guide to the allowable time step in an explicit difference scheme (to be defined later) the time step is taken such that the initial data mesh points of the difference scheme at least fall within the domain of dependence. In this case, shown in the lower left part of Figure 7, we have the disconcerting situation that the procedure is stable, but data which physically cannot influence the flow at Point P are being used to determine the solution at P. Such idiosyncracies of the finite difference procedure frequently arise, and they are simply reminders that difference equations have their own peculiar features quite apart from those of the corresponding differential equations.

TYPES OF DIFFERENCE SCHEMES

Let us consider next the elemental marching step. Here the

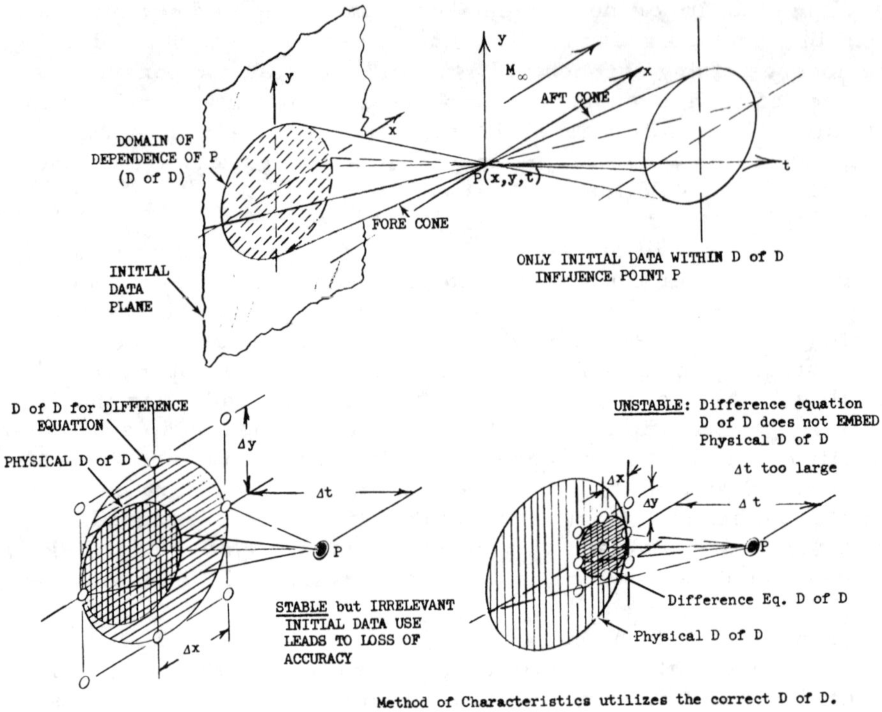

Figure 7. Domain of Dependence (D of D) and the Allowable Time Step

solution is advanced in time from the initial data as a Taylor expansion where the time derivatives are obtained from the flow equations. In the format of the flow equations given in Figure 4, the time derivative of the dependent vector W can be expressed explicitly in terms of space derivatives $F_x(W)$ and $G_y(W)$.

As the first alternative in Figure 3 in the selection of the difference scheme, one has the option of an explicit or an implicit scheme. In the former the derivatives F_x and G_y are evaluated using only the known initial data. The resulting difference equation for W then contains only a single unknown value of W at the new time which then can be determined explicitly. In this case there will be a CFL restriction on the allowable time step that is invariably much more stringent than that necessary to resolve the flow unsteadiness.

On the other hand, in an implicit difference scheme, the space derivatives F_x and G_y are usually evaluated on both the initial as well as the new time plane. As a result, the difference equation will involve the unknown vector W at more than one point at the new time. Such implicitness will then require the simultaneous consideration of the difference equations for every mesh point and hence the inversion of large matrices. Offsetting this is the significant advantage of implicit difference schemes of being unconditionally stable. The restriction on the time step here would then be dictated only by the scale of the flow unsteadiness.

In Ref. [1] we have chosen to use an explicit scheme, in fact, the second order Lax-Wendroff scheme shown in Figure 8. Again this choice was dictated, not by an objective selection process, but by our past experience. The choice of the second order difference scheme for the present transonic problem has some basis since first order schemes are found to be simply inadequate from the accuracy point of view, while the added accuracies achieved by the higher order schemes have been generally found to be not

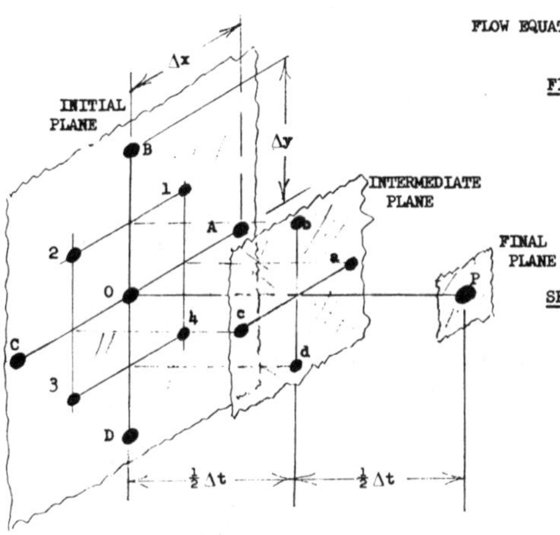

FLOW EQUATION: $\frac{\partial W}{\partial t} = -\frac{\partial F}{\partial x} - \frac{\partial G}{\partial y}$; $F = F(W)$; $G = G(W)$

FIRST STEP: Determine Intermediate Values W_a etc.

$$W_a = \tfrac{1}{4}(W_A + W_1 + W_0 + W_4) - \tfrac{1}{2}\Delta t\left(\frac{F_A - F_0}{\Delta x} + \frac{G_1 - G_4}{\Delta y}\right)$$

Determine $F_a = F(W_a)$ etc.

SECOND STEP:

$$W_P = (1-\mu)W_0 + \underbrace{\tfrac{1}{4}(W_A + W_B + W_C + W_D)\mu}_{\text{Diffusive Damping}} - \Delta t\left(\frac{F_a - F_c}{\Delta x} + \frac{G_b - G_d}{\Delta y}\right)$$

DIFFUSIVE DAMPING COEFFICIENT μ USED FOR EXAMPLE TO CRITICALLY DAMP OVERSHOOTS AT SHOCKS

CFL CONDITION: $(|u| + c)\frac{\Delta t}{\Delta x} < 2^{-\frac{1}{2}}$

($\mu = 0$ and $\Delta x = \Delta y$)

Figure 8. Lax-Wendroff Two-Step Difference Scheme (Explicit Scheme of Second-Order Accuracy)

commensurate with the added complexities.

SAMPLE SECOND ORDER EXPLICIT DIFFERENCE SCHEMES [2, 3]

Let us first examine the Lax-Wendroff difference scheme in Figure 8 more closely. Here in the first step the unknown W at the intermediate time step $\frac{1}{2}\Delta t$(Pts. a, b, c and d) are first determined as well as the corresponding values of F(W) and G(W). The latter values are then used to form F_x and G_y and hence W_t at the midpoint between Pts. O and P, which is then used to determine W_P in the second step of Figure 8. It is to be noted that in the second step diffusive damping has been added by expressing the value of W at Pt. O as a weighted sum of W_O and the average of the values of W at neighboring Pts. A, B, C and D. Here the diffusive effect is introduced by bringing into play the influence of the neighboring points in the latter averaging process. Supplementary diffusive damping is used (usually tenatively) to suppress instabilities in localized trouble spots as well as to suppress finite difference "wiggles" occurring for example at mesh subdomain interfaces, or, in the case of the basic Lax-Wendroff scheme ($\mu = 0$), at shock waves. The amounts of diffusive damping here are generally sufficiently small to retain the second order accuracy.

Another frequently used explicit scheme of second order is due to MacCormack of NASA-Ames shown in Figure 9. This is a predictor-corrector scheme in which the predictor step is used to obtain a first guess of the solution which is then used in the corrector step to obtain the final solution. It involves significantly fewer arithmetic operations than the Lax-Wendroff scheme since it utilizes data already generated at adjacent mesh points. The main deficiency of the MacCormack scheme is the use of the one-sided differences rather than central differences to approximate the derivatives. Such a deficiency would be serious in regions where large values of F_{xx} or G_{yy} arise, as in the shock profile, where neither the forward nor the backward differences are adequate approximations to the central difference, leading to a poor first approximation in the predictor step that cannot be fully rectified in the corrector step.

Finally, a time-split difference scheme given by MacCormack is shown in Figure 10. The method of time splitting has been pioneered by the Russian researchers who labeled the procedure as the method of fractional steps. In the specific scheme of Fig. 10

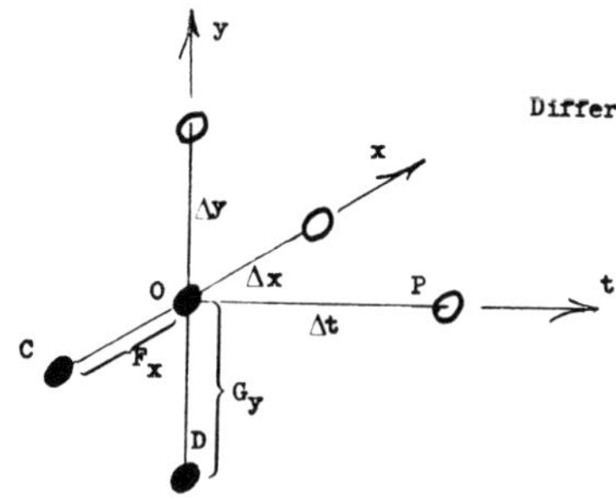

Differential Equation: $W_t = -F_x - G_y$

PREDICTOR STEP

$$\bar{W}_P = W_O - \Delta t\left[\frac{F_O - F_C}{\Delta x} + \frac{G_O - G_D}{\Delta y}\right]$$

Determine $\bar{W}$ also at points a & b

From $\bar{W}$ determine F & G

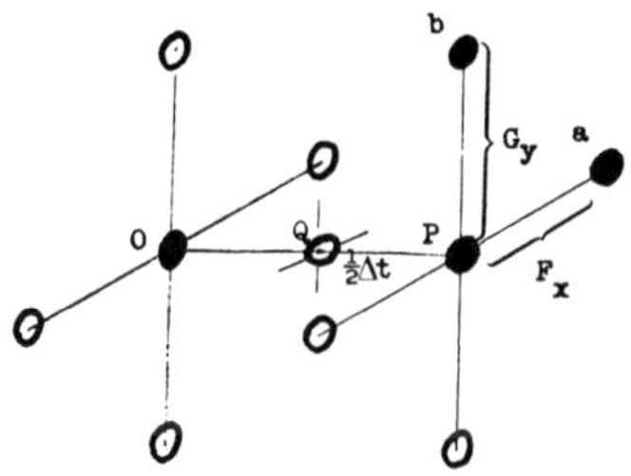

CORRECTOR STEP

$$W_P = \underbrace{\tfrac{1}{2}(W_O + \bar{W}_P)}_{W_Q} - \tfrac{1}{2}\Delta t\left[\frac{\bar{F}_a - \bar{F}_P}{\Delta x} + \frac{\bar{G}_b - \bar{G}_P}{\Delta y}\right]$$

EMPIRICALLY DERIVED CFL CONDITION

$$\Delta t \leq \min.\left\{|u|\Delta x^{-1} + |v|\Delta y^{-1} + c\,(\Delta x^{-2} + \Delta y^{-2})^{\frac{1}{2}}\right\}^{-1}$$

Figure 9. MacCormack's Predictor-Corrector Difference Scheme (Second Order, Explicit)

the basic time step of $2\Delta t$ is split into 2 equal parts, each fractional step comprising a MacCormack type predictor-corrector scheme with, however, one of the space derivatives suppressed. With the suppression of a space derivative together with the use of one-sided differences, in a given fractional step, one must expect serious inaccuracies within the profile of an oblique shock as well as in steep gradient regions as about the leading edge. The above scheme with $2\Delta t$ as the basic time step is of first order accuracy. However, if in the next time step the order of the operators L_1 and L_2 are reversed, second order accuracy then will be achieved.

Needless to say, there are many other successful second order difference schemes, both explicit as well as implicit. A suitable difference scheme for any specific problem must be judged, not only by its arithmetic precision and the economy of its makeup, but by its ability to handle criticle flow features as shock waves.

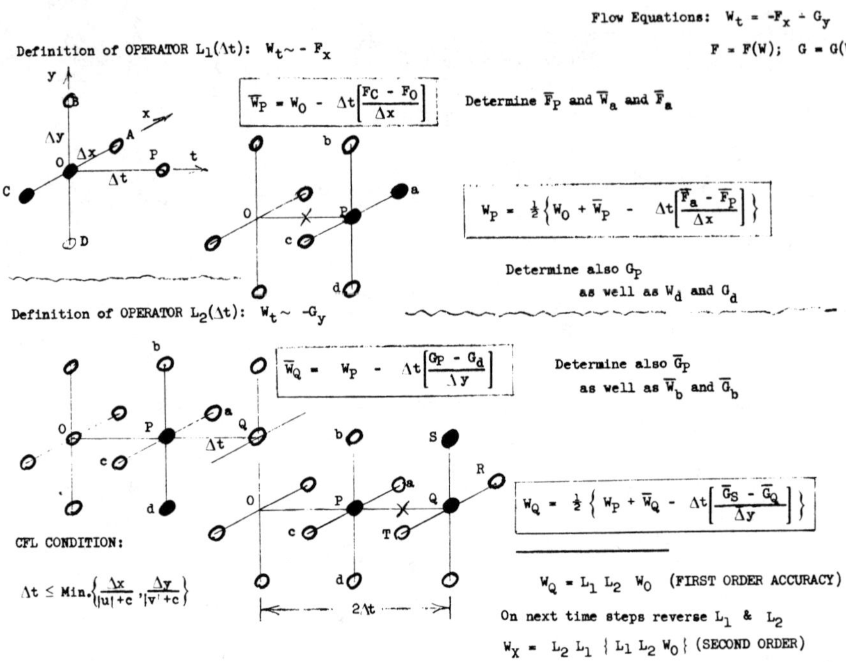

Figure 10. Explicit Time-Split Difference Scheme (Following MacCormack) (Method of Fractional Steps)

THE AIRFOIL BOUNDARY CONDITION

In the present problem the boundary condition at the airfoil is the forcing function that drives the flow unsteadiness. As such it should be treated numerically with at least the same integrity as any other aspect of the problem. Of the many procedures that have been employed, only that utilizing the characteristics or the surface piston waves can be recommended. These procedures have been fully validated by numerous successful examples. (See, for example, Refs. [1] and [4])

The procedure using the method of characteristics requires little elaboration so far as presenting its essential features. One simply projects the characteristic forecone back in time from the point on the boundary at the new time onto the initial data plane, interpolating between mesh points to obtain the required initial data on the appropriate bicharacteristics. With the latter the compatibility conditions in a suitable form are used together with the boundary condition to determine the solution at the new time. An iterative procedure is used to improve the accuracy. If the required interpolations are carried out carefully, one can use a simple cartesian mesh system about the airfoil.

The method of surface piston waves used in Ref. [1] is also a proven procedure, and its essentials are illustrated in Figure 11 for a simplified boundary. In the first step a suitable one-sided difference scheme is used to determine a tentative solution at P. In general the resulting normal velocity U_n is not zero. In the second step a planar isentropic piston wave is emitted from P whose strength is chosen such that through the wave the normal velocity is reduced to zero. The tangential velocity U_t is kept invariant across the wave, while the density and pressure changes across the wave correspond to the isentropic changes due to the

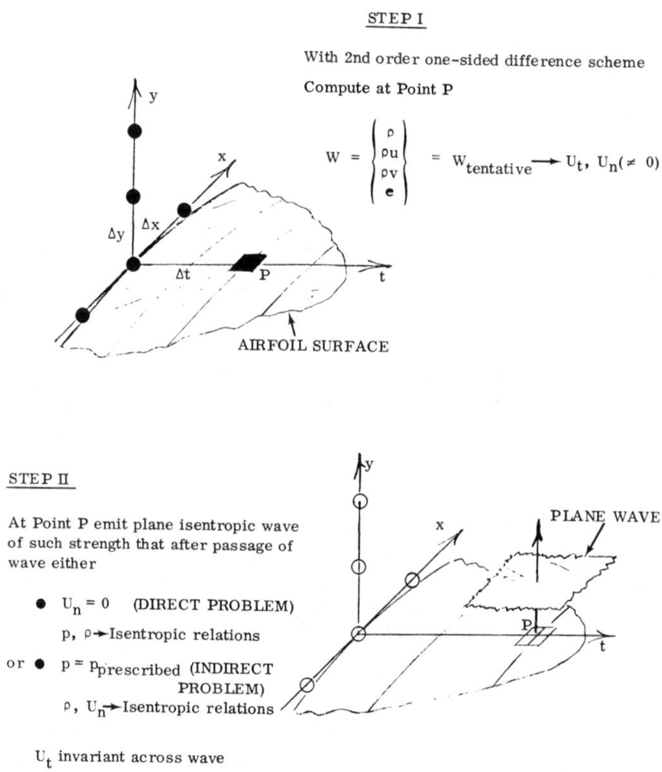

Figure 11. Procedure to Fulfill Condition at Airfoil Surface

change in the normal velocity across the wave.

Here the use of locally planar piston waves is a legitimate mechanism to input the effects of the moving boundary element only so long as the mesh spacing is sufficiently small compared to the local radius of curvature of the surface. The effect of the piston waves can then be considered as a substitute for that portion of the domain of dependence lost "below" the surface.

The complicating aspect of the above procedure is that a curvilinear orthogonal mesh subsystem conformal to the airfoil surface is required such as that shown earlier in Figure 5. This then requires an overlap and a data transfer routine with the surrounding cartesian mesh.

In Ref. [1] the piston wave procedure was simplified by imposing the boundary condition, not on the moving airfoil surface, but on the airfoil surface at a stationary mean location. Because of the importance of the airfoil boundary condition, the accuracy of the above approximation will be checked in the near future by fixing the curvilinear mesh subsystem to the moving airfoil.

THE TREATMENT OF SHOCK WAVES

There is a dichotomy of procedures to treat embedded shock waves. In the first class, the shock-fitting procedure, the shock wave is treated as a discontinuity across which the shock jump conditions are explicitly imposed. In the second class, the shock capturing procedure, numerical viscosity is utilized to spread the shock discontinuity, and the resulting continuous flow is then treated in a normal fashion using, however, a refined mesh about the shock.

Consider first the shock-fitting procedure, for example, developed by Moretti [5]. In this procedure a moving mesh system is employed which is continuously aligned with the moving shock. The conditions upstream of the shock are first determined in the usual manner using, however, one-sided spatial differences biased upstream of the shock. The shock velocity is approximated; and the upgraded shock velocity and conditions downstream of the shock are determined using the shock jump conditions supplemented by the characteristic compatibility condition for a characteristic arriving at the shock point. An iterative procedure then follows.

In a general unsteady transonic flow, logic must be further provided to identify the birth or demise of a shock to signal the switching to the proper computer subroutine. Here prominent features of the shock profile or the crossing of characteristics are used to tag the shock. Needless to say, careful screening must be provided to avoid false alarms.

Shock-fitting procedures obviously require much computer logic, and most probably the interactive guidance of the fluid dynamicist as well. A refined mesh about the shock, however, is not required, although this advantage may be largely negated by the need to employ a constantly changing mesh aligned with the shock.

In the floating shock-fitting procedure also derived by Moretti (Ref. [5]) the changing mesh is dispensed with, and the shock wave is allowed to float over a fixed background mesh system. The basic mesh point star for the difference scheme is now composed of a point on the shock together with its appropriate neighbors on the fixed mesh. With these irregular mesh stars the procedure then essentially proceeds as previously. Thus, in the floating shock-fitting procedure a continuously changing shock mesh star replaces the changing mesh configuration of the first procedure. Care must be exercised with such a continuously distorting mesh star to maintain the proper domain of dependence and to avoid extreme distortions of the mesh star so as not to trigger instabilities.

Let us pass on to the shock capture procedure whose essential features are given in Figure 12. As a result of the numerical viscosity inherent in the flow difference equations, shock waves, as well as other flow discontinuties such as slip lines, acquire a profile; and the flow as a whole becomes continuous. A sufficiently refined mesh must then embed the shocks so that the unphysical subdomain contained within the shock structure does not preempt excessively the physical domain.

In the shock capture procedure the fulfillment of the shock jump conditions is automatic provided the flow equations are used in the proper conservation form as reshown in Figure 12. The use of the difference equations in this conservation form insures the telescoping of the truncation errors along a given integration path. That is, if the marching process is carried out sequentially starting at a mesh point just ahead of the shock to a point just downstream of the shock, the large truncation errors occurring within the shock profile are identically cancelled upon arrival at

- o Shocks rendered continuous by numerical viscosity
- o Refined Mesh must embed shock profile
- o Equations must be used in PROPER CONSERVATION FORM (Telescoping of Truncation Errors)

$$\begin{Bmatrix} \rho \\ \rho u \\ \rho v \\ e \end{Bmatrix}_t = - \begin{Bmatrix} \rho u \\ \rho u^2 + p \\ \rho uv \\ u(p+e) \end{Bmatrix}_x - \begin{Bmatrix} \rho v \\ \rho uv \\ \rho v^2 + p \\ v(p+e) \end{Bmatrix}_y$$

Conservation of MASS
Conservation of MOMENTA
Conservation of ENERGY

<u>Tailoring of Lax-Wendroff Shock Overshoots by Diffusive Damping</u>

NO DIFFUSIVE DAMPING

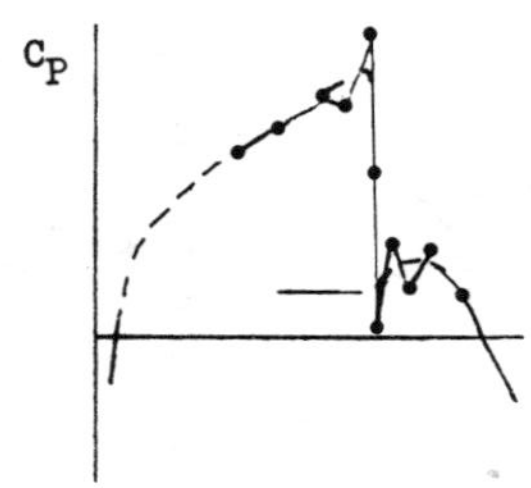

EXCESSIVE DAMPING

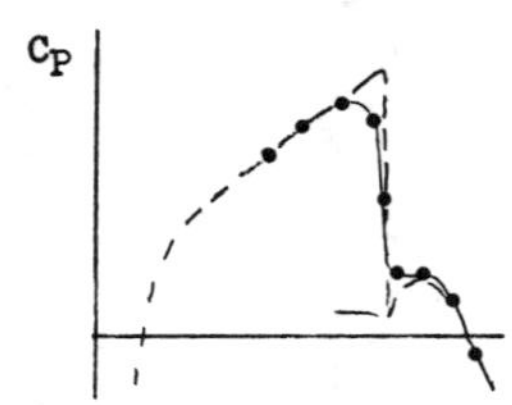

TAILORED DAMPING

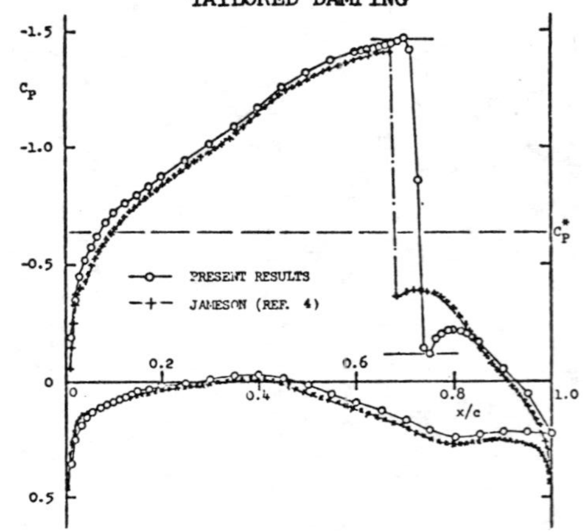

Figure 12. Automatic Shock Capture Procedure

the post-shock mesh point.

Thus, with no additional logic, shock configurations, no matter how complex, are automatically captured, that is, properly located fulfilling the proper shock jump conditions. A suitably refined mesh about the shocks must, however, be used.

In the case of the basic Lax-Wendroff difference scheme, a smooth shock profile is seldom achieved, with wiggles or overshoots occurring as shown in the lower left part of Figure 12. However, by the judicious addition of diffusive damping (selection of a suitable value of μ in Figure 8), wiggles and overshoots can be eliminated as shown in the tailored shock in the lower right part of Figure 12.

In the author's experience the shock-capture procedure has proven to be fully satisfactory. It is difficult to rationalize the added program logic in shock-fitting procedures when, with no additional program logic but with some increase in the number of mesh points, acceptable results can be achieved using the shock capture procedure.

AN EXAMPLE (REF. [1])

The example to be presented was computed using the Lax-Wendroff second order explicit difference scheme, the automatic shock capture routine, the piston wave boundary condition procedure, and the hybrid mesh configuration of Figure 5. The data transfer process used between mesh subdomains as well as the selection process for the time step in a given mesh subdomain and the attendant task of synchronizing the time are shown in Figure 13. The details here should be self-explanatory. In the calculations frequently wiggles occur in the solution about the interfaces of the different mesh subdomains due to the differences of the truncation errors. In such cases the diffusive damping is used to smoothen the solution.

The example to be considered concerns the NACA 64A-410 airfoil oscillating harmonically in pitch about the midchord at a free stream Mach number of 0.72. The amplitude of the pitch oscillation is 2° about a mean incidence of 2°. The reduced frequency is 0.2; that is, the airfoil must travel a distance of 10π chords at the free stream velocity to complete one cycle of the oscillation.

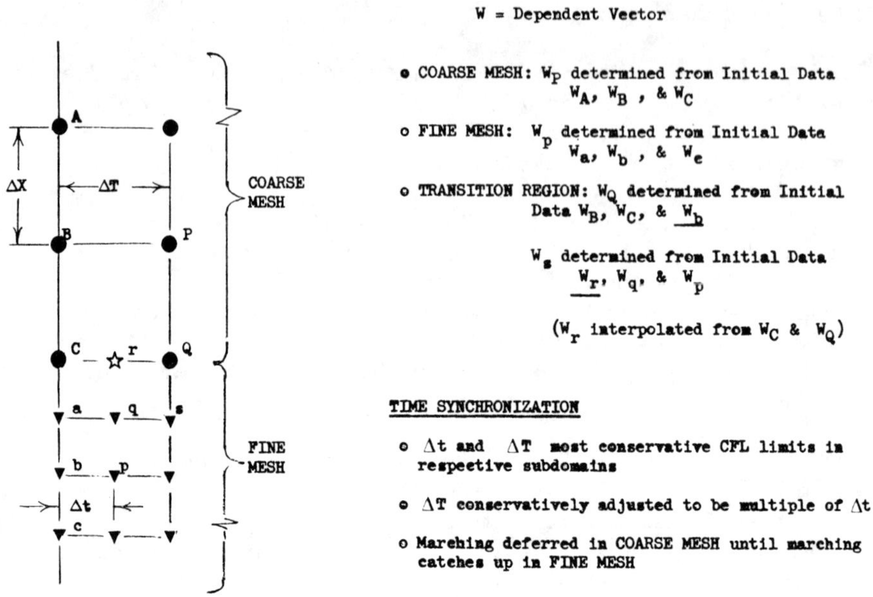

Figure 13. Transition Between Mesh Subsystems and Time Synchronization

In Figure 14 we show the resulting instantaneous chordwise pressure distributions at various phases of the cycle, while in Figure 15 we have plotted the time histories of the shock location as well as the local surface pressures at several locations on the airfoil. It is interesting to note here that the shock wave becomes stuck at its most downstream location for nearly a quarter of the cycle. Local pressure histories are closely sinusoidal except at $x/c = 0.7$ on the upper surface where the pressure trace is more akin to a square wave than to a sine wave as a result of the passage of the shock over this point. Such non-sinusoidal traces would invalidate time linearizations.

The above calculations required approximately 2 hours on the CDC 7600 computer.

CONCLUDING REMARKS AND SUGGESTIONS

There is no question that the procedure of Ref. [1] yields reasonable inviscid unsteady solutions. Computing times required, however, are unquestionably large, and it would be of interest to

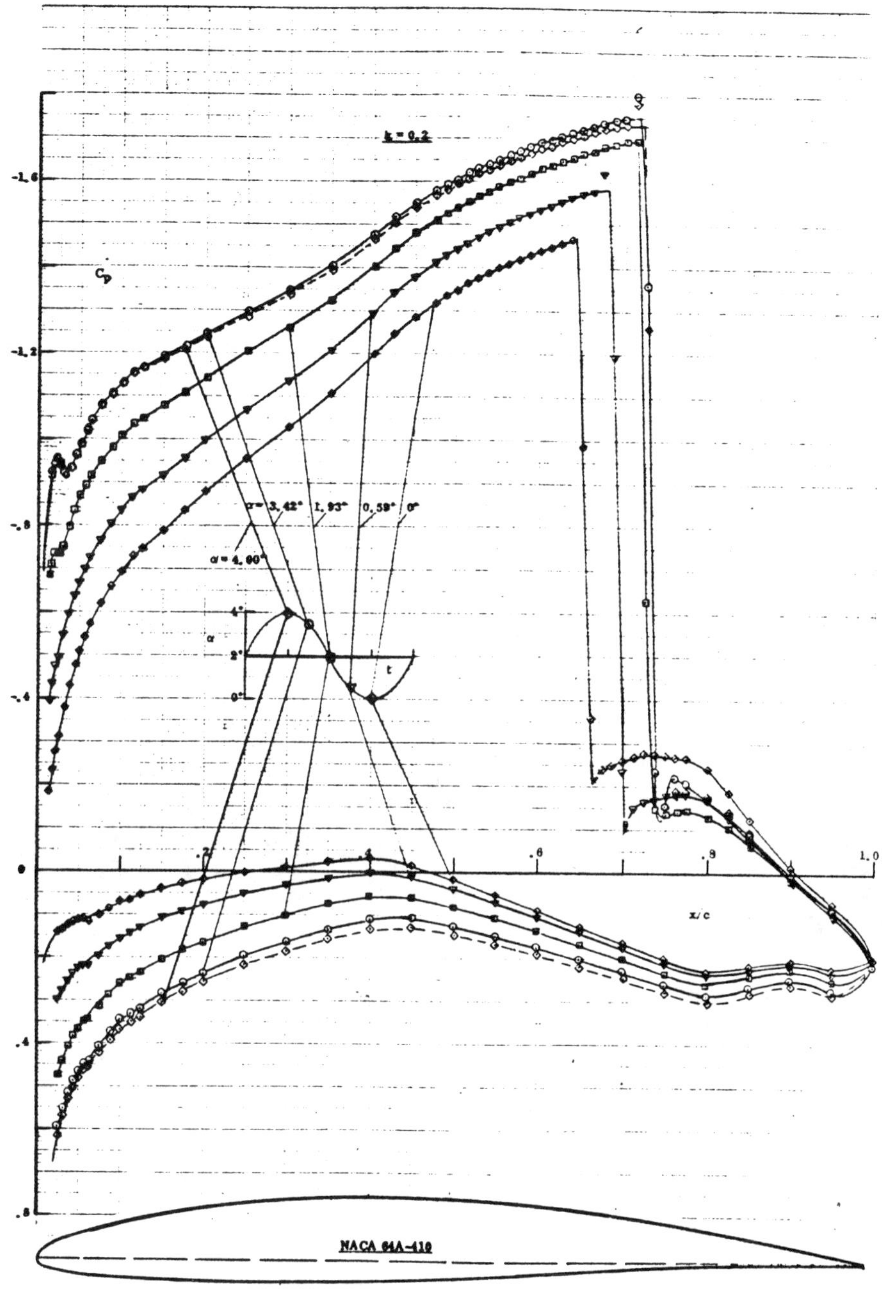

Figure 14. Unsteady Pressure Distributions for k = 0.2

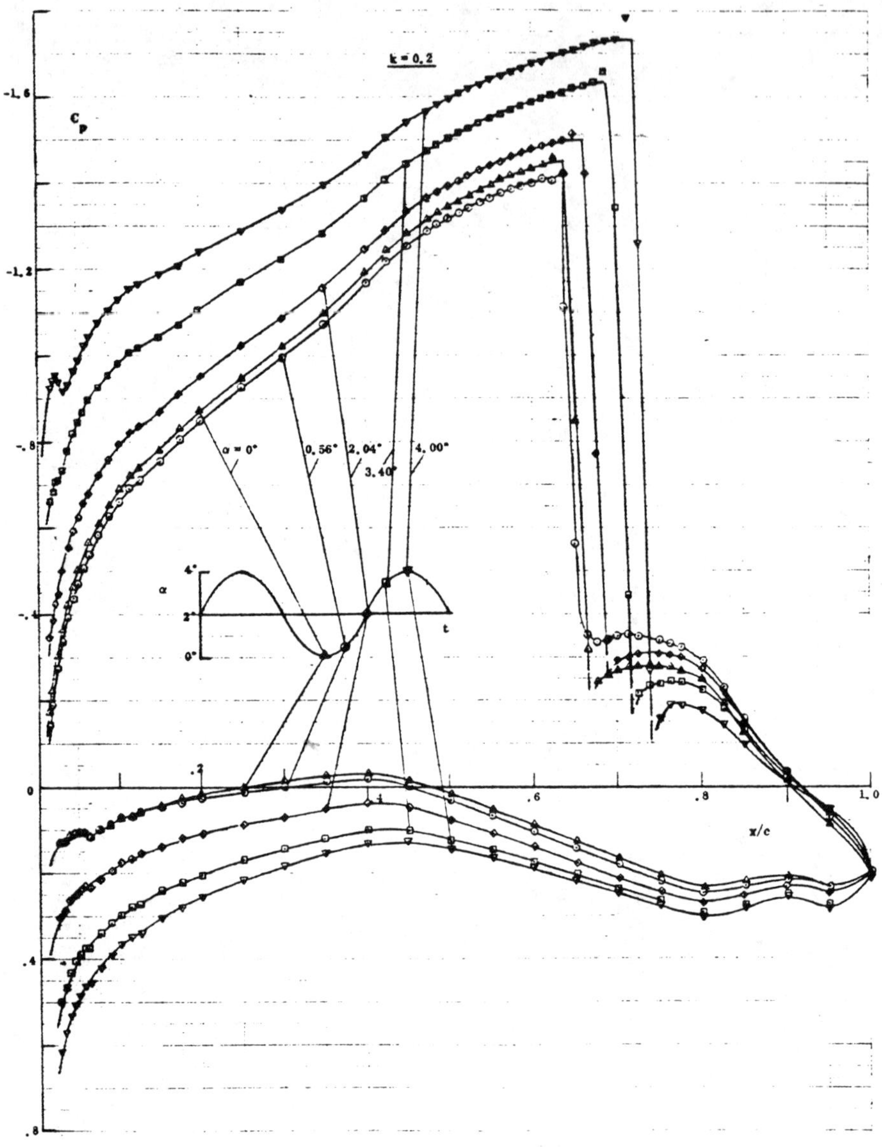

Figure 14. Concluded

conclude the review by examining the sources that are responsible.

The primary cause of the large computing time can be traced to the use of an explicit differencing scheme. Here the CFL limit

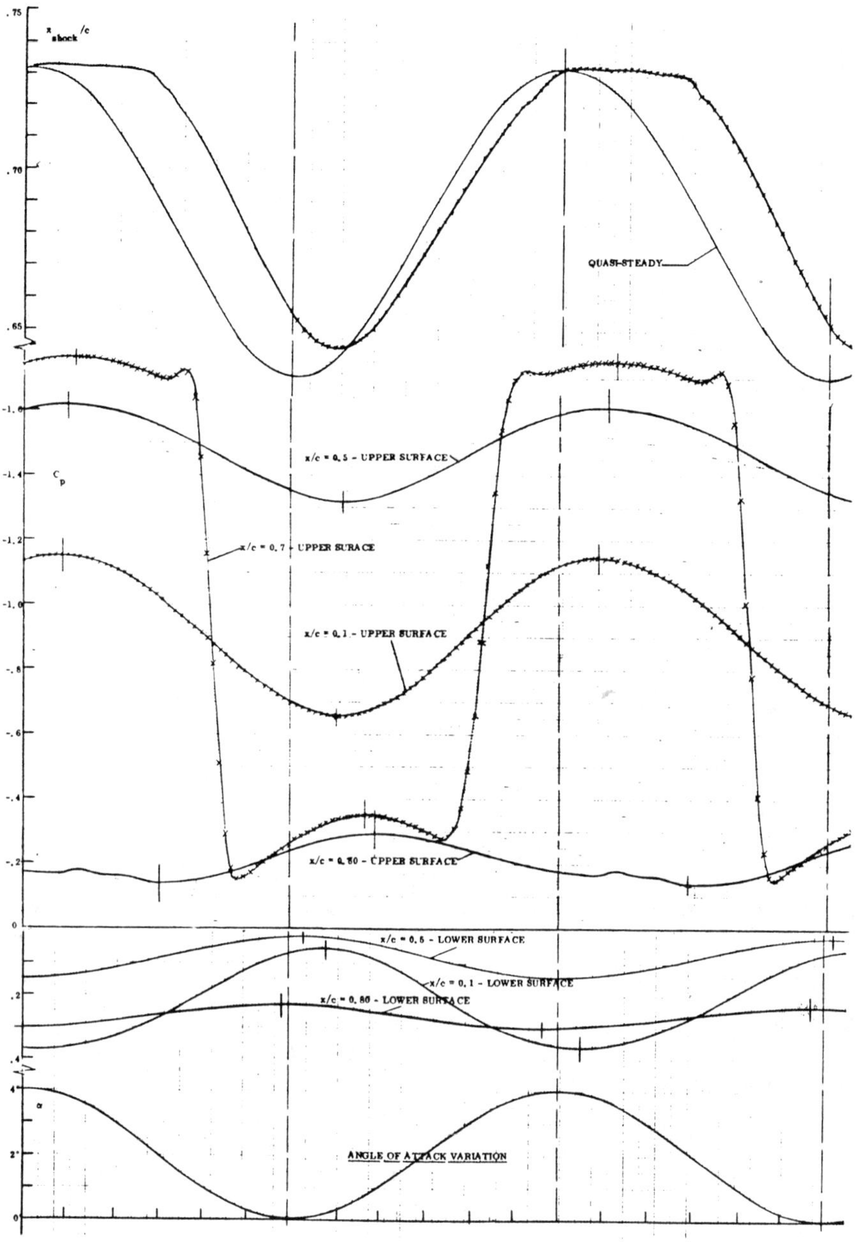

Figure 15. Time Variation of Local C_p's and Shock Location for k=0.2

on the time step is at least several orders smaller than that required to achieve the required accuracy of the time marching procedure. Moreover, the use of an excessively refined mesh rapidly increases the computer time, not only by the increase of the number of mesh points, but by the corresponding reduction in the allowable time step. In the above example a significant fraction of the total number of mesh points was used to cover the leading edge region. Undoubtedly, a coarser mesh here should suffice. On the other hand, the use of the fine mesh about the shock should be more than justified by precluding the complexities of the added logic required in the alternative shock-fitting procedure.

Clearly, the most promising means to reduce the computer time significantly is the use of the implicit difference scheme. Here the task will be to evolve efficient means to invert large matrices. The use of the velocity potential should also be explored since in this case four first order equations with four unknowns can be replaced by a single second order equation with a single unknown with a generally tolerable error in the shock jump condition. In the case of advanced computers as the CDC 7600 computer significant reduction of computer time can further be accomplished in a straightforward manner by writing the computer code to be in harmony with the characteristics of the computer.

Finally, we must not forget that important viscous effects have been omitted. Meaningful unsteady results can be obtained only when such effects have been incorporated.

REFERENCES

[1] Magnus, R. and Yoshihara, H., Finite Difference Calculations of the NACA 64A-410 Airfoil Oscillating Sinusoidally in Pitch at M = 0.72, Convair Report CASD-NSC-74-004 (1974).

[2] Richtmyer, R. and Morton, K., Difference Methods for Initial-Value Problems, Second Edition, Interscience Publishers (1967), pp. 360-8.

[3] MacCormack, R., Survey of Computational Methods for Three-Dimensional Supersonic Inviscid Flows with Shocks, AGARD Report LS-64 (1973).

[4] Abbett, M., Boundary Condition Calculation Procedures for Inviscid Supersonic Flows, Proceedings of the AIAA Computational Fluid Dynamics Conference (1974).

[5] Moretti, G., Thoughts and Afterthoughts about Shock Computations, PIBAL Report 72-37 (1972).

DISCUSSION OF THE YOSHIHARA PAPER

MC CROSKEY: It appeared to me in watching your movie that the unsteadiness in the flow, as might be evidenced by the phase lag of the flow field to develop, seemed much more pronounced at the low incidence - the bottom of the cycle - than at high incidence - at the top of the cycle. Is this a valid observation and, if so, does it have any implications for the kinds of flow that might be quasi-steady and which ones are purely unsteady?

YOSHIHARA: By the bottom of the cycle you mean for the lower angle of attack?

MC CROSKEY: When the incidence seemed approximately zero, it seemed to me the flow field took longer to follow the adjustments. There seemed to be more phase lag around zero incidence than at the top. But maybe it was just an impression from the movie.

YOSHIHARA: Yes - I have no comment. I think you should realize that any point on the airfoil - how it affects the flow - it's got to propagate and it's got to buck the big wind that's coming by and if you have a local embedded supersonic region, it's got to go around that. So in the slow adjustment process, any time the supersonic region is extensive laterally, at least that path of adjustment is very much delayed. These kinds of things feed in, but whether it shows up for what you said, I don't know.

JAMESON: What was the computer time?

YOSHIHARA: It was listed there as two hours. I'm not concerned about computing time at this stage; some people are, but I'm not.

MORETTI: Some people is me! Of course!

YOSHIHARA: Yes! - because quite frankly, my point of view is that I'm not going to do a dozen cases, and I want to do something correctly, you see - so it took two hours.

VOICE IN AUDIENCE: On what?

YOSHIHARA: On the 7600 - that's a lot of computing time.

MORETTI: Two hours is all right, but I don't understand how many cycles you do in two hours.

YOSHIHARA: What we do is 6000 points. We start the flow from an initial steady flow at the mean incidence so there is a transient time before you come to the oscillatory state. We go through that transient stage - about a 1000 time steps on the coarse mesh, not the fine mesh - and then we go through two oscillation cycles to make sure we've reached the stationary state. Two hours.

ROBERTS: I noticed that while your airfoil was oscillating, I think, toward the high angle of attack, there was a very small recompression region near the leading edge, and of course, this is a completely inviscid solution, isn't that right? You see some of that on real airfoils and cascades and I wonder if you'd comment on that.

YOSHIHARA: Yes. The characteristic of the pressure distribution that the gentleman is alluding to is the fact that this is quite typical of the 64A 410 airfoil - namely as you expand over the nose it overexpands then recompresses, then expands. Now, in a real viscous flow you won't see that because you get flow separation and it just completely hides that. It is actually there and the only reason that you see it is, by taking a very fine mesh - you take a coarser mesh, and it is completely lost.

YOSHIHARA: Thank you, Gino - I can say you were almost unbiased!

MORETTI: Ah, well, no! - My bias will come in writing! I have to say one thing quickly, and that is that about a year ago at Tucson, Dr. Yoshihara presented something similar and I made some - well, I presented, something different, which was my estimate of running time in unsteady flows and so we had a little bit of a discussion there. And I promised Dr. Yoshihara to come out with another interpretation of this problem which in my opinion should be faster. I must say publicly, as I said publicly last year at this sort of thing that I didn't do that yet and so I've nothing to say. So if I never do it, I have to apologize to him publicly for promising something and not doing it!

FOUR ISSUES IN THE COMPUTATION OF TRANSONIC FLOWS IN TURBOMACHINERY

DAVID A. OLIVER
Massachusetts Institute of Technology

ABSTRACT

The development of fully three-dimensional inviscid transonic flow prediction in real turbomachine rotors with full consideration of thickness, stagger, and the complex geometry of the rotor has been proceeding at the MIT Gas Turbine Laboratory. The computational theory and its implementation include multi-dimensional shock waves. The codes developed at MIT are described in "Computational Aspects of the Prediction of Multi-Dimensional Transonic Flows in Turbomachinery," by D. A. Oliver and P. Sparis, in Aerodynamic Analyses Requiring Advanced Computers (NASA SP-347, 1975) and "Three-Dimensional Flow Calculation for a Transonic Compressor Rotor," by W. T. Thompkins and D. A. Oliver, to be presented at the 47th PEP Meeting on Through Flow Calculations in Axial Turbomachinery.

This work has raised four fundamental issues which are discussed in the present report:

1. The use of time-dependent versus steady state iterative solution techniques
2. The behavior of non-linear instability and the use of local damping operations in the transonic regime
3. Treatment of complex blade geometry and high order accuracy boundary conditions
4. The use of continuum versus finite element methods.

Critical comments with some illustrations will be directed to each of these issues.

DISCUSSION OF THE OLIVER PAPER

ADAMSON: In the supersonic part of the flow above the sonic cylinder, one would expect shocks to come off the leading edge and perhaps these shocks should reflect back and forth. That would mean a nonuniform pressure distribution along the blades. I wondered - is the reason we do not see this a result of the fact that because of the artificial viscosity the shock is smeared out so much that you don't get the proper reflections? If not, what is the reason for it?

OLIVER: I mentioned that the mesh resolution in the front - if you recall the slide that shows the distribution of the mesh - I mentioned at the time that I showed that, that we were purposely cutting out the mesh in front, which means we wipe out the possibility of resolving the appropriate leading edge shocks. We have done 2-D calculations in which we have run the mesh upstream and we have been able to begin to show some of the 2-D shocks that are running upstream off the blades. And in this 3-D calculation ... you can't see it.

ADAMSON: I was worried about the shock that would reflect and then come back and effect the pressure downstream.

OLIVER: You're thinking of the passage shock. But in this wide open throttle calculation, contrary to the one that is loaded, the shocks out at the tip are extremely weak if nonexistent - just a strong shock sitting down close to the hub.

MC CUNE: Dave, what did you finally end up doing about the boundary conditions upstream and downstream? You remember we had long debates about this. How did you finally settle it?

OLIVER: The upstream boundary condition is one of imposed uniform flow corresponding to the operating conditions - almost. The almost part is that when that flow interacts with the blades, there are signals that will come upstream in the subsonic region. We let those waves escape with a one-dimensional characteristic construction at the inlet. Now that's not quite correct because the waves that come upstream are three-dimensional. They don't quite fully escape properly with a one-dimensional characteristics construction, but they do pretty well. On the downstream boundary, we have used two boundary conditions. In the calculation that I showed first today, which is one of the unthrottled case, we used a condition of zero axial gradient. Now, that's wrong because the streamlines are not axial. It would become increasingly more correct if we put on a zero gradient along the streamlines. So we do distort the structure of the flow in the vicinity of the exit plane. We've done two-dimensional experiments in which we have changed the boundary conditions at the exit plane, and tried to observe the impact in the blades, and close to the blades, and immediately behind the blades. We satisfied ourselves that we have very little influence because of the far downstream location of our exit boundary. In the second calculation which is for the loaded case, for which I just showed that one slide showing the warpage of the stream surfaces, there's a more drastic downstream condition which is used, which is the condition of azimuthal uniformity. So that the theta variation is wiped out completely, and what is put on downstream, is the experimentally measured pressure distribution. So we have an experimentally measured pressure distribution that is placed on the downstream boundary, but the theta average pressure distribution. Therefore, the structure in the theta direction is wiped out, but the radial structure of pressure is maintained.

COMPARISON OF A FINITE DIFFERENCE METHOD WITH A TIME-MARCHING METHOD FOR BLADE TO BLADE TRANSONIC FLOW CALCULATIONS

J. M. THIAVILLE
SNECMA

INTRODUCTION

Two formulations are generally used for computing the transonic flow through compressor and turbine cascades: First, a time-dependent formulation of the equations of motion allows the flow field to be identically calculated in subsonic and supersonic regions. Second, a finite difference formulation of the transonic potential flow is solved by a relaxation process using noncentered finite differences in the supersonic regions.

Both formulations have been developed on an IBM 370-145 computer and applied to three examples: a compressor and two turbine two-dimensional cascades. The comparison with experimental results are presented in the present paper.

TIME-DEPENDENT METHOD

The time-dependent method used is the one developed by P. W. McDonald [1] in which an integral representation of the equations of motion is applied to finite area elements. The numerical process assures that the conservation laws of mass and momentum are satisfied at each time step of the calculation. A damping procedure was employed to ensure stability and accuracy with a relatively coarse mesh: less than 1000 elements were used and the residual damping error was approximately 3% on the stagnation pressure. The method permits a natural appearance of weak shock waves and a choked turbine cascade but the flow is assumed isentropic. The finite area integral method was preferred to the transient finite difference formulations [2] for the present cases as it saves computation time and remains well adapted to cascades presenting high turning angles and limited supersonic pockets.

FINITE-DIFFERENCE METHOD

The finite-difference method used was developed by T. S. Luu ([3] and [4]). The calculation mesh is created by the streamlines and equipotential lines of the incompressible flow solution. The transonic potential flow is computed in the thus created orthogonal mesh using a relaxation procedure including Jameson's rotated difference scheme and an accelerated convergence method. The outlet Mach number is increased step by step from incompressible to the design value so that the same computation gives the cascade performance for several Mach numbers. The conservation of mass is not fully respected and the suction-side and pressure-side streamlines are prolonged to downstream infinity so that the velocity is not uniform at the cascade outlet.

COMPARISON WITH EXPERIMENTAL RESULTS

The first selected example is a turbine rotor tip section cascade. The outlet Mach number is 0.89, the trailing edge thickness normalized to throat area equals 6% and the suction side maximum Mach number nearly unity. The turning angle is 80°. Figure 1 shows a very good agreement between both numerical methods and test results.

The second example (Figure 2) is a compressor hub section stator cascade presenting a 45° turning angle with an inlet Mach number of 0.82. The agreement with experiment is equally very good. However, the time-dependent method seems to predict the suction side supersonic pocket better than the finite difference method does. The latter method had the outlet velocity imposed and so the non-conservativity may have introduced a small difference in the inlet computed Mach number.

The last example is shown in Figures 3 and 4 and concerns a turbine inlet guide vane hub section. The outlet flow angle was 15° and the trailing edge thickness was 15% of the throat area.

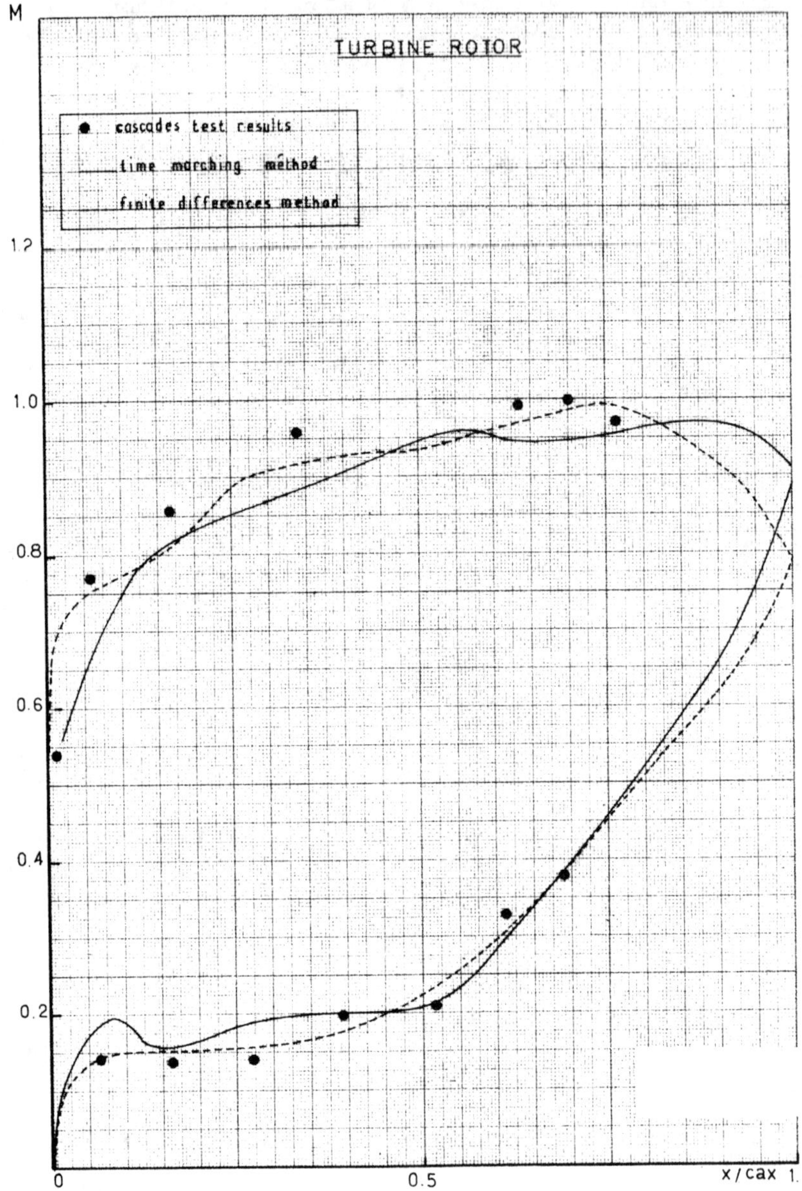
M
TURBINE ROTOR
cascades test results
time marching method
finite differences method
1.2
1.0
0.8
0.6
0.4
0.2
0
0.5
x/cax 1.

Figure 1

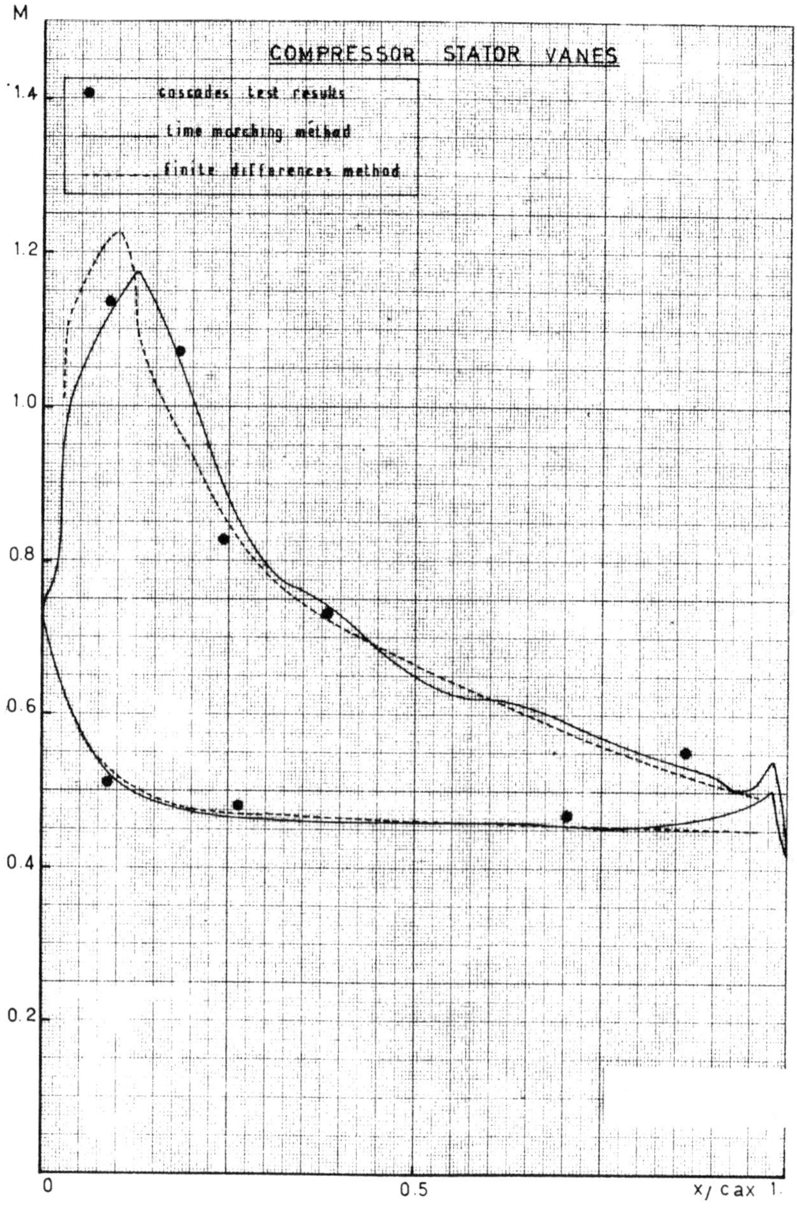
M
COMPRESSOR STATOR VANES
cascades test results
time marching method
finite differences method
1.4
1.2
1.0
0.8
0.6
0.4
0.2
0
0.5
x/cax 1.

Figure 2

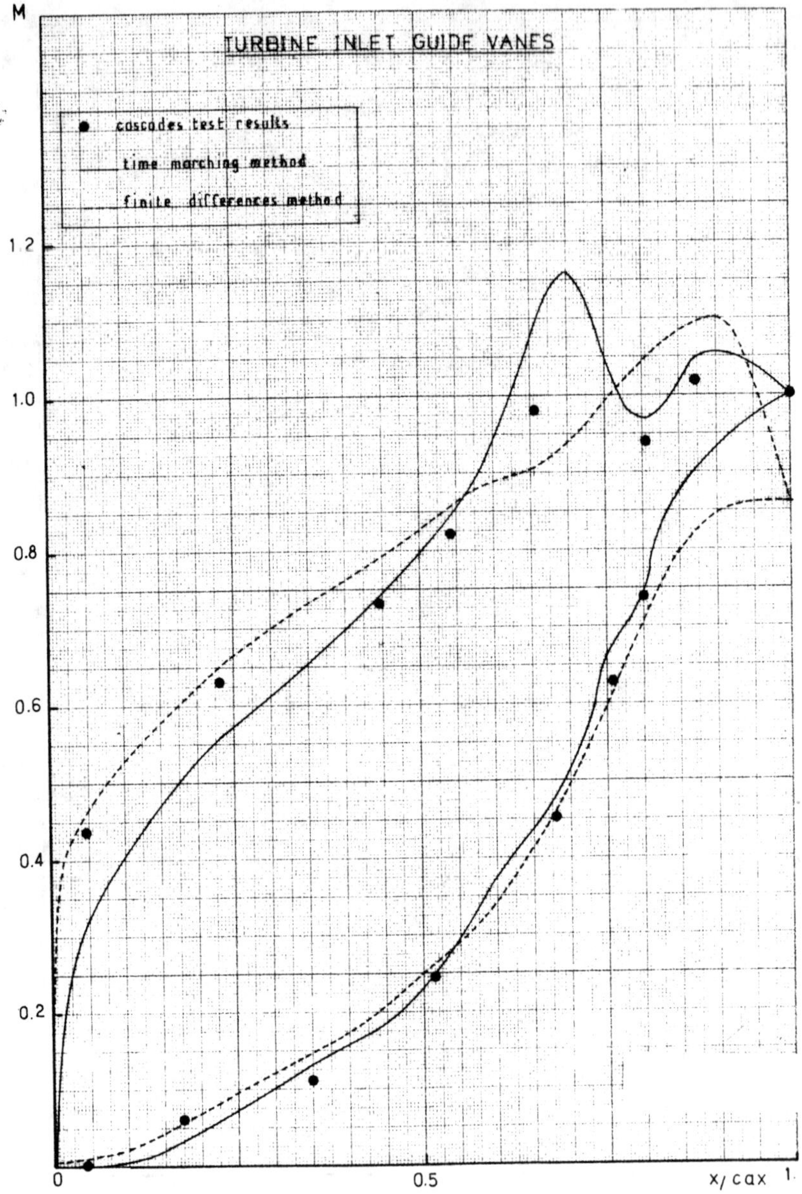

M
TURBINE INLET GUIDE VANES
cascades test results
time marching method
finite differences method
1.2
1.0
0.8
0.6
0.4
0.2
0
0.5
x/cax
1.

Figure 3

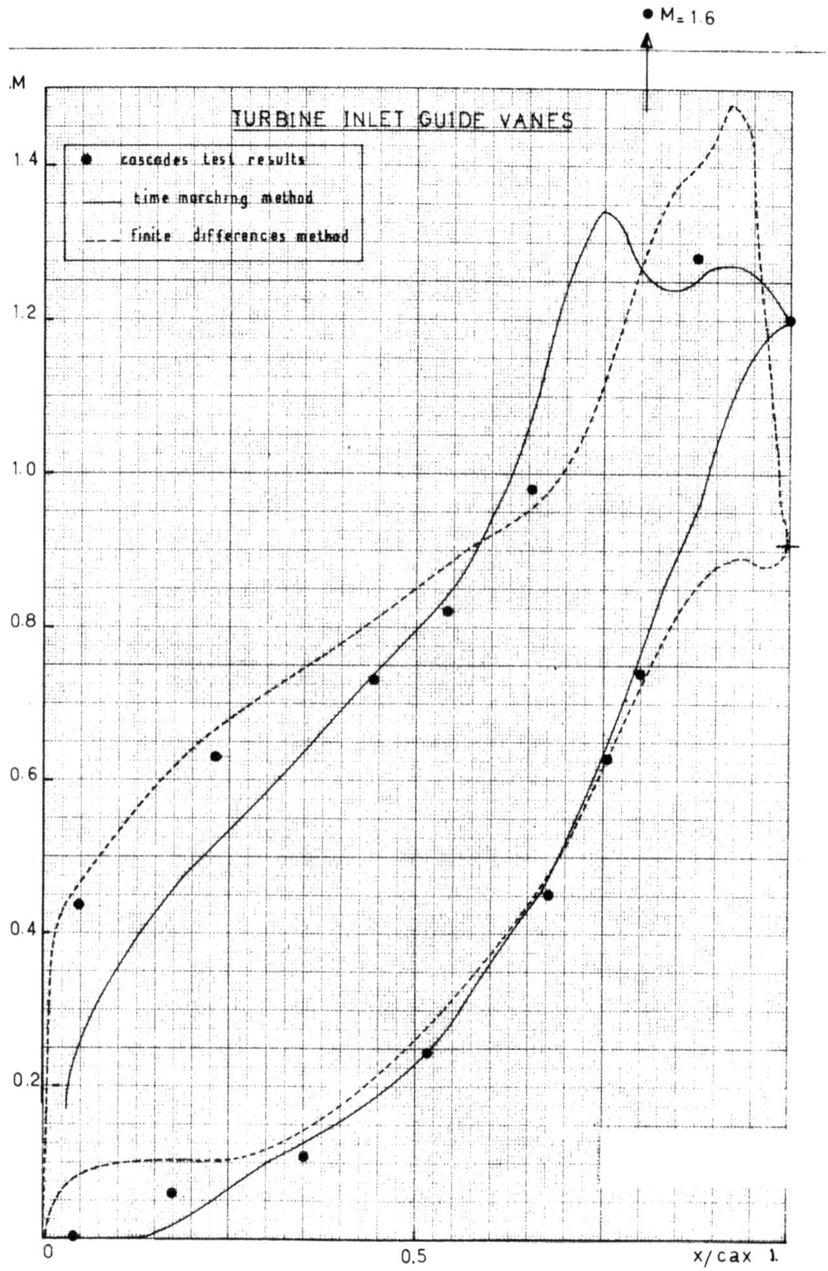

M= 1.6
M
TURBINE INLET GUIDE VANES
cascades test results
time marching method
finite differences method
1.4
1.2
1.0
0.8
0.6
0.4
0.2
0
0.5
x/cax
1.

Figure 4

Several remarks may be made concerning Figure 3 for an outlet Mach number equal to unity:

The finite difference method seems to give somewhat better results for the low suction side Mach numbers. However, it is not sure whether this is due to the method itself. Maybe the finite element forms used by the McDonald method could have been better adapted to the profile geometry.

The time-marching method gives a good prediction of the sonic point and supersonic region on the suction side. The way the downstream streamlines are prolonged in the Luu method introduces an error in the value of the trailing edge velocity with a non-uniform computed outlet Mach number.

At an outlet Mach number of 1.2 (Figure 4) the above mentioned remarks are shown more clearly:

The suction side region near the leading edge is better predicted by the finite difference method while the suction side sonic point and minimum pressure point are better located by the time-dependent method. Both methods were unable to predict the maximum Mach number of 1.6 and the corresponding shock wave because it is produced by the effect of the trailing edge thickness, which is not well accounted for in both calculations. In McDonald's computation the mesh used is too coarse and in Luu's computation there is no trailing edge effect at all regarding the prolonged streamlines.

In all cases pressure side velocities are in very good agreement with experiment.

The computation time was of the order of half an hour for the time-marching method while the finite difference method consumed 15 minutes for the calculation of the flow field around the same cascade and for several Mach numbers.

CONCLUDING REMARKS

The accuracy of the McDonald method seems to be better at high Mach numbers than that of the finite difference method. The latter method gives better results in the subsonic regions of the flow. The accuracy could have been improved for the results presented in Figure 4 by using a transient finite difference calculation method which could predict the strong trailing edge shock wave. However this last method which we developed according to Gopalakrisnan's work, used an orthogonal grid badly adapted to the high turning turbine cascades and needed a strong smoothing procedure to insure stability.

The effects of radius variation and streamtube thickness with axial distance (three-dimensional cascades) are now being introduced to our calculation methods but no examples were yet available for the SQUID Workshop.

ACKNOWLEDGEMENTS

The author would like to thank T. S. Luu (L. I. M. S. I., Paris-Orsay) and M. Couston (V. K. I., Rhodes-St-Genese) who developed the calculation methods used in the present paper.

REFERENCES

[1] McDonald, P. W., "The computation of transonic flow through two-dimensional gas turbine cascades," ASME Paper No. 71-GT-89 (1971).

[2] Gopalakrisnan, S. and Bozzola, R., "Computation of shocked flows in compressor cascades," ASME Paper No. 72-GT-31 (1972).

[3] Luu, T. S. and Coulmy, G., "Calcul de l'écoulement transsonique avec choc à travers une grille d'aubes." A. T. M. A. (1975).

[4] Luu, T. S. and Worms, G., "Calcul des grilles d'aubes transsoniques avec choc. Confrontation avec les essais en soufflerie." A. A. A. F. Poitiers (1975).

DISCUSSION OF THE THIAVILLE PAPER

OLIVER: Could you give some idea of the estimate of computer time for the two methods?

THIAVILLE: Yes. Time-marching methods, integral methods - which we use in that example - 500 iterations was large mesh and then after 500 iterations, with a small mesh, with a double mesh. And it used half an hour in an IBM 370/145; that is to say, the slow computer. In front of a CDC, for example, it is thirty times slower. In a CDC 7600 we would like to say that it could be less than one minute.

MORETTI: This is a time-marching method?

THIAVILLE: Time-marching method.

MORETTI: Now the other was a relaxation, right? Of course, finite difference, but ...

THIAVILLE: Yes, a relaxation, of course, and we have a local ponderation.

MORETTI: But that would be much faster?

THIAVILLE: Yes, but we used to calculate a lot of Mach numbers in the same calculation, and we used with a finite difference method, a quarter of an hour on the same computer, but we had twenty Mach number points.

ROTATIONAL TRANSONIC INTERNAL FLOWS

EUGENE F. BROWN

Virginia Polytechnic Institute and State University

For several years now I have been involved in the development of transonic computational methods for predicting the performance of aircraft exhaust nozzles. I am here today to share with you the results of my most recent research which I believe will be of interest to those of you involved in making transonic blade-to-blade calculations. The unique feature of this work is that the acknowledged computational efficiency of relaxation methods is retained while permitting rotational flow effects to be included.

When this work was begun almost three years ago we applied the relaxation scheme pioneered by Murman to the full potential flow equation. This was much the same approach used by South and Jameson [1] in their airfoil calculations with the exception that, because of the largely axial character of the flow, we did not find it necessary to use rotated differences. The use of the potential function, however, has rather significant limitations from the standpoint of both nozzle and turbomachinery applications. This is because the true nature of the flow in a nozzle or a turbomachinery passage is likely to be rotational, that is, to possess significant radial gradients of total temperature and total pressure. In a turbomachine the rotationality of the flow is imparted by a number of effects including distorted inflow, flow separation, and the radial dependence of the energy delivered by the compressor.

The most obvious way of introducing the rotationality of the flow into the analysis is to return to the primitive-variable formulation of the problem. We were reluctant to do this, however, because of the additional storage requirements and increased computational time widely known to accompany this approach. We decided instead to reformulate the problem in terms of a velocity function and an auxiliary rotation function by defining the axial and radial velocity components by

$$u \equiv \tilde{\phi}_x + F$$

and

$$v \equiv \tilde{\phi}_r,$$

where $\tilde{\phi}$ is the velocity function and F is the rotation function. These two functions can be shown to obey the following two equations

$$[c^2 - (\tilde{\phi}_x + F)^2](\tilde{\phi}_{xx} + F_x) + [c^2 - \tilde{\phi}_r^2]\tilde{\phi}_{rr} - (\tilde{\phi}_x + F)\tilde{\phi}_r(2\tilde{\phi}_{xr} + F_r)$$
$$+ \frac{c^2}{r}\tilde{\phi}_r = 0 \tag{1}$$

$$F_r = \frac{1}{u\gamma P}(1 - \frac{\gamma - 1}{2} M^2)P_r \tag{2}$$

where P is the total pressure, u is the axial velocity component, M is the local velocity divided by the speed of sound at stagnation conditions, and the subscripts indicate partial differentiation with respect to the subscripted variable. In these equations it has been assumed that the flow is inviscid (but not irrotational). The first of these equations is readily recognized as being identical in form to the potential equation (except for the presence of F and its derivatives). The second equation has been derived from Crocco's theorem. Because of the similarity of the first equation to the potential equation it is solved for the velocity function using standard relaxation methods. The values of F and its derivatives needed to carry out this calculation are obtained from the second equation. In the second equation the required radial gradient of total pressure is obtained by a streamline tracking procedure from the values of the total pressure supplied at the inlet. This is possible since in the case of an inviscid, non-thermally conducting gas, the total pressure is constant along streamlines.

Figure 1 shows the results of a representative calculation of the flow in an axisymmetric turbofan exhaust nozzle (bypass duct)

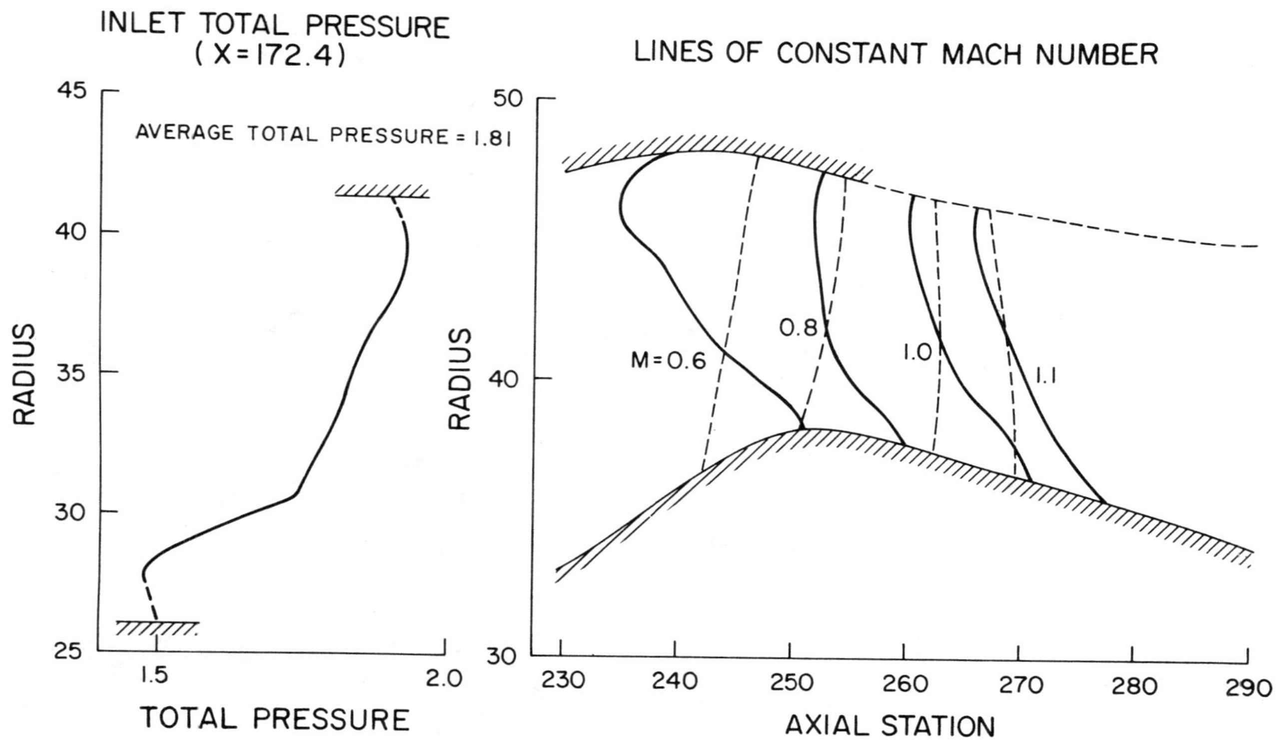

Figure 1. Turbofan Bypass Duct

made with the most recent version of our program. This figure shows the radial variation of total pressure at the inlet and the resulting lines of constant Mach number. In addition, dashed constant Mach number lines are shown which correspond to uniform inlet conditions. Clearly, the flow in the transonic region is strongly affected by the rotationality of the flow. On the inner wall the lines of constant Mach number are shifted strongly downstream while on the outer surface of the nozzle they are shifted upstream. Comparison of the static pressure distributions on the inner and outer walls also reveals the significance of the rotationality of the flow.

Although this method has not as yet been applied to turbomachinery calculations it is not difficult to imagine that the nozzle walls shown in the preceding figure represent the suction and pressure surfaces of blades. Therefore, these results suggest that the rotationality of the flow may be an important consideration in turbomachinery calculations as well and that the method we have developed to account for these effects in propulsion nozzles may also be a useful computational tool for transonic turbomachinery flows.

REFERENCE

[1] South, J. C., Jr. and A. Jameson (1973), "Relaxation Solutions for Inviscid Axisymmetric Transonic Flow Over Blunt or Pointed Bodies," Proceedings of the AIAA Computational Fluid Dynamics Conference, pp. 8-17.

ACKNOWLEDGMENT

This work was supported by the Douglas Aircraft Company, McDonnell Douglas Corporation, Independent Research and Development (IRAD) Program.

DISCUSSION OF THE BROWN PAPER

JAMESON: Would you show the second slide - no the next one - yes, here we are. That formula that I can't quite read for F_r - could you elaborate on that a bit for us?

BROWN: Sure. I'm sorry that you can't read it. Shall I read it?

(Does so.)

JAMESON: I can't be reading it right, because M_o divided by M_o is one.

BROWN: Yes! The leading term reduces to $1/u\gamma P_o$. M_o/M_o was introduced to simplify the computer program. It, of course, cancels out.

MURMAN: Shouldn't that be the partial derivative of P_o with respect to the stream function in the expression for F_r?

BROWN: Yes. But, the fact that the total pressure is constant along streamlines permits the stream function derivative to be written in terms of the flow direction and the radial derivative. (Please note that the symbol ψ in this paper is used for the transformed radial coordinate and not the stream function.)

MORETTI: I see that perplexities now are not allowed because they take too much time!

MELNIK: The total pressure is a function of the stream function; it only appears as a function of the radial direction through its dependence on the stream function.

BROWN: Yes, that's correct.

APPLICATION OF A MULTI-LEVEL GRID METHOD TO TRANSONIC FLOW CALCULATIONS

JERRY C. SOUTH, JR.*

NASA Langley Research Center

ACHI BRANDT**

Weizmann Institute of Science

SUMMARY

A multi-level grid method has been studied as a possible means of accelerating convergence in relaxation calculations for transonic flows. The method employs a hierarchy of grids, ranging from very coarse (e.g., 8 x 2 mesh cells) to fine (e.g., 128 x 32); the coarser grids are used to diminish the magnitude of the smooth part of the residuals, hopefully with far less total work than would be required with, say, optimal SLOR iterations on the finest grid. The method was applied to the solution of the transonic small-disturbance equation for the velocity potential in the conservation form. Nonlifting transonic flow past a parabolic-arc airfoil is the example studied, with meshes of both constant and variable step size.

INTRODUCTION

The multi-level grid method, for accelerating convergence in relaxation calculations, has been shown to be very efficient for

*Assistant Head, Theoretical Aerodynamics Branch

**Professor of Mathematics, currently on leave at IBM Research Center, Mathematics Department, Yorktown Heights, New York.

solving elliptic problems with Dirichlet boundary conditions. For background and historical material, see references [1] to [4]. In [5], Brandt gives an extensive discussion and analysis of the method, together with several different procedures for applying the method. The idea of the method is based on the fact that in many typical elliptic boundary-value problems, the error is composed of a discrete spectrum of wave lengths, which range from the width of the region down to the width of a mesh cell. The short wave-length components of the error are usually diminished quite rapidly in a relaxation calculation, while the long wave-length components diminish very slowly. After only a few iterations the residual will be smooth, since the short wave-length error components have been eliminated; and thus the residual can be represented accurately on a coarser mesh. An equation called the "residual" equation is then solved on the coarser mesh, and the resulting correction is added to the last approximation on the fine mesh, yielding a significant improvement with very little work.

Since relaxation methods are currently the most attractive for obtaining numerical solutions to transonic aerodynamics problems, the question arises as to whether a multi-level, or multi-grid (MG) method can be used in a mixed flow with shock waves. In this paper we report some early results using the MG method to solve a simple transonic problem: we consider the transonic small-disturbance equation for the velocity potential, for nonlifting flow past a parabolic-arc airfoil.

PROBLEM DESCRIPTION

The transonic small-disturbance equation for the velocity potential can be written in conservation form as:

$$p_x + q_y = 0 \tag{1}$$

where

$$p = [K - \frac{(\gamma+1)}{2} M_\infty^2 \phi_x]\phi_x \tag{2}$$

$$q = \phi_y \tag{3}$$

$$K = (1 - M_\infty^2)/\tau^{2/3} \tag{4}$$

Equation (1) is to be solved subject to the boundary conditions that the disturbance potential, ϕ, vanishes at infinity and the flow is tangent to the airfoil surface, in the interval $|x| \leq 1/2$; i.e., at y = 0,

$$\begin{aligned} \phi_y &= F'(x) \quad \text{for } |x| \leq 1/2 \\ &= 0 \qquad\quad \text{for } |x| > 1/2 \end{aligned} \tag{5}$$

where F(x) is the (upper surface) thickness distribution function. τ is the usual thickness ratio, and γ, M_∞ and K are the ratio of specific heats, free-stream Mach number, and transonic similarity parameter, respectively. The form of Eqs (1) to (5) is a correctly-scaled transonic similarity form, in that all quantities are of order 1. Physical quantities, denoted by a "hat" symbol are related to the scaled quantities as follows:

$$\begin{aligned} \hat{\phi} &= c\tau^{2/3}\phi \\ \hat{x} &= cx \\ \hat{y} &= c\tau^{-1/3}y \\ \hat{t}(x) &= 2c\tau F(x) \end{aligned} \tag{6}$$

where c is the airfoil chord length and $\hat{t}$ is the total thickness distribution of the symmetric airfoil.

Equation (1) is of hyperbolic or elliptic type depending on whether

$$U = K - (\gamma + 1)M_\infty^2 \phi_x \tag{7}$$

is negative or positive, respectively.

Finite-Difference Equations

Murman's conservative difference scheme [6] can be conveniently presented in terms of Jameson's "switching function" [7] as follows:

$$(1 - \mu_{ij}) P_{ij} + \mu_{i-1,j} P_{i-1,j} + Q_{ij} = 0 \tag{8}$$

where

$$P_{ij} = U_{ij} \frac{\phi_{i+1,j} - 2\phi_{ij} + \phi_{i-1,j}}{\Delta x^2} \tag{9}$$

$$U_{ij} = K - (\gamma+1)M_\infty^2 \frac{\phi_{i+1,j} - \phi_{i-1,j}}{2\Delta x} \tag{10}$$

$$Q_{ij} = \frac{\phi_{i,j+1} - 2\phi_{ij} + \phi_{i,j-1}}{\Delta y^2} \tag{11}$$

and where

$$\begin{aligned} \mu_{ij} &= 0 \qquad \text{if } U_{ij} > 0 \\ &= 1 \qquad \text{if } U_{ij} \leq 0 \end{aligned} \tag{12}$$

It should be noted here that, in the interest of simplicity, we have presented only the constant-step-size (unstretched grid) form of the difference equations. In the case of a stretched grid, the conservative difference equations cannot be factored into the nice form given above, but this presents no real difficulty. The actual computer program is written for a stretched grid, with the identity transformation (constant step size) included as a special case.

Vertical Line Relaxation

A vertical line relaxation scheme for solving Eq. (8) by iteration can be written as:

$$AT_{i,j-1} + BT_{ij} + CT_{i,j+1} = R_{ij} + DT_{i-1,j} + ET_{i-2,j} \tag{13}$$

where

$$T_{ij} = \phi^+_{ij} - \phi_{ij} \tag{14}$$

ϕ^+ denotes a "new" value of ϕ, obtained during the latest iteration sweep, while ϕ is the value from the previous sweep. R_{ij}, which is the left-hand side of Eq. (8), is evaluated with "old" values of ϕ_{ij}, as are the iteration coefficients A through E, which are given in the appendix.

Multi-Grid Approach

Let us introduce a sequence of grids $G_1, G_2, \ldots, G_m$, where for simplicity, $h_k = 2h_{k+1}$, and h_k represents the step size of the G_k grid. We can represent the iteration operator (e.g., Eq. (13)) on the finest grid G_M as:

$$L_M(\phi_M) = f_M \tag{15}$$

where ϕ_M is the exact discrete solution on the G_M grid. We can write

$$\phi_M = u_M + v_M \tag{16}$$

where u_M is the approximate solution and v_M is the error. Then we have the <u>residual equation</u>:

$$\begin{aligned} \bar{L}_M(v_M) &= f_M - L_M(u_M) \\ &= -R_M \end{aligned} \tag{17}$$

where R_M is the residual of the approximation u_M on the G_M grid. $\bar{L}_M$ is in general different from L_M in the nonlinear case, which complicates matters. Nevertheless, if R_M is smooth, the error will be smooth, and the residual equation (17) can be solved on a coarser grid. Thus, for example, we can write

$$\bar{L}_{M-1}(w_{M-1}) = I_M^{M-1}(R_M) \tag{18}$$

where w is an approximation to the error v_M on the G_{M-1} grid, and I_k^ℓ denotes interpolation from the G_k to G_ℓ. After solving the problem (18) (usually with homogeneous boundary conditions), we interpolate the function w_{M-1} back onto the G_M mesh, and thus form an improved approximation:

$$(u_M)_{new} = (u_M)_{old} + I_{M-1}^M(w_{M-1}) \,. \tag{19}$$

In the complete MG algorithm, the solution of Eq. (18) is also performed by relaxation; and if the convergence rate falls below a prescribed level, we can apply a similar procedure, backing up to the G_{M-2} grid level, and so on, until we arrive at G_1, if necessary. The G_1 grid is so coarse that a direct solution could be used economically, but we have used iteration here also.

Full Approximation. In the general nonlinear case, the form of the operator $\bar{L}$ can be quite complicated - more so than the original operator, L - and thus applications to, say, the full potential equation may be tedious to program. It turns out that for the transonic small disturbance equation, the job is simple, and our first program did use the exact expression for $\bar{L}$ in an efficient way. However, there is an equivalent, easier method for solving the residual equation, which we call the full approximation method, as follows:

Suppose we add to both sides of Eq. (18) the function

$$L_{M-1}(u_M) - f_{M-1} = \tilde{R}_{M-1} \tag{21}$$

Then, since

$$\bar{L}_{M-1}(w_{M-1}) + L_{M-1}(u_M) \simeq L_{M-1}(\phi_M),$$

we have

$$L_{M-1}(\phi_M) \simeq \tilde{R}_{M-1} - I_M^{M-1}(R_M) \tag{22}$$

We can now use the original operator on all the grids, which greatly simplifies the programming. The right-hand side of Eq. (22) is the difference between the residuals of u_M calculated with the coarse- and fine-grid operators. Note that when the solution converges on the G_M grid, then

$$R_M \rightarrow 0 \tag{23a}$$

$$I_M^{M-1}(R_M) \rightarrow 0 \tag{23b}$$

but $\tilde{R}_{M-1}$ will remain finite, since ϕ_M is a solution on the G_M grid; $\tilde{R}_{M-1}$ is essentially the truncation error of the L_{M-1} operator.

After Eq. (22) is solved to sufficient accuracy, we determine the function

$$w_{M-1} = \phi_M - I_M^{M-1}(u_M) \tag{24}$$

by subtraction at all points of the grid G_{M-1}, and then interpolate w_{M-1} to the G_M grid as before in Eq. (19).

More explicit details of the method will be deferred to a forthcoming report.

RESULTS AND DISCUSSION

In order to estimate the efficiency of the method, a work unit can be defined as the amount of computational effort required for one relaxation sweep on the (finest) G_M grid. Thus, a relaxation sweep on the G_k grid costs

$$n_w = (1/4)^{M-k}$$

work units, for example. Likewise, when we calculate the residuals for the G_k grid, we perform these calculations at the points of the G_{k-1} grid, i.e., 1/4 as few points; hence each residual calculation costs less than 1/4 the effort of a relaxation sweep on the G_k grid, or approximately $(1/4)^{M-k+1}$. Note that this is an overestimate, since the tridiagonal system (13) is not inverted, nor do we calculate the iteration coefficients during the residual calculations. On the other hand, we did not count the work of interpolation in Eq. (19), for example, or any other "overhead" of that type.

An overall estimate of efficiency can be given by the number

$$a = \{ \| R_{M, n_w} \| / \| R_{M, 1} \| \}^{1/n_w} \tag{25}$$

where

$\| R_{M, 1} \|$ = norm of R_M after first sweep on G_M

$\| R_{M, n_w} \|$ = norm of R_M after n_w work units

and

$$\| R_M \| = (\Delta x \, \Delta y \, \Sigma_{ij} \, R_M^2)^{1/2} \tag{26}$$

Hence the norm we use is the root mean square of the residual on G_M. This number is typically about 5 to 10 times smaller than the maximum norm in the transonic problems. We consider an approximate solution to be converged when

$$\| R_M \| < C/(\text{no. of grid points}) \tag{27}$$

where the prescribed constant C is typically chosen as 1 so as to estimate the nominal truncation error.

Unstretched Grids

In the case of a grid with constant steps in both directions, the present MG method performed quite well. In the following some typical results are summarized. All of the figures shown are copies of the screen display, on a remote computer terminal, of an abbreviated history of the MG runs. The first integer is the grid level M, corresponding to G_M in our text. The next three "E"- format numbers are:

1. $\max_{ij} |R_{ij}|$ (see Equation (13)).
2. $||R_M||$ (see Equation (26)).
3. $\max_{ij} |T_{ij}|$ (see Equation (14)).

The two integers following 1. above are the i, j location where the $\max_{ij} |R_{ij}|$ occurred. The last two numbers in a row are the number of work units, n_W, and the number of supersonic points. One row is printed for each relaxation sweep on the finest (G_M) grid, but not for the coarser grids. However, each time the calculation "backs up" to a coarser grid, the words RESCAL are printed and the value of $\max_{ij} |R_{ij}|$ and $||R_M||$ are printed, together with the grid level (L) which has just been relaxed. Note that these norms correspond to $I_M^{M-1}(R_M)$ in the right-hand side of Eq. (22). In all MG runs shown, a relaxation factor of 1.0 was used on all grids. Likewise all MG runs in these examples used 5 levels of grids (M=5), with G_1 being 4 x 2 mesh cells in the x- and y-directions, respectively, and G_5 being a 64 x 32. We have done 6 levels, with G_6 being 128 x 64 with no deterioration in MG performance.

<u>Laplace's Equation</u>. To show just how fast the MG method works for a nice, smooth, elliptic problem, we present a run in Figures 1 and 2 for Laplace's equation with the prescribed normal derivative equal to sin πX along $y = 0$. In Figure 1, the convergence history is shown for G_4, a 32 x 16 grid, and according to Eq. (25), we achieved a = .540. Now because of the smoothness of the solution, it may be expected that interpolating the converged G_4 solution onto G_5 will give a very good starting approximation for G_5. This is true, as can be seen on Figure 2, where the G_5 grid was started with the interpolated G_4 solution. For G_5, we obtained a = .583, but the efficiency of the two <u>combined</u> levels is more like a = .46!

```
-----CONVERGENCE CRITERION FOR L1= 4 GRID IS EPS=1.783E-03

----------NO X-STRETCH----------
   XL= 2.00, DX=6.250E-02

----------NO Y-STRETCH----------
   YL= 1.00, DY=6.250E-02

  4   4.535E+01    9 17   1.148E+01    5.457E-02    1.0   0
  4   2.181E+01    9 17   6.209E+00    3.170E-02    2.0   0
 RESCAL. L=4, RMAX= 8.085E+00, RL2= 3.059E+00
 RESCAL. L=3, RMAX= 2.619E+00, RL2= 1.489E+00
 RESCAL. L=2, RMAX= 5.215E-01, RL2= 3.927E-01
  4   4.077E+00   23 17   1.026E+00    5.184E-03    4.6   0
  4   1.981E+00   23 17   5.233E-01    2.874E-03    5.6   0
 RESCAL. L=4, RMAX= 7.216E-01, RL2= 2.496E-01
 RESCAL. L=3, RMAX= 1.917E-01, RL2= 9.714E-02
 RESCAL. L=2, RMAX= 2.410E-02, RL2= 1.456E-02
  4   4.043E-01   23 17   1.018E-01    4.691E-04    8.3   0
  4   1.753E-01   23 17   4.521E-02    2.329E-04    9.3   0
  4   9.351E-02   22 17   2.581E-02    1.392E-04   10.3   0
 RESCAL. L=4, RMAX= 3.563E-02, RL2= 1.211E-02
 RESCAL. L=3, RMAX= 9.977E-03, RL2= 4.209E-03
 RESCAL. L=2, RMAX= 2.228E-03, RL2= 1.311E-03
  4   2.147E-02   22 17   4.921E-03    2.450E-05   12.9   0
  4   9.032E-03   21 17   2.014E-03    1.217E-05   13.9   0
  4   4.645E-03   20 17   1.167E-03    7.044E-06   14.9   0

     PLOT OF CPBAR FOR LEVEL 4

         CPSTAR=I1.0 *INTERRUPTED*
```

Figure 1. MG Solution of Laplace's Equation with Smooth Boundary Conditions. 32 x 16 Grid.

Figure 3 shows the convergence history for the same problem, using SLOR all the way on G_5, achieving convergence in $n_w = 141$ yielding a = .924.

<u>Nonlinear, Subcritical Flow ($M_\infty = 0.7$).</u> In Figure 4 is shown the history of a nonlinear, but subcritical flow solved by MG. Here the convergence rate is the same or better, on G_5, as it was for Laplace's equation with smooth boundary conditions discussed previously (i.e., a = .549). In this case, however, the Neumann boundary condition is an "N-wave" - far from smooth - and hence we can conclude that discontinuous boundary conditions do not deteriorate MG performance. SLOR, with $\omega = 1.85$, achieved a = .868.

<u>Supercritical Flow ($M_\infty = .85$).</u> Figure 5 illustrates the history for a typical supercritical flow with a moderate-sized supersonic

```
-----CONVERGENCE CRITERION FOR L1= 5 GRID IS EPS=4.662E-04

---------NO X-STRETCH----------
  XL= 2.00, DX=3.125E-02

---------NO Y-STRETCH----------
  YL= 1.00, DY=3.125E-02

 5   4.042E-01  18 33   7.043E-02    1.230E-04    1.0    0
 5   1.901E-01  17 33   3.685E-02    6.806E-05    2.0    0
RESCAL. L=5, RMAX= 6.517E-02, RL2= 1.715E-02
RESCAL. L=4, RMAX= 1.912E-02, RL2= 7.399E-03
RESCAL. L=3, RMAX= 4.332E-03, RL2= 2.409E-03
 5   3.742E-02  47 33   7.307E-03    1.054E-05    4.6    0
 5   1.504E-02  48 33   3.713E-03    4.707E-06    5.6    0
RESCAL. L=5, RMAX= 4.820E-03, RL2= 1.650E-03
RESCAL. L=4, RMAX= 1.464E-03, RL2= 8.060E-04
RESCAL. L=3, RMAX= 7.393E-04, RL2= 5.503E-04
RESCAL. L=2, RMAX= 1.938E-04, RL2= 1.751E-04
 5   4.992E-03  46 33   8.927E-04    1.531E-06    8.3    0
 5   2.372E-03  45 33   4.637E-04    8.477E-07    9.3    0

    PLOT OF CPBAR FOR LEVEL 5

        CPSTAR=-1.000E+03

I    I *INTERRUPTED*
```

Figure 2. MG Solution of Laplace's Equation with Smooth Boundary Conditions. 64 x 32 Grid Initialized by Solution of Figure 1.

region. Since the G_5 grid has 2145 grid points, the supersonic region, with 124 points, occupies about six percent of the grid. For this case, a = .593. The same case using SLOR all the way converged in n_w = 68, using a relaxation factor of 1.85, and the a = .855.

<u>Highly Supercritical Flow (M_∞ = .95).</u> Figure 6 illustrates the history for a highly supercritical flow, where the shock wave is at the trailing edge of the airfoil. Note the final number of supersonic points (355) is established after 38 work units. It is typical that, at that point, the MG method begins to work best, since most of the high frequency error components have been eliminated. For this case, a = .858, achieving convergence in 67.6 work units. The same case was converged with SLOR all the way in 228 work units, with a = .957.

-----CONVERGENCE CRITERION FOR L1= 5 GRID IS EPS=4.662E-04

----------NO X-STRETCH----------
XL= 2.00, DX=3.125E-02

----------NO Y-STRETCH----------
YL= 1.00, DY=3.125E-02

5	1.514E+02	20	33	3.125E+01	8.961E-02	1.0	0
5	5.225E+01	19	33	1.826E+01	4.414E-02	2.0	0
5	3.595E+01	19	33	1.506E+01	3.161E-02	3.0	0
5	2.841E+01	19	33	1.314E+01	2.527E-02	4.0	0
5	2.368E+01	18	33	1.172E+01	2.115E-02	5.0	0
5	2.030E+01	18	33	1.058E+01	1.818E-02	6.0	0
5	1.768E+01	17	33	9.612E+00	1.586E-02	7.0	0
5	1.557E+01	17	33	8.768E+00	1.399E-02	8.0	0
5	1.390E+01	16	33	8.017E+00	1.240E-02	9.0	0
5	1.230E+01	16	33	7.341E+00	1.106E-02	10.0	0
5	1.097E+01	16	33	6.726E+00	9.878E-03	11.0	0
5	9.821E+00	15	33	6.163E+00	8.845E-03	12.0	0
5	8.789E+00	15	33	5.646E+00	7.919E-03	13.0	0
5	7.960E+00	14	33	5.168E+00	7.083E-03	14.0	0
5	7.155E+00	45	33	4.728E+00	6.445E-03	15.0	0
5	6.598E+00	44	33	4.321E+00	5.945E-03	16.0	0
5	6.120E+00	44	33	3.947E+00	5.516E-03	17.0	0
5	5.639E+00	43	33	3.603E+00	5.128E-03	18.0	0
5	5.309E+00	43	33	3.289E+00	4.787E-03	19.0	0
5	4.973E+00	42	33	3.003E+00	4.485E-03	20.0	0
5	4.667E+00	41	33	2.743E+00	4.210E-03	21.0	0
5	4.403E+00	41	33	2.510E+00	3.972E-03	22.0	0
5	4.161E+00	40	33	2.301E+00	3.754E-03	23.0	0
5	3.940E+00	40	33	2.116E+00	3.555E-03	24.0	0
5	3.746E+00	39	33	1.953E+00	3.380E-03	25.0	0
5	3.564E+00	38	33	1.810E+00	3.216E-03	26.0	0
5	3.395E+00	38	33	1.686E+00	3.064E-03	27.0	0
5	3.244E+00	37	33	1.579E+00	2.927E-03	28.0	0
5	3.101E+00	36	33	1.486E+00	2.798E-03	29.0	0
5	2.966E+00	35	33	1.406E+00	2.675E-03	30.0	0
5	2.836E+00	34	33	1.335E+00	2.558E-03	31.0	0
5	2.712E+00	33	33	1.274E+00	2.446E-03	32.0	0
5	2.594E+00	32	33	1.219E+00	2.339E-03	33.0	0
5	2.480E+00	31	33	1.169E+00	2.236E-03	34.0	0
5	2.370E+00	30	33	1.122E+00	2.137E-03	35.0	0
5	2.264E+00	29	33	1.078E+00	2.041E-03	36.0	0
5	2.160E+00	28	33	1.036E+00	1.948E-03	37.0	0
5	2.060E+00	27	33	9.956E-01	1.857E-03	38.0	0
5	1.962E+00	26	33	9.552E-01	1.769E-03	39.0	0
5	1.867E+00	25	33	9.151E-01	1.683E-03	40.0	0
5	1.773E+00	24	33	8.751E-01	1.599E-03	41.0	0
5	1.681E+00	23	33	8.351E-01	1.516E-03	42.0	0
5	1.591E+00	22	33	7.950E-01	1.434E-03	43.0	0
5	1.502E+00	21	33	7.550E-01	1.354E-03	44.0	0

Figure 3. SLOR Solution of Laplace's Equation with Smooth Boundary Conditions. 64 x 32 Grid. (Continued)

5	1.413E+00	20	33	7.150E-01	1.274E-03	45.0	0
5	1.327E+00	19	33	6.752E-01	1.196E-03	46.0	0
5	1.241E+00	18	33	6.358E-01	1.118E-03	47.0	0
5	1.156E+00	17	33	5.969E-01	1.042E-03	48.0	0
5	1.072E+00	16	33	5.587E-01	9.658E-04	49.0	0
5	9.884E-01	15	33	5.212E-01	8.909E-04	50.0	0
5	9.064E-01	14	33	4.848E-01	8.170E-04	51.0	0
5	8.255E-01	13	33	4.494E-01	7.440E-04	52.0	0
5	7.458E-01	12	33	4.153E-01	6.722E-04	53.0	0
5	6.675E-01	11	33	3.825E-01	6.016E-04	54.0	0
5	5.907E-01	10	33	3.512E-01	5.324E-04	55.0	0
5	5.156E-01	9	33	3.214E-01	4.647E-04	56.0	0
5	4.424E-01	8	33	2.933E-01	3.987E-04	57.0	0
5	3.947E-01	16	33	2.669E-01	3.556E-04	58.0	0
5	3.547E-01	17	33	2.422E-01	3.198E-04	59.0	0
5	3.218E-01	20	33	2.194E-01	2.901E-04	60.0	0
5	2.940E-01	20	33	1.984E-01	2.651E-04	61.0	0
5	2.690E-01	21	33	1.793E-01	2.425E-04	62.0	0
5	2.459E-01	21	33	1.621E-01	2.216E-04	63.0	0
5	2.244E-01	21	33	1.467E-01	2.022E-04	64.0	0
5	2.223E-01	20	33	1.455E-01	2.003E-04	65.0	0
5	2.025E-01	20	33	1.346E-01	1.826E-04	66.0	0
5	1.876E-01	20	33	1.248E-01	1.691E-04	67.0	0
5	1.743E-01	20	33	1.157E-01	1.571E-04	68.0	0
5	1.621E-01	20	33	1.072E-01	1.461E-04	69.0	0
5	1.508E-01	20	33	9.943E-02	1.360E-04	70.0	0
5	1.403E-01	20	33	9.221E-02	1.265E-04	71.0	0
5	1.305E-01	20	33	8.552E-02	1.176E-04	72.0	0
5	1.212E-01	20	33	7.931E-02	1.092E-04	73.0	0
5	1.124E-01	20	33	7.356E-02	1.013E-04	74.0	0
5	1.042E-01	20	33	6.822E-02	9.389E-05	75.0	0
5	9.653E-02	20	33	6.327E-02	8.700E-05	76.0	0
5	8.944E-02	20	33	5.868E-02	8.061E-05	77.0	0
5	8.288E-02	20	33	5.442E-02	7.470E-05	78.0	0
5	7.682E-02	20	33	5.047E-02	6.924E-05	79.0	0
5	7.122E-02	20	33	4.680E-02	6.419E-05	80.0	0
5	6.604E-02	20	33	4.340E-02	5.953E-05	81.0	0
5	6.125E-02	20	33	4.024E-02	5.521E-05	82.0	0
5	5.681E-02	20	33	3.731E-02	5.120E-05	83.0	0
5	5.269E-02	20	33	3.460E-02	4.749E-05	84.0	0
5	4.887E-02	20	33	3.208E-02	4.405E-05	85.0	0
5	4.532E-02	20	33	2.974E-02	4.085E-05	86.0	0
5	4.202E-02	20	33	2.757E-02	3.788E-05	87.0	0
5	3.896E-02	20	33	2.555E-02	3.512E-05	88.0	0
5	3.612E-02	20	33	2.369E-02	3.256E-05	89.0	0
5	3.348E-02	20	33	2.195E-02	3.018E-05	90.0	0
5	3.103E-02	20	33	2.035E-02	2.797E-05	91.0	0
5	2.875E-02	19	33	1.886E-02	2.592E-05	92.0	0
5	2.664E-02	19	33	1.747E-02	2.401E-05	93.0	0
5	2.468E-02	19	33	1.619E-02	2.225E-05	94.0	0
5	2.286E-02	19	33	1.500E-02	2.061E-05	95.0	0
5	2.118E-02	19	33	1.390E-02	1.909E-05	96.0	0
5	1.961E-02	19	33	1.288E-02	1.768E-05	97.0	0
5	1.816E-02	19	33	1.193E-02	1.637E-05	98.0	0
5	1.682E-02	19	33	1.106E-02	1.516E-05	99.0	0
5	1.558E-02	19	33	1.024E-02	1.404E-05	100.0	0

Figure 3. (Continued)

```
5   1.443E-02  19 33   9.489E-03   1.301E-05  101.0   0
5   1.336E-02  19 33   8.791E-03   1.205E-05  102.0   0
5   1.230E-02  20 33   8.145E-03   1.116E-05  103.0   0
5   1.147E-02  20 33   7.545E-03   1.034E-05  104.0   0
5   1.063E-02  20 33   6.990E-03   9.578E-06  105.0   0
5   9.847E-03  20 33   6.476E-03   8.875E-06  106.0   0
5   9.126E-03  20 33   6.000E-03   8.226E-06  107.0   0
5   8.460E-03  20 33   5.559E-03   7.625E-06  108.0   0
5   7.844E-03  20 33   5.151E-03   7.070E-06  109.0   0
5   7.266E-03  19 33   4.772E-03   6.549E-06  110.0   0
5   6.730E-03  20 33   4.422E-03   6.066E-06  111.0   0
5   6.236E-03  20 33   4.097E-03   5.620E-06  112.0   0
5   5.778E-03  20 33   3.796E-03   5.208E-06  113.0   0
5   5.354E-03  20 33   3.518E-03   4.826E-06  114.0   0
5   4.962E-03  20 33   3.260E-03   4.472E-06  115.0   0
5   4.598E-03  20 33   3.021E-03   4.144E-06  116.0   0
5   4.261E-03  20 33   2.799E-03   3.840E-06  117.0   0
5   3.949E-03  20 33   2.594E-03   3.559E-06  118.0   0
5   3.659E-03  20 33   2.404E-03   3.298E-06  119.0   0
5   3.391E-03  20 33   2.228E-03   3.056E-06  120.0   0
5   3.142E-03  20 33   2.065E-03   2.832E-06  121.0   0
5   2.912E-03  20 33   1.913E-03   2.625E-06  122.0   0
5   2.699E-03  20 33   1.773E-03   2.432E-06  123.0   0
5   2.501E-03  20 33   1.643E-03   2.254E-06  124.0   0
5   2.317E-03  20 33   1.523E-03   2.088E-06  125.0   0
5   2.147E-03  20 33   1.411E-03   1.935E-06  126.0   0
5   1.989E-03  20 33   1.307E-03   1.793E-06  127.0   0
5   1.843E-03  20 33   1.211E-03   1.661E-06  128.0   0
5   1.703E-03  20 33   1.122E-03   1.540E-06  129.0   0
5   1.583E-03  20 33   1.040E-03   1.427E-06  130.0   0
5   1.467E-03  20 33   9.637E-04   1.322E-06  131.0   0
5   1.359E-03  20 33   8.930E-04   1.225E-06  132.0   0
5   1.259E-03  20 33   8.275E-04   1.135E-06  133.0   0
5   1.167E-03  20 33   7.668E-04   1.052E-06  134.0   0
5   1.081E-03  20 33   7.105E-04   9.747E-07  135.0   0
5   1.002E-03  20 33   6.584E-04   9.032E-07  136.0   0
5   9.235E-04  20 33   6.101E-04   8.369E-07  137.0   0
5   8.604E-04  20 33   5.653E-04   7.755E-07  138.0   0
5   7.973E-04  20 33   5.239E-04   7.186E-07  139.0   0
5   7.388E-04  20 33   4.854E-04   6.659E-07  140.0   0
5   6.846E-04  20 33   4.498E-04   6.170E-07  141.0   0

   PLOT OF CPBAR FOR LEVEL 5

        CPSTAR=-1.000E+0I3
*INTERRUPTED*
```

Figure 3. (Concluded)

Stretched Grids

We found quickly that vertical line relaxation alone is not the best way to relax the solution in the MG mode in the case of a stretched grid. A possible explanation for this is that all of the high-frequency error components are not rapidly damped by vertical line relaxation in a general stretched mesh, where the mesh aspect ratio varies from very small to very large values. For if we consider the line relaxation algorithm for Laplace's equation, with a local mesh aspect ratio equal to A, and a relaxation factor ω, the amplification factor is:

```
-----CONVERGENCE CRITERION FOR L1= 5 GRID IS EPS=4.662E-04

----------NO X-STRETCH----------
  XL= 2.00, DX=3.125E-02

----------NO Y-STRETCH----------
  YL=  .50, DY=1.563E-02

  5   2.560E+02  17 33   1.934E+01    3.658E-02    1.0   0
  5   3.366E+01  48 33   7.515E+00    1.982E-02    2.0   0
  5   4.624E+01  47 33   5.035E+00    1.362E-02    3.0   0
  5   3.105E+01  46 33   4.132E+00    1.044E-02    4.0   0
 RESCAL. L=5, RMAX= 2.340E+01, RL2= 3.918E+00
 RESCAL. L=4, RMAX= 1.053E+01, RL2= 2.654E+00
 RESCAL. L=3, RMAX= 2.919E+00, RL2= 1.131E+00
  5   7.191E+00  16 33   4.797E-01    7.600E-04    6.7   0
  5   2.431E+00  48 33   1.936E-01    4.398E-04    7.7   0
  5   1.224E+00  47 33   1.409E-01    3.043E-04    8.7   0
 RESCAL. L=5, RMAX= 7.230E-01, RL2= 1.290E-01
 RESCAL. L=4, RMAX= 2.337E-01, RL2= 8.850E-02
 RESCAL. L=3, RMAX= 7.744E-02, RL2= 3.357E-02
 RESCAL. L=2, RMAX= 1.914E-02, RL2= 1.003E-02
  5   3.394E-01  16 33   2.275E-02    3.581E-05   11.4   0
  5   1.192E-01  49 33   1.021E-02    2.003E-05   12.4   0
  5   6.405E-02  48 33   6.139E-03    1.345E-05   13.4   0
  5   4.275E-02  49 33   4.452E-03    9.966E-06   14.4   0
 RESCAL. L=5, RMAX= 3.290E-02, RL2= 3.831E-03
 RESCAL. L=4, RMAX= 8.416E-03, RL2= 1.801E-03
 RESCAL. L=3, RMAX= 1.737E-03, RL2= 6.368E-04
  5   9.887E-03  49 33   8.715E-04    1.358E-06   17.3   0
  5   3.697E-03   6 33   3.332E-04    5.895E-07   18.3   0

    PLOT OF CPBAR FOR LEVEL 5

        CPSTAR=-8I.673 *INTERRUPTED*
```

Figure 4. MG Solution of Parabolic-Arc Airfoil. $M_\infty = 0.7$, $\tau = 0.1$, 64 x 32 Grid.

$$g(\theta_x, \theta_y; A, \omega) = \frac{A[2(1-\omega) + \omega e^{i\theta_x}]}{A(2 - \omega e^{-i\theta_x}) + 2\omega(1 - \cos\theta_y)} \tag{28}$$

If $A = (\Delta y/\Delta x)^2$ is large, we have a problem, for then, with $\theta_x = 0$, and $\theta_y = \pi/2$, we have

$$g(0, \frac{\pi}{2}, A, \omega) = \frac{A(2-\omega)}{A(2-\omega) + 2\omega} \tag{29}$$

and if $\omega = 1$, we see that

$$|g| \to 1 \quad \text{as} \quad A \to \infty .$$

```
-----CONVERGENCE CRITERION FOR L1= 5 GRID IS EPS=4.662E-04

---------NO X-STRETCH----------
   XL= 2.00, DX=3.125E-02

---------NO Y-STRETCH----------
   YL=  .46, DY=1.451E-02
 5   2.758E+02  17 33   2.007E+01    5.148E-02    1.0    0
 5   3.596E+01  16 33   6.192E+00    2.763E-02    2.0    0
 5   4.482E+01  49 33   4.164E+00    1.964E-02    3.0    0
 5   2.919E+01  16 33   3.457E+00    1.552E-02    4.0    0
RESCAL. L=5, RMAX= 2.008E+01, RL2= 3.304E+00
RESCAL. L=4, RMAX= 9.996E+00, RL2= 2.277E+00
RESCAL. L=3, RMAX= 2.871E+00, RL2= 1.060E+00
 5   1.683E+01  16 33   1.027E+00    1.020E-02    6.7  122
 5   4.587E+00  45 33   3.800E-01    2.929E-03    7.7  124
 5   2.562E+00  47 33   2.085E-01    8.709E-04    8.7  124
 5   1.670E+00  49 33   1.649E-01    6.444E-04    9.7  123
RESCAL. L=5, RMAX= 1.420E+00, RL2= 1.528E-01
RESCAL. L=4, RMAX= 4.032E-01, RL2= 7.270E-02
RESCAL. L=3, RMAX= 7.971E-02, RL2= 3.097E-02
 5   1.023E+00  40 28   6.943E-02    6.537E-04   12.6  123
 5   1.252E-01  48 33   1.565E-02    1.385E-04   13.6  124
 5   9.050E-02  49 33   8.809E-03    6.942E-05   14.6  124
 5   7.234E-02  49 33   6.728E-03    4.278E-05   15.6  124
RESCAL. L=5, RMAX= 5.714E-02, RL2= 6.025E-03
RESCAL. L=4, RMAX= 1.174E-02, RL2= 2.668E-03
RESCAL. L=3, RMAX= 4.874E-03, RL2= 1.610E-03
 5   2.454E-02  41 32   1.620E-03    2.058E-05   18.8  124
 5   8.248E-03  41 32   5.571E-04    7.043E-06   19.8  124
 5   4.966E-03  41 32   3.842E-04    4.435E-06   20.8  124

     PLOT OF CPBAR
```

Figure 5. MG Solution of Parabolic-Arc Airfoil. $M_\infty = 0.85$, $\tau = 0.1$, 64 x 32 Grid.

Clearly, choosing $\omega \simeq 2$ alleviates the problem, but then other high-frequency components are retarded, i. e., for $\theta_x = \pi/2$ and $\theta_y = 0$,

$$\left| g\left(\frac{\pi}{2}, 0; A, \omega\right)\right| = \sqrt{\frac{4(1-\omega)^2 + \omega^2}{4 + \omega^2}} \tag{30}$$

which approaches 1 as ω nears 2.0. A solution to this problem is to sweep in all directions alternatively (forward, backward, up, and down, in a general problem), but of course special care must be taken in supersonic regions.

Figure 7 shows an MG run with vertical line relaxation for the $M_\infty = .95$ flow, with the grid stretched to infinity in both the x- and y-directions. A logarithmic stretch was used, with thirty

```
----CONVERGENCE CRITERION FOR L1= 5 GRID IS EPS=4.662E-04


----------NO X-STRETCH-----------
   XL= 2.00, DX=3.125E-02


----------NO Y-STRETCH-----------
   YL= 2.00, DY=6.250E-02

  5    6.400E+01  17 33    9.670E+00    7.392E-02     1.0     0
  5    1.024E+02  16 33    8.136E+00    8.362E-02     2.0     0
 RESCAL. L=5, PMAX= 3.265E+02, RL2= 2.911E+01
 RESCAL. L=4, PMAX= 5.370E+01, RL2= 1.002E+01
  5    8.886E+02  48 33    6.405E+01    3.193E-01     7.1   263
  5    1.223E+03  53 33    1.213E+02    1.851E-01     8.1   257
  5    4.512E+02  53 33    2.751E+01    8.244E-02     9.1   259
  5    2.526E+02  54 33    1.559E+01    8.460E-02    10.1   261
  5    7.746E+01  56 33    6.815E+00    3.350E-02    11.1   261
  5    7.205E+01  56 33    6.527E+00    2.270E-02    12.1   260
 RESCAL. L=5, PMAX= 5.651E+01, RL2= 7.330E+00
 RESCAL. L=4, PMAX= 1.339E+01, RL2= 2.663E+00
  5    2.165E+02  54 31    1.806E+01    5.433E-02    15.8   289
  5    8.696E+01  51 33    7.351E+00    1.994E-02    16.8   290
  5    3.332E+01  52 32    2.582E+00    9.188E-03    17.8   291
  5    1.927E+01  52 32    1.520E+00    5.171E-03    18.8   291
  5    1.060E+01  52 32    1.071E+00    2.910E-03    19.8   291
 RESCAL. L=5, RMAX= 6.095E+00, RL2= 1.037E+00
 RESCAL. L=4, RMAX= 1.121E+00, RL2= 4.073E-01
 RESCAL. L=3, RMAX= 5.433E-01, RL2= 2.971E-01
  5    9.944E+01  51 33    1.013E+01    7.621E-02    24.0   349
  5    2.423E+02  56 27    1.503E+01    3.845E-02    25.0   349
  5    5.182E+01  56 27    3.275E+00    1.437E-02    26.0   349
  5    8.438E+00  57 27    6.046E-01    2.213E-03    27.0   349
  5    2.967E+00  51 33    3.543E-01    1.190E-03    28.0   349
  5    2.497E+00  52 33    3.048E-01    9.618E-04    29.0   349
 RESCAL. L=5, RMAX= 2.153E+00, RL2= 3.416E-01
 RESCAL. L=4, RMAX= 7.052E-01, RL2= 1.563E-01
  5    2.168E+01  51 33    1.754E+00    9.381E-03    32.7   352
  5    2.833E+01  52 32    1.743E+00    7.565E-03    33.7   352
 RESCAL. L=5, RMAX= 9.679E-01, RL2= 1.414E-01
 RESCAL. L=4, RMAX= 2.172E-01, RL2= 5.301E-02
 RESCAL. L=3, RMAX= 3.632E-02, RL2= 2.092E-02
  5    1.490E+01  51 33    1.545E+00    5.100E-03    37.9   355
  5    2.485E+00  52 32    1.812E-01    9.018E-04    38.9   355
  5    1.298E+00  52 32    9.746E-02    3.815E-04    39.9   355
  5    4.763E-01  51 33    5.710E-02    1.879E-04    40.9   355
  5    3.061E-01  53 33    4.373E-02    1.203E-04    41.9   355
 RESCAL. L=5, RMAX= 2.637E-01, RL2= 4.582E-02
 RESCAL. L=4, RMAX= 1.086E-01, RL2= 2.461E-02
  5    3.617E+00  51 33    3.424E-01    1.369E-03    45.6   355
  5    6.573E-01  51 33    4.942E-02    2.354E-04    46.6   355
  5    3.493E-01  52 32    2.300E-02    9.997E-05    47.6   355
  5    1.287E-01  52 32    1.225E-02    4.495E-05    48.6   355
```

Figure 6. MG Solution of Parabolic-Arc Airfoil. $M_\infty = 0.95$, $\tau = 0.1$, 64 x 32 Grid. (Continued)

```
 5   7.147E-02  51 33   8.543E-03   2.778E-05   49.6  355
 5   5.496E-02  53 33   7.056E-03   2.037E-05   50.6  355
RESCAL. L=5, RMAX= 4.636E-02, RL2= 7.939E-03
RESCAL. L=4, RMAX= 1.684E-02, RL2= 4.349E-03
RESCAL. L=3, RMAX= 2.934E-03, RL2= 1.674E-03
 5   1.060E+00  54 31   1.260E-01   3.408E-04   55.0  355
 5   2.358E-01  52 32   1.706E-02   6.924E-05   56.0  356
 5   1.429E-01  52 32   8.246E-03   4.379E-05   57.0  356
 5   5.666E-02  52 32   4.008E-03   1.755E-05   58.0  356
 5   2.093E-02  52 32   2.516E-03   6.462E-06   59.0  356
 5   1.370E-02  54 31   1.959E-03   4.476E-06   60.0  356
RESCAL. L=5, RMAX= 9.397E-03, RL2= 1.800E-03
RESCAL. L=4, RMAX= 4.238E-03, RL2= 9.833E-04
 5   1.185E-01  54 31   1.133E-02   3.775E-05   63.6  356
 5   2.752E-02  52 32   1.895E-03   8.125E-06   64.6  356
 5   1.736E-02  52 32   9.430E-04   5.340E-06   65.6  356
 5   7.437E-03  52 32   4.773E-04   2.345E-06   66.6  356
 5   3.398E-03  52 32   3.140E-04   1.082E-06   67.6  356

   PLOT OF CPBAR FOR LEVE
```

Figure 6. (Concluded)

percent of the grid points in the x-direction on the airfoil chord. Note that the maximum residual tends to occur far above the airfoil (small values of j), where $\Delta y/\Delta x$ is large. For this case, a = .936. The same case, solved by SLOR takes about 382 cycles to converge (a = .974). Some benefit is still achieved from the MG mode of operation, even though the MG performance is far worse than what we believe can be obtained by a better relaxation algorithm.

Since this last case is a particularly interesting flow, we have included some pictures of the output for the pressure distribution along y = 0 (Figure 8), a chart of the Mach numbers in the computational plane (Figure 9), and an isobar plot (Figure 10). Note in Figure 8 that an <u>oblique</u> shock occurs at the trailing edge, followed by a nearly-constant velocity supersonic zone in the wake, then a normal shock in the wake about 1/2-chord behind the trailing edge, and finally a very slow recovery to free-stream conditions. The airfoil lies between I = 24 and 42 (x < .5). Figures 9 and 10 show the "fishtail" shock pattern more clearly. In Figure 9, only odd values of J are printed in order to fit the picture on the screen. J = 1 corresponds to infinity, as do I = 1 and 65. The values of I are the first column of integers, and the Mach numbers x 100 are shown in the array. Flow is from top to bottom in the picture, with the line y = 0 (and the airfoil surface) on the left (J = 33, see bottom row of integers indicating the value of J). The isobar plot, Figure 10, uses integers for supersonic flow values. The triangular region of nearly-constant velocity between the oblique

```
------CONVERGENCE CRITERION FOR L1= 5 GRID IS EPS=4.662E-04


   ------X-STRETCH FUNCTIONS------
         XO=.300, AX= 8.077E-01

 I      XX       SX(I)       SXH(I)

 2   -3.346 3.309E-02 5.668E-02
33    0.000 6.190E-01 6.189E-01


    -----Y-STRETCH FUNCTIONS------
         YO= .300, AY= 3.749E-01

 J        YY          ZZ(J)       SY(J)       SYH(J)

 2       1.553        3.346        .082        .122
33       0.000        0.000       1.334       1.333
 5    1.594E+02   24  33    1.309E+01    1.409E-01     1.0     0
 5    8.563E+01   23  33    5.101E+00    1.727E-01     2.0    29
 5    2.375E+02   42  33    1.708E+01    1.584E-01     3.0    79
 5    2.660E+02   40  32    2.801E+01    8.949E-02     4.0   103
 5    3.055E+01   41  32    7.625E+00    3.740E-02     5.0   105
 5    2.795E+01   41  33    2.454E+00    1.601E-02     6.0   106
 5    1.377E+01   40  32    1.823E+00    1.898E-02     7.0   110
RESCAL. L=5, RMAX= 1.058E+01, RL2= 1.606E+00
RESCAL. L=4, RMAX= 2.451E+00, RL2= 7.702E-01
PESCAL. L=3, RMAX= 1.005E+00, RL2= 4.308E-01
 5    4.729E+01   41  33    3.782E+00    6.239E-02    11.5   596
 5    9.541E+00   54   5    9.508E-01    3.237E-02    12.5   580
 5    2.378E+00   55   5    2.404E-01    9.204E-03    13.5   581
 5    9.867E-01   54   8    1.826E-01    7.228E-03    14.5   583
PESCAL. L=5, RMAX= 8.616E-01, RL2= 1.749E-01
RESCAL. L=4, RMAX= 4.822E-01, RL2= 1.509E-01
 5    2.474E+01   55   8    2.446E+00    9.443E-02    18.7   666
 5    1.381E+01   55   9    1.081E+00    4.467E-02    19.7   662
 5    1.854E+00   55   9    1.552E-01    6.989E-03    20.7   663
 5    9.410E-01   55  13    7.922E-02    4.535E-03    21.7   662
 5    2.615E-01   23  33    6.413E-02    3.568E-03    22.7   661
RESCAL. L=5, RMAX= 4.717E-01, RL2= 6.442E-02
PESCAL. L=4, RMAX= 1.569E-01, RL2= 5.730E-02
RESCAL. L=3, RMAX= 7.464E-02, RL2= 4.028E-02
RESCAL. L=4, RMAX= 8.033E-02, RL2= 2.405E-02
RESCAL. L=3, RMAX= 2.327E-02, RL2= 1.312E-02
 5    1.375E+01   57   8    1.451E+00    1.298E-01    29.1   700
 5    1.107E+01   58   4    7.928E-01    6.186E-02    30.1   707
 5    2.089E+00   58   4    1.259E-01    1.617E-02    31.1   708
 5    3.088E-01   58   6    4.061E-02    2.903E-03    32.1   708
 5    4.475E-01   56  13    4.273E-02    2.210E-03    33.1   708
 5    3.364E-01   55  25    3.696E-02    1.819E-03    34.1   707
RESCAL. L=5, RMAX= 1.494E-01, RL2= 2.505E-02
RESCAL. L=4, RMAX= 6.038E-02, RL2= 1.865E-02
```

Figure 7. MG Solution of Parabolic-Arc Airfoil. $M_{\infty} = 0.95$, $\tau = 0.1$, 64 x 32 Stretched Grid. (Continued)

```
RESCAL. L=3, RMAX= 1.944E-02, RL2= 1.001E-02
 5   2.009E+00  57  9   2.567E-01   1.259E-02   38.1  703
 5   1.752E+00  58  8   9.614E-02   5.882E-03   39.1  701
 5   1.873E-01  57  8   2.712E-02   1.405E-03   40.1  701
 5   1.456E-01  28  2   2.310E-02   9.147E-04   41.1  701
RESCAL. L=5, RMAX= 5.464E-02, RL2= 1.357E-02
RESCAL. L=4, RMAX= 1.867E-02, RL2= 8.053E-03
RESCAL. L=3, RMAX= 1.311E-02, RL2= 6.848E-03
 5   1.396E+00  57  8   1.621E-01   8.903E-03   45.5  703
 5   6.808E-01  59  4   4.034E-02   3.074E-03   46.5  706
 5   8.706E-01  58  4   4.040E-02   5.677E-03   47.5  705
 5   1.283E-01  58  4   1.715E-02   7.932E-04   48.5  705
 5   1.081E-01  28  2   1.541E-02   1.099E-03   49.5  706
RESCAL. L=5, RMAX= 3.900E-02, RL2= 8.531E-03
RESCAL. L=4, RMAX= 1.446E-02, RL2= 6.519E-03
RESCAL. L=3, RMAX= 9.752E-03, RL2= 5.055E-03
 5   5.614E-01  54 33   8.403E-02   3.645E-03   53.8  709
 5   2.291E-01  56 25   2.407E-02   9.221E-04   54.8  709
 5   9.146E-02  27  2   1.327E-02   4.715E-04   55.8  709
 5   8.569E-02  27  2   1.183E-02   4.428E-04   56.8  709
RESCAL. L=5, RMAX= 2.281E-02, RL2= 5.526E-03
RESCAL. L=4, RMAX= 9.945E-03, RL2= 4.334E-03
RESCAL. L=3, RMAX= 6.692E-03, RL2= 3.509E-03
 5   7.281E-01  55 33   1.049E-01   5.649E-03   61.6  719
 5   1.462E-01  57  9   1.481E-02   6.466E-04   62.6  718
 5   7.145E-02  27  2   1.128E-02   5.998E-04   63.6  718
RESCAL. L=5, RMAX= 2.158E-02, RL2= 4.738E-03
RESCAL. L=4, RMAX= 6.124E-03, RL2= 3.065E-03
RESCAL. L=3, RMAX= 4.357E-03, RL2= 2.242E-03
 5   1.879E-01  57  9   2.522E-02   1.427E-03   67.8  719
 5   6.357E-02  27  2   8.833E-03   3.249E-04   68.8  718
 5   5.932E-02  27  2   7.698E-03   3.037E-04   69.8  718
RESCAL. L=5, RMAX= 1.193E-02, RL2= 2.734E-03
RESCAL. L=4, RMAX= 4.773E-03, RL2= 1.924E-03
RESCAL. L=3, RMAX= 3.028E-03, RL2= 1.610E-03
 5   3.033E-01  57  9   5.058E-02   2.332E-03   74.6  719
 5   7.834E-02  57  9   8.555E-03   3.260E-04   75.6  718
 5   4.857E-02  27  2   6.385E-03   2.479E-04   76.6  718
RESCAL. L=5, RMAX= 8.888E-03, RL2= 2.191E-03
RESCAL. L=4, RMAX= 3.680E-03, RL2= 1.485E-03
RESCAL. L=3, RMAX= 2.341E-03, RL2= 1.244E-03
 5   2.257E-01  57  9   3.685E-02   1.609E-03   81.1  719
 5   5.824E-02  57  9   6.729E-03   2.412E-04   82.1  719
 5   3.974E-02  27  2   5.176E-03   2.027E-04   83.1  719
RESCAL. L=5, RMAX= 7.538E-03, RL2= 1.813E-03
RESCAL. L=4, RMAX= 2.996E-03, RL2= 1.191E-03
RESCAL. L=3, RMAX= 1.819E-03, RL2= 9.642E-04
 5   1.457E-01  57  9   2.421E-02   1.034E-03   87.7  720
 5   3.728E-02  57  9   5.112E-03   1.766E-04   88.7  720
 5   3.238E-02  27  2   4.148E-03   1.650E-04   89.7  720
RESCAL. L=5, RMAX= 6.119E-03, RL2= 1.426E-03
RESCAL. L=4, RMAX= 2.435E-03, RL2= 9.612E-04
RESCAL. L=3, RMAX= 1.465E-03, RL2= 7.765E-04
```

Figure 7. (Continued)

```
RESCAL. L=4, RMAX= 1.969E-03, RL2= 7.824E-04
RESCAL. L=3, RMAX= 1.181E-03, RL2= 6.246E-04
 5    3.651E-02  57  9    1.449E-02    6.139E-04  100.7  720
 5    2.300E-02  26  2    3.269E-03    1.165E-04  101.7  720
 5    2.141E-02  26  2    2.710E-03    1.088E-04  102.7  720
RESCAL. L=5, RMAX= 4.034E-03, RL2= 9.262E-04
RESCAL. L=4, RMAX= 1.593E-03, RL2= 6.344E-04
RESCAL. L=3, RMAX= 9.540E-04, RL2= 5.041E-04
 5    6.792E-02  57  9    1.133E-02    4.760E-04  107.3  720
 5    1.963E-02  26  2    2.622E-03    9.453E-05  108.3  720
 5    1.733E-02  26  2    2.187E-03    8.824E-05  109.3  720
RESCAL. L=5, RMAX= 3.269E-03, RL2= 7.460E-04
RESCAL. L=4, RMAX= 1.286E-03, RL2= 5.146E-04
RESCAL. L=3, RMAX= 7.709E-04, RL2= 4.067E-04
RESCAL. L=3, RMAX= 1.113E-03, RL2= 4.737E-04
 5    9.352E-02  57  9    1.658E-02    7.312E-04  114.3  720
 5    2.239E-02  57  9    2.619E-03    9.790E-05  115.3  720
 5    1.411E-02  26  2    1.920E-03    7.158E-05  116.3  720
RESCAL. L=5, RMAX= 3.647E-03, RL2= 7.748E-04
RESCAL. L=4, RMAX= 1.256E-03, RL2= 5.132E-04
RESCAL. L=3, RMAX= 5.571E-04, RL2= 2.927E-04
 5    2.460E-02  56 33    4.848E-03    2.291E-04  120.2  720
 5    1.231E-02  26  2    1.557E-03    6.230E-05  121.2  720
 5    1.146E-02  26  2    1.390E-03    5.818E-05  122.2  720
RESCAL. L=5, RMAX= 2.354E-03, RL2= 4.517E-04
RESCAL. L=4, RMAX= 8.740E-04, RL2= 3.136E-04
RESCAL. L=3, RMAX= 4.934E-04, RL2= 2.688E-04
 5    6.133E-02  57  9    1.048E-02    4.774E-04  127.0  720
 5    3.962E-01  56 14    1.748E-02    6.610E-04  128.0  721
 5    4.219E-01  56 14    1.839E-02    7.587E-04  129.0  720
 5    3.143E-02  56 14    1.618E-03    5.677E-05  130.0  720
 5    3.927E-01  56 14    1.734E-02    6.599E-04  131.0  721
 5    4.229E-01  56 14    1.840E-02    7.569E-04  132.0  720
 5    3.114E-02  56 14    1.502E-03    5.770E-05  133.0  720
 5    3.944E-01  56 14    1.740E-02    6.642E-04  134.0  721
 5    4.245E-01  56 14    1.847E-02    7.585E-04  135.0  720
 5    3.147E-02  56 14    1.457E-03    5.953E-05  136.0  720
 5    5.648E-03  26  2    8.710E-04    2.913E-05  137.0  720
 5    3.923E-01  56 14    1.737E-02    6.594E-04  138.0  721
 5    4.251E-01  56 14    1.849E-02    7.597E-04  139.0  720
 5    3.150E-02  56 14    1.412E-03    6.021E-05  140.0  720
 5    4.627E-03  26  2    8.095E-04    2.400E-05  141.0  720
 5    4.418E-03  26  2    7.791E-04    2.295E-05  142.0  720
RESCAL. L=5, RMAX= 2.993E-03, RL2= 6.473E-04
RESCAL. L=4, RMAX= 1.417E-03, RL2= 5.199E-04
RESCAL. L=3, RMAX= 8.070E-04, RL2= 4.333E-04
 5    3.601E-01  56 14    2.067E-02    6.375E-04  146.8  721
 5    2.046E-02  57  9    1.685E-03    7.053E-05  147.8  721
 5    5.232E-03  58  7    8.345E-04    4.539E-05  148.8  721
 5    3.538E-03  26  2    7.334E-04    3.089E-05  149.8  721
RESCAL. L=5, RMAX= 2.582E-03, RL2= 5.282E-04
RESCAL. L=4, RMAX= 1.042E-03, RL2= 4.074E-04
RESCAL. L=3, RMAX= 4.392E-04, RL2= 2.205E-04
 5    2.770E-02  57  9    5.012E-03    2.303E-04  153.7  721
 5    6.247E-03  57  9    6.371E-04    3.000E-05  154.7  721
 5    3.060E-03  26  2    4.179E-04    1.558E-05  155.7  721


                  PLOT OF CPBAR FOR LEVEL 5
         I
                      CPSTAR=-9.004E-02

               II     *INTERRUPTED*
```

Figure 7. (Concluded)

CPSTAR=-9.004E-02

I	XX	CP
2	-3.346	.0209
3	-2.774	.0332
4	-2.433	.0444
5	-2.187	.0548
6	-1.993	.0647
7	-1.832	.0744
8	-1.694	.0841
9	-1.572	.0940
10	-1.462	.1041
11	-1.362	.1147
12	-1.270	.1258
13	-1.184	.1378
14	-1.104	.1507
15	-1.028	.1650
16	-.956	.1811
17	-.887	.1995
18	-.821	.2212
19	-.758	.2477
20	-.696	.2816
21	-.637	.3288
22	-.579	.4055
23	-.522	.5629
24	-.467	.5193
25	-.413	.2761
26	-.359	.1276
27	-.307	.0172
28	-.254	-.0855
29	-.203	-.1816
30	-.152	-.2582
31	-.101	-.3234
32	-.050	-.3834
33	0.000	-.4399
34	.050	-.4939
35	.101	-.5458
36	.152	-.5959
37	.203	-.6447
38	.254	-.6923
39	.307	-.7391
40	.359	-.7852
41	.413	-.8308
42	.467	-.6319
43	.522	-.3765
44	.579	-.3424
45	.637	-.3425
46	.696	-.3351
47	.758	-.3272
48	.821	-.3202
49	.887	-.3137
50	.956	-.3075
51	1.028	-.3015
52	1.104	-.2956
53	1.184	-.2895

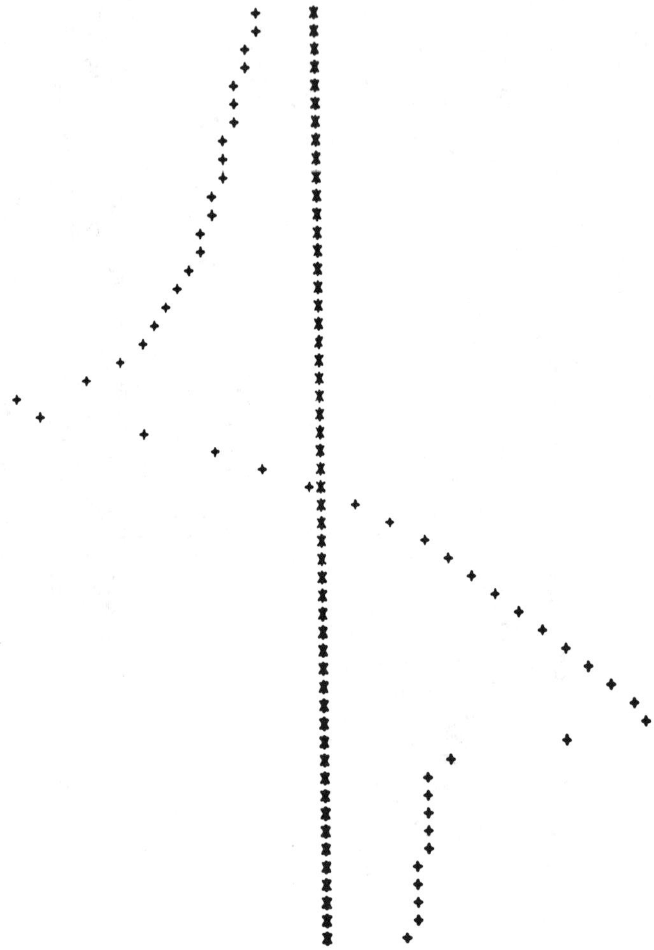

Figure 8. Pressure Distribution Along y=0 for Solution of Figure 7. (Continued)

```
54  1.270  -.2834                            *      +
55  1.362  -.1238                            *+
56  1.462   .0701                     +      *
57  1.572   .1197                   +        *
58  1.694   .1266                   +        *
59  1.832   .1309                   +        *
60  1.993   .1317                   +        *
61  2.187   .1283                   +        *
62  2.433   .1196                   +        *
63  2.774   .1039                    +       *
64  3.346   .0803                     +      *

CHART OF CIP-P *INTERRUPTED*
```

Figure 8. (Concluded)

shock at the trailing edge and the normal shock in the wake is clearly evident.

A summary of all these results is shown in Figure 11.

CONCLUDING REMARKS

The multigrid (MG) method for accelerating relaxation calculations has been shown to be applicable to transonic flow with embedded shock waves. In this paper, vertical line relaxation was used for solving the nonlinear, conservative difference equation modelling the small-disturbance equation for the velocity potential. The multigrid approach appears to work about three to five times faster than optimal SLOR on unstretched grids of moderate size (64 x 32). The relative advantage of MG to SLOR increases as the grid gets finer, since the MG convergence rate is nearly independent of mesh size.

On stretched grids, the present MG method slows down, being only about twice as fast as SLOR. It is felt that the reason for this is clear, the indicated remedy being alternating-direction relaxation sweeps.

	86	36	S7	87	87	87	88	88	88	89	89	90	90	91	92	94	
14	85	96	86	86	86	86	87	87	88	88	88	89	89	90	91	92	94
15	85	85	85	85	85	86	86	87	87	88	88	89	89	90	91	92	94
16	84	84	84	84	85	85	86	86	87	87	88	89	89	90	91	92	94
17	82	82	83	83	84	85	85	86	87	87	88	89	89	90	91	92	94
18	81	81	82	82	83	84	85	86	87	87	88	89	89	90	91	92	94
19	79	79	80	81	83	84	85	86	86	87	88	89	90	90	91	92	94
20	77	77	79	81	82	83	85	86	86	87	88	89	90	90	91	92	94
21	73	75	78	80	82	83	85	86	87	88	88	89	90	91	91	92	94
22	68	72	76	79	82	83	85	86	87	88	89	89	90	91	92	93	94
23	54	70	76	80	82	84	85	86	87	88	89	90	90	91	92	93	94
24	53	72	78	81	83	85	86	87	88	89	89	90	91	91	92	93	94
25	77	78	81	83	85	86	87	88	89	89	90	91	91	92	92	93	94
26	87	85	85	86	87	87	88	89	89	90	91	91	92	92	93	93	94
27	94	90	89	89	89	89	90	90	90	91	91	92	92	93	93	94	94
28	99	95	93	92	91	91	91	91	92	92	92	92	93	93	93	94	94
29	104	101	98	96	94	94	93	93	93	93	93	93	93	93	94	94	94
30	108	105	102	99	97	96	95	95	94	94	94	94	94	94	94	94	94
31	111	108	105	103	101	99	97	96	96	95	95	95	95	95	95	95	94
32	114	111	108	105	103	101	100	99	97	97	96	96	95	95	95	95	94
33	117	113	110	108	105	104	102	101	99	98	97	97	96	96	96	95	94
34	119	116	113	110	108	106	104	102	101	100	99	98	97	97	96	96	94
35	122	118	115	112	110	108	106	104	103	101	100	99	98	97	97	96	94
36	124	121	117	114	112	110	108	106	104	103	102	100	99	98	97	96	94
37	126	123	120	117	114	112	109	107	106	104	103	102	100	99	98	97	94
38	128	125	122	119	116	113	111	109	107	106	104	103	101	100	99	97	94
39	130	127	124	121	118	115	113	111	109	107	105	104	102	101	100	98	94
40	132	129	126	123	120	117	115	112	110	108	107	105	103	102	100	98	94
41	134	131	127	125	122	119	117	114	112	110	108	106	104	103	101	99	94
42	125	130	129	126	124	121	118	116	114	111	109	107	105	104	102	100	94
43	114	126	129	127	125	123	120	118	115	113	111	109	107	105	103	100	94
44	112	119	126	127	126	124	122	119	117	114	112	110	108	106	103	101	94
45	112	114	121	125	126	125	123	121	118	116	114	111	109	107	104	101	94
46	112	112	116	122	125	126	124	122	120	118	115	113	110	108	105	102	94
47	112	112	113	117	122	124	125	123	121	119	117	114	111	109	106	103	94
48	111	111	112	114	118	122	124	124	123	120	118	115	113	110	107	103	94
49	111	111	111	112	114	118	122	123	123	122	119	117	114	111	108	104	94
50	111	111	111	111	112	115	119	122	123	122	121	118	116	113	109	105	94
51	110	110	110	110	111	112	115	119	121	122	121	120	117	114	110	106	94
52	110	110	110	110	110	111	112	115	119	121	122	120	118	115	112	107	94
53	110	110	110	110	110	110	111	112	115	119	121	121	119	117	113	108	94
54	109	109	109	109	110	110	110	111	112	115	118	120	120	118	114	109	94
55	101	101	102	102	102	103	103	105	107	110	114	118	119	118	115	110	94
56	90	90	91	91	91	92	92	93	94	97	101	108	115	118	116	111	94
57	87	87	87	87	87	87	87	87	86	86	87	91	98	107	114	112	94
58	87	87	87	87	87	87	86	86	86	85	84	83	83	89	97	103	94
59	87	87	87	87	87	86	86	86	86	85	85	84	83	82	83	90	94
60	87	87	87	87	87	86	86	86	86	86	85	85	84	84	84	86	94
61	87	87	87	87	87	87	87	87	86	86	86	86	86	86	86	87	94
62	87	87	87	87	87	87	87	87	87	87	87	87	87	87	88	89	94
63	88	88	88	88	88	88	88	88	88	88	89	89	89	89	89	90	94
64	90	90	90	90	90	90	90	90	90	90	90	90	90	91	91	91	94
	33	31	29	27	25	23	21	19	17	15	13	11	9	7	5	3	1

I *INTERRUPTED*

Figure 9. Mach Number Chart for Solution of Figure 7.

Future investigations will include the alternating sweeps, and the extension of the method to lifting flows.

```
 2   PPPPPPPPPPPPPPPPPPPPPPPPPPPPMMMM
 3   PPPPPPPPPPPPPPPPPPPPPPPPPPPPPPPM
 4   PPPPPPPPPPPPPPPPPPPPPPPPPPPPPPPM
 5   PPPPPPPPPPPPPPPPPPPPPPPPPPPPPPPM
 6   PPPPPPPPPPPPPPPPPPPPPPPPPPPPPPPM
 7   PPPPPPPPPPPPPPPPPPPPPPPPPPPPPPPM
 8   PPPPPPPPPPPPPPPPPPPPPPPPPPPPPPPM
 9   PPPPPPPPPPPPPPPPPPPPPPPPPPPPPPPM
10   PPPPPPPPPPPPPPPPPPPPPPPPPPPPPPPM
11   XXXXXXXXXXPPPPPPPPPPPPPPPPPPPPPM
12   XXXXXXXXXXXXXXPPPPPPPPPPPPPPPPPM
13   XXXXXXXXXXXXXXXXXPPPPPPPPPPPPPPM
14   XXXXXXXXXXXXXXXXXXPPPPPPPPPPPPPM
15   XXXXXXXXXXXXXXXXXXXPPPPPPPPPPPPM
16   XXXXXXXXXXXXXXXXXXXPPPPPPPPPPPPM
17   DDDDDXXXXXXXXXXXXXXXPPPPPPPPPPPM
18   DDDDDDDDXXXXXXXXXXXXPPPPPPPPPPPM
19   DDDDDDDDDXXXXXXXXXXXPPPPPPPPPPPM
20   LLLLDDDDDXXXXXXXXXPPPPPPPPPPPPPM
21   CCLLLDDDDDXXXXXXXXXPPPPPPPPPPPPM
22   STCLLLDDDDXXXXXXXXPPPPPPPPPPPPPM
23   SSTCLLDDDXXXXXXXXPPPPPPPPPPPPPPM
24   SSCLLDDDXXXXXXXXPPPPPPPPPPPPPPPM
25   LLLDDDDXXXXXXXPPPPPPPPPPPPPPPPPM
26   XXXXXXXXXXXXXPPPPPPPPPPPPPPPPPMM
27   PPPPPPPPPPPPPPPPPPPPPPPPPPPPPPMM
28   MMMMPPPPPPPPPPPPPPPPPPPPPPPPPMMM
29   000MMMMMMPPPPPPPPPPPPPPPPPPMMMMM
30   110000MMMMMMMMMMMMMMMMMMMMMMMMMM
31   1111000000MMMMMMMMMMMMMMMMMMMMMM
32   2211110000000MMMMMMMMMMMMMMMMMMM
33   2221111100000000MMMMMMMMMMMMMMMM
34   3222211111000000000MMMMMMMMMMMMM
35   3322222111111000000000MMMMMMMMMM
36   3333222211111110000000000MMMMMMMM
37   43333222221111111000000000MMMMMM
38   4433333222221111110000000000MMMMMM
39   44443333222222111111100000000MMMM
40   5444433333222222111111100000000MMMM
41   55444443333322222221111110000000MMM
42   44444443333322222221111110000000MM
43   2344444444333332222221111110000000MM
44   2233444444443333322222211111000000MM
45   2222334444444333333222222111110000MM
46   22222333334444333333222222111110000M
47   11122223333333333333222221111000M
48   1111122223333333333332222211111000M
49   111111122222333333333332222211100M
50   1111111122222333333333332222211100M
51   11111111112222233333333332222221110M
52   111111111111122222333333332222221110M
53   1111111111111111222223333333322222110M
54   11111111111111111122222233333222110M
55   00000000000000000111222223333222211M
56   PPPPPPPPPPPPPPPPPMMMM0011222222211M
57   XXXXXXXXXXXXXXXXXXXXXPPMM0112221M
58   XXXXXXXXXXXXXXXXXXXXXXXDXXPPM000M
59   XXXXXXXXXXXXXXXXXXXXXXXXXXDDDXXPPM
60   XXXXXXXXXXXXXXXXXXXXXXXXXXXXXXPM
61   XXXXXXXXXXXXXXXXXXXXXXXXXXXXXXPM
62   XXXXXXXXXXXXXXXXXXXXXXXXXXXXXPPM
63   PPPPPPPPPPPPPPPPPPPPPPPPPPPPPPPM
64   PPPPPPPPPPPPPPPPPPPPPPPPPPPPPPPM

   5   2.693E-03   26   2    3.638E-04    1.378E-05     2.0   721
 1$Z
 0TK       = .45
```

Figure 10. Isobar Chart for Solution of Figure 7.

SUMMARY OF MULTIGRID RESULTS
64 X 32 CELLS

Problem Description		*Effective Spectral Radius	
		MG	SLOR
Unstretched Grid	Laplace's Eq., Smooth B.C.'s	.583 (.46 Combined Levels)	.924
	Parabolic Airfoil, M_∞ = .70	.549	.868
	" " " .85	.593	.855
	" " " .95	.858	.957
Stretched Grid	" " " .95	.936	.974

*Eff. Spec. Rad. = (Final Error/Initial Error) $^{1/(\text{Work Units})}$

Figure 11. Summary of Multigrid Results

ACKNOWLEDGEMENTS

During the course of this work, Professor Antony Jameson of the Courant Institute of Mathematical Sciences, New York University, also carried out research on the multigrid method. He showed independently that the "Full Approximation" approach would work. Our many discussions have been beneficial.

This research, partially supported by NASA Grant NGR-47-102-001, was initiated while Dr. Brandt was visiting ICASE (Institute for Computer Applications in Science and Engineering) at Langley Research Center.

APPENDIX

Iteration Coefficients

We have used various choices for iteration coefficients in Eq. (13). The coefficients used to make the calculations presented in this paper are simply based on the Newton linearization of Eqs. (8), (9), and (11). They are as follows:

First define: (dropping the j index, since all quantities are evaluated at the same j)

$$b_{i+\frac{1}{2}} = [K - (\gamma+1)M_{\infty}^{2} \frac{\phi_{i+1} - \phi_{i})}{\Delta x}] \Delta x^{-2} \tag{A1}$$

Then we have

$$\overline{U}_i = \frac{1}{2}(b_{i+\frac{1}{2}} + b_{i-\frac{1}{2}}) = U_i \Delta x^{-2} \tag{A2}$$

$$A = C = -\Delta y^{-2} \tag{A3}$$

$$B = 2\Delta y^{-2} + 2(1-\mu_i)\overline{U}_i/\omega - \mu_{i-1} b_{i-\frac{1}{2}} \tag{A4}$$

$$D = (1-\mu_i)\, b_{i-\frac{1}{2}} - 2\mu_{i-1}\overline{U}_{i-1} \tag{A5}$$

$$E = \mu_{i-1}\, b_{i-\frac{3}{2}} \tag{A6}$$

where

$$\begin{aligned} \mu_i &= 0 \quad \text{if } U_i > 0 \\ &= 1 \quad \text{if } U_i \leq 0 \end{aligned} \tag{A7}$$

REFERENCES

[1] Fedorenko, R. P. (1962), A Relaxation Method for Solving Elliptic Difference Equations. USSR Computational Mathematics and Mathematical Physics, vol. 1, pp. 1092-1096.

[2] Fedorenko, R. P. (1964), The Speed of Convergence of One Iterative Process. USSR Computational Mathematics and Mathematical Physics, 4:3, pp. 227-235.

[3] Bakhvalov, N. S. (1966), On the Convergence of a Relaxation Method with Natural Constraints on the Elliptic Operator. USSR Computational Mathematics and Mathematical Physics, 6:5 , pp. 101-135.

[4] Brandt, Achi (1972), Multi-Level Adaptive Technique (MLAT) for Fast Numerical Solution to Boundary Value Problems. Proceedings of the Third International Conference on Numerical Methods in Fluid Mechanics, vol. 1, pp. 82-89.

[5] Brandt, Achi: Multi-Level Adaptive Techniques (MLAT). Part I. The Multi-Grid Method. IBM Report to be published.

[6] Murman, E. M. (1973), Analysis of Embedded Shock Waves Calculated by Relaxation Methods. Proceedings of AIAA Computational Fluid Dynamics Conference, Palm Springs, California, July 19-20, pp. 27-40.

[7] Jameson, Antony (1975), Transonic Potential Flow Calculations Using Conservation Form. Proceedings AIAA 2nd Computational Fluid Dynamics Conference, Hartford, Connecticut, June 19-20, pp. 148-161.

DISCUSSION OF THE SOUTH-BRANDT PAPER

CASPER: You use a successive line overrelaxation, so you're using a double ordered technique, I guess. If you use a semi-direct method where the linearized problem was solved directly, would you still expect the same sort of behavior that the relaxation iteration would tail off as it does here?

SOUTH: Well, semi-direct in the relaxation method - I guess you mean the fast solver; that's not a relaxation method.

CASPER: Well, all right. The question I'm asking is: Is this behavior that you're showing a part of the linear iteration or the nonlinear iteration? If you're using a direct method to solve the linearized problem, would this Fedorenko method be attractive?

SOUTH: Yes, when you come up to the operation count; I guess Tony Jameson can answer that better than I. You pay a price for using a direct solution. You have a rather high operation count, and, if you look at the asymptotic count, if you have an n x n square mesh, the fast solver costs you roughly $n^2 \log n$ operations whereas the advertized operation count is asymptotically just n^2 - you don't have a log n on the Fedorenko method. But the question is, for reasonable sized grids, is log n bigger than that big constant on the Fedorenko method? But, anyway, it is competitive with the fast solver; Tony Jameson has used it and thinks that they're about the same. And the multigrid is more flexible.

OLIVER: I have two very short but precise questions. One: Is there an either practical or precise optimal criterion for when to cut at one grid and go to the next? The second question is: By stretch, do you mean nonuniform spacing in one direction, or where the aspect ratio of the grid is different - where Δx, say, does not equal Δy?

SOUTH: The problems we had were in the completely stretched grids and stretched to infinity in all directions, for the second question. But you can simulate the same effect just by choosing a constant mesh, but a weird aspect ratio. There are just certain points of the flow that just don't damp rapidly. The long wavelengths don't damp rapidly like you need to for this to work. The first question was, when do you back up, right? It's not too sensitive to the choice, but you can measure pretty well with it slowing down, like if the current residual - maximum residual - is greater than 7/10 of the previous calculation, then you say, back up to the coarser grid - that kind of thing.

DIXON: What do you do in the transonic calculation when your fine grid has a fairly sharply defined shock and you want to go back up to the coarser grid which would seem to smear it out?

SOUTH: It's no problem, because you're not really solving a different function on the coarse grid, you're using the coarse grid essentially to solve for the error in the problem and eventually the error on all grids goes to zero. That's part of the principle of the method.

MORETTI: That seems to be the beauty of the method.

SOUTH: Yes.

THOMPSON: Doesn't your error function ever shift sign?

SOUTH: It satisfies a mixed type equation. Wait a minute, do you want to know, did the error <u>function</u> switch sign? I'm sure it does.

THOMPSON: When it switches sign, then, don't you get into problems with that, if you're coming at it from above or coming at it from below as you shift your grid? Isn't that going to throw the whole thing off?

SOUTH: No.

APPLICATION OF TIME-DEPENDENT FINITE VOLUME METHOD TO TRANSONIC FLOW IN LARGE TURBINES

C. L. FARN AND D. K. WHIRLOW

Westinghouse Research Laboratories

ABSTRACT

A discussion is given of our experience with the time-dependent finite volume technique in the calculation of transonic flows in turbines. Comparisons are made with other techniques, and a number of examples are given for flows in the blade-to-blade surface and the hub-to-shroud surface, and for the unsteady flow generated by upstream wakes. The examples given serve to demonstrate the power and versatility of the method.

1. INTRODUCTION

The flow field in axial flow turbomachines represents one of the most challenging tasks for aerodynamic analysts because it is three dimensional and unsteady in nature. In practice, these difficulties are bypassed by solving flow fields on the blade-to-blade and hub-to-shroud surfaces separately. However, for large turbines used in central power stations, to obtain numerical solutions for flows on these two surfaces is by no means a simple task because of the following three factors. First, the volume of the fluid expands tremendously as it travels from the inlet of the turbine to the exit. A volumetric expansion ratio of 1:1000 is not uncommon for large steam turbines. Consequently, the hub and shroud of a typical low-pressure turbine are highly flared. Second, in order to generate as much power as possible per turbine, nearly all of the modern low-pressure turbines are designed to operate in the fully choked transonic (subsonic-supersonic) flow region. Third, with the increasingly high cost of fuel, the available energy of the working fluid must be utilized to its full extent. As a result, the exhaust pressure at the low-pressure end of a large steam turbine plant is very low (say 1 psia) and the exhaust volumetric flow is extremely large (say 2×10^8 ft^3/hr). In order to accommodate such a large volumetric flow, the blades of the last rotating row are very long and the blade profile varies from the impulse type at the root to the reaction type at the tip.

For blade-to-blade problems, the existing methods can be classified into four categories according to their formulations and numerical approximations. They are (a) steady-state formulation and streamline curvature method [1]-[3], (b) steady-state formulation and combined method [4], [5], (c) time-dependent, differential formulation and finite difference approximation [6], [7], and (d) time-dependent, integral formulation and finite volume approximation [8].

Great numerical convergence difficulties were experienced in applying Method (a) to the fully-choked transonic turbine cascades and the agreement between theoretical and experimental results deteriorated in the supersonic flow region. The method works nicely for impulse blades with convergent passages in which the flow is well guided and the supersonic region is relatively small. For reaction blades and nozzles, the exit velocity is well into the supersonic region and the jet deflection becomes very significant. Obviously, this method is not adequate. Furthermore, solutions obtained by Method (a) correspond to the limiting-load condition, the flow expands fully in the blade passages and all shocks are expelled out of the passages. For partially loaded blades, it is not clear how the shocks can be computed in this method.

In an attempt to improve the accuracy of Method (a) in the supersonic region, Wilkinson [4] replaced the streamline curvature solution in this region by the characteristics solution. However, the difficulties in numerical convergence still remained. Decuypere [5] seemed to overcome both convergence and accuracy problems by dividing the flow field into three regions according to the local Mach number ($M < 1$, ≈ 1 and > 1) and then computing flow fields in these regions simultaneously. The streamline curvature method was used in the subsonic region. In the sonic and supersonic regions, power series and characteristics methods were adopted respectively. This scheme is very rigorous from the mathematical point of view. However, in order to apply the scheme to a given blade passage, one has first to guess the three flow regions and join the solutions computed by the aforementioned methods on the common boundaries of these flow regions. It is evident that the process is too tedious to be practical. Furthermore, because of the use of method of characteristics, it is doubtful that Method (b) can be applied to the flow field on an arbitrary blade-to-blade surface of revolution with variable stream channel height.

The shortcomings associated with Methods (a) and (b) can be avoided by using a time-dependent formulation. Two different schemes exist in the open literature. One, Method (c), is based on the governing equations in differential form and they are approximated by finite difference equations. The other, Method (d), utilizes the governing equations in integral form and the finite volume approximation is used. In both schemes, the approximate governing equations are solved numerically by the time-marching techniques. Methods (c) and (d) can be applied to an arbitrary blade-to-blade surface of revolution with variable stream channel height. In addition, they are capable of handling both limiting and partial load cases. The only major difficulty for these two methods is the numerical instability associated with the different time-marching techniques. This difficulty is generally overcome by using artificial viscosity or numerical damping. The artificial viscosity or numerical damping has little effect on the accuracy of the theoretical pressure distributions on the blade surfaces for limiting load cases (see Figures 2 and 3). For partial load cases, the oblique shocks in blade passages are somewhat smeared (see Figure 4). However, the errors in the theoretical pressure ahead and behind the shock seem to cancel each other and do not significantly affect the prediction of the overall blade loadings, at least, for turbine cascades. Comparatively, Method (c) is superior in the numerical stability, while Method (d) is easier in coding and takes less computation time per time step. In addition, blades with very complex geometry may be too difficult to handle for Method (c).

Now, let us turn our attention to the hub-to-shroud problems. All of the published research works in this area are based on the steady-state formulation. However, different flow models were used in the formulation. Roughly speaking, they can be divided into three groups. They are streamwise vorticity model [11], [13], circumferentially average flow model [10], [16] and S2 surface flow model [15]. The first two models are self-explanatory, and the detailed description of the last model may be found in [9].

Traditionally, different schemes for solving hub-to-shroud problems are classified into three categories according to the numerical methods used in computing the solutions. They are (e) streamline curvature method [10]-[14], (f) matrix through flow method [15], [16], and (g) finite element method [17]. Methods (e) and (f) are very familiar to many industrial users, and relative merits of these two methods were discussed extensively in [18] through [20]. It is worth noting that Method (f) can be used only for flow field computations while Method (e) may be used for both flow field and row by row computations. When it is used for row by row computations, only one grid line (or computation station) is needed in each of the free spaces between two adjacent blade rows and no grid lines are required within the blade rows [12], [14]. Application of the finite element method to the hub-to-shroud problems is a relatively new development. Judging from the results reported in [17], Method (g) seems to yield excellent solutions for a partially choked transonic compressor (single stage).

Of all the references [10]-[20] discussed in the previous paragraph, only two investigated fully-choked transonic flows [12] and [14]. In both of these studies, Method (e) was used to compute row by row solutions for large low-pressure steam turbines. Since no grid lines were assigned within the blade rows, the difficulties due to numerical convergence in the sonic region were artificially circumvented. The price for this circumvention, in our opinion, is considerable. In fact, the effects of variation in blade geometry can only be accounted for, in the row by row solutions, by empirical correlations. As correctly pointed out by Davis and Millar [20], for computation of fully-choked transonic flow fields, methods using the time-dependent equations may make both streamline curvature and matrix through flow methods obsolete.

2. GOVERNING EQUATIONS FOR TIME-DEPENDENT FINITE VOLUME FORMULATION

We, at Westinghouse, are interested in finding a method which can be applied both to blade-to-blade and hub-to-shroud problems for large power station turbines. It must be capable of overcoming three difficulty factors depicted in the first paragraph of Section 1. Subsequent discussion in the same section indicates that Method (d), which is based on the time-dependent, integral formulation and the finite volume approximation, is best suited for the purpose.

The governing equations used in Method (d) are given as follows,

$$V \frac{\partial \rho}{\partial t} + \int_A \rho \, (\vec{w} \cdot \hat{n}) \, dA = 0 \tag{1}$$

$$V \frac{\partial}{\partial t} (\rho v_z) + \int_A (\rho v_z)(\vec{w} \cdot \hat{n}) dA = - \int_A p(\hat{n} \cdot \hat{z}_k) dA \tag{2}$$

$$V\left[\frac{\partial}{\partial t} (\rho v_r) - \frac{\rho v_\theta^2}{r_k} \right] + \int_A (\rho v_r)(\vec{w} \cdot \hat{n}) dA = - \int_A p(\hat{n} \cdot \hat{r}_k) dA \tag{3}$$

$$V\left[\frac{\partial}{\partial t} (\rho v_\theta) + \rho v_r \left(\Omega + \frac{v_\theta}{r_k}\right)\right] + \int_A (\rho v_\theta)(\vec{w} \cdot \hat{n}) dA = - \int_A p(\hat{n} \cdot \hat{\theta}_k) dA \tag{4}$$

$$V \frac{\partial}{\partial t} (\rho e) + \int_A (\rho h_t)(\vec{w} \cdot \hat{n}) dA = - \int_A p(\vec{v}_A \cdot \hat{n}) dA \tag{5}$$

$$e = u + \frac{v^2}{2} \qquad\qquad h_t = e + \frac{p}{\rho} \qquad (6),(7)$$

$$u = \text{fn}(p, \rho) \tag{8}$$

The eight dependent variables to be solved from the above equations are ρ, v_z, v_r, v_θ, e, u, h_t, and p.

Equations (1) and (5) are the continuity and energy equations respectively. Equations (2) through (4) are three components of the momentum equation. Equation (6) is merely the definition of the specific total energy and Eq. (7), the definition of the specific total enthalpy. Equation (8) represents a property relation for the gas medium.

It is worth noting that, in the above formulation, the control volume V has a rigid surface with the area A and is rotating about the axis of the turbine under study at a constant angular speed Ω. For blade-to-blade problems, Ω is set equal to zero for stator blades and to the angular speed of the rotor for rotor blades. For hub-to-shroud problems, Ω is equal to zero everywhere except inside of the rotating rows where Ω is again set equal to the angular speed of the rotor. During the derivations of Eqs. (2) through (5), effects of heat transfer and viscosity were neglected.

3. APPLICATION

The governing equations given in Section 2 together with the appropriate boundary conditions are applied to three categories of transonic flow problems in turbomachinery. They are (a) flow field on the blade-to-blade surface, (b) flow field on the hub-to-shroud surface, and (c) interference between moving blade rows. For the first two categories, the final solutions are independent of time. For the last category, the final solution is periodic in time and its period is, of course, equal to that of the disturbance imposed upstream or downstream of the blade row under investigation.

3.1 Flow Field on the Blade-to-Blade Surface

Three blade profiles were analyzed by the time-dependent, finite volume method. They are designated by GTRB, STRB, and STSN. Profile GTRB represents a typical gas turbine rotor blade section and STRB, a typical steam turbine rotor blade section (reaction type). The pitch to chord ratios for GTRB and STRB are 0.71 and 0.90 respectively. These two profiles were tested by Sieverding [4] and the end walls of the cascade test sections were parallel to each other for both blades. Profile STSN is a typical steam turbine stator nozzle. Its pitch to chord ratio is equal to 0.50. Although the endwalls of the test section for STSN are parallel to each other in the inlet and exit regions, they form a divergent channel within the cascade. The half angle of the divergence is 7.5°.

Figures 1 and 2 show the theoretical and experimental blade surface pressure distributions for GTRB at the subsonic exit and limiting load conditions respectively. It is worth noting that the blade passage is not choked for the former case but is fully choked for the latter. For both cases, the agreement between theoretical and experimental results is excellent.

The theoretical blade surface pressure distribution for Profile STRB at the limiting load and STSN at a partial load are given in Figures 3 and 4 respectively. Also shown in these figures are the corresponding experimental data. In general, the agreement is very good. However, for Profile STSN, the theoretical pressure recovery across the oblique shock on the suction surface of the blade is somewhat smeared due to the numerical damping discussed in Section 1.

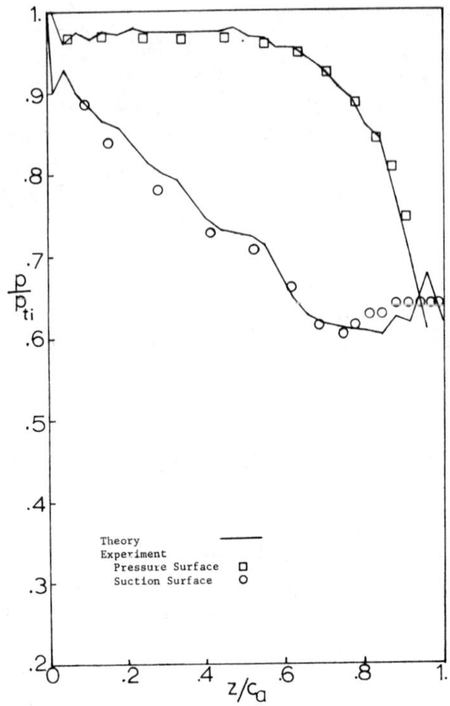

Figure 1. Comparison of Theoretical and Experimental Blade Surface Pressure Distributions, Blade Profile GTRB, $\beta_i = 30^o$, $p_e/p_{ti} = 0.681$.

3.2 Flow Field on the Hub-to-Shroud Surface

Two cases were studied by the present method. The first case is a fictitious stator row with cylindrical hub and shroud. The blades in the stator row are parallel symmetric airfoils and the stagger angle is zero. In fact, the passage on a given cylindrical surface formed by any two adjacent blades is a two-dimensional convergent-divergent nozzle. Thirty-four radial and eleven axial grid lines were used in computation. The No. 11, 15, and 23 radial grid lines coincide with the leading edge, throat, and trailing edge of the stator row, respectively. Numerical results corresponding to two pressure ratios (p_e/p_{ti} = 0.141 and 0.815) were obtained. Since the inlet flow is assumed to be axial and no geometrical factors would cause the flow to move in the radial and tangential directions, the numerical solution on any cylindrical

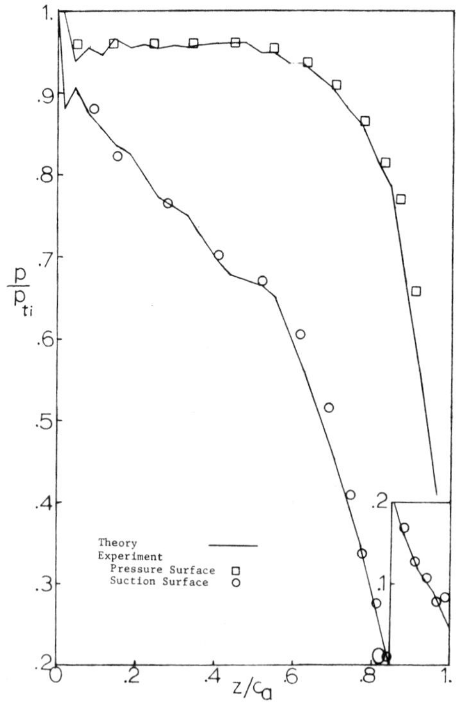

Figure 2. Comparison of Theoretical and Experimental Blade Surface Pressure Distributions, Blade Profile GTRB, $\beta_i = 30^o$, $p_e/p_{ti} = 0.120$.

surface should agree well with the theoretical solution computed according to one-dimensional (1-D) gasdynamics. Comparison of the numerical and 1-D theoretical static pressure distributions on the mid-span cylindrical surface is shown in Figure 5 for $p_e/p_{ti} = 0.141$ and in Figure 6 for $p_e/p_{ti} = 0.815$. For both cases, the agreement is excellent. It is worth noting that, although the exit velocity is subsonic for $p_e/p_{ti} = 0.815$, the stator row is fully choked. Furthermore, a strong normal shock exists in the divergent part of the nozzle and causes significant losses in total pressure. The exit to inlet total pressure ratio (p_{te}/p_{ti}) is equal to 0.929 according to the present method and 0.928 based on 1-D gasdynamics.

The second case analyzed by the present method is the last stage of a large low-pressure steam turbine. The hub and shroud

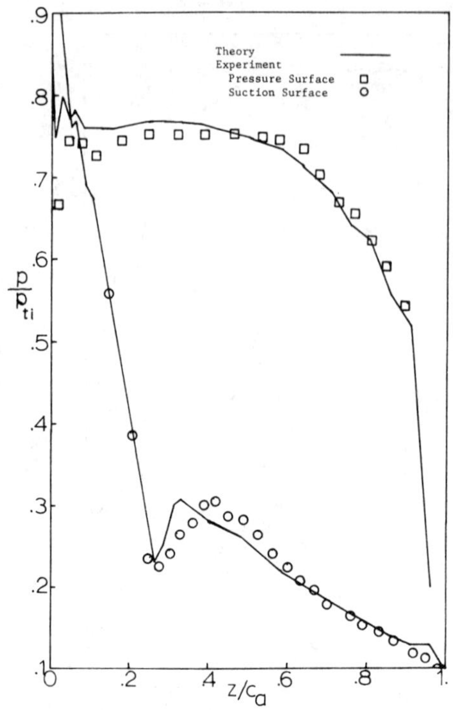

Figure 3. Comparison of Theoretical and Experimental Blade Surface Pressure Distributions, Blade Profile STRB, $\beta_i = -66^o$, $p_e/p_{ti} = 0.128$.

of the stage are highly flared and the blades of the rotating row are tapered and twisted. Figure 7 shows the theoretical static pressure distributions along the blade height at the inlet, exit of the stage and at the interstage. It is interesting to note that, as indicated by the numerical solutions within the blade rows, the rotating row is fully choked and the stationary row, partially choked.

3.3 Interference Between Moving Blade Rows

The time-dependent finite volume technique is also applicable to the problem of interaction between blade rows. For example, if the exit conditions of a stator can be measured by an intrastage probe, the unsteady flow, and the forces and moments, can be calculated for the following rotor row. The measurements can be

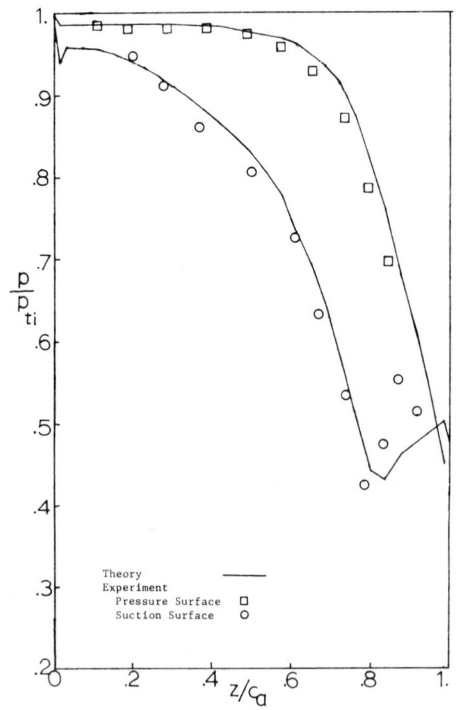

Figure 4. Comparison of Theoretical and Experimental Blade Surface Pressure Distributions, Blade Profile STSN, $\beta_i = 0^o$, $p_e/p_{ti} = 0.480$.

made on a stator cascade, but the result is only an approximation because the effect of the downstream rotor blade on the stator flow field is absent. Nonetheless, the approximation is useful because stator cascade data are available whereas intrastage measurements are very sparse. With this method, one is able to overcome some of the limitations which have been necessary in classical unsteady aerodynamics. Some of the principal features and limitations of the method are:

Features.

1. The flow is truly transonic and is not limited to small fluctuations about Mach one. The test case given below is for an inlet Mach number of approximately 0. 7 and an exit Mach number of 1. 18.

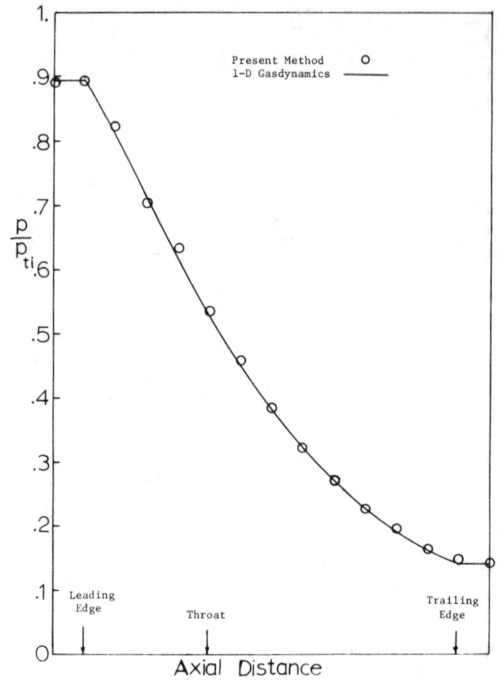

Figure 5. Theoretical Pressure Distributions in the Axial Direction at the Mid Span of a Fictitious Stator Row, $p_e/p_{ti} = 0.141$.

2. The upstream disturbance may be arbitrary save for the fact that it must be periodic. The magnitude of the fluctuations need not be small compared with the mean flow, and the period of the disturbance need not be equal to the period of the rotor blade passage.

3. The blades are real blades with camber, thickness and stagger.

Limitations.

1. Simultaneous disturbances with different periods cannot be accommodated economically.

2. No viscous effects are present.

3. Only two-dimensional cascades can be handled at present. There is no provision for flared casings.

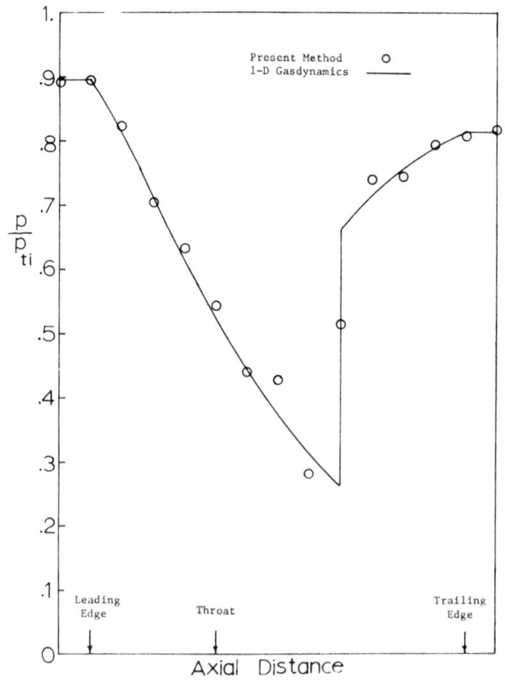

Figure 6. Theoretical Pressure Distributions in the Axial Direction at the Mid Span of a Fictitious Stator Row, $p_e/p_{ti} = 0.815$.

4. Only ideal gases are considered.

5. The blade positions are fixed, so flutter is not considered.

The unsteady case represents a modification of the normal approach (see Section: Flow Field on the Blade-to-Blade Surface). It is less general in that it is limited to ideal gases and strictly two-dimensional flow, but the computation scheme has been modified to allow arbitrary periodic disturbances along the inlet plane. The principal modification in the unsteady approach is that the computation is made periodic with respect to the upstream disturbance rather than with respect to the cascade pitch.

A sample case is given in Figures 8 and 9. The blade passages are formed by steam turbine rotor blade sections of a form found near the tip. Figure 8 shows the unsteady inlet conditions presented to the rotor. These conditions were derived from measurements

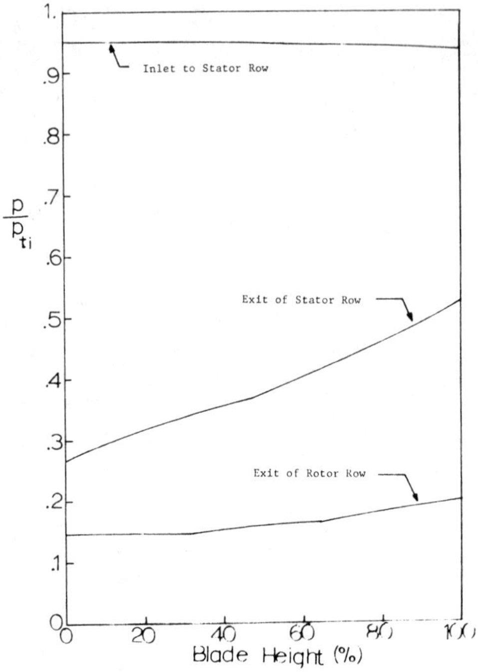

Figure 7. Theoretical Static Pressure Distributions in the Radial Direction for the Last Stage of a Low-Pressure Turbine.

made on a typical steam turbine stator tested in an air cascade. The absolute stator exit conditions were converted to relative rotor inlet conditions and a harmonic analysis was done. The mean values were altered to agree with the known mean steady rotor flow, but the relative harmonic content of the disturbance was retained. Figure 9 shows the fluctuations of the axial and tangential forces on a particular blade of the cascade. The fluctuation for this case is some ± three to four percent about the mean, and the frequency is essentially that of the fundamental. The forces are quite sinusoidal in appearance, the higher frequency components having been washed out by the combination of integration around the blade and the numerical damping inherent in the system.

Although the magnitudes of the fluctuating forces are not very high, they are significant for power station turbines which undergo many cycles and where fatigue could be a problem.

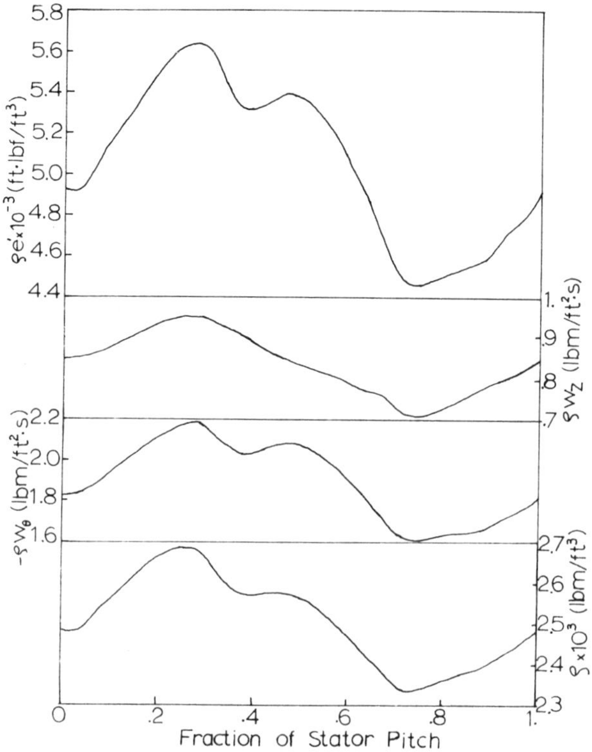

Figure 8. Variation of Relative Inlet Variables Along One Stator Pitch.

Steady periodic flow appears in general to be attained by 8000 time steps, so 10,000 steps are usually used for safety. For the 495 node points used in this example (45 axially by 11 peripherally), the computer run time for 10,000 steps is one hour on a UNIVAC 1106 (approximately three minutes on a CDC 7600).

CONCLUSIONS

We have found the time-dependent finite volume technique to be exceptionally versatile for the computation of transonic flows. In addition to obviating the usual problem of mixed elliptic/hyperbolic mathematical domains, one can use the technique for difficult flow geometry, non-ideal working fluids and unsteady boundary conditions.

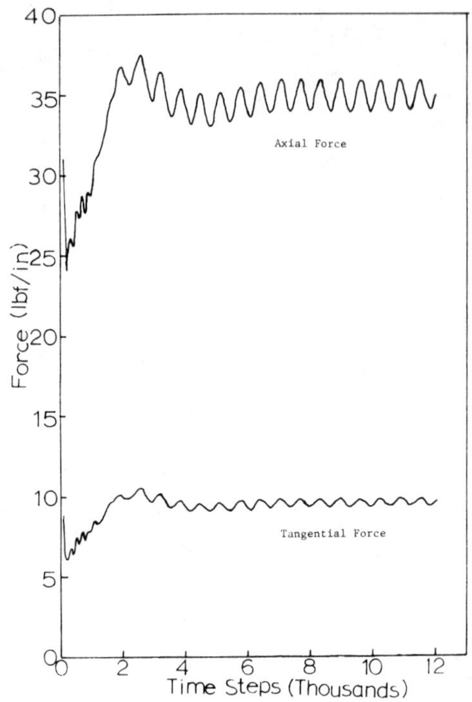

Figure 9. Fluctuating Axial and Tangential Forces Plotted Against Number of Time Steps.

The current versions of the computer programs, particularly those for the hub-to-shroud surface and unsteady flow, do have limitations, but these are due to computer storage limitations rather than to anything inherent in the method itself.

As for the method itself, there are difficulties when the Mach number is too low, but the method is usable in the high subsonic range, and below that other methods are applicable. Also, our experience has been with turbine blades; the behavior of the method when confronted with a compressor is unknown. Certainly that case would be a most interesting extension of the technique.

REFERENCES

[1] Bindon, J. P. and Carmichael, A. D. (1971), "Streamline Curvature Analysis of Compressible and High Mach Number Cascade Flows," J. Mech. Engrg. Sci., 13, 344-357.

[2] Bindon, J. P. and Day, I. J. (1974), "Experimental Evaluation of a Streamline Curvature Method for Flared Subsonic Cascades and for Transonic Cascades," Proc. Inst. Mech. Engrs., 188, 215-220.

[3] Wilkinson, D. H. (1972), "Calculation of Blade-to-Blade Flow in a Turbomachine by Streamline Curvature," ARC R. & M., No. 3704,

[4] Sieverding, C. H. (1973), "Experimental Data on Two Transonic Turbine Blade Sections and Comparison with Various Theoretical Methods," VKI Lecture Series 59 (Transonic Flows in Turbomachinery).

[5] Decuypere, R. H. (1973), "Method for Calculating the Steady Irrotational Isentropic Flow in a Two-Dimensional Supercritical Turbine Cascade," Inst. Mech. Engrs., Conference Publication 3, 196-203.

[6] Gopalakrishnan, S. and Bozzola, R. (1972), "Computation of Shocked Flows in Compressor Cascades," ASME Paper No. 72-GT-31.

[7] Kurzrock, J. W. and Novick, A. S. (1975), "Transonic Flow Around Rotor Blade Elements," ASME Paper No. 75-FE-24.

[8] McDonald, P. W. (1971), "The Computation of Transonic Flow Through Two-Dimensional Gas Turbine Cascades," ASME Paper No. 71-GT-89.

[9] Wu, C. H. (1952), "A General Theory of Three-Dimensional Flow in Subsonic and Supersonic Turbomachines of Axial-, Radial-, and Mixed-Flow Types," NACA TN-2604.

[10] Smith, L. H., Jr. (1966), "The Radial-Equilibrium Equation of Turbomachinery," J. of Engineering for Power, Trans. ASME, Series A, 88, 1-12.

[11] Novak, R. A. (1967), "Streamline Curvature Computing Procedures for Fluid-Flow Problems," J. of Engineering for Power, Trans. ASME, Series A, 89, 478-490.

[12] Herzog, J. (1970), "Calculation of Flow Distribution in Large Radius Ratio Stages of Axial Flow Turbines and Comparison of Theory and Experiment," NASA SP-304, Part II, 565-580.

[13] Araki, T. (1973), "Three-Dimensional Flow Through Turbines," Inst. Mech. Engrs., Conference Publication 3, 38-45.

[14] Renaudin, A. and Somm, E. (1970), "Quasi-Three-Dimensional Flow in a Multi-stage Turbine Calculation and Experimental Verification," Flow Research on Blading, Edited by L. S. Dzung, Elsevier Publishing Co., 51-87.

[15] Marsh, H. (1968), "A Digital Computer Program for the Through-Flow Fluid Mechanics in an Arbitrary Turbomachine Using a Matrix Method," ARC R. & M., No. 3509.

[16] Biniaris, S. (1974), "The Calculation of the Quasi-Three-Dimensional Flow in an Axial Gas Turbine, ASME Paper No. 74-GT-67.

[17] Hirsch, Ch. and Warzee, G., (1974), "A Finite Element Method for Flow Calculations in Turbomachines," University of Brussels, Report V. U. B. -STR-5.

[18] Smith, D. J. L. and Barnes, J. F. (1968), "Calculation of Fluid Motion in Axial Flow Turbomachines," ASME Paper No. 68-GT-12.

[19] Lindley, D., Westmoreland, J. and Whybrow, J. F. T. (1969-1970), "Some Problems in the Application of Meridional Flow Programs to the Design of Large Steam Turbines," Proc. Inst. Mech. Engrs., 184, Part 3G (II), 38-48.

[20] Davis, W. R. and Millar, D. A. J. (1974), "A Comparison of the Matrix and Streamline Curvature Methods of Axial Flow Turbomachinery Analysis, From a User's Point of View," ASME Paper No. 74-WA/GT-4.

NOMENCLATURE

A	=	surface area of the control volume
c_a	=	axial chord of the blade
e	=	specific total energy, see Eq. (6)
e'	=	specific relative total energy $(= u + w^2/2)$
h_t	=	specific total enthalpy, see Eq. (7)
$\hat{n}$	=	unit vector normal to the surface of the control volume and pointing outward
p	=	static pressure
p_t	=	total pressure
r	=	radial variable
$\hat{r}$	=	unit vector in the radial direction
t	=	time variable
u	=	specific internal energy
v	=	magnitude of flow velocity
v_r	=	r-component of flow velocity
v_z	=	z-component of flow velocity
v_θ	=	θ-component of flow velocity
$\vec{v}_A$	=	local velocity of the surface of the control volume
V	=	control volume
w	=	magnitude of $\vec{w}$
w_z	=	z-component of $\vec{w}$
w_θ	=	θ-component of $\vec{w}$
$\vec{w}$	=	velocity of the flow relative to the surface of the control volume
z	=	axial variable
$\hat{z}$	=	unit vector in the axial direction
β	=	flow angle $[= \tan^{-1}(v_\theta/v_z)]$
ρ	=	density
$\hat{\theta}$	=	unit vector in the azimuthal direction
Ω	=	angular speed of the control volume
$(\ )_e$	=	exit condition
$(\ )_i$	=	inlet condition
$(\ )_k$	=	condition at a point representing the k-th control volume

DISCUSSION OF THE FARN-WHIRLOW PAPER

MURMAN: The agreement you showed between the inviscid calculations and the experiments was quite good. Is there any reason? It seems that there should be a strong viscous effect on a blade that is that loaded, but you show very good results. Can you comment on that?

FARN: These are turbine blades. The pressure distribution is favorable. We did not try a compressor.

MURMAN: I see - O.k.

DODGE: [Remark in background - not at microphone - not recorded.]

FARN: I'm still not convinced that the time-dependent method will work well for compressors because I haven't seen any comparisons between theoretical and test data.

MORETTI: Speaking of comparison, what was Figure 9? I didn't understand it.

FARN: Figure 9 is all theoretical results.

MORETTI: When you say theoretical, you mean ... your computations?

FARN: Numerical, yes.

MORETTI: You don't show those as dots as in the other figures.

FARN: Well, these are the integrated blade force - with time. This is time dependent. The rotor row sees the wake of the stator and then experiences the unsteady force - so this is a periodic force on the rotor blade.

ERDOS: In the rotor-stator interaction case, in the last one, are you solving the entire periphery, or a blade-to-blade passage?

FARN: We solve a blade-to-blade passage. You have the modified boundary conditions on the side. It's not a regular periodic condition. There is a phase angle difference.

MORETTI: But - those fluctuations, you showed in the last figures - Are they due to the ... ?

FARN: They are due to the wake of the upstream row. If you use uniform flow, then you get steady loading. The corresponding figure will be a straight line, no wiggles.

MCNALLY: How do you take that viscous effect into account when you have the rotor-stator interaction case? I thought this was an inviscid solution. How does the wake come in?

FARN: Theoretically, if you traverse in a turbine, you find varying properties - and then you transform them into rotating coordinates. It's the use of the input.

MORETTI: If I understand, it's not a wake in a viscous sense.

FARN: That's right.

MORETTI: And neither, it seems to me, a wake in a sense of shed vorticity, because you have some smearing of quantities there. So it's just input data.

FARN: It's just a nonuniform flow field.

FINITE-DIFFERENCE CALCULATIONS OF THREE-DIMENSIONAL TRANSONIC FLOW THROUGH A COMPRESSOR BLADE ROW, USING THE SMALL-DISTURBANCE NONLINEAR POTENTIAL EQUATION* **

WILLIAM J. RAE

Calspan Corporation

INTRODUCTION

Axial-flow compressors and turbines have been operated for many years in the transonic range, i.e., under conditions where the resultant of the axial approach velocity of the flow and the circumferential velocity of the blades is close to the local sound speed somewhere along the blade span. Under these circumstances, an observer riding on one of the blades sees a steady transonic flow, in which the subsonic relative approach flow may accelerate to supersonic speed when passing through the row of blades (see Figure 1). Standard methods of calculating turbomachinery flow fields cannot be used when this condition occurs, and the designer must resort to a variety of analytical extrapolation methods, and a judicious choice of blade-element data [1]. In the past, these design procedures were adequate for the development of many

---

*This paper contains the combined results of two oral presentations on work in progress, which were entitled, "On the Formulation of the Small-Disturbance Nonlinear Potential Equation" and "Finite-Difference Calculations of Three-Dimensional Transonic Flow Through a Compressor Blade Row."

**This work was supported by the Air Force Office of Scientific Research.

generations of axial-flow machines.

In recent years, however, increasing demands for fuel economy and compression per stage, with decreased weight and noise generation, have placed much greater emphasis on the need for more accurate solutions to turbomachinery flowfield problems [2, 3]. However, available methods for predicting turbomachinery flowfields are severely limited in situations where locally supersonic zones occur [4]. Analytical solutions are restricted to linearized cases [5-13], and the finite-difference relaxation methods that are in widespread use for wholly subsonic flows [14-18] cannot be used when supercritical regions are present. Thus, the only approaches open are either to accept some further degree of approximation, such as that of a shock-free velocity gradient solution [15], or to perform a lengthy time-dependent calculation, whose large-time limit yields the steady-state solution [19-21]. These solutions, moreover, are restricted to two-dimensional flows.

The state of the art of prediction methods for transonic turbomachinery flows stands in marked contrast to the case of isolated airfoils in transonic flow, where the progress made during the past five years has made it possible to handle cases having a wide range of complexity [22-28].

PROBLEM FORMULATION

The basic purpose of this research has been to learn about the interaction between three-dimensionality and transonic nonlinearity in a compressor flow. In order not to complicate the problem unduly, and to maximize the applicability of isolated-airfoil computational techniques, the small-disturbance approximation has been adopted from the outset. Thus, the compressor blades are taken to be thin and lightly loaded, and to be essentially aligned with the helical streamline paths seen by a blade-fixed observer as the resultant of the axial through-flow and the angular velocity of the blades (see Figure 1). For the range of transonic flow being considered, the Mach number of the resultant flow may vary from a subsonic value near the hub to a supersonic value near the tip.

The basic technical approach is to seek numerical solutions to the finite-difference counterparts of the equations of motion, subject to appropriate boundary conditions and periodicity conditions. A number of subproblems had to be solved before the finite-

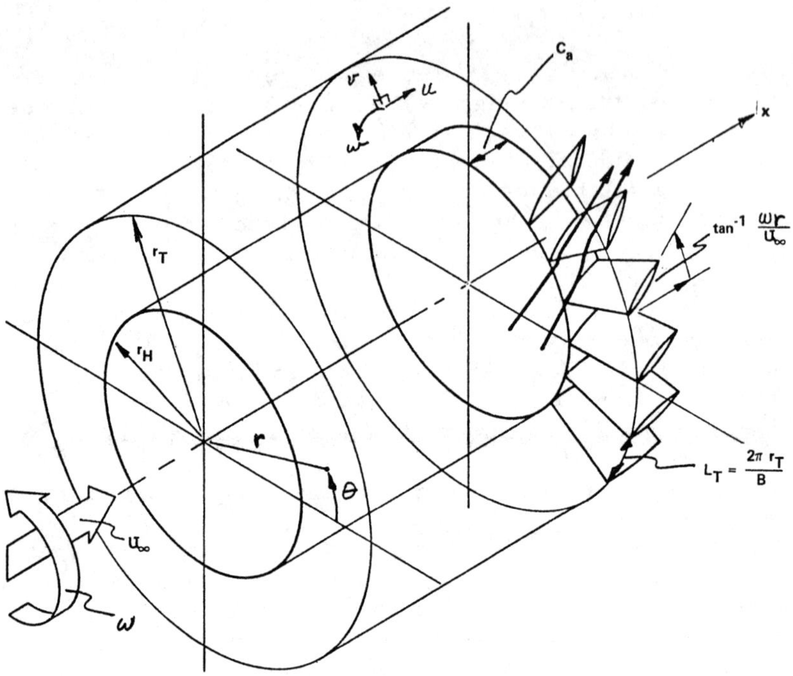

Figure 1. Blade Fixed Coordinates. Rotor is Stationary in a Helical Approach Flow

difference solutions could be found; among the more important of these were development of the basic equations, adaptations of isolated-airfoil techniques so as to account for periodicity and locally supersonic flow, and finding a stable method of iterating in the radial direction. These items are described briefly in the paragraphs below.

Basic Equations

The small-disturbance nonlinear form of the potential equation that is appropriate for transonic flow in a blade row is [10]

$$\{1 - M_\infty^2(1+\rho^2) - (\gamma+1)M_\infty^2\phi_z\}\phi_{zz} + \rho^2\phi_{zz} - 2(1+\rho^2)\phi_{\zeta z}$$

$$+ \frac{(1+\rho^2)^2}{\rho^2}\phi_{\zeta\zeta} + (1+\rho^2)(\phi_{\rho\rho} + \frac{1}{\rho}\phi_\rho) = 0$$

where

$$z = \frac{\omega x}{U_\infty}, \quad \rho = \frac{\omega r}{U_\infty}, \quad \zeta = \theta - z; \quad M_\infty = \frac{U_\infty}{a_\infty}$$

and where the perturbation velocities are

$$u/U_\infty = \left.\frac{\partial\phi}{\partial z}\right)_{\rho,\theta} = \left.\frac{\partial\phi}{\partial z}\right)_{\rho,\zeta} - \left.\frac{\partial\phi}{\partial\zeta}\right)_{z,\rho}$$

$$v/U_\infty = \left.\frac{\partial\phi}{\partial\rho}\right)_{z,\theta} = \left.\frac{\partial\phi}{\partial\rho}\right)_{z,\zeta}$$

$$w/U_\infty = \left.\frac{1}{\rho}\frac{\partial\phi}{\partial\theta}\right)_{z,\rho} = \left.\frac{1}{\rho}\frac{\partial\phi}{\partial\zeta}\right)_{z,\rho}$$

In these coordinates, the blades lie in the helical surfaces defined by

$$0 \leq z \leq \frac{\omega c_a}{U_\infty}, \quad \rho_H \leq \rho \leq \rho_T \quad \text{and} \quad \zeta = \frac{2j\pi}{B}, \ j=0, 1, \ldots, B-1,$$

where B is the number of blades. In the small-disturbance approximation, the boundary conditions at the blade surfaces are satisfied in these helical surfaces. Thus, the use of the ζ-coordinate greatly facilitates enforcement of the boundary conditions.

In addition, these coordinates have the advantage that the periodicity conditions are easily enforced, i.e.,

$$\phi_\zeta(z,\rho,\zeta) = \phi_\zeta(z,\rho,\zeta + \frac{2j\pi}{B})$$

This ease in enforcing both the periodicity and boundary conditions is not shared by coordinate systems which are aligned with the local streamline direction:

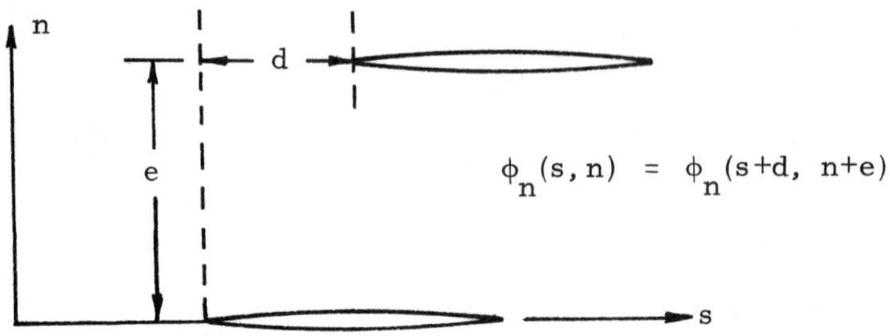

Similitude Considerations

The equations presented in the previous section contain the leading nonlinear terms that must be retained when the difference of the relative Mach number from one is small. The specific combination of terms retained implies an assumption that

$$\frac{u_s}{w_o} = O(1 - M_0^2); \quad M_0^2 \equiv \{U_\infty^2 + (\omega r)^2\}/a_\infty^2$$

However, the boundary conditions imposed at the blade surfaces also require that u_s and u_n, or some combination of them, will have orders of magnitude that are dictated by the small parameters that measure the blade geometry and loading. Because these components are connected by the equations of motion, the small parameters must then be related to each other, in order that the equations be consistently ordered in the limit where these parameters approach zero. This section contains a discussion of the geometric and loading parameters, and their relation to $1 - M_0^2$ in the small-disturbance limit.

The blade thickness and lift distributions are taken to be characterized by the small parameters τ and α, where $\tau/\alpha = O(1)$. These parameters mainly affect the orders of the velocity components u_s and u_n, in a manner discussed below. The radial component v, however, is ordered by the axial rates of change of the annular radii, dr_H/dx and dr_T/dx. In the present work, both of these are assumed small compared with τ. Thus the basic ordering problem is treated in a quasi two-dimensional manner, involving only u_s and u_n.

The velocity-component equations from which the potential equation is derived are rewritten with the s- and n-coordinates made dimensionless by the local chord length

$$\left\{1 - M_0^2 - (\gamma+1)M_0^2 \frac{u_s}{w_o}\right\} \frac{\partial (u_s/w_o)}{\partial (s/c)} + \frac{\partial (u_n/w_o)}{\partial (n/c)} =$$

$$= (\gamma-1)M_0^2 \frac{u_s}{w_o} \frac{\partial (u_n/w_o)}{\partial (n/c)} + 2M_0^2 \frac{u_n}{w_o} \frac{\partial (u_n/w_o)}{\partial (s/c)}$$

$$\frac{\partial (u_s/w_o)}{\partial (n/c)} = \frac{\partial (u_n/w_o)}{\partial (s/c)}$$

The next step is to establish the dependence of the velocity components on the parameter τ. This dependence is influenced by the solidity of the blade row, c/L_T. At very low solidity, the dependence approaches that of the isolated airfoil case, i.e.,

$$\frac{u_n}{w_o} = \frac{dn}{ds} = \tau\, g'(s/c)$$

where g' is of unit order. If the equations above are divided by β_0^4 and β_0^3, respectively, where $\beta_0^2 = 1 - M_0^2$, it follows that the solution has the familiar transonic similitude:

$$\frac{u_s/w_o}{\beta_0^2}\,,\quad \frac{u_n/w_o}{\beta_0^3} = \text{fcns}\left(\frac{s}{c}\,,\ \frac{\beta_0 n}{c}\,;\ K\right)$$

where

$$K \equiv [(\gamma+1)M_0^2\,\tau]^{2/3}/\beta_0^2 = O(1)$$

This ordering requires that $\beta_0 n/c$ be of unit order, a requirement that cannot be satisfied if the solidity is of unit order and $\beta_0 \to 0$.

When the solidity is of unit order, a new phenomenon comes into play, namely that the flow between the blades begins to act like a transonic channel flow, in which the distance over which the transition through sonic conditions occurs is controlled by the effective radius of curvature of the channel. This influence was first pointed out by Ackeret and Rott [33]. The implications on the ordering of the flow variables has been studied by Hall [34], who observed that the axial extent of the transonic zone (which affects the ordering of the streamwise coordinate) is governed by the radius of curvature of the channel throat.

The adaptation of Hall's analysis to the present problem leads to the following conclusions. The normal velocity component at the blade surface is still given by the equation above, but it is no longer true that u_n/w_o is of order τ, since the surface slope function itself vanishes at the sonic location. In order to establish the orders, it is necessary to use the irrotationality condition:

$$\left.\frac{\partial(u_s/w_o)}{\partial(n/c)}\right|_{n=0} = \left.\frac{\partial(u_n/w_o)}{\partial(s/c)}\right|_{n=0} = \tau\, g''(s/c)$$

where surface curvature $g''(s/c)$ is of unit order, and does not vanish in the region of near-sonic flow. Since $n/c = O(1)$, it can be concluded that $u_s/w_o = O(\tau)$. In order to deduce the orders of u_n/w_o and s/c, the first of the velocity-component equations shown above is differentiated with respect to s/c, and the irrotationality condition is used to eliminate u_n:

$$[\beta_0^2 - (\gamma+1)M_0^2 \frac{u_s}{w_o}] \frac{\partial^2(u_s/w_o)}{\partial(s/c)^2} + \frac{\partial^2(u_s/w_o)}{\partial(n/c)^2} = O\left\{\frac{\partial(u_s/w_o)}{\partial(s/c)}\right\}^2$$

From the known orders of u_s/w_o and n/c, it is clear that we must have

$$\beta_0^2 = O(\tau), \qquad \frac{s}{c} = O(\tau^{1/2})$$

The solution is then of the form

$$\frac{u_s/w_o}{\beta_0^2}, \frac{u_n/w_o}{\beta_0^3} = \text{fcns}\left(\frac{s}{c\sqrt{\tau}}, \frac{n}{c}; \frac{\beta_0^2}{\tau}, \frac{\alpha}{\tau}, \frac{c}{L_T}\right)$$

where all of the parameters are of unit order. It can now be verified that the nonlinear terms on the right-hand side are smaller by a factor τ than those retained.

FINITE-DIFFERENCE SOLUTIONS

Adaptation of Isolated-Airfoil Finite-Difference Techniques

The discussion given above concerning coordinate system's is included to expose the nature of the difficulties that must be faced in adapting the finite-difference schemes of the isolated-airfoil case to the special circumstances of a turbomachine blade row. In particular, if one seeks to align the streamwise coordinate with the stream direction, it then becomes very awkward to satisfy the periodicity and boundary conditions.

The skew coordinates z, ζ remove both of these problems, but they introduce the complication that the grid points used in the finite-difference method must be inside the fore Mach cone extending upstream of the grid point at which the solution is being calculated (Figure 2). The remedy to this problem can be found in a series of papers by Jameson [29-31], who developed an alteration in the method of differencing that accounts properly for the rotation of the forward Mach cone. Jameson's method was originally developed for cases where the misalignment between the

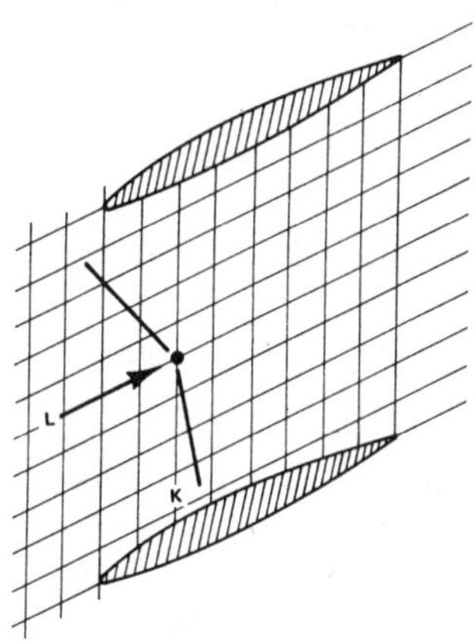

Figure 2. Coordinate System and Finite-Difference Grid, Showing Forward Mach Cone at Point K, L.

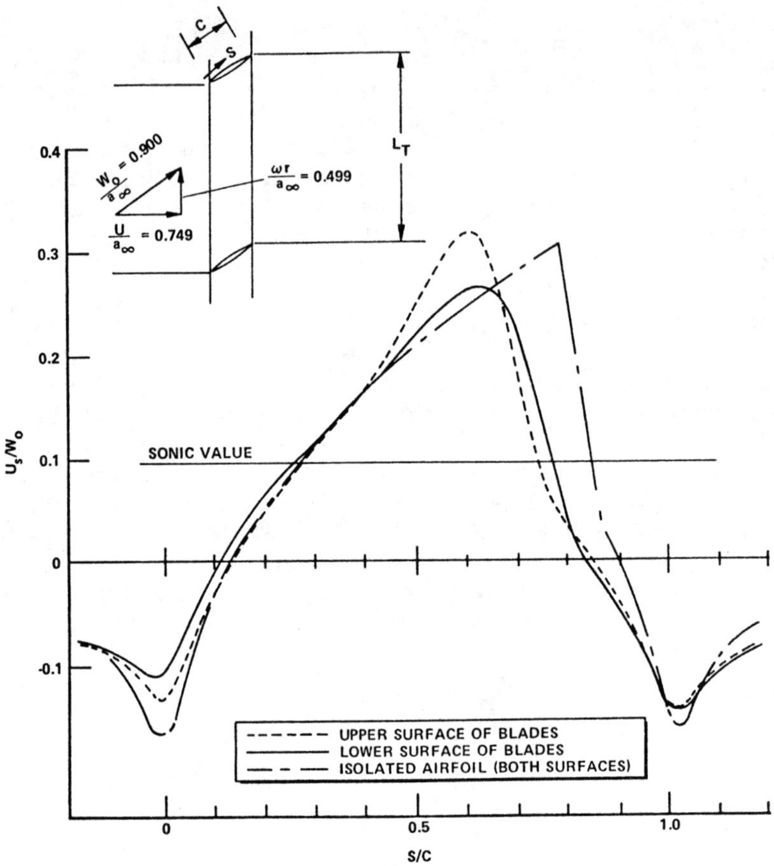

Figure 3. Blade-Surface Velocity Distributions in a Subsonic, Supercritical Cascade Flow. 6% Thickness/Chord Ratio, Double Parabolic-Arc Blades

coordinate lines and the flow direction is caused by large deflections in the flow. There is a subtle difference in the present case, where the flow deflections are always small, and the misalignment is caused by the skew nature of the coordinate system itself.

Figure 3 presents the results of a calculation made with this program, of the flow in a cascade at a relative approach Mach number of 0.9, over a set of 6% thick symmetric parabolic-arc blades at a stagger angle of 33.69 degrees, and with a solidity c/L_T of 0.277. This blade shape was chosen because an isolated-airfoil calculation, developed under another AFOSR-supported program at Calspan, was available for it. A low value of the solidity was chosen, so as to approximate a nearly isolated condition. The results given in Figure 3 are the distribution of the streamwise component of the perturbation velocity, normalized by the resultant magnitude of the approach velocity. (Multiplying the ordinates of Figure 3 by -2 would give the pressure-coefficient distribution.) The differences between the upper- and lower-surface distributions indicate that the blades were not completely isolated. This is further supported by the fact that a small amount of circulation (corresponding to a negative section lift coefficient of about 10^{-3}) had to be applied, in order to satisfy the Kutta condition at the trailing edge.

The agreement between these cascade results and those of the isolated airfoil are taken as a validation of the computational method. The principal difference lies in the steepness of the shock transition. In most relaxation solutions for the flow over isolated airfoils, this transition is much steeper, and is completed in four to six grid points, which typically extend over 5% of the airfoil chord. This degree of definition is inadequate for the turbomachinery case, where this length might be as large as 10-15% of the blade-to-blade spacing in a high-solidity blade row.

The steepness of the shock transition is affected by the stagger angle of the blades. Figure 4 shows results for the same set of blades as in Figure 3, with the same values of c/L_T and w_0, except that the stagger angle is reduced to 5.711 degress. It is clear that the shock transition is much steeper at the lower stagger angle. However, practical blading designs have values of ωr that are typically two to three times U_∞; thus, the stagger angles will in general be even greater than that shown in Figure 3, and constitute a major problem that must be addressed.

Radial Iteration

The computer code used for the above calculations was further developed, to do the three-dimensional case. The sequence used is to solve first the difference equations at a given radius, treating the radial derivatives as known quantities, and then to move to the next larger radius, and repeat the process from hub to tip.

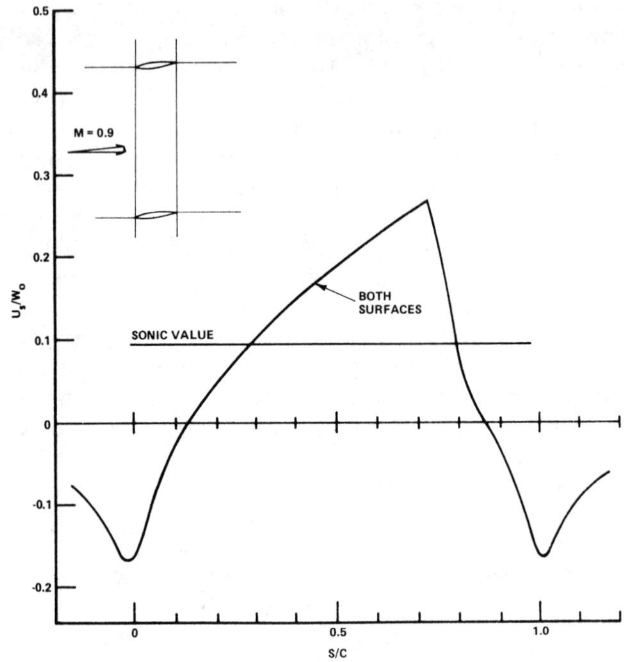

Figure 4. Blade-Surface Velocity Distributions in a Subsonic, Supercritical Cascade Flow. 6% Thickness/Chord Ratio, Double Parabolic-Arc Blades

As new values of the solution become available at various radii, they can then be used to update the radial derivatives. This process works successfully in isolated-airfoil cases [32], but had never been tried in a mixed-flow situation, where subsonic and supersonic flows exist side-by-side. It was found that this method worked stably, and successful demonstration calculations were made with it.

The computer program is capable of handling arbitrary blade geometry and loading, and will operate in either of two modes: given the blade geometry, it will find the loading (the direct problem) or, given the blade loading and thickness distributions, it will find the camber-line shape required (the indirect problem).

Two calculations have recently been done with this code, for the purpose of demonstrating its capabilities. The blade row used for both of these calculations is shown in Figure 5. It had thirty

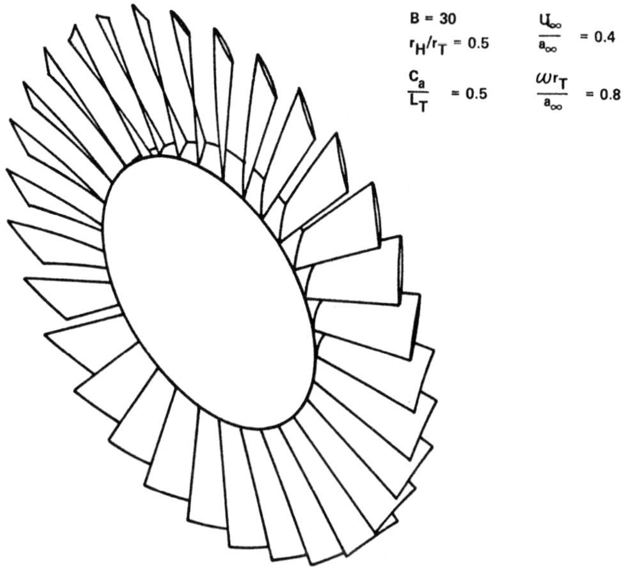

Figure 5. Blade Row Used for Demonstration Calculations

blades with a hub-to-tip radius ratio of 0.5 and a solidity c_a/L_T of 0.5. Both calculations used an axial Mach number of 0.4, and an angular Mach number at the tip of 0.8, making the resultant Mach number at the tip 0.894. The grid used for these calculations had 40 points in the z-direction, 20 points in the ζ-direction, and 10 points radially, for a total of 8000 grid points. Constant grid spacings were used in the ρ- and ζ-directions; a variable grid was used in the z-direction, with points more closely spaced in the vicinity of the blades (14 of the 40 points lay between the leading- and trailing-edge stations). The grid extended from $2c_a$ upstream of the leading-edge station to $2c_a$ downstream of the trailing-edge station.

The relaxation procedure began by finding the solution at the hub. Line relaxation was used, with all points on the line z = constant, ρ = constant updated simultaneously. This line is then swept from upstream to downstream, and the sweep is repeated a number of times, typically the same as the number of grid points in the z-direction, in order for information to be carried from one end of the grid to the other. The solution then proceeds to the next radius, where the process is repeated. After this first sequence of radial iterations has reached the tip radius, it is then repeated. The number of repetitions is typically the same as the number of radial surfaces in the grid.

The calculations reported below used 10 radial sweeps, with 40 axial sweeps for each radial one. Thus, each point in the grid was visited 400 times, for a total of 3.3×10^6 point-updatings. The program was written in FORTRAN IV and, using the H compiler, was run on the IBM 370/168 of the Martin-Marietta Corporation Great Lakes Data Center. It was found that the computing time required was 36 microseconds per grid point per visitation, exclusive of the time required for the output. Thus, each of the two calculations reported below took about two minutes, plus the time used in the output. (The latter time requires an additional 5 to 10 seconds, depending on the amount of output desired.)

In the first demonstration calculation, the blade geometry was specified. The blades were taken to have parabolic-arc distributions of thickness and camber:

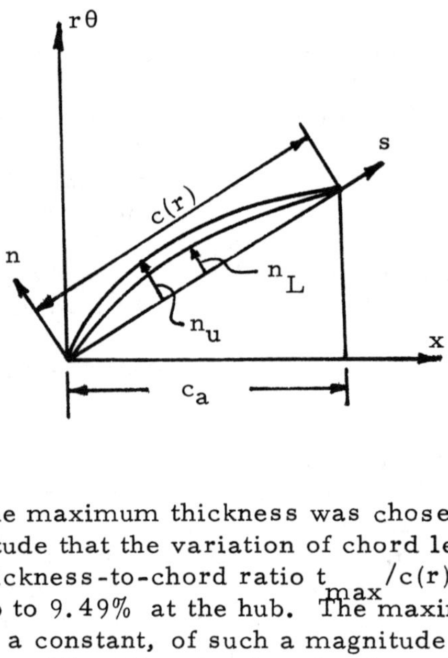

$$n_{u,L}(s) = h(s) \pm \frac{1}{2} t(s)$$

$$t(s) = \frac{4t_{max}(r)}{[c(r)]^2} \{ s[c-s] \}$$

$$h(s) = \frac{4h_{max}(r)}{[c(r)]^2} \{ s[c-s] \}$$

The maximum thickness was chosen as a constant, of such a magnitude that the variation of chord length with radius produces a thickness-to-chord ratio $t_{max}/c(r)$ which caries from 6% at the tip to 9.49% at the hub. The maximum camber was also chosen as a constant, of such a magnitude as to make the camber vary from 4% of the chord at the tip to 6.33% of the chord at the hub. The blade shapes at the hub and tip are shown in Figure 6.

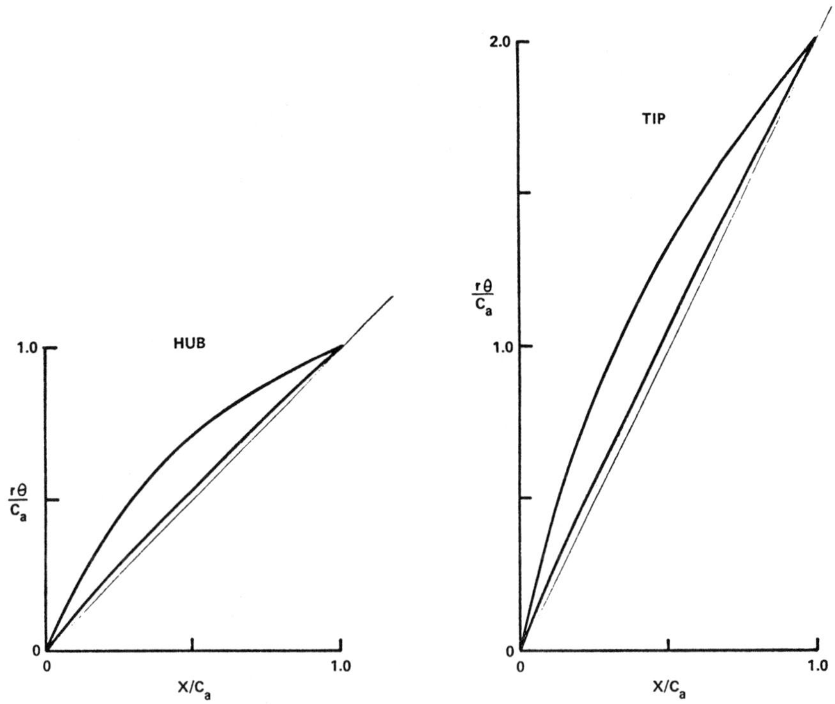

Figure 6. Blade Geometry Specified for First Demonstration Calculation

The results of this calculation show that the flow accelerates to locally supersonic conditions over the outer half of the span. Contours of the local Mach number are shown in Figures 7a and 7b for radial stations near the hub and near the tip.

The second demonstration calculation was done using the computer program in the design mode. The thickness distribution was chosen to be the same as used in the above calculation, while the loading distribution had the form:

$$\Delta C_p \equiv \frac{p_L - p_u}{\frac{1}{2}\rho_\infty U_\infty^2}$$

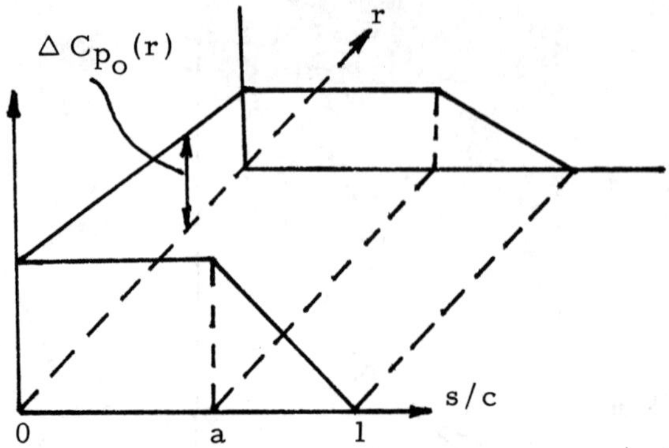

where the value of $\Delta C_{p_o}(r)$ varied linearly between prescribed

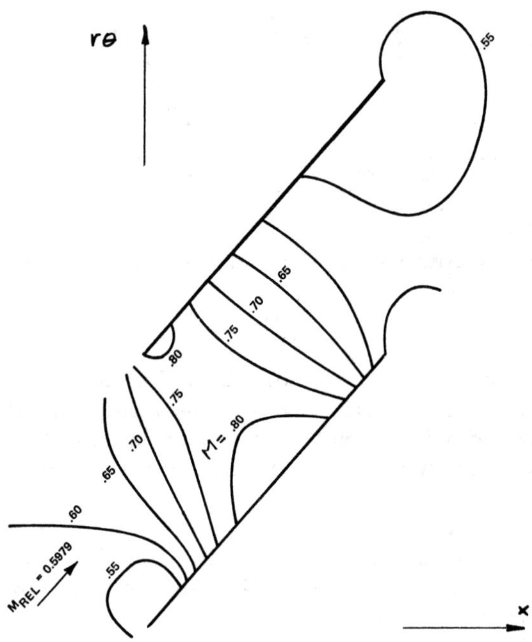

Figure 7a. Contours of the Local Mach Number. First Demonstration Calculation. $r = \frac{10}{9}\, r_{HUB}.$

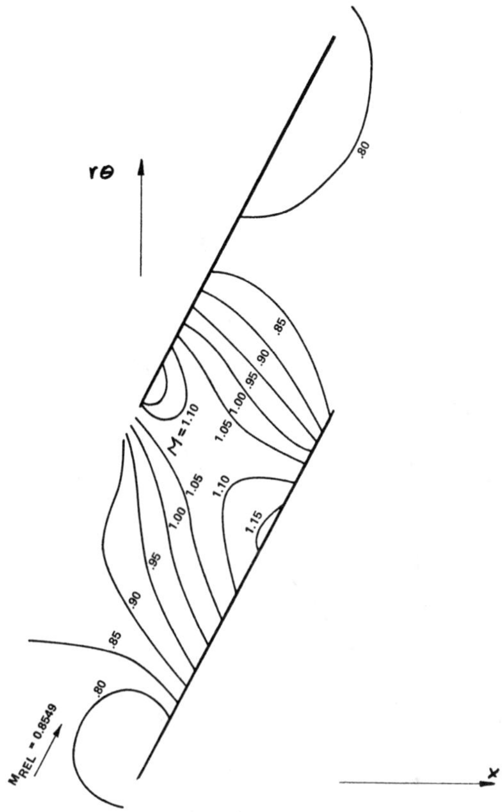

Figure 7b. Contours of the Local Mach Number. First Demonstration Calculation. $r = \frac{17}{9} r_{HUB}$

values at the hub and tip. The values used were a = 0.6, ΔC_{p_o} = 0.5 at the hub, and 1.0 at the tip. The flow resulting from these specifications was everywhere subsonic; contours of the local Mach number at two radial stations are given in Figures 8a and 8b. The blade shape required to achieve this loading is shown in Figure 9 for the hub and tip. It is interesting to note that an angle of attack makes up the largest portion of the blade shape, with only a modest contribution from camber.

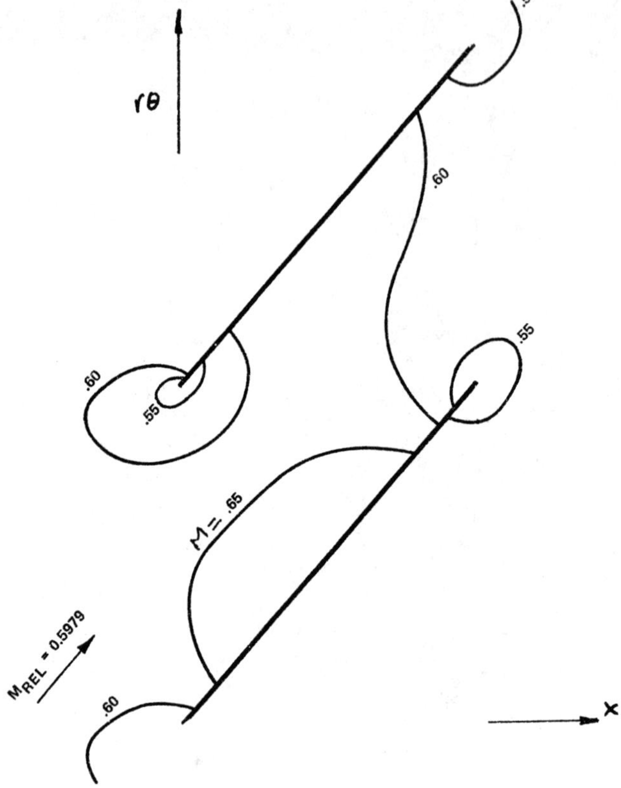

Figure 8a. Contours of the Local Mach Number. Second Demonstration Calculation. $r = \frac{10}{9} r_{HUB}$

The significance of these results is that real turbomachinery flows can be calculated, which include three-dimensionality and transonic zones, at a very modest computer cost. To our knowledge, these results are the first of their kind to have been found.

This work will be described in greater detail in References [35] and [36].

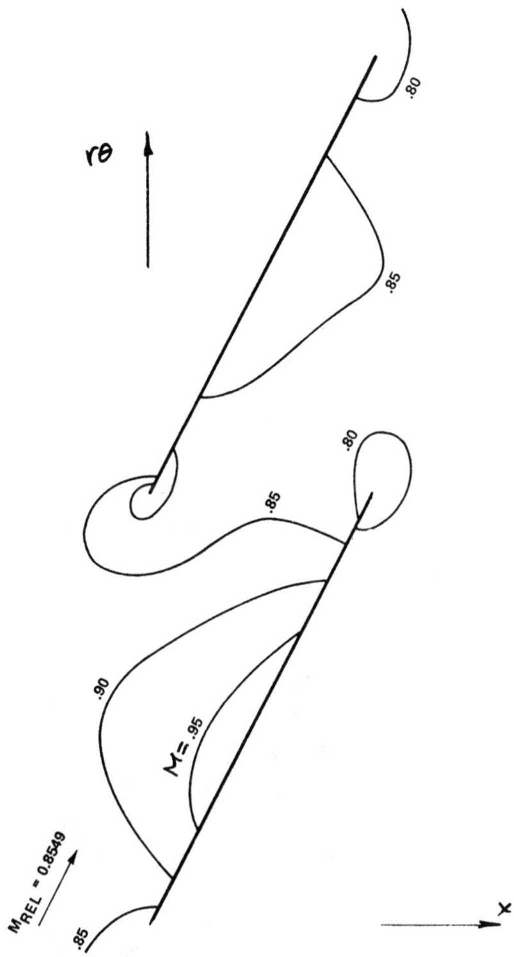

Figure 8b. Contours of the Local Mach Number. Second Demonstration Calculation. $r = \frac{17}{9} r_{HUB}$

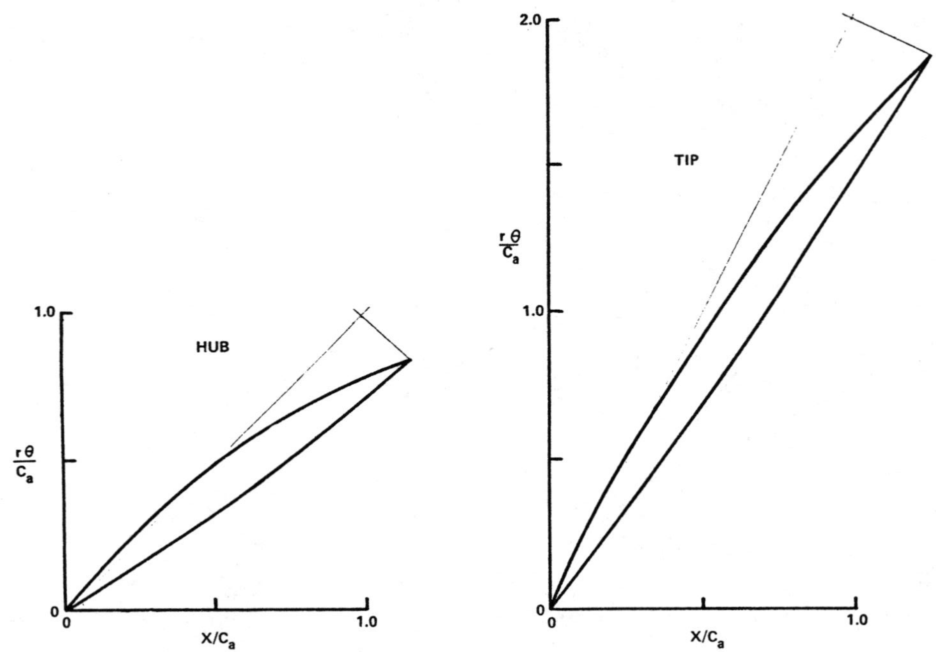

Figure 9. Blade Geometry Found from Second Demonstration Calculation

REFERENCES

[1] Johnsen, I. A. and Bullock, R. O., eds. (1965), Aerodynamic Design of Axial-Flow Compressors, NASA SP-36.

[2] Hartmann, M. J., Benser, W. A., Hauser, C. H. and Ruggeri, R. S., (1971), "Fan and Compressor Technology," pp. 1-36 of Aircraft Propulsion, Proceedings of a conference held at NASA Lewis Research Center, November 18-19. NASA SP-259.

[3] Benser, W. A., Bailey, E. E. and Gelder, T. F. (1974), "Holographic Studies of Shock Waves Within Transonic Fan Rotors," Paper presented at the ASME 19th International Gas Turbine Conference, Zurich, Switzerland, March 30 - April 4. Available as NASA TM X-71430.

[4] Gostelow, J. P.. (1973), "Review of Compressible Flow Theories for Airfoil Cascades," Jour. of Engineering for Power, Trans. ASME(A) 95, pp. 281-292.

[5] McCune, J. E., (1958), "The Three-Dimensional Flow Field of an Axial Compressor Blade Row - Subsonic, Transonic, and Supersonic," Ph. D. thesis, Cornell University.

[6] McCune, J. E., (1958), "A Three-Dimensional Theory of Axial Compressor Blade Rows - Application in Subsonic and Supersonic Flows," J. Aero. Sci., 25, pp. 544-560.

[7] McCune, J. E., (1958), "The Transonic Flow Field of an Axial Compressor Blade Row," Journal of the Aerospace Sciences, 25:10, pp. 616-626.

[8] Erickson, J. C., Lordi, J. A., and Rae, W. J., (1971), "On the Transonic Aerodynamics of a Compressor Blade Row," Calspan Corporation Report AI-3003-A-1.

[9] Okurounmu, O. and McCune, J. E. (1970), "Three-Dimensional Vortex Theory of Axial Compressor Blade Rows at Subsonic and Transonic Speeds," AIAA Journal, 8:7, pp. 1275-1283.

[10] Okurounmu, O. and McCune, J. W., (1974), "Lifting Surface Theory of Axial Compressor Blade Rows: Part I - Subsonic Compressor, Part II - Transonic Compressor, AIAA Journal, 12.

[11] Lordi, J. A. (1971), "Report on a Study of Noise Generation by a Rotating Blade Row in an Infinite Annulus," U.S. Air Force Office of Scientific Research Scientific Report AFOSR TR-71-1485. (Also available as Calspan Corporation Report AI-2386-A-1).

[12] Lordi, J. A. (1971), "Noise Generation by a Rotating Blade Row in an Infinite Annulus," AIAA Paper No. 71-617, AIAA 4th Fluid and Plasma Dynamics Conference, Palo Alto, California, June 21-23.

[13] Lordi, J. A., Homicz, G. F. and Rehm, R. G. (1973), "Theoretical Studies on Fan Noise Generation by a Transonic Compressor Blade Row," Calspan Corporation Report AI-2386-A-02.

[14] Katsanis, T., (1968), "Computer Program for Calculating Velocities and Streamlines on a Blade-to-Blade Stream Surface of a Turbomachine," NASA TN D-4525.

[15] Katsanis, T., (1969), "FORTRAN Program for Calculating Transonic Velocities on a Bade-to-Blade Stream Surface of a Turbomachine," NASA TN D-5427.

[16] Katsanis, T. (1967), "A Computer Program for Calculating Velocities and Streamlines for Two-Dimensional, Incompressible Flow in Axial Blade Rows," NASA TN D-3762.

[17] Katsanis, T. and McNally, W. D. (1969), "FORTRAN Program for Calculating Velocities and Streamlines on a Blade-to-Blade Stream Surface of a Tandem Blade Turbomachine," NASA TN D-5044.

[18] Katsanis, T. and McNally, W. D. (1969), "Revised FORTRAN Program for Calculating Velocities and Streamlines on a Blade-to-Blade Stream Surface of a Turbomachine," NASA TM X-1764.

[19] McDonald, P. W. (1971), "The Computation of Transonic Flow Through Two-Dimensional Gas Turbine Cascades," ASME Paper No. 71-GT-89.

[20] Gopalakrishnan, S. and Bozzola, R. (1972), "Computation of Shocked Flows in Compressor Cascades," ASME Paper No. 72-GT-31.

[21] Kurzrock, J. W. and Novick, A. S. (1973), "Transonic Flow Around Compressor Rotor Blade Elements. Volume I: Analysis," Air Force Aero Propulsion Laboratory Report AFAPL TR-73-69.

[22] Krupp, J. A. (1971), "The Numerical Calculation of Plane Steady Transonic Flows Past Thin Lifting Airfoils," Boeing Scientific Research Laboratories Report D180-12958-1.

[23] Murman, E. M., and Cole, J. D. (1971), "Calculation of Plane Steady Transonic Flows," AIAA Journal, 9, pp. 114-121.

[24] Murman, E. M. and Krupp, J. A. (1971), "Solution of the Transonic Potential Equation Using a Mixed Finite Difference System," Proceedings of the Second International Conference on Numerical Methods in Fluid Dynamics, published as Vol. 8 of Lecture Notes in Physics, Springer-Verlag, pp. 199-206.

[25] Krupp, J. A. and Murman, E. M. (1972), "Computation of Transonic Flows Past Lifting Airfoils and Slender Bodies," AIAA Journal, 10, pp. 880-886.

[26] Bailey, F. R. and Steger, J. L. (1973), "Relaxation Techniques for Three-Dimensional Transonic Flow About Wings," AIAA Journal, 11, pp. 318-325.

[27] Steger, J. L. and Lomax, H. (1971), "Generalized Relaxation Methods Applied to Problems in Transonic Flows," Proceedings of the Second International Conference on Numerical Methods in Fluid Dynamics, published as Vol. 8 of Lecture Notes in Physics, Springer-Verlag, pp. 193-197.

[28] Murman, E. M., (1973), "Analysis of Embedded Shock Waves Calculated by Relaxation Methods," Proceedings of the AIAA Computational Fluid Dynamics Conference, July, pp. 27-40.

[29] Jameson, A. (1974), "Iterative Solution of Transonic Flows Over Airfoils and Wings, Including Flows at Mach 1," Communications on Pure and Applied Mathematics, 27, pp. 283-309.

[30] Jameson, A., (1973), "Numerical Calculation of the Three-Dimensional Transonic Flow over a Yawed Wing," Proceedings of the AIAA Computational Fluid Dynamics Conference, July, pp. 18-26.

[31] South, J. C. and Jameson, A. (1973), "Relaxation Solutions for Inviscid Axisymmetric Transonic Flow over Blunt or Pointed Bodies," Proceedings of the AIAA Computational Fluid Dynamics Conference, July, pp. 8-17.

[32] Ballhaus, W. F. and Bailey, F. R. (1972), "Numerical Calculation of Transonic Flow About Swept Wings," AIAA Paper 72-677.

[33] Ackeret, J. and Rott, N., (1949), "Ueber die Stromungen von Gasen durch ungestaffelte Profilgitter," Schweizerische Bauzeitung, 67, pp. 40-41, 58-61.

[34] Hall, I. M., (1962), "Transonic Flow in Two-Dimensional and Axially Symmetric Nozzles," Quarterly Journal of Mechanics and Applied Mathematics, 15, pp. 487-508.

[35] Rae, W. J. (to be published), "Nonlinear Small-Disturbance Equations for Three-Dimensional Transonic Flow Through a Compressor Blade Row."

[36] Rae, W. J. (to be published), "Numerical Relaxation Solutions for Three-Dimensional Transonic Flow Through a Compressor Blade Row in the Nonlinear Small-Disturbance Approximation."

DISCUSSION OF THE FIRST RAE PAPER

MC CUNE: I was just wondering, Bill, what the philosophy that underlies this is - you want to stick to small perturbations and do better within the framework of small perturbations?

RAE: That's right. The major motivation was to shed light, I would hope, on the question of how it can be that a supersonic and a subsonic flow can exist side by side, and to examine that interaction.

MC CUNE: Yes ... and I think that's an important question to investigate, but from the practical point of view you will still be limited to light loading.

RAE: Absolutely. There's an open question as to how far you go.

MC CUNE: You still, for example, can't satisfy radial equilibrium, on the average, right?

RAE: Well, I'm not sure to what extent the approximation abandons that, but it is surely true, and I watched with great interest your introduction of the Beltrami equation, because the principal penalty I think we pay is for locating the vortex sheet.

MC CUNE: That's one thing that using the Beltrami equation allows you to do.

RAE: Our vortex sheets are still located in the helical surface.

DISCUSSION OF THE SECOND RAE PAPER

SICHEL: I noticed that sonic contour you had looked something like the one that Tom Adamson reported on yesterday in the simplified 3-D calculation. Could you say something about how your inlet velocity profile compared to the simple shear profile that Tom and Oliver talked about.

RAE: The inlet profile is, of course, not a linear shear; in fact, it isn't a shear at all. It's the resultant of U_∞ and ωr, and in the small disturbance nonlinear approximation there isn't any term that sometimes arises in rectilinear shear flow approximations, that corresponds to the convection of a circulation perturbation. Moreover, it was restricted to a case where the approach Mach number was uniformly subsonic. I indicated earlier my reservations about the boundary conditions at upstream ∞. This program has built into it, zero perturbations at upstream ∞. My belief is that those must be replaced by a radiation condition when there's a supersonic - when the sonic circle lies inside the plane. That's not to say the program wouldn't run; it'll certainly handle the supercritical case, but I'm not comfortable with the way the boundary conditions are applied. Does that answer your question?

PLATZER: How would you satisfy the initial conditions if you had a mixed subsonic/supersonic approach flow?

RAE: I don't know the answer exactly. My suspicion is that what's happening in the real physics is that on the supersonic portions of the blades near the tips, Mach waves, in this approximation, are launched which make their way upstream in some contra-rotating helical path and are diffracted as they go, and give rise to a nonzero perturbation at a number of chord lengths upstream. How do describe their radial variation or their azimuthal variation, I am not sure; I suspect that the answer lies in an application of the linear theory to look at the flow details of that station, and I suspect that having a fix on what the physics looks like, one ought to be able to use a radiation boundary condition in somewhat the same way that supersonic flows over 2-D airfoils can make use of a radiation condition at the upper boundary of the grid. But that's strictly my speculation at the moment.

TRANSONIC RELAXATION METHODS

PAUL R. DODGE

Garrett-Airesearch Manufacturing Company of Arizona

INTRODUCTION

Methods for solving transonic flow problems in turbomachinery are relatively new. To date, they have been based on adaptations from external aerodynamic methods. The oldest is the method of characteristics (MOC). The problem with this method, of course, is that it is not really a transonic analysis, but rather a supersonic analysis. All supersonic cascades and rotor sections have been designed and tested [1] using MOC methods. Applications to inlet regions of cascades have also been made [2].

The application of time-dependent methods to cascades by Gopalakrishnan [3] and McDonald [4] resulted in the first all transonic solutions. These methods suffer from being relatively slow. Smoothing introduced to speed convergence and maintain stability also introduced significant errors for some configurations. Recently, effort has been applied to relaxation methods. These methods, pioneered for external aerodynamics by Murman [5], offer possibilities of significant increases in efficiency and accuracy.

Cascade geometries introduce significant complications into a flow picture. The presence of adjacent blade rows seriously complicates boundary conditions, introducing such constraints as periodicity and unique incidence angles to the boundary conditions ahead of and behind the blade row. An isolated airfoil can always pass the flow, because an infinite area is available. This is not always so with a cascade, since area as well as surface curvature affects the flow solution. Cascades can support larger diffusions than isolated airfoils; consequently, stronger and more varied shock structures can be expected. Inlet Mach numbers to transonic cascades are normally higher than anything considered to be transonic for isolated airfoils. Exit Mach numbers on the other hand may well be below sonic. As a result, methods totally adequate for isolated airfoils may not be suitable for cascades. Thus, some adaptation is necessary when applying relaxation methods to cascades. The following section describes in brief two stages of transonic development and the results of each. The first is a straight-forward adaptation of an isolated airfoil relaxation method. The second is a new non-orthogonal method that has proven more successful in solving a wider, more complete range of transonic cascade problems.

ORTHOGONAL RELAXATION

The starting point for any relaxation calculation is some form of potential equation. For external aerodynamics, a perturbation potential equation is often used. For cascade aerodynamics, some of the advantages of this approach rapidly dissipate due to large variations in Mach number between inlet and exit. A potential equation suitable for cascade calculations given by Eq. (1).

$$\frac{1}{a^2} \nabla\phi \cdot \{\nabla (\frac{\nabla\phi \cdot \nabla\phi}{2})\} - \nabla^2\Phi - \frac{1}{b} \nabla b \cdot \nabla\Phi = 0. \qquad (1)$$

As is the case for all isentropic potential equations, this form exhibits a jump solution that does not satisfy all the founding equations (continuity, energy, and momentum). However, for realistic jump conditions, the error between normal jumps is small. This is illustrated by Figure 1 from [6]. The error between forms is negligible below the separation limit. Higher pressure ratio shocks introduce such severe viscous effects that the error in jump conditions is not noticed.

When constructing differences for Eq. (1) numerous consistent forms can be supplied. However, as discovered by Murman and Cole [5], realistic supersonic results are only obtained when the difference star includes only upstream influence. Consequently, a switch in difference stars must be made at the sonic point. In addition, Murman [7] more recently has pointed out the need of constructing difference equations, right at a jump, that faithfully

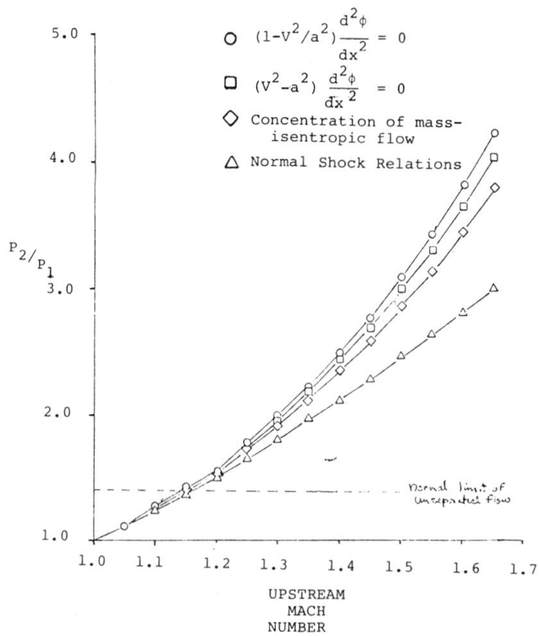

Figure 1. Static Pressure Ratio Across a Jump.

model the analytically expected jump. The errors associated with not fulfilling this requirement, however, do not appear large when compared to the error between ideal inviscid flow and the real viscous flows exhibited by actual cascades.

The first application of these techniques to cascades was reported by Dodge [8]. A numerically generated orthogonal grid system was utilized. Periodicity conditions were applied in front of and behind the blade row. A typical grid system is shown in Figure 2. The results of surface Mach number distributions are shown for several supercritical compressor cascades in Figures 3 through 10. In general, when the cascade is well below choke, results are good. As flow is increased, two effects take over. The first is the viscous effect, primarily due to shock wave-boundary layer interactions. Also evident is the occurrence of choke points well below what one would expect on a one-dimensional area basis, often so low as to be below the observed experimental choke point that contains viscous effects. As a result of several numerical studies, this defect was diagnosed to be caused by the

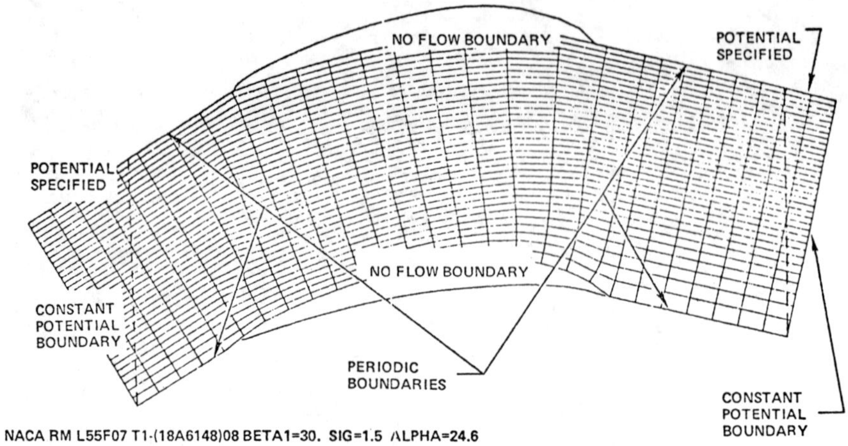

Figure 2. . Grid System With Boundary Conditions.

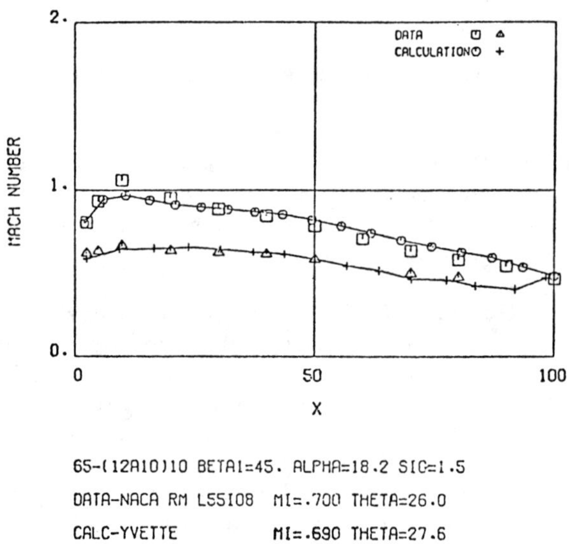

Figure 3.

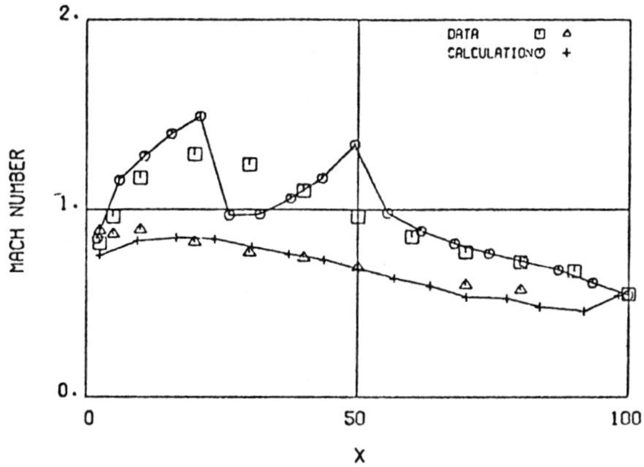

65-(12A10)10 BETA1=45. ALPHA=18.2 SIG=1.5

DATA-NACA RM L55I08 MI=.792 THETA=25.3

CALC-YVETTE MI=.796 THETA=27.6

Figure 4.

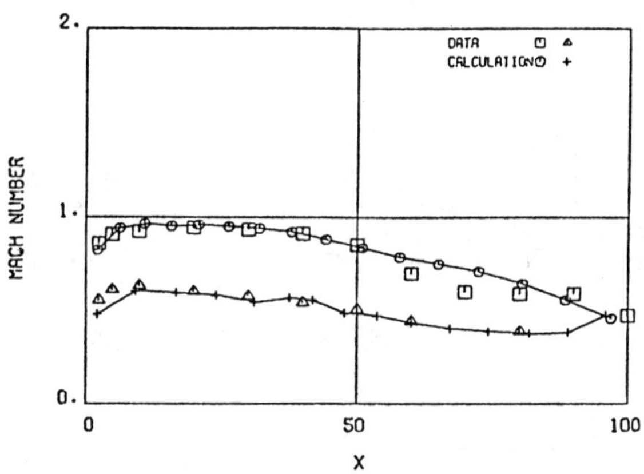

65-(12A2I8B)10 BETA1=60. ALPHA=14. SIG=1.

DATA--RM L55I08 MI=.746 THETA=16.4

CALC--YVETTE MI=.744 THETA=19.

Figure 5.

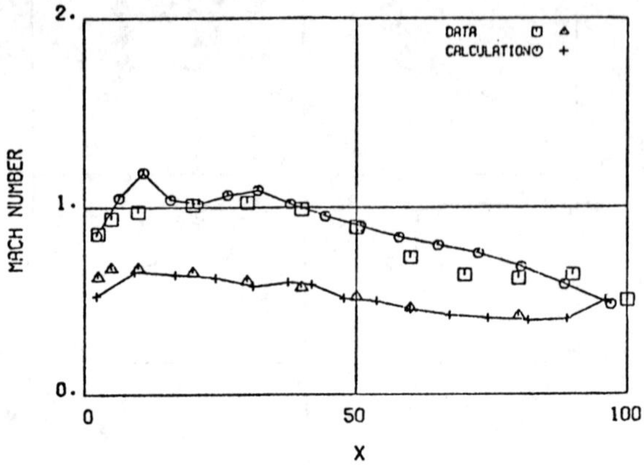

65-(12A2I8B)10 BETA1=60. ALPHA=14. SIG=1.

DATA--RM L55I08 MI=.793 THETA=16.5

CALC--YVETTE MI=.793 THETA=19.

Figure 6.

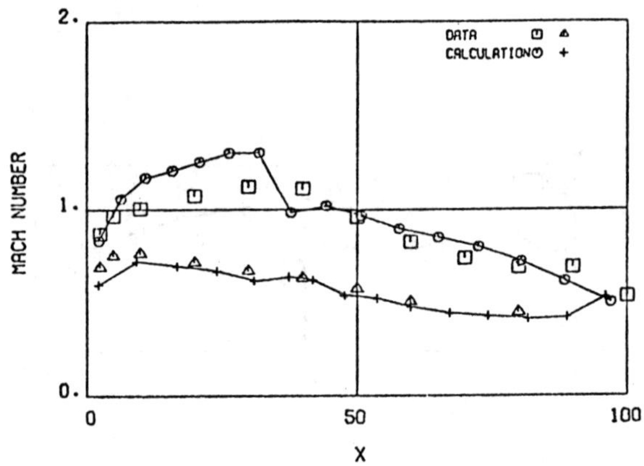

65-(12A2I8B)10 BETA1=60. ALPHA=14. SIG=1.

DATA--RM L55I08 MI=.842 THETA=17.4

CALC--YVETTE MI=.839 THETA=19.

Figure 7.

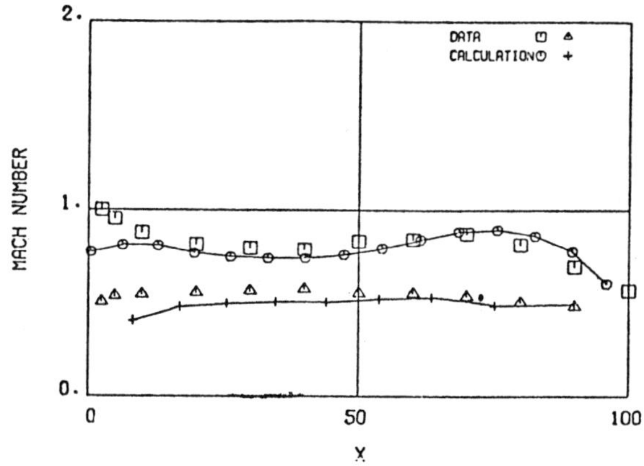

T1-(18A6I4B)08 BETA1=30. ALPHA=24.6 SIG=1.5
DATA-NACA RM L55F07 MI=.658 THETA=42.3
CALC-YVETTE MI=.654 THETA=44.

Figure 8.

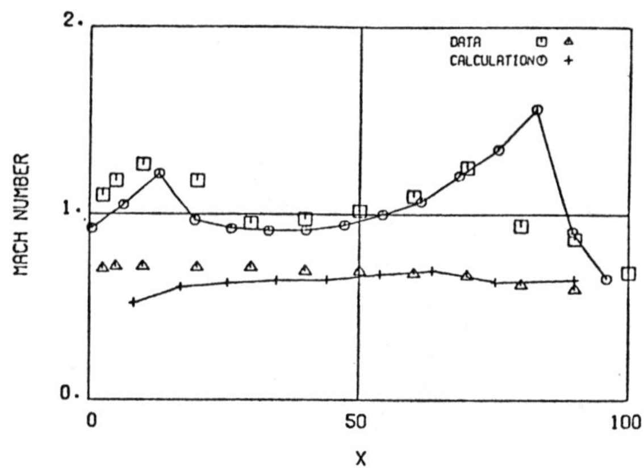

T1-(18A6I4B)08 BETA1=30. ALPHA=24.6 SIG=1.5
DATA-NACA RM L55F07 MI=.799 THETA=42.1
CALC-YVETTE MI=.799 THETA=44.

Figure 9.

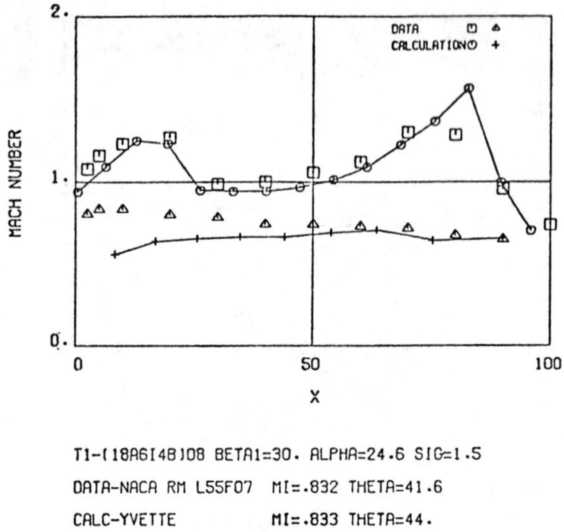

Figure 10.

lack of agreement between the difference star and characteristic regions. The difference star utilized was implicit. Its region of influence is shown in Figure 11. At high Mach number, only the region inside the characteristic cone should affect the potential;

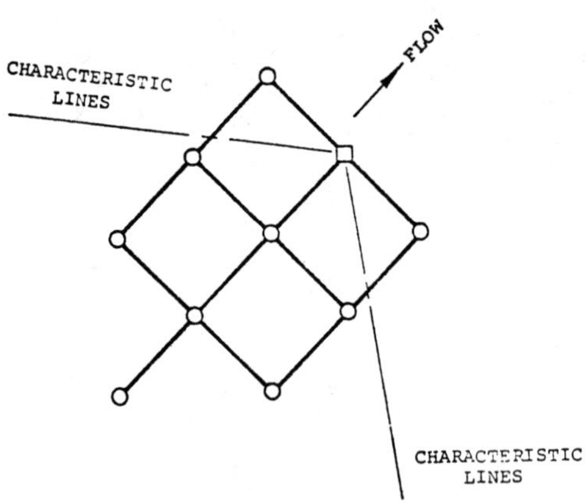

Figure 11. Difference Star for a Supersonic Node.

however, the difference equation introduces an effect from upstream points. This produces a strong wall-to-wall influence not present in the differential equation. Sonic bubbles appear too large, resulting in indications of choking flow well below the actual occurrence.

NON-ORTHOGONAL RELAXATION

A cure for this difficulty was developed subsequently, as reported in [9]. Only a brief outline of the method will be presented here. It is based on a non-orthogonal grid system similar to that generated by a normal method of characteristics. Equation (1) can be transformed into Eq. (2).

$$AH + BE + C \frac{d\phi}{d\xi} + D \frac{d\phi}{d\zeta} = 0 \tag{2}$$

H is a hyperbolic operator

$$H = \frac{d^2\phi}{d\zeta d\xi} + \frac{d^2\phi}{d\xi d\zeta} \tag{3}$$

E is the elliptic operator

$$E = \frac{d^2\phi}{d\xi^2} + \frac{d^2\phi}{d\zeta^2} \tag{4}$$

The coefficients A, B, C, and D are functions of the angle the grid makes with the flow direction (see Figure 12), the local Mach number, and the b width distributions. In general, when converged, the coefficient B of the elliptic operator is zero when the flow is supersonic. The hyperbolic operator coefficient, A, is zero when the flow is subsonic. The hyperbolic operator uses the difference star shown in Figure 13a and the elliptic shown in Figure 13 b. The result is a natural transonic relaxation system always matching regions of influence of the difference equation to the differential equation. The disadvantage of this approach is that the grid must be reconstructed as needed.

Results for several supersonic and supercritical cases are shown in Figures 14-21. Final grid systems are shown in Figures 16 and 20. Choke flows are within a few percent of the one

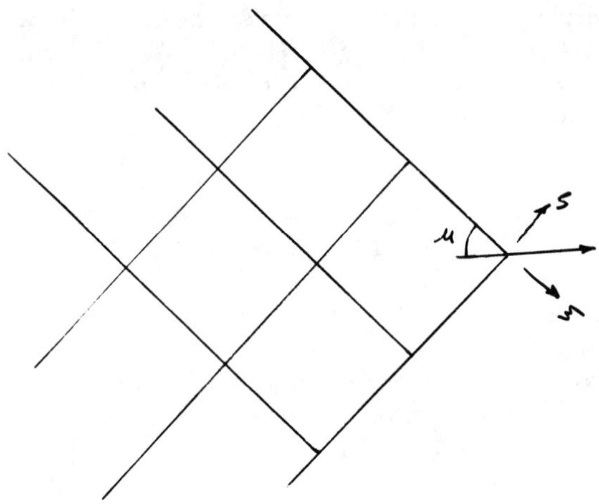

Figure 12. Non-Orthogonal Grid System.

dimensional values. General agreement is good until viscous effects become overpowering.

CONCLUSIONS

A relaxation method based on a non-orthogonal grid system has been developed that provides realistic solutions to cascades with either subsonic or supersonic inlet and exit Mach numbers. Accurate choked flows are predicted, run times are substantially better than those of time-dependent methods, and a more accurate and distinct shock structure is apparently produced. However, for supersonic inlet cascades, viscous effects become extremely important, and adversely influence the agreement of results.

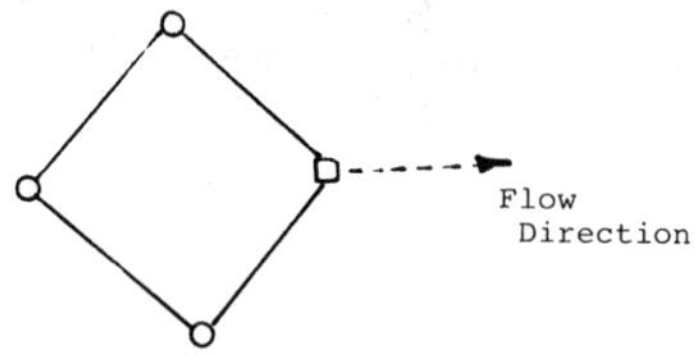

Figure 13a. Hyperbolic Difference Star.

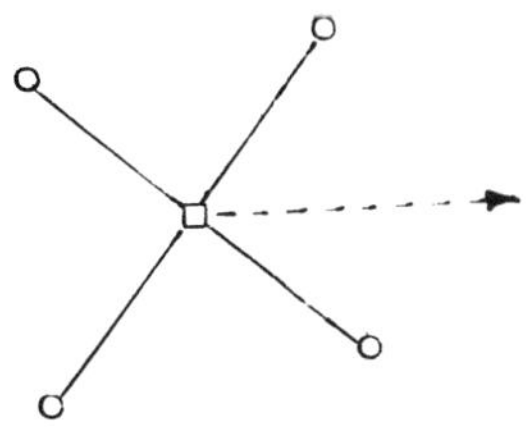

Figure 13b. Elliptical Difference Star.

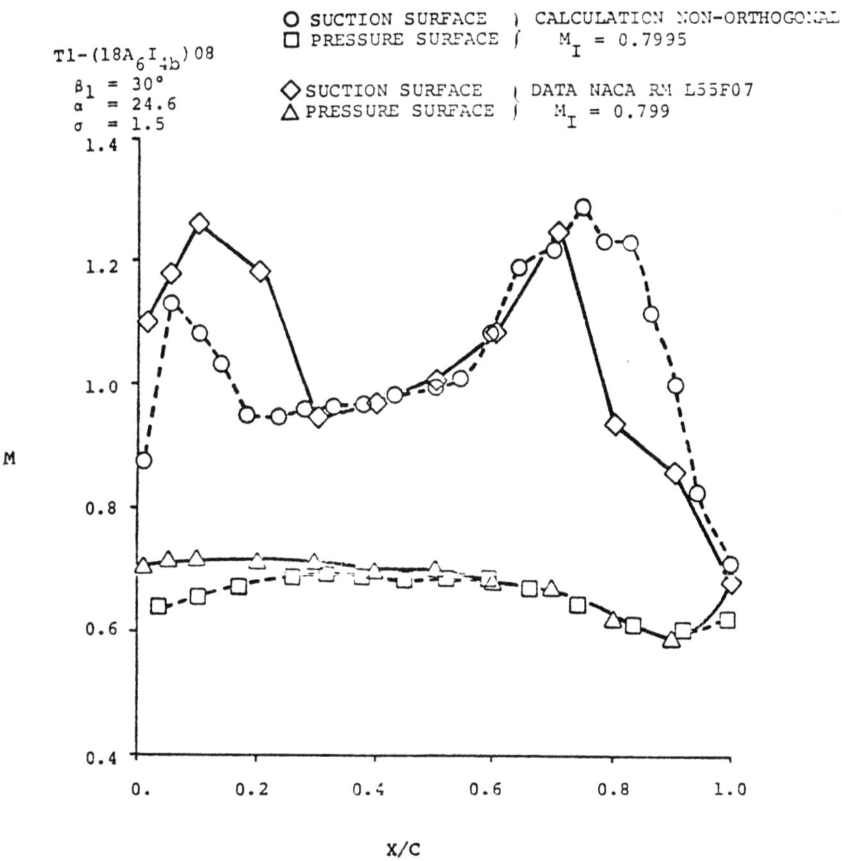

Figure 14. Surface Mach Number vs. Distance.

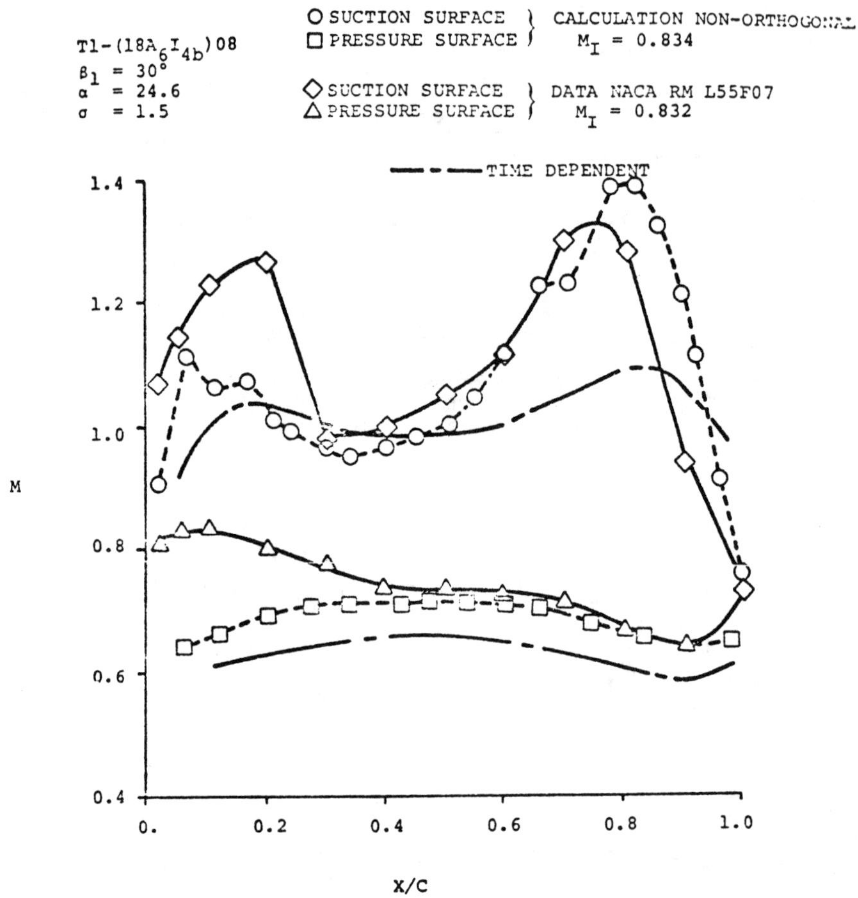

Figure 15. Surface Mach Number vs. Distance.

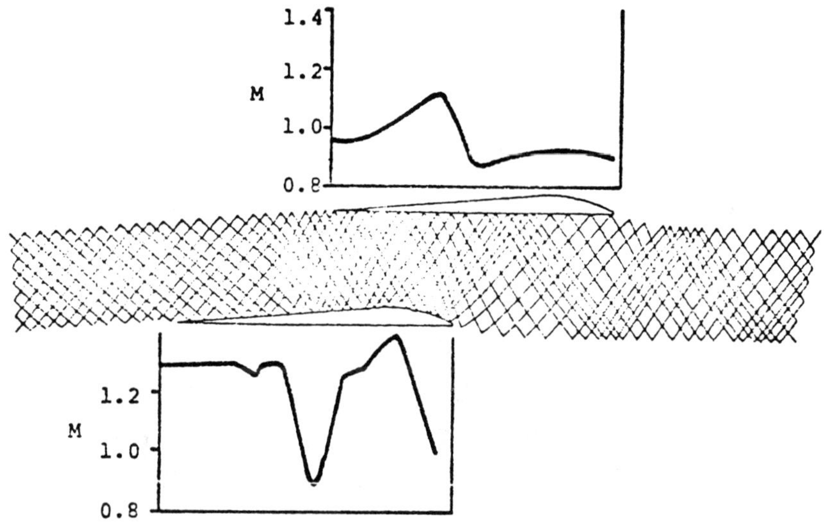

M_E = 0.800, M_1 = 1.28

STAGGER ANGLE 55°

β_1 = 60°

σ = 1.43

Figure 16. Grid System and Calculated Surface Mach Numbers.

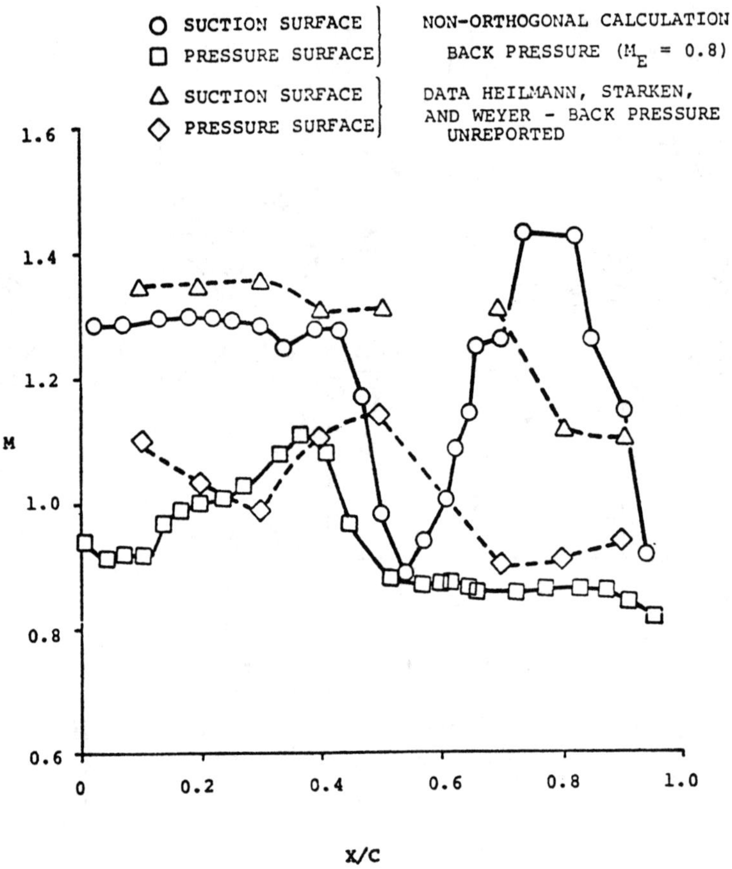

Figure 17. Surface Mach Numbers vs. Distance (CW1 Blade β_s = 55°, M_1 = 1.28).

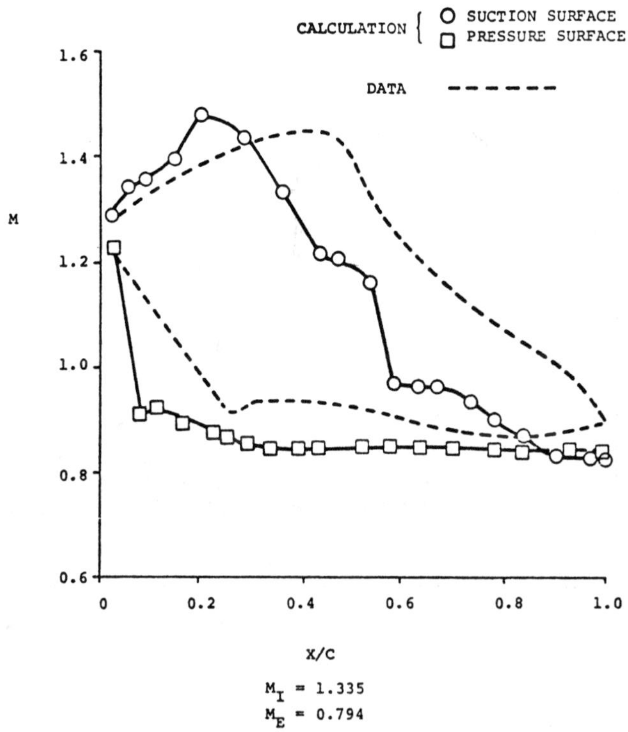

Figure 18. V.K.I. 9.5° Camber DCA Blade.

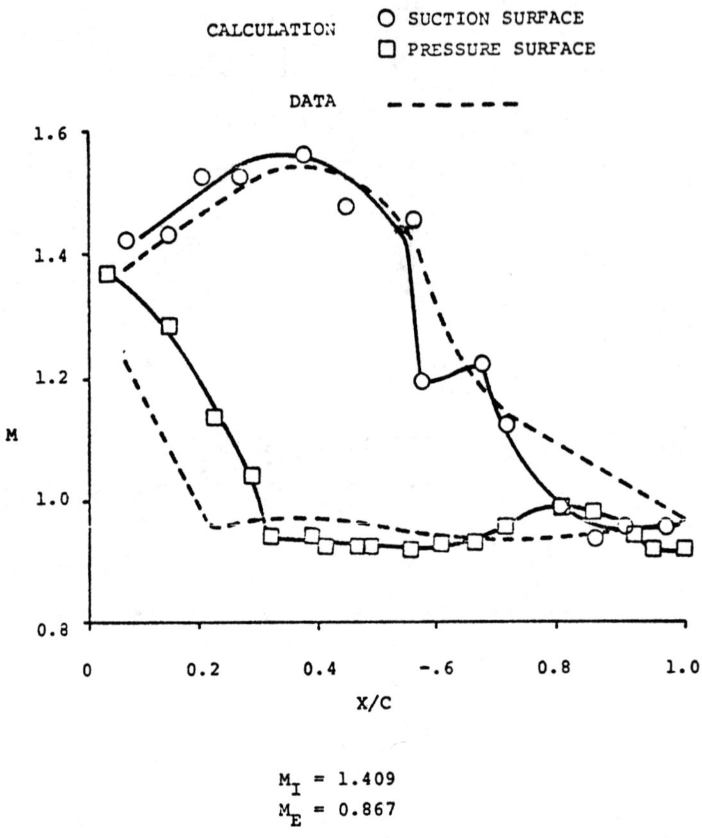

Figure 19. V.K.I. 9.5° Camber DCA Blade.

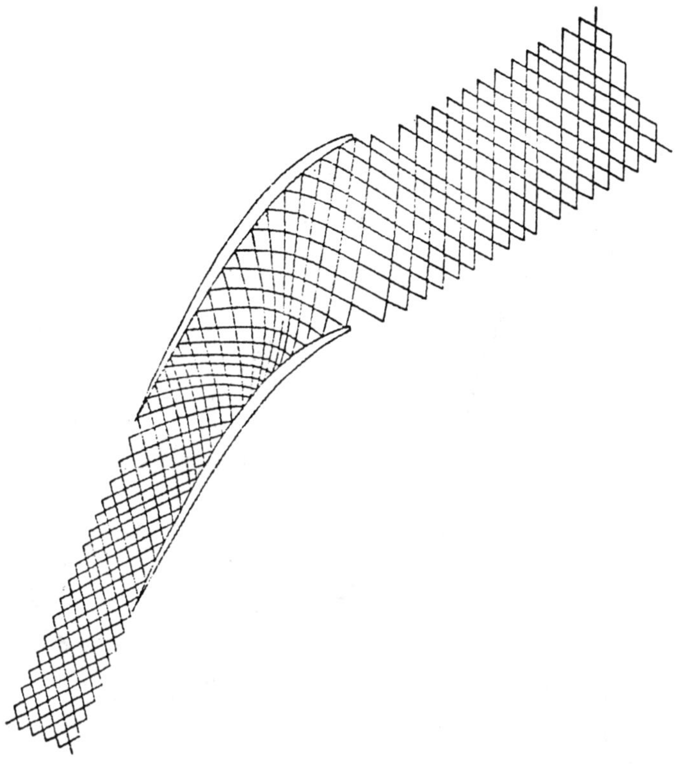

Figure 20. Calculated Grid System, ARL Cascade.

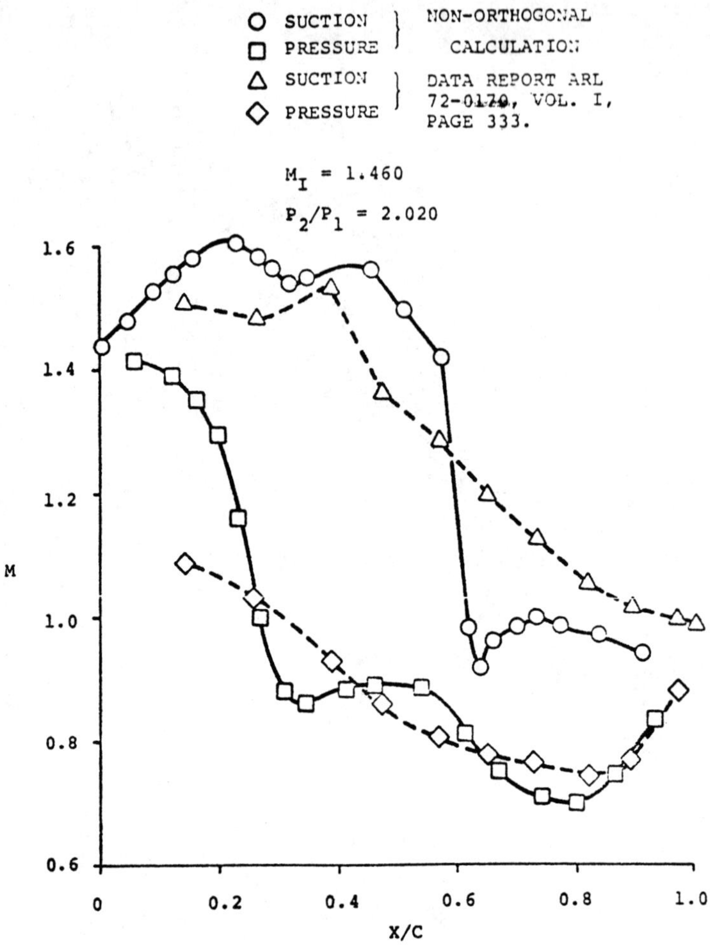

Figure 21. Surface Mach Number Distributions, ARL Cascades.

REFERENCES

[1] Erwin and Vitale, "Rotor Design of a High Tip Speed Low Loading Transonic Fan," AIAA 72-83, January 17-19, (1972).

[2] Starken, H. and Lichtfuss, J.J., "Some Experimental Results of Two-Dimensional Compressor Cascades at Supersonic Inlet Velocities," J. of Engineering for Power of ASME, July (1970).

[3] Gopalakrishnan, S., "Time Dependent Methods," presented at Lecture Series 59 of von Karman Institute, May (1973).

[4] McDonald, P. W., "The Computations of Transonic Flow Through Two-Dimensional Gas Turbine Cascades," ASME Paper, 71-GT-89.

[5] Murman, Earll M. and Cole, Julian D., "Calculations of Plane Steady Transonic Flows," AIAA Journal, vol. 9, pp. 114, January (1971).

[6] Dodge, Paul R., "Transonic Relaxation," von Karman Institute 1976 Lecture Series 3, February 2-6, (1976).

[7] Murman, Earll M., "Analysis of Embedded Shock Waves Calculated by Relaxation Methods," Proceedings of the AIAA Computational Fluid Dynamics Conference, July (1973).

[8] Dodge, Paul R., "A Transonic Relaxation Method for Cascade Flow Systems," von Karman Institute Lecture Series 59, May (1973).

[9] Dodge, Paul R., "A Non-Orthogonal Numerical Method for Solving Transonic Cascade Flow," to be presented at the ASME Gas Turbine Conference, March (1976).

DISCUSSION OF THE DODGE PAPER

MORETTI: I'm completely confused about this computational grid you have. If, in the elliptic zone you have to add points which are ahead and behind the point you are computing and the other - you have all the points behind, what happens when the Mach number is exactly one? Shouldn't you have points which lie on a perpendicular to the flow direction and therefore shouldn't you have a grid there which squashes into nothing?

DODGE: Yes, in theory; but it's a region actually where we use a mixture of both elliptic and hyperbolic - from a region of about 0.85 Mach number to about 1.1, which corresponds roughly to a 60° Mach angle; there's a mixture of both elliptic and hyperbolic. So there is really in that particular region some upstream influence. It does not seem to cause any particular inaccuracies or difficulties in the calculation.

MORETTI: In other words, does it make Tony Jameson particularly unhappy?

DODGE: Well, I don't know about that! Maybe it would make you unhappy from a theoretical standpoint but not from a practical standpoint.

RAE: I noticed that in all the grids that you showed, that the periodic points don't seem to lie on a surface of a constant value of one of the coordinates. Could you tell me how you enforced the periodicity of, say, velocity components?

DODGE; That's two different questions, depending on which method you're talking about. In the orthogonal method, it's simply a matter of interpolation, as you pointed out. In the non-orthogonal method, the actual total solution is between two complete surfaces. It's a complete channel solution. Now, the surfaces upstream and downstream of the blade are adjusted, when the major updates are made, are adjusted in shape, to attempt to get periodicity conditions upstream and downstream of the blade. You can do this so far as you match the unique incidence condition at the inlet and the unique exit condition at the exit.

RAE: I take it you've never encountered a problem in that bringing information forward, so to speak, violates any zone of silence condition.

DODGE: In the supercritical cases that's not a problem because that's not a region where your supersonic bubble is.

RAE: Right. When will you have the unique incidence effect?

DODGE: In the non-orthogonal situation, the differencing star at the boundaries was either centered or backward depending on whether the node is supersonic - and since all the walls are solid walls, it's a complete channel calculation. The calculation part - you get no problem. And then, everytime the grid system is updated, at the same time, the slip line or streamline shapes - whatever you want to call them - ahead and behind the blade, are updated in shape, to try to come up with a periodicity condition ahead and behind. Subsonically, that's done by a method that Wilkinson has applied to curvature of streamlines and supersonically it takes a little advantage of characteristics theory and a few things like that.

CALCULATION OF TRANSONIC POTENTIAL FLOWFIELDS ABOUT COMPLEX, THREE-DIMENSIONAL CONFIGURATIONS*

D. A. CAUGHEY

Cornell University

ANTONY JAMESON

New York University

ABSTRACT

Methods for extending iterative, finite-difference calculations of transonic potential flowfields to complex three-dimensional configurations are discussed. One particularly attractive approach is to use relatively simple conformal mappings in combination with shearing transformations to generate computational domains that are nearly-conformally mapped from the physical space in one family of coordinate surfaces, and which map the complex boundaries to grid surfaces. The application of such a method to a general wing-body combination or to a multi-bladed fan is discussed. A transformation to map the wing-fuselage or fan-hub combination to a convenient computational domain is proposed. The transformation is useful in its own right for treating the two-dimensional problems of flow past a profile in a wind tunnel or through a cascade. Some results of preliminary calculations are presented.

*This work was supported by NASA under Grants NGR 33-016-167 and NGR 33-016-201. The computations were performed at the ERDA Mathematics and Computing Laboratory, New York University, under Contract E(11-1)-3077.

INTRODUCTION

In recent years, iterative solutions of finite-difference approximations to the transonic potential equation have met with remarkable success. These methods have been used to predict the inviscid, mixed flowfields about airfoils [1], wings [2], and simple wing-body combinations [3] within the framework of small-disturbance theory; they have also been applied to calculate the inviscid, transonic flowfields about airfoils [4, 5], bodies of revolution [5, 6], nacelles [7, 8], and oblique [9] and swept [10] wings using the full potential equation.

There are two computational advantages of using small-disturbance theory. First, the equation is somewhat simpler, and its type is completely determined by the coefficient of the ϕ_{xx} term, allowing the construction of relatively simple type-dependent differencing schemes. Second, the boundary conditions can usually be transferred to some mean surface which can often be chosen as a coordinate plane in a simple cartesian system. For complex geometries, this is a great simplification, since the treatment of boundary conditions in cases when the boundary surface does not coincide with grid points is generally either very complicated and time-consuming or inaccurate.

When using the full potential equation the use of such mean surface approximations is generally inconsistent with the accuracy of the equation itself. The alternatives are (1) to use interpolation formulas to apply the boundary conditions at the grid points nearest the boundary surfaces, or (2) to use coordinate transformations which reduce the boundaries to coordinate surfaces. At the present time it is not clear which approach is better for the handling of complex geometrical shapes. The interpolation schemes require additional complexity in the difference codes to treat the variety of mesh-boundary intersections that may occur; the transformation method adds complexity to the equations themselves, and may require additional storage for transformation derivatives. Our approach is the latter. In this case the boundary conditions are satisfied exactly (in a finite-difference sense) on the boundary surfaces; it is also hoped that since the transformations need be calculated only once at the beginning of the solution, the method will compare favorably in terms of speed with the grid-interpolation methods.

PROPOSED ANALYSIS

Conformal transformations have proved a useful tool for the finite-difference calculation of two-dimensional flowfields. A preliminary transformation is performed which maps the body boundary onto some canonical curve (e.g., a profile onto the unit circle), and a finite-difference grid is set up in convenient coordinates in the mapped plane for which the canonical curve is a coordinate line. A difference approximation to the transformed equation is then solved by relaxation, with the boundary conditions applied along the appropriate coordinate lines.

Much of the advantage of this approach is lost when we go to problems in three dimensions, because there is no generalization of conformal mapping in this case. Also, if we perform separate conformal mappings to canonical curves in each of some family of coordinate surfaces, we lose orthogonality, so that numerous mapping derivatives have to be calculated. The labor of determining the map functions at each point of the grid, and the storage required to save them for repeated use during the iterative solution of the difference equations, can then become excessive.

An attractive alternative approach has been used by Jameson for calculating the three-dimensional flow past yawed [9] and swept [10] wings. The same basic idea has been applied to the flow past axisymmetric inlet nacelles by Caughey and Jameson [7]. The method consists of applying a simple conformal transformation (which can usually be generated by elementary functions) in each of one family of coordinate surfaces which almost maps the boundary surfaces to coordinate planes. A shearing transformation is then introduced to complete the mapping of the boundaries to coordinate surfaces. This final transformation renders the coordinate system non-orthogonal, but if the initial conformal mapping is carefully chosen, the shearing is everywhere slight, and the weak non-orthogonality seems not to cause any stability problems.

For the analysis of flow past an isolated, three-dimensional wing, a convenient mapping is the square root transformation applied in planes containing the wing section. If the branch point of the transformation is located just inside the leading edge of the profile at each spanwise station, this has the effect of mapping the wing surface to a shallow bump, which can then be reduced to a plane by a simple shearing transformation. The square-root transformation is particularly nice because the mapping modulus and

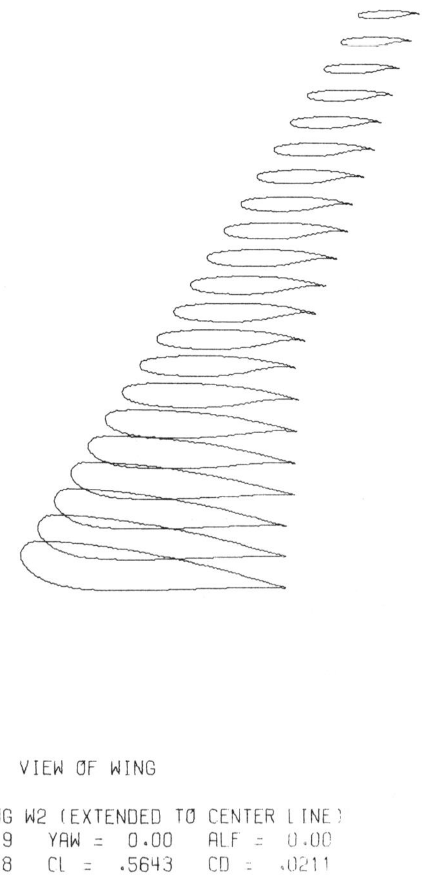

Figure 1a. Geometry of Swept Wing.

its derivatives can all be calculated from the coordinates in the mapped plane using only algebraic (non-transcendental) functions. Thus, there is little to be gained by storing the mapping derivatives; rather they can be rapidly calculated each time they are needed. The result of a typical calculation using this method is presented in Figure 1.

To treat more complex configurations in a similar manner, we need to reduce all boundary surfaces to shallow bumps by simple mappings. An example of the next level of complexity we have in mind is shown in Figure 2. We consider a wing mounted on a

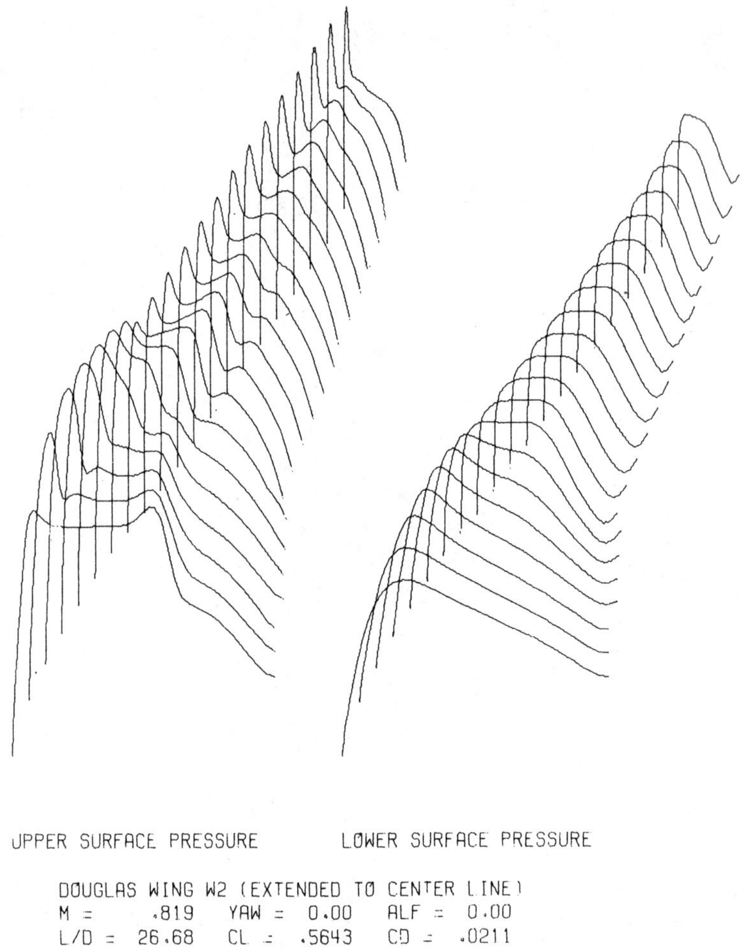

Figure 1b. Upper and Lower Surface Pressure Distributions on Swept Wing.

fuselage having a circular cross-section of varying radius denoted by R(x). We assume the flow is symmetric about the vertical plane containing the fuselage centerline, so that we may apply a symmetry condition there and consider the flow only in the half space.

We first define a singular line, just inside the leading edge of the wing, which will later be used as the branch point in a

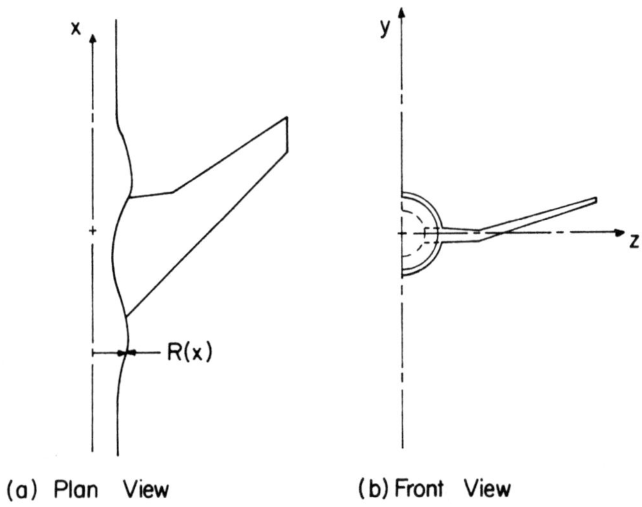

Figure 2. Geometry of Wing-Fuselage Combination.

conformal map to "unwrap" the wing surface. The location of this singular line can be denoted as

$$x = x_s(z),$$

$$y = y_s(z),$$

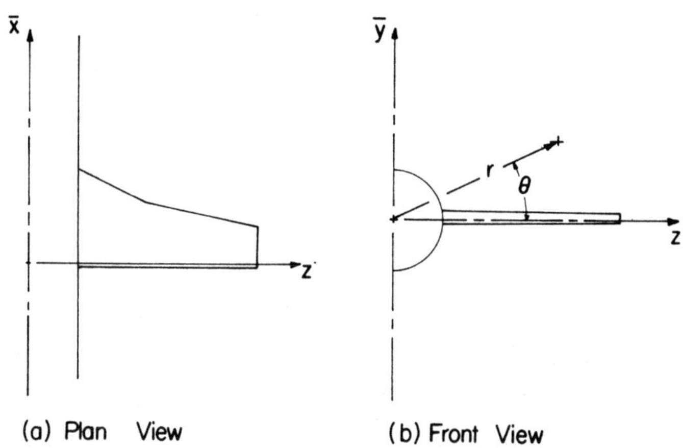

Figure 3. Normalized Geometry of Wing-Fuselage Combination

where x'_s, y'_s will correspond to the local sweep and dihedral, respectively, of the singular line. If we then introduce the coordinates

$$\bar{x} = x - x_s(z),$$

$$\bar{y} = y - y_s(z),$$

and finally

$$r = \sqrt{\bar{y}^2 + z^2/R(x)}$$

$$\theta = \tan^{-1}(\bar{y}/z),$$

the geometry will be transformed to that shown in Figure 3. A surface of r = constant will then look like Figure 4. It is in these surfaces, then, that we wish to introduce our nearly-conformal transformations. Note that the geometry of Figure 4 corresponds in the two-dimensional case to the flow past a profile in a solid-walled wind tunnel or, if the symmetry conditions at $\theta = \pm\pi/2$ are replaced by periodicity conditions, to the flow past one blade in an infinite, two-dimensional cascade.

The transformation which is the generalization of Jameson's square-root mapping to this case can be shown to be (after a

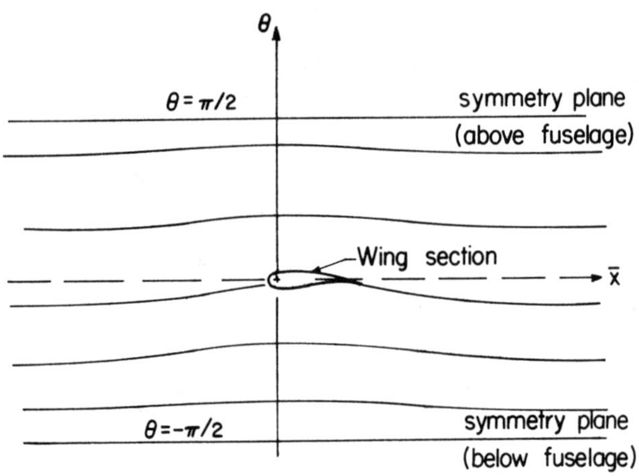

Figure 4. Surface of r = constant ($1 < r < r_{tip}$).

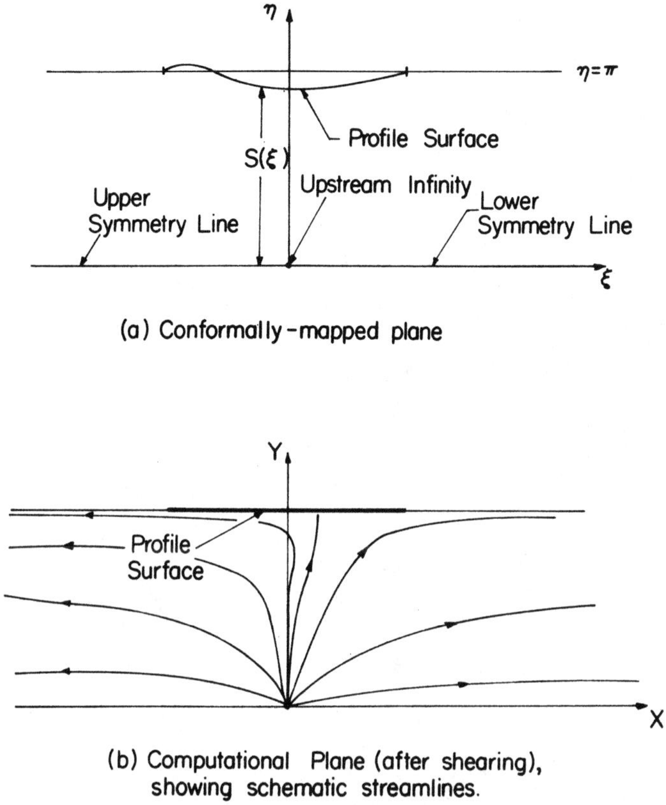

Figure 5. Nearly-Conformal Mapping of r = constant Surfaces.

rescaling of θ and a shift in the origin)

$$\bar{x} + i\theta = \log\{1. - \cosh \zeta\}, \tag{1}$$

where $\zeta = \xi + i\eta$. A schematic representation of the ζ-plane is shown in Figure 5a. The upper and lower symmetry lines (tunnel walls) map to the negative and positive real axes, respectively. The profile maps to a slight bump, near the line $\eta = \pi$. If we let $S(\xi)$ be the width of the infinite strip in the ζ-plane, then the final shearing transformation

$$X = \xi$$

$$Y = \eta / S(X)$$

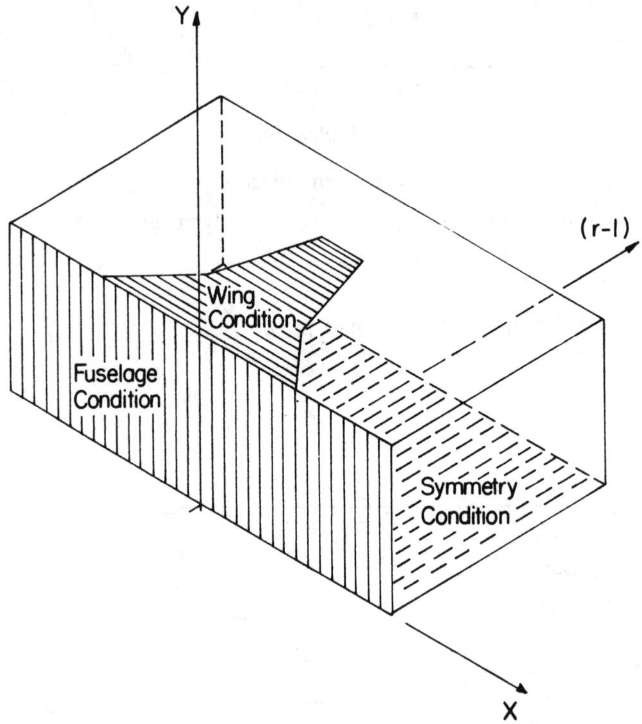

Figure 6. Sketch of Boundaries in Computational Domain.

reduces the strip to one of constant width, as shown in Figure 5b.

A complete sketch of the three-dimensional coordinate system is shown in Figure 6. The space is infinite in the directions of $\pm$ X, r, but the computational domain can be rendered a finite rectangular parallelepiped by suitable stretching transformations in these directions.

Finally, it can be mentioned that the above-described coordinate system is a logical one to use for the problem of flow through a fan (or propeller) or a three-dimensional cascade. Figure 7 shows the type of geometry that might be treated in this way. In this case the range of θ treated would be limited to the interval $[-2\pi/N, 2\pi/N]$, where N is the number of fan (or cascade blades, and the symmetry condition applied on these planes would be replaced by the requirement of periodicity.

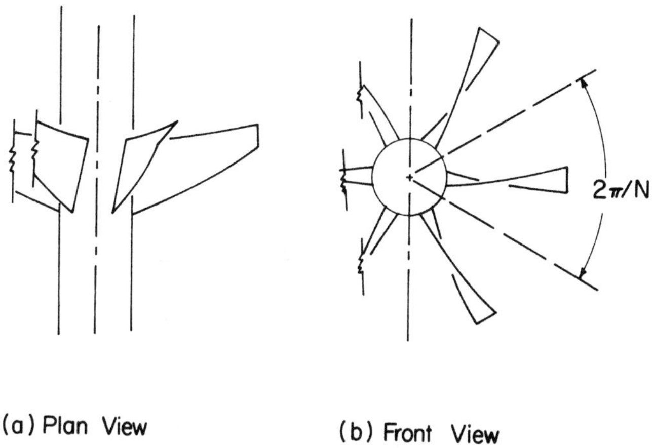

Figure 7. Geometry of an N-Bladed Fan.

A fundamental difference between this problem and that of the wing-fuselage combination is that the onflow would now be rotational in the reference frame rotating with the blades. For cases in which the onflow is irrotational in an absolute frame, however, a reduced potential can be introduced to describe the velocity field in terms of the gradient of a single scalar potential plus a constant rotational component. (See, e.g., Vavra [11].)

PRELIMINARY RESULTS

To demonstrate the efficacy of the coordinate system just described for the calculation of three-dimensional flows, some preliminary two-dimensional calculations have been performed using the mapping of Eq. (1). These particular calculations have been performed using the symmetry condition on the lines $\theta = \pm \pi/2$, and hence correspond to the flow past a profile in a non-ventilated, or solid-wall wind tunnel.

A sketch of the finite-difference mesh produced by this transformation is shown in Figure 8. The sketch shows the point distribution for a very crude grid containing 32 x 8 mesh cells, and is for the geometry of the original Korn airfoil [4] in a tunnel having a total height six times the airfoil chord.

To avoid difficulties at infinity, the singular part of the

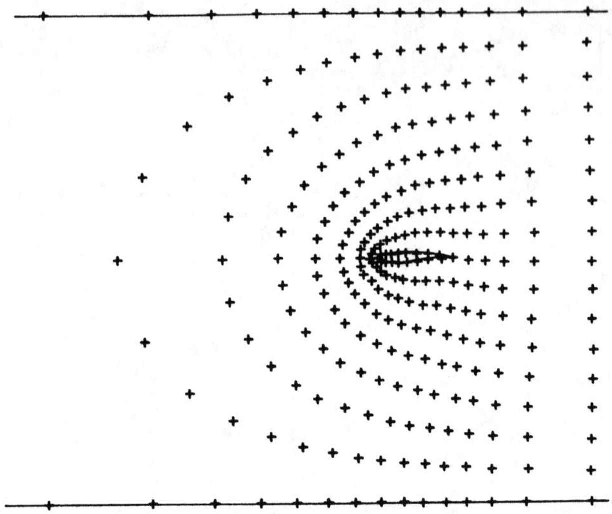

Figure 8. Representative Mesh Distribution in Physical Plane (32 x 8 Grid).

velocity potential is removed, and the calculation is performed in terms of the reduced potential

$$G = \phi - \bar{x},$$

where, from Eq. (1),

$$\bar{x} = \log(\cosh \xi - \cos \eta).$$

If we define

$$T = 1/S(\xi),$$

and

$$U = \phi_\xi = \frac{\sinh \xi}{\cosh \xi - \cos \eta} + G_X + Y_\xi G_Y,$$

$$V = \phi_\eta = \frac{\sin \eta}{\cosh \xi - \cos \eta} + T\, G_Y,$$

the planar potential equation becomes

$$AG_{XX} + BG_{XY} + CG_{YY} + DG_{Y} + E = 0, \tag{2}$$

with

$$A = a^2 - U^2/h^2,$$

$$B = -2\left\{\frac{U(TV + Y_\xi U)}{h^2} - Y_\xi a^2\right\},$$

$$C = a^2(T^2 + Y_\xi^2) - \frac{(TV + Y_\xi U)^2}{h^2}$$

$$D = (a^2 - U^2/h^2)\eta\ T'' - 2T'\frac{UV}{h^2},$$

$$E = \frac{-U^2 + V^2}{h^2}\,\frac{1 - \cosh\xi\cos\eta}{(\cosh\xi - \cos\eta)^2} + 2\frac{UV}{h^2}\frac{\sinh\xi\sin\eta}{(\cosh\xi - \cos\eta)^2}$$

$$\frac{-(U^2 + V^2)}{h^4}\,\frac{U\sinh\xi\cos\eta + V\cosh\xi\sin\eta}{(\cosh\xi - \cos\eta)^2},$$

where a is the local speed of sound, and h^2 is the square of the modulus of the map function

$$h^2 = \left|\frac{d(\bar{x} + i\theta)}{d(\xi + i\eta)}\right|^2 = \left|\frac{\cosh\xi + \cos\eta}{\cosh\xi - \cos\eta}\right|.$$

The profile shape downstream of the trailing edge is continued smoothly to infinity in the computational domain, and allowance is made for a constant jump in potential across this cut in the physical plane. The magnitude of this jump is determined by the Kutta condition at the trailing edge of the profile.

A finite-difference form of Eq. (2) is solved using the rotated differencing scheme first suggested by Jameson [9]. The equation is solved subject to the conditions that

$$G_Y = \frac{S' \left\{ \dfrac{\sinh \xi}{\cosh \xi - \cos \eta} + G_X \right\} - \dfrac{\sin \eta}{\cosh \xi - \cos \eta}}{T(1 + S'^2)}$$

on the profile surface,

$$G_Y = 0$$

on the symmetry lines, and

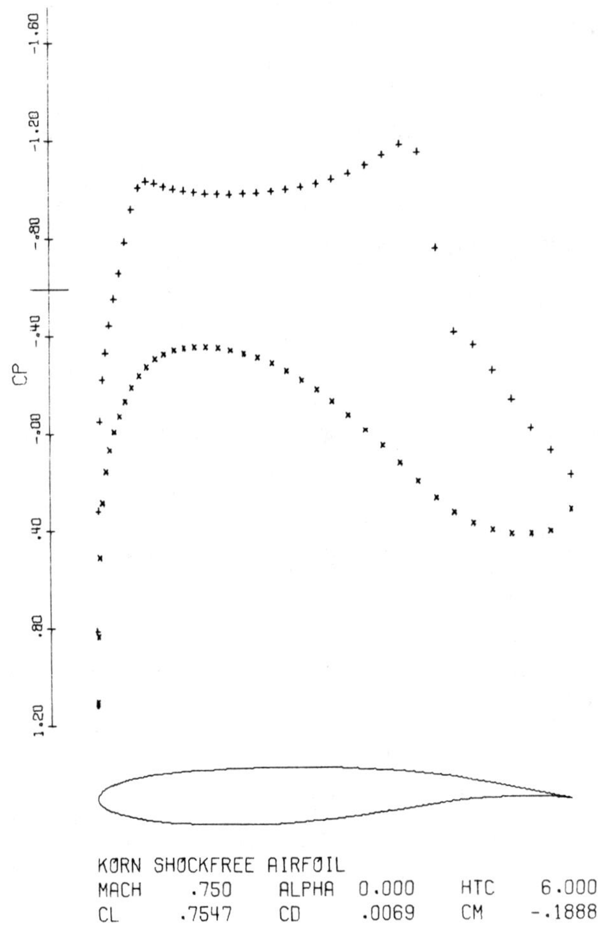

Figure 9a. Pressure Distribution on Korn Airfoil in a Wind Tunnel. h/c = 6.0.

$$G_X = 0$$

at downstream infinity. The problem for G is thus a purely Neumann one, and the value of G is allowed to float to an arbitrary level during the course of the iteration.

A comparison of two such solutions is shown in Figure 9. The results are for the flow at $M_\infty = 0.75$ past the Korn airfoil at zero angle of attack, and are calculated on a grid containing 128 x 32 mesh cells in the X and Y directions, respectively. One

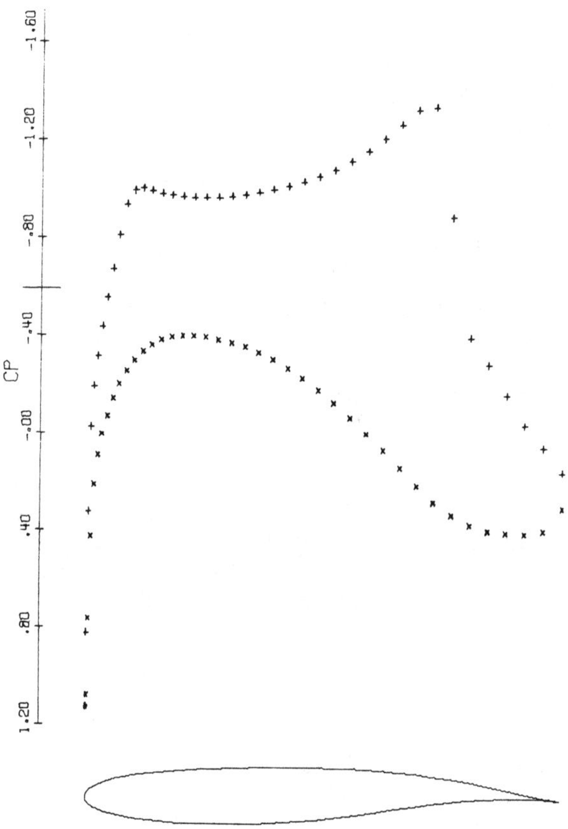

KØRN SHØCKFREE AIRFØIL

MACH	.750	ALPHA	0.000	HTC	4.000
CL	.7980	CD	.0177	CM	-.2235

Figure 9b. Pressure Distribution on Korn Airfoil in a Wind Tunnel. h/c = 4.0.

calculation is for the profile in a tunnel having a total height six times the airfoil chord; the other in a tunnel having a total height four times the airfoil chord.

A REMARK ON TWO-DIMENSIONAL CASCADE CALCULATIONS

While the basic mapping of Eq. (1) can be used to perform two-dimensional cascade or wind-tunnel calculations as demonstrated in the preceding section, it is not necessarily the best coordinate system for these problems. For the numerical calculation of flows past isolated profiles, the most accurate results to date have been performed in coordinate systems which conformally map the exterior of the profile to the interior of the unit circle [4, 5], as mentioned in the introduction. This has the advantage of allowing a large number of grid points to be placed on the airfoil surface, and of easily allowing concentrations of the points at both the leading and trailing edges of the profile (the latter being especially important for aft-cambered airfoils which have large gradients there).

The analogous mapping for the cascade problem would be to map the physical geometry to a ring. One possible mapping to perform this transformation can be visualized in two steps as shown in Figure 10. First we map the infinite strip with to a plane with two slits on the real axis, e.g., by

$$z = -\log w,$$

where $z = x + iy$ and $w = u + iv$ are the complex variables in the physical and mapped planes, respectively. Then the mapping of these slits to concentric circles can be expressed in terms of the Elliptic Integral (see Kober [12])

$$\frac{\sigma}{\bar{r}} = \exp\left\{ \int_{\infty}^{\zeta} \frac{ds}{\sqrt{4s(s-1)(s-e_2)}} \right\},$$

where e_2 is the value of u at the image of the profile trailing edge in the w plane, and $\bar{r}$ is the radius of the inner circle of the ring. The profile would be mapped to a nearly circular contour near the outer circle by this sequence of transformations. This contour could be mapped conformally to an exact circle by an iterative scheme similar to that used to map an isolated profile to a circle,

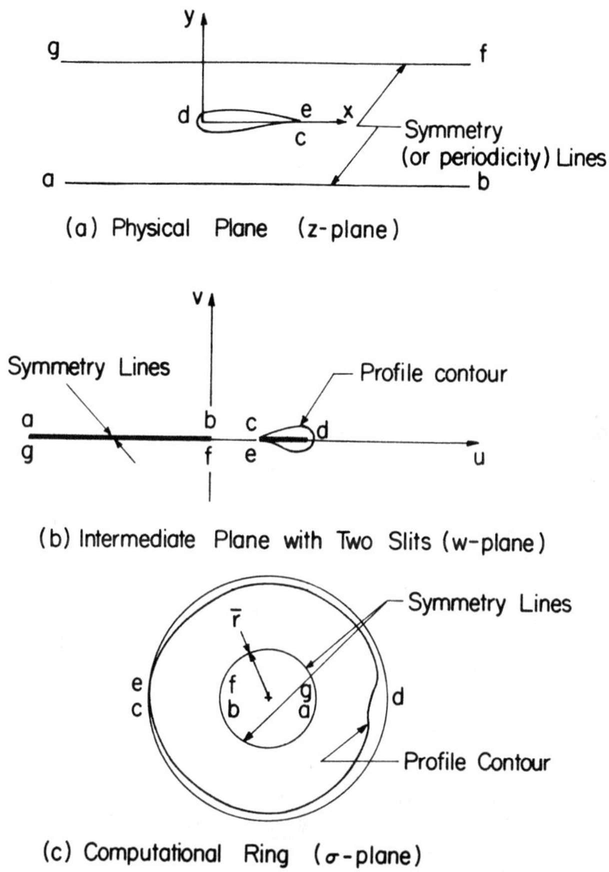

Figure 10. A Mapping for Two-Dimensional Cascade Calculations.

or else a simple shearing transformation could be used to make it a coordinate line.

CONCLUSION

Some ideas for handling complex three-dimensional geometries in the finite-difference calculation of inviscid transonic flow patterns have been presented. An attractive approach seems to be to use nearly conformal maps in a family of coordinate surfaces which reduce the boundaries to those of the computational grid.

A convenient mapping for use in calculations involving wing-fuselage combinations, fans, or three-dimensional cascades has been introduced, and sample results demonstrating its success in two-dimensional problems have been presented. Possible conformal mappings for the two-dimensional cascade (or wind tunnel) problem are also briefly discussed. These ideas have already proven effective in the calculation of transonic flows past swept and yawed wings. On the basis of these successes, we may expect the further application of these ideas to result in solutions with adequate accuracy for a variety of important engineering problems.

Finally, it should be noted that three-dimensional calculations require a substantial amount of computer time. For example, the swept wing calculation of Figure 1 requires about 75 minutes on the CDC 6600 (or 15-20 minutes on the CDC 7600). There are, however, a number of promising possibilities for accelerating the rates of convergence of the iterative schemes. For example, extrapolated relaxation [7], alternating direction methods [13], fast elliptic-solvers used in conjunction with relaxation [7, 14], and a multi-grid method [15] have all been successfully applied to transonic problems. Thus, there appears to be the prospect of substantial reductions in the cost of these calculations.

REFERENCES

[1] E. M. Murman and J. D. Cole (1971), "Calculation of Plane, Steady, Transonic Flows," AIAA J., 9, 114.

[2] W. F. Ballhaus and F. R. Bailey (1972), "Numerical Calculations of Transonic Flow about Swept Wings," AIAA Paper 72-677, Presented at AIAA 5th Fluid and Plasma Dynamics Conference, Boston, Mass., June 26-28.

[3] F. R. Bailey and W. F. Ballhaus (1972), "Relaxation Methods for Transonic Flow about Wing-Cylinder Combinations and Lifting Swept Wings," presented at Third International Conference on Numerical Methods in Fluid Dynamics, Paris, France, July 3-7.

[4] P. R. Garabedian and D. G. Korn (1972), "Analysis of Transonic Airfoils," Comm. Pure Appl. Math., 24, 841.

[5] Antony Jameson (1971), "Transonic Flow Calculations for Airfoils and Bodies of Revolution," Grumman Aerodynamics Report 370-71-1.

[6] J. C. South, Jr. and Antony Jameson (1973), "Relaxation Solutions for Inviscid Axisymmetric Transonic Flow Over Blunt or Pointed Bodies," Proceedings of AIAA Computational Fluid Dynamics Conference, pp. 8-17.

[7] D. A. Caughey and Antony Jameson (1976), "Accelerated Iterative Calculation of Transonic Nacelle Flowfields," AIAA Paper 76-100, Presented at AIAA 14th Aerospace Sciences Meeting, Washington, D.C., January 26-28.

[8] B. G. Arlinger (1975), "Calculation of Transonic Flow Around Axisymmetric Inlets," AIAA Paper 75-80, presented at AIAA 13th Aerospace Sciences Meeting, Pasadena, California, January 20-22.

[9] Antony Jameson (1974), "Iterative Solution of Transonic Flows over Airfoils and Wings, Including Flows at Mach 1," Comm. Pure Appl. Math., 27, 283.

[10] Antony Jameson and D. A. Caughey (to appear), "Numerical Calculation of the Transonic Flow Past a Swept Wing."

[11] M. H. Vavra (1960), Aero-Thermodynamics and Flow in Turbomachinery, John Wiley and Sons.

[12] H. Kober (1957), Dictionary of Conformal Representations, Dover.

[13] Antony Jameson, "An Alternating-Direction Method for the Solution of the Transonic Small-Disturbance Equation," New York University ERDA Report C00-3077-96.

[14] Antony Jameson, "Accelerated Iterative Schemes for Transonic Flow Calculations using Fast Poisson-Solvers," New York University ERDA Report C00-3077-82.

[15] Jerry C. South, Jr. and Achi Brandt (1976), "Application of a Multi-Level Grid Method to Transonic Flow Calculations," Project SQUID Workshop on Transonic Flow Problems in Turbomachinery, Monterey, California.

DISCUSSION OF THE CAUGHEY-JAMESON PAPER

DODGE: I am a little concerned on when you are talking about the fan, what your boundary conditions are in your two periodicity walls.

JAMESON: If we have the kind of coordinate systems I've suggested, there is a periodic boundary condition at the appropriate place.

DODGE: There is a flow across the wall.

JAMESON: Yes. In the case of a wing cylinder, symmetry says there's no flow across the wall, but when you go to the fan you introduce the periodic conditions but there doesn't seem to be any reason why that should create difficulty. The coordinates stretch out to infinity, so you get the far field taken care of, hopefully.

CHENG: I have two remarks or questions, one related to your acceleration algorithm, another more on aerodynamics. The first one - in your acceleration algorithm, the work according to our experience, that is, the work of Dr. Hafez and myself, we find that using the simplest kind of algorithm like that implied by what you said, that is, taking out the largest eigenvalue, we find it's not consistently working for the transonic small disturbance problem. We have to include the second largest eigenvalue to account for that. In fact, we are quite surprised at the success of your example that makes use of the formula based on the largest single eigenvalue. I would like you to comment on the fact that - what is the reason that you find using the higher order formula not practical.

JAMESON: I did try - Dave Caughey and I tried - a higher order formula and that didn't seem to work very well in the experiments we made. I think the question of whether this will work depends on whether you really have got a dominant eigenvector appearing in the error distribution and that depends pretty much on the nature of your relaxation scheme, the coordinate system, and a number of other factors. Also it depends on the over relaxation patch. We know that if you were doing Laplace's equation in a square and used the optimum relaxation factor, then every eigenvalue would have the same magnitude and the extrapolation idea wouldn't be too good. However, it seems that in a typical transonic calculation, that you try to raise the relaxation factor to the value that would be optimal for a Laplace's equation, that they have a habit of diverging, and therefore we appear very often to be running at a much lower relaxation factor which seems to result - it certainly does in the circle plane, because I printed out error distributions that showed this in a dominant eigenvectoral period. Should that occur, then the extrapolation will certainly do some good - but you can't say that it's necessarily going to occur in every calculation you try, but I believe it's worth experimenting with in these 3-D calculations. Even if you only get a factor of 2, it's still interesting, isn't it? 75 minutes down to 40 minutes on the 6600 would be a worthwhile savings; it's worth a try - not much programming involved.

CHENG: Thank you - the second question has to do with the aerodynamics - that is, one of the important aspects in the 3-D effect of the high aspect ratio wing, is the effect corresponding to Prandtl's upwash correction. Now, that depends quite a lot on how we take care of the far field. Now, I understand there are difficulties in the storage problem and in the conversion problem in this program. I just would like to hear your comment on how you handle that, and that type of thing.

JAMESON: All right - well - firstly, these coordinate systems are stretched out to infinity. When you get to infinity in a 3-D lifting situation, there will be a disturbance in the downstream infinity plane - the Treftz plane; however, if you look at the parabolic coordinate system, it turns out that the parabolic coordinates expand in such a way that the disturbance in the Trefftz plane is contracted down to a single line coincident with the vortex sheet. That's just something that's a property of that coordinate system, so that therefore it's quite appropriate to apply a zero disturbance potential in that particular coordinate system. If we go to the fan calculation, I think the downstream boundary conditions you may apply would have to be thought about rather carefully, because you haven't got the same expansion there, and I think you would certainly need to trail vortex sheets behind the blade, of course - and that should still have a disturbance that would be visible in the coordinate system I've described at the downstream infinity. However, I think you would find that the appropriate equation would be Laplace's equation in the far downstream plane - which could be solved in the course of the relaxation technique. So I don't believe there's any major difficulty there. I do compute induced drag with that swept wing calculation and it has the right type of behavior, though the program does seem to underestimate the induced drag, at the present time.

CALCULATION OF SUPERCRITICAL FLOW PAST A DOUBLE WEDGE BY TELENIN'S METHOD

KEUN SICK CHANG AND MAURICE HOLT

University of California

Chattot (1976) applied the method of Telenin to calculate supercritical flow past a symmetrical double wedge. The physical plane is shown in Figure 1. The flow is subsonic everywhere except in a small supersonic region attached to the shoulder of the wedge CTD where CT is the sonic line; this region is terminated on the downstream side by the shock DT.

To simplify the solution of the problem we transform it into the hodograph plane. In this plane the straight lines bounding the edge transform into straight lines in the same directions. The point C-D is singular and represents a Prandtl Meyer expansion which transforms into a characteristic CD_1 (see Figure 1). The shock wave (S) transforms into curves $D_2T(S_2)$ and $TD_1(S_1)$, the shapes of which are unknown at the start of the calculation. The unknown in the hodograph plane is the stream function ψ which satisfies Chaplygin's equation. The point at infinity in the physical plane transforms into the point A_∞, F_∞. This is a singular point in the neighborhood of which the behavior of ψ is known.

Telenin's method is applied along three directions, BC, BA_∞, ED_2, each extending up to the characteristic CD (and its continuation). The boundary conditions are $\psi = 0$ along the three lines shown and along the characteristic through C. To the right of the curve D_2D_1 the transformation of this solution on the physical plane is triple valued and a shock wave D_2D_1 is fitted step by step to exclude this region. A full account of the work is given in Chattot (1976).

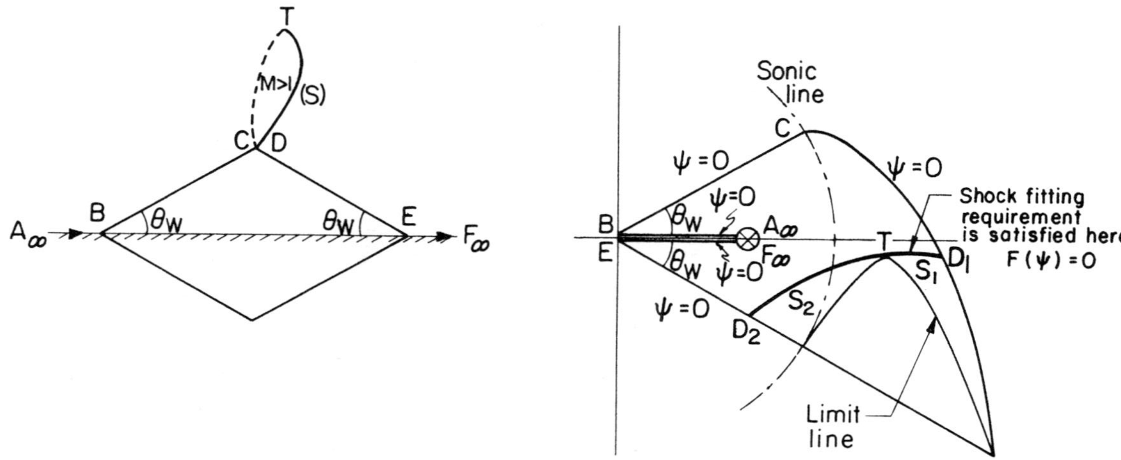

Figure 1. Flow Past a Double Wedge With No Angle of Attack

For a wedge at very small angle of attack the zero streamline in both the physical and the hodograph planes is shown in Figures 2 and 3. In both figures, the displacement of the forward stagnation point to a position on the lower wedge surface and the local separation bubble near the wedge vertex are neglected. As pointed out in Vincenti and Wagoner (1954) and in Guderley and Yoshihara (1953), this is justifiable at sufficiently small angles of attack. These figures cover the upper and lower surfaces, respectively. In asymmetric flow, we have to consider two Riemann surfaces with common branch line and branch point between the two. The hodograph boundary for each Riemann surface is certainly changed from that of the symmetric flow but it does not present any additional difficulty.

A question arises, however, concerning the assumption that the zero streamline emanating from the stagnation point at the trailing edge is straight in the hodograph plane. As a matter of fact, this part of the streamline in the physical plane would have a curvature which is slight for a small angle of attack and which vanishes as the streamline becomes parallel to the other streamlines at infinity. In the hodograph plane, the streamline $E'F'_{\infty}$ is not straight but is slightly curved by an unknown amount. It seems reasonable, however, that we assume that the method suggested by Chattot is not too sensitive to the small boundary perturbation given locally, especially at $\phi = \pm\pi$, which lies in the subsonic region (see Eq. (6.2.1) of Chattot (1976)). Thus it is expected that the solution obtained by solving the boundary value problems in the hodograph plane for the boundaries shown in Figures 2 and 3 do in fact approximate the exact solution well in the region of primary concern, namely, near the supersonic pockets terminated by recompression shocks.

REFERENCES

Chattot, Jean-Jacques, J. Fluid Mechanics (1976) to appear.

Guderley, G. and Yoshihara, H. (1953), J. Aero. Sci., 20, 757-768.

Vincenti, W. G. and Wagoner, C. B. (1954), NACA, TR 1180.

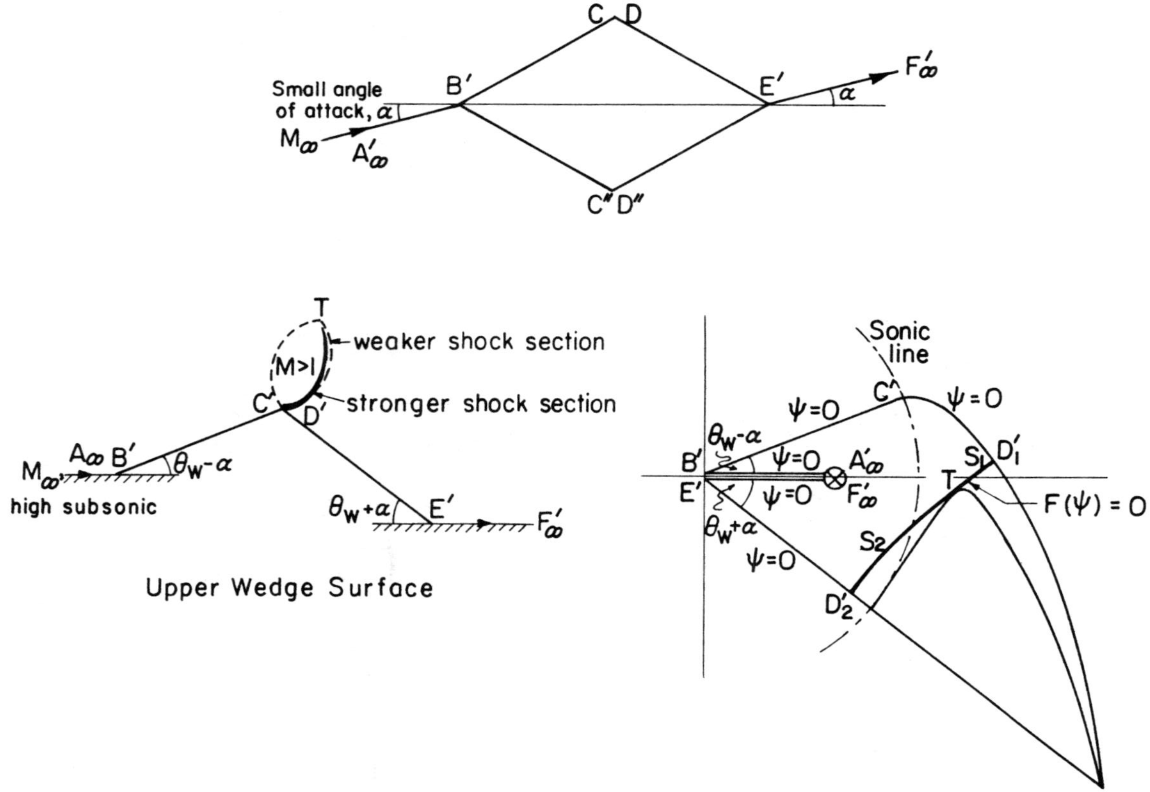

Figure 2. Flow Past a Double Wedge With Angle of Attack

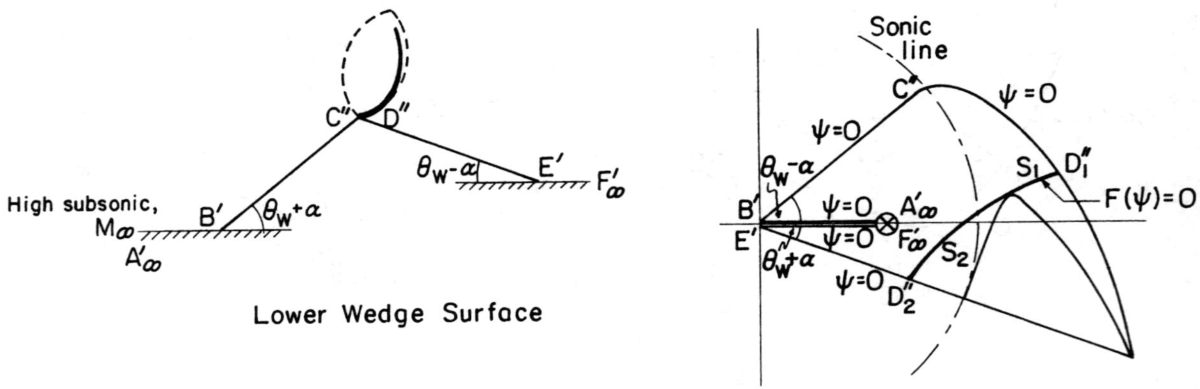
High subsonic,
M_∞
A'_∞
B'
$\theta_W+\alpha$
C''
D''
$\theta_W-\alpha$
E'
F'_∞
Lower Wedge Surface
Sonic line
C''
$\psi=0$
$\psi=0$
$\theta_W-\alpha$
B'
E'
$\psi=0$
A'_∞
$\psi=0$
F'_∞
$\theta_W+\alpha$
$\psi=0$
D''_2
S_2
S_1
D''_1
$F(\psi)=0$

Figure 3.

DISCUSSION OF THE CHANG-HOLT PAPER

CHENG: There's one interesting piece of information perhaps you might be able to point out and I believe that it will be of significance in the study of boundary layer separation near the trailing edge. Namely, it is that: what is the singular behavior of your pressure coefficient in approaching the trailing edge, in terms of physical variables? Since you already have the analytical behavior in hodograph variables, can you say something about its behavior? The reason being that, assuming any power law behavior of the trailing edge or leading edge, you might be able to reduce the transonic small disturbance equation to a self-similar solution form - that is, an ordinary differential equation - but that system degenerates in the case of a wedge and one cannot find much information out of that approach.

HOLT: I don't know what the answer is, but certainly it's in the - I guess we could extract it from the paper.

CHENG: You could work it out from that, couldn't you?

HOLT: Yes. Now, we solved the exact equations, you realize that? Except, we assume the shock is weak or of moderate strength - which I think is reasonable.

CHENG: I'm not referring to the shock; I'm referring to the trailing edge behavior.

HOLT: I know. But we did solve the exact equations.

SOUTH: For the wedge at angle of attack - you have the unsettling problem of the stagnation point possibly being slightly below the sharp nose, and then the compressible flow wants to expand to sonic velocity right at the apex of the wedge. Do you intend to wrestle with this problem - the microscopic detail, of the exact solution there?

HOLT: Well, we certainly intend to move the stagnation point away from the leading edge of the - the corner of the wedge. It seems as a first start that we would treat the flow around the corner in an incompressible way because it's going to be a very local effect. Maybe when we can do things like interaction with boundary layer programs; then we could get more ambitious in the whole thing.

MELNIK: I should refresh your memory. I don't know if you know of the work, but many years ago, when Vincenti was at AMES, Vincenti and Wagoner looked at a problem very similar to that, and I think it might have been a wedge - he did a relaxation solution - and it was clear that the flow as it expands around the sharp leading edge, exceeds the limiting velocity, and you have a pocket. They interpreted it, in the inviscid flow, as a pocket of zero density, and with a line separating that. So it's clear if you want physical interpretation, you've got to put a boundary layer in, or some rounding.

HOLT: Yes. I think that - to put in a free streamline coming from that edge would be feasible; you could map that in the hodograph plane, and you know where it is.

SUPERCRITICAL CASCADE DESIGN*

DAVID G. KORN
New York University

INTRODUCTION

Recently, there has been a revival of interest in the area of transonic flow. This has been inspired by several factors, in particular the energy crisis and its demand for fuel conservation. Improved aerodynamic design is one of the most promising ways to increase turbomachinery efficiency and thereby reduce fuel consumption. It is natural to look for methods to design blades which will be as efficient as possible. Performance improves as the speed increases until shocks appear. Therefore we look for designs which delay the onset of shock formation.

This paper is a review of my research efforts during the past several years in an attempt to achieve better aerodynamic designs for wing and blade cross-sections. Also, I will outline our current work [4]. A design generated by our new code is presented.

COMPLEX EXTENSION

The impetus for our work on the design problem has been our ability to compute analytic solutions to second order partial equations in two dimensions by the method of complex extension [3]

* Work sponsored by NASA under Grant NGR 33-016 - 167 and by ERDA under Contract E(11-1)-3077 with New York University.

regardless of whether the equation is elliptic, hyperbolic, or mixed. For the case of an inviscid isentropic fluid in the x, y plane we have the second order partial differential equation

$$(c^2 - u^2)\phi_{xx} - 2uv\phi_{xy} + (c^2 - v^2)\phi_{yy} = 0 \qquad (1)$$

for the velocity potential ϕ where u and v are velocity components in the x and y direction respectively and c^2 is a function of $u^2 + v^2$ which can be computed from Bernoulli's equation

$$\frac{u^2 + v^2}{2} + \frac{c^2}{\gamma - 1} = \text{const.} \qquad (2)$$

where γ depends on the fluid and is approximately 1.4 for air.

The method of complex extension can be carried out efficiently on these equations by introducing characteristic coordinates ξ and η as independent variables. Introducing the stream function ψ, the equations of motion for x, y, ϕ and ψ, can be written in terms of ξ and η as

$$\lambda_- y_\xi + x_\xi = 0; \qquad \lambda_+ y_\eta + x_\eta = 0$$
$$\phi_\xi = \tau_+ \psi_\xi; \qquad \phi_\eta = \tau_- \psi_\eta \qquad (3)$$

where $\tau_{\pm}$ and $\lambda_{\pm}$ are analytic functions of ξ and η. The relationship between characteristic coordinates ξ and η and the hodograph variables u and v can be expressed explicitly modulo an analytic function which maps a characteristic coordinate onto itself. The selection of this mapping function will be discussed later.

The classical method of characteristics can be used to generate solutions to (3) using complex arithmetic for all quantities [5] A discussion of the paths of integration needed for the analytic continuation into the hyperbolic region can be found in [1] and [6].

SINGULAR SOLUTIONS

While many improvements have been made in the speed and accuracy of computing solutions with the method of complex characteristics, most of the effort has gone into finding ways of superimposing solutions to solve the desired physical problem. The first consideration for the solution we generate is that it have the appropriate behavior near infinity in the physical plane. For an airfoil this means that the flow becomes uniform in the neighborhood of infinity. For the cascade there is both an inlet velocity and an exit velocity. The singularity for an airfoil can be represented by a pole at a point in the characteristic coordinate plane while two logarithmic singularities are needed to represent the flow in the neighborhood of infinity for the cascade case. For large blade spacings these logarithms are close and behave like a dipole. In the limit they approach the singularity for the airfoil case. These singular solutions can be computed to high resolution by an expansion of the form

$$Z = Z_1 S_1 + Z_2 S_2 + Z_3 \tag{4}$$

where Z is a symbolic notation referring to either x, y, ϕ or ψ and S_1 and S_2 are the appropriate singularities. We can compute Z_1 and Z_2 as solutions to the differential equations (3) with characteristic initial conditions similar to those of the Riemann function. The differential equation for Z_3 can be found by plugging Z into equations (3). The representation of the singularity is discussed in more detail in [1] for the isolated airfoil and in [5] for the cascade.

INITIAL DATA

The aspect of the design which has been the most difficult to implement in a satisfactory way has been the selection of initial data needed for a desired objective. We have at our disposal the analytic mapping function for the characteristic coordinates and an analytic function for the characteristic initial value problem. In all but our most recent work the mapping function was chosen merely to require that the profile be single valued in the characteristic coordinate plane. This could be achieved in most cases by a quadratic function which opened up the typically two sheeted hodograph plane. A two parameter family of such transformations, expressed in terms of the branch point, was all that was needed. For more highly cambered blades we found it helpful to add additional branch points thereby increasing the degree of the polynomial.

In all but our most recent work the initial function was defined by

$$Z_3(\xi, 0) = \sum_{k=1}^{m} \alpha_k f_k(\xi) \tag{5}$$

where the f_k are elementary functions such as powers and logarithms and the α_k are complex parameters. The location of the logarithms must also be specified. In our earliest work the user selected these parameters more or less by trial and error. For each choice a solution was computed which often did not even resemble an airfoil. Moreover, the parameters were non-orthogonal so that simple rules for adjusting parameters could not be found.

Later versions used a least square technique to select the linear parameters α_k in (5). The user specified the shape of the desired streamline in the characteristic plane. This proved to be an enormous help and many airfoil sections were designed with this program [2]. There were three drawbacks to the method. First, the problem was still overdetermined so that good agreement between the prescribed shape in the characteristic plane and the solution was not always possible. Very often a non-physical solution was generated which had holes in the streamline in the characteristic plane.

Second, the user had to know how to express the desired design objective in terms of a curve in the characteristic plane. Few had the patience to train themselves for this task. The final drawback was that practical designs often took many runs. For cascade design which has more degrees of freedom than the isolated airfoil, approximately 30 runs were used to design a typical case. Besides the cost of computer time, it placed too high a demand on the user to make the appropriate runs.

WORK IN PROGRESS

All of the drawbacks outlined in the previous section have been overcome in a satisfactory manner by the technique described in [4]. To use the code one must supply only the desired speed distribution as a function of arclength and the critical speed. In a single run a shape is computed which will have the prescribed speed distribution for subsonic flow and a nearby speed distribution in the transonic case.

To achieve this we have formulated a pair of boundary value problems whose simultaneous solution leads to the desired shape. The mapping of the characteristic coordinate plane onto the hodograph plane is determined by prescribing the speeds at evenly spaced points along the unit circle which are obtained from the input distribution. The power series for the mapping is determined by trigonometric interpolation using Fast Fourier transform techniques. Secondly we require that the stream function vanish on the unit circle in the characteristic plane. This is achieved by using the method of complex extension to compute a family of solutions to equations (3) using the powers of ξ as characteristic initial data and by selecting the weights by interpolation through evenly spaced points of the circle.

The two problems are solved in an iterative fashion in alternate steps. At each iteration cycle we compute a valid profile. Successive profiles produce closer agreement to the input distribution than the last. For transonic flow the problem is overdetermined and the blade shape which results has a slightly modified pressure distribution. Closure of the profile is controlled by simple adjustments to the input pressure distribution.

The code is long and complicated but a typical run requires only 3 or 4 iterations which takes about 5 to 10 minutes on a CDC 6600. Figure 1 is an example of a cascade airfoil which has been designed with this code.

More work still needs to be done in order to make the code more reliable and easy to use for engineering applications. Also, for high solidity design it may be necessary to change the basic domain from the circle to the ellipse.

REFERENCES

[1] Bauer, F., Garabedian, P. and Korn, D. (1972): Supercritical Wing Sections, Lecture Notes in Economics and Mathematical Systems, vol. 66, Springer-Verlag, New York.

[2] Bauer, F., Garabedian, P., Korn, D. and Jameson, A. (1975): Supercritical Wing Sections II, Lecture Notes in Economics and Mathematical Systems, vol. 108, Springer-Verlag, New York.

[3] Garabedian, P.R. (1964): Partial Differential Equations, Wiley, New York.

[4] Korn, D. and Garabedian, P. (to appear): "A Systematic Method for Computer Design of Supercritical Airfoils in Cascade," Comm. Pure Appl. Math.

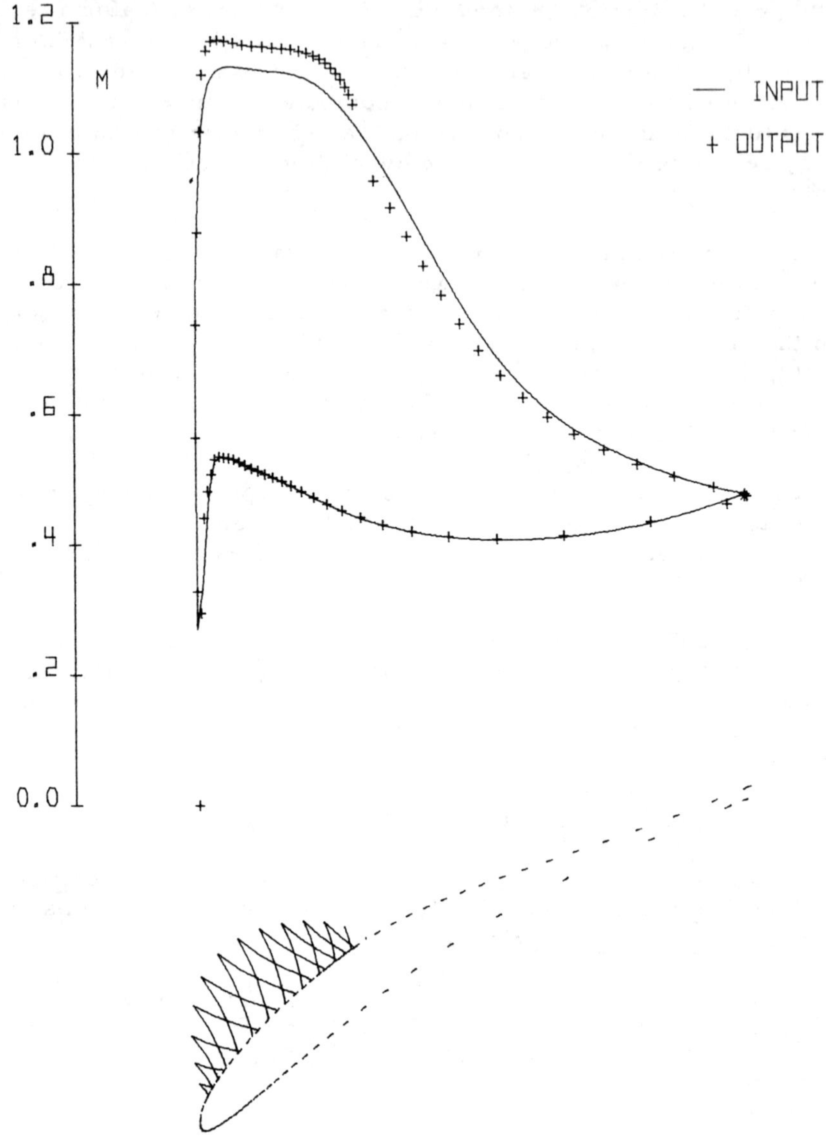

Figure 1. Cascade Airfoil with $M_1 = 0.72$, $M_2 = 0.47$, $G/C = 1.2$

[5] Korn, D.G. (1975): "Numerical Design of Transonic Cascades," ERDA Research and Development Report COO-3077-72, Courant Institute of Mathematical Sciences, New York University.

[6] Korn, D.G. (1975): "Transonic Design in Two-Dimensions," Lecture Notes in Mathematics, vol. 430, Springer-Verlag, New York, pp. 271-288.

DISCUSSION OF THE KORN PAPER

SOUTH: Do you envision any limitation on the size of the supercritical region or the extent of supercritical flow for this large turning angle cascade blade?

KORN: There are a lot of interesting questions that occur and some of them were touched on a little bit this morning - and in fact, in the first talk by Oswatitsch there was - I guess it was a turbine blade that was presented where you had a supersonic zone that extended from the one blade all the way onto the - connecting onto the other blade. Now that had a supersonic exit. We don't have the capability of doing a supersonic exit. But we do expect that there will be some point where the supersonic zone from one blade will extend up to the other blade. The question is, really, does such a flow really exist. Maybe I should stress at this point - what's really needed on this is some very good experimental work - to be able to see how well it performs. After all, I would say the supercritical airfoil has been a success, and we suspect that the same technology should work for the supercritical blade; however, we are in a lower Reynolds number regime - we're well aware of this - and transition does play a bigger role. So it really needs some good cascade test - and transonic testing for highly loaded blades is not easy to get.

JAMESON: For your new effort I think you might just say - what do you project are the computer times for the new method? How long does it take?

KORN: Well, typical runs now are taking - to get the type of convergence where you don't see the difference between the input distribution and the one that you compute, you need to take about five iterations of approximately two minutes each - that's on a 6600 - so, on the order of ten minutes. It's certainly not a very costly program. In fact, every iteration produces an airfoil - and they're converging 2:1 - which describes a pressure distribution.

BALLHAUS: Can you impose constraints like on the particular lift coefficient you're looking for, or the volume of the airfoil?

KORN: The constraints that we impose now are the inlet Mach number and also the turning angle so in a sense we're imposing the lift as it is now. So, effectively we are because we are imposing the inlet and exit conditions which does give you the circulation.

BALLHAUS: Can't you also impose constraints in that you impose a pressure distribution?

KORN: When we give the pressure distributions, all we would get is the critical speed. The lift is determined by just integrating q·ds. This gives you the circulation.

A NAVIER-STOKES SOLUTION OF THE THREE-DIMENSIONAL VISCOUS COMPRESSIBLE FLOW IN A CENTRIFUGAL COMPRESSOR IMPELLER

JAMES L. HARP, JR.
Thermo Mechanical Systems Company

INTRODUCTION

During the past two years Thermo Mechanical Systems Co. has been developing a computer code for calculating steady, three-dimensional, viscous, compressible flow fields in centrifugal compressor impellers. The program can handle subsonic, transonic, and supersonic flow at the compressor inlet.

The work has been funded by the NASA Lewis Research Center with Dr. Ted Katsanis serving as Technical Monitor.

IMPELLER PROBLEM

A set of finite difference analogs of the full three-dimensional compressible Navier-Stokes equations was developed and programmed. In addition to three-dimensionality and compressibility, the following were included:

1. Centrifugal force
2. Coriolis force
3. Transition and turbulence
4. Impeller tip clearance

5. Arbitrary impeller geometry

In the present effort a two-dimensional time-dependent computer code was utilized to calculate the three-dimensional steady flow within the impeller blading. The numerical method is an explicit time marching scheme in two spatial dimensions. Details of the method are presented in [1] and [2].

A schematic of a centrifugal compressor impeller is shown in Figure 1. Initially, an inviscid solution is generated on the hub blade-to-blade surface by the method of Katsanis and McNally [3].

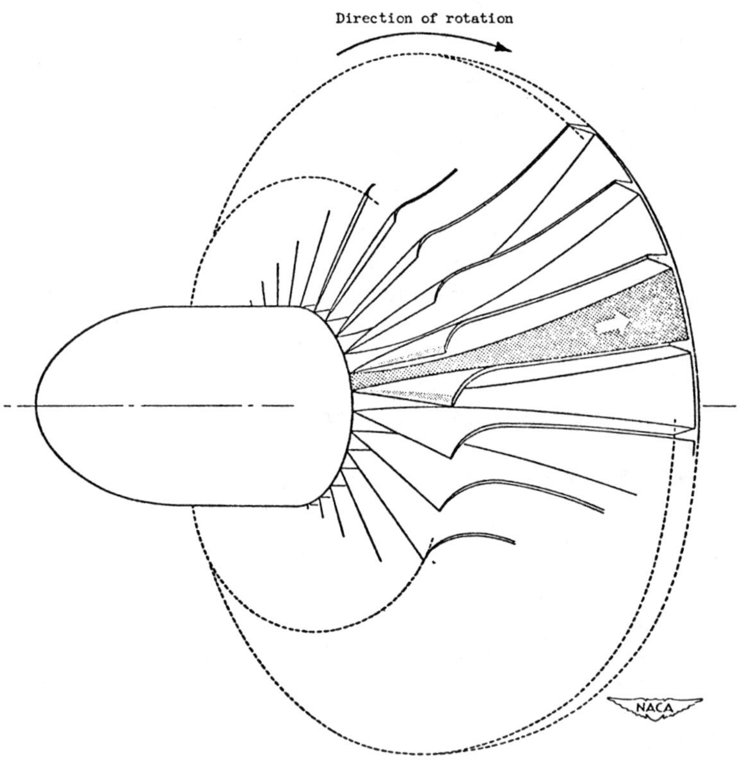

Figure 1. Passage Between Blades in Impeller of Typical Centrifugal Compressor.

This solution is shown by the shaded surface in Figure 1. Starting with the known inviscid solution, the viscous effects are calculated through iteration. After the initial blade-to-blade surface is calculated, a second blade-to-blade surface is calculated a very small distance above the initial surface, and a small increment in time later. Then a third surface is calculated, and so on. This mode of marching accounts for two very important fluid mechanical effects that occur in impellers.

1. Upstream Influence Effects

The flow in the impeller is primarily subsonic relative to the moving blades; hence, downstream conditions influence upstream conditions. Since each blade-to-blade surface extends from inducer inlet to discharge, the downstream flow can influence the upstream flow as the blade-to-blade surface moves from the hub to shroud.

2. Blade Boundary Layer Separation

Separations, which occur on the blade surfaces, produce vortices whose axes are normal to the blade-to-blade surfaces. Thus, the vortices themselves are contained in the blade-to-blade surface and are easily calculable.

In order to get a complete flow field solution, it is also necessary to compute in cross-sectional surfaces which move from the inlet to the discharge of the impeller. This mode of marching accounts for three additional fluid mechanical effects that occur in impellers.

1. Channel Corner Vortices

At the junction of the blades and the hub, vortices may form whose axes are generally normal to the cross-sectional surfaces; hence, the corner vortices would be contained in these surfaces and are easily calculable.

2. Shroud Effects

Relative to the blades, the shroud imposes a moving boundary condition. The effects of this moving boundary condition may induce separation in the neighborhood of the shroud. This separation is calculable in cross-sectional surfaces since each surface contains the shroud vortices.

3. Blade Tip Clearance Effects

Since the shroud and blade tip are contained in each cross-sectional plane, spillage in the tip clearance region is calculable in this mode of marching.

Therefore, to properly solve for a complete impeller flow field, an iteration procedure with both modes of marching is required. Starting from an inviscid solution as the "zeroth" iterate, we determine the first viscous iterate by marching blade-to-blade surfaces which move from the hub to the shroud. Based on the first viscous iterate we determine a second viscous iterate by marching in cross-sectional planes which move from the inducer to the discharge. In this way, the five principal impeller fluid-mechanical effects, described above, can be accounted for. The second iterate will be a complete solution to the three-dimensional, compressible, Navier-Stokes equations for flow in a centrifugal impeller.

The blade-to-blade mode of marching has been developed in this current research effort, and two problems have been run: (1) a radial impeller and (2) a backswept impeller.

RADIAL IMPELLER

The specifications and operating conditions for the radial impeller are shown in Figure 2. The impeller was 6 inches in diameter, and operated at a rotational speed of 38,600 rpm which gave

<u>RADIAL IMPELLER</u>

1.	Impeller diameter -	6.0 inches
2.	Rotational speed -	38,600 rpm
3.	Tip speed -	1010 fps
4.	Inlet total temperature -	536° R
5.	Inlet total pressure	861 psfa
6.	Specific heat ratio -	1.667
7.	Gas constant -	38.73 ft/^{o}R
8.	Reynolds number at discharge -	5000
	Laminar flow conditions	

Figure 2

a tip speed of 1010 fps. An artificial gas was made up which had the thermodynamic properties of Argon and a Reynolds number of 5000 so that laminar flow could be maintained throughout the impeller.

Initially, an inviscid flow field was determined for the hub-to-blade surface, and the velocity vector plot of the results is shown in Figure 3. The symbols indicate the pressure and suction blade surfaces. The arrows indicate the direction of flow and the length of arrow indicates the relative velocity. Figure 3 indicates the velocity is relatively low on the pressure blade surface and relatively high on the suction blade surface. Since the flow is inviscid, no boundary layer profile is indicated.

Figure 4 presents the viscous solution on a surface which is 32 percent of the distance from the hub to the shroud. A large vortex has formed on the suction surface near the impeller exit. The vortex causes the flow velocity in the remaining portion of the channel to increase which results in a reduced impeller pressure recovery.

BACKSWEPT IMPELLER

After completing the radial impeller calculation, a turbulence model, based on the mixing length theory, was incorporated into the program.

A backswept impeller problem was then run with the turbulence model incorporated. The specifications and operating conditions of the backswept impeller are shown in Figure 5. The impeller was 6.28 inches in diameter, and operated at a rotational speed of 75,000 rpm which gave a tip speed of 2055 fps. The design pressure ratio was 8:1. The Reynolds number was large enough to provide for transition and turbulence.

In Figure 6 a velocity vector plot of the inviscid blade-to-blade solution on the hub is shown. The velocities in the impeller inlet are relatively uniform, whereas at the impeller discharge, the velocities are low on the pressure surface and high on the suction surface.

Figure 7 shows a viscous solution on a blade-to-blade surface 19 percent of the distance from the hub to shroud. There is no separation at the impeller exit, but there is a small separation

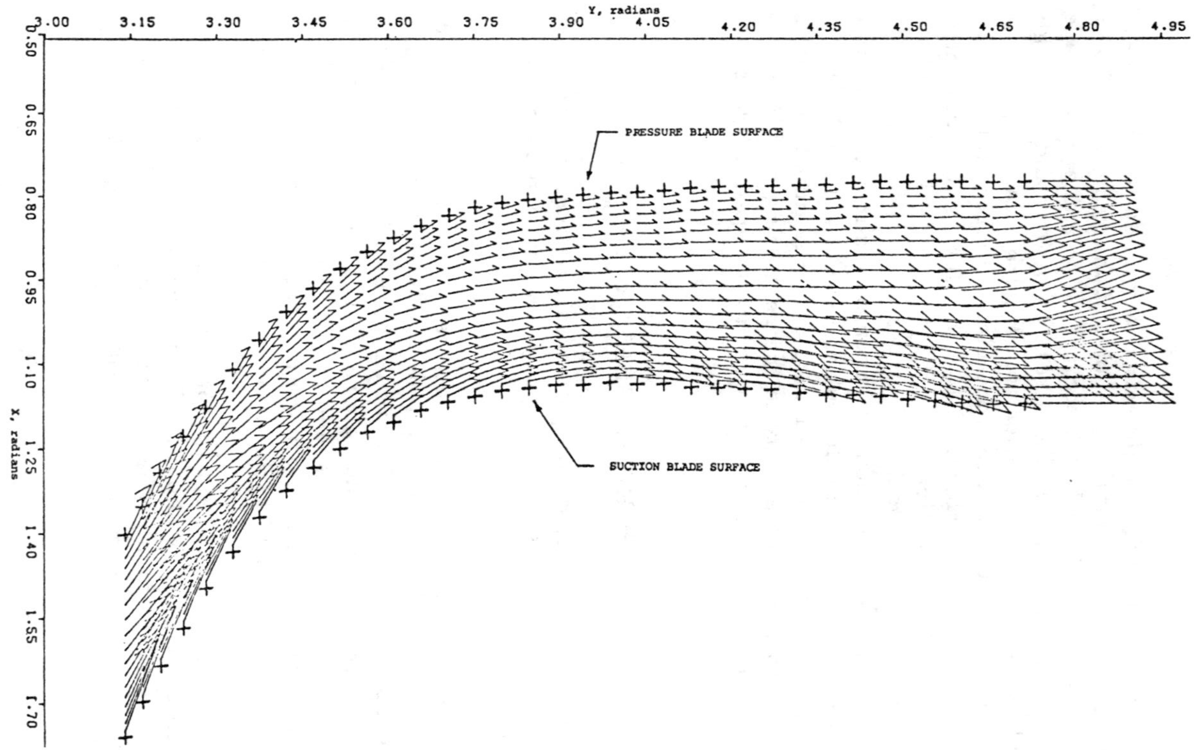

Figure 3. Relative Velocity Vector Plot of the Inviscid Impeller Flow Field on the Hub Blade-to-Blade Surface of Revolution; z = 1.22524 Radians; + Symbols Indicate the Pressure and Suction Blades.

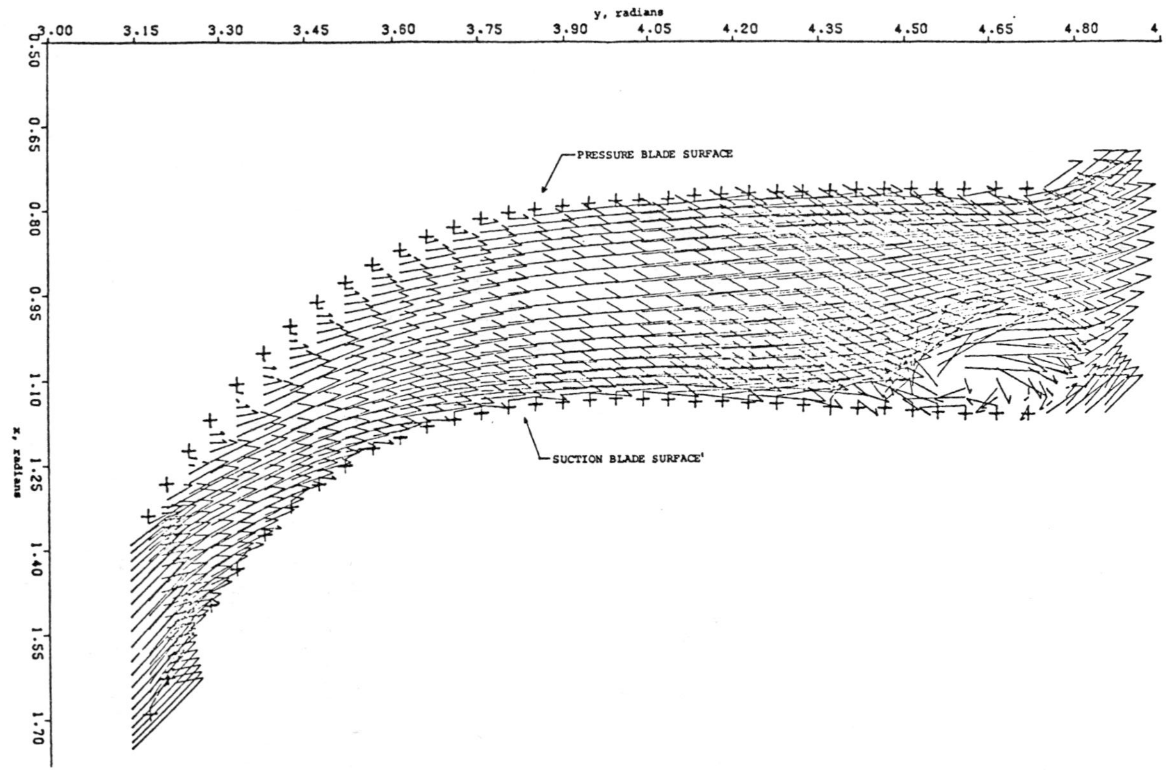

Figure 4. Relative Velocity Vector Plot of the Viscous Impeller Flow Field on the Elliptic Blade-to-Blade Surface of Revolution at $z = -1.1991$ Radians; + Symbols Indicate the Pressure and Suction Blade Surface; Surface Moved 32% from Hub to Shroud.

BACKSWEPT IMPELLER CONDITIONS

1.	Impeller diameter -	6.28 inches
2.	Rotational speed -	75,000 rpm
3.	Tip speed -	2055 ft/sec
4.	Design pressure ratio -	8:1
5.	Inlet total temperature -	$519^{o}R$
6.	Inlet total pressure -	2117 lb/ft^2
7.	Discharge Reynolds Number -	1.43×10^6

Figure 5

on the suction surface at the impeller inlet.

Figure 8 shows an enlargement of the inlet portion of the previous figure. There is a definite separation on the suction surface at the inlet, but this flow reattaches as the flow proceeds downstream. The buildup of the boundary layer on the pressure surface as the flow moves downstream is clearly indicated.

Figure 9 shows a velocity vector plot for a surface 72 percent of the distance from hub to shroud. The flow now has become relatively uniform except for the boundary layer at both blade surfaces.

Figure 10 shows a pressure-ratio contour plot for a surface 98 percent of the distance from hub-to-shroud. The pressure ratio near the impeller exit is about 4. 5. Since this is an 8:1 compressor, the remaining pressure rise would occur in the diffuser. It is seen that very large pressure gradients occur between the suction and pressure surfaces midway through the impeller.

Figure 11 shows a pressure coefficient plot for a surface 98 percent of the distance between the hub and shroud. Pressure coefficient is defined as the local total pressure divided by the ideal total pressure. It is seen in Figure 11 that by the time the flow has passed two-thirds of the way through the impeller, viscous effects are observed completely across the channel.

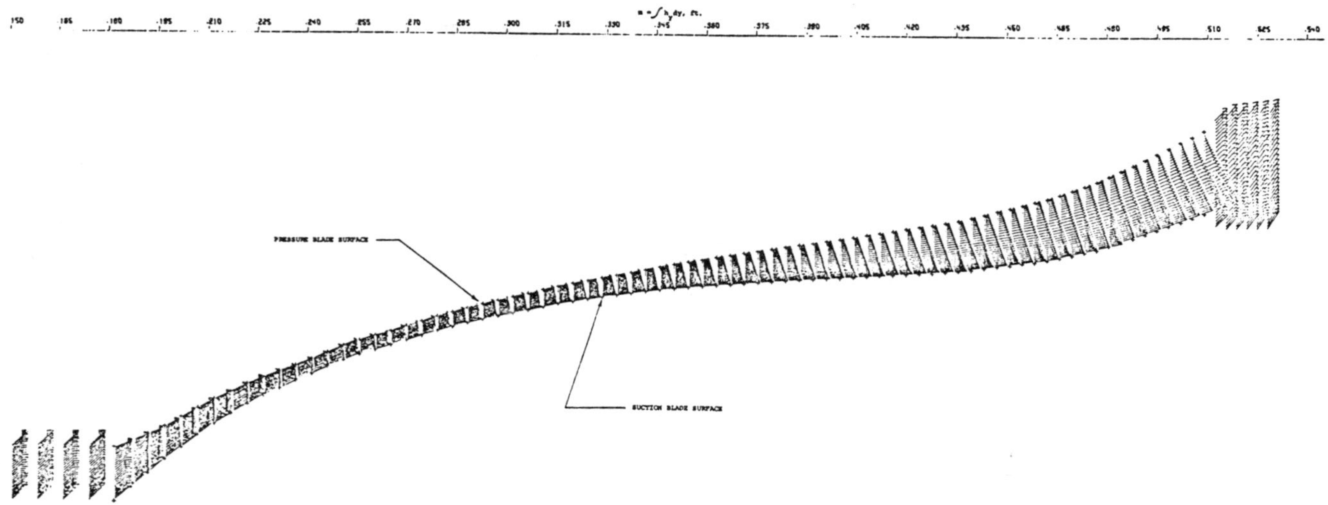

Figure 6. Relative Velocity Vector Plot of the Inviscid Impeller Flow Field on the Hub Blade-to-Blade Surface of Calculation; $z = 0$; + Symbols Indicate the Pressure and Suction Blade Surfaces.

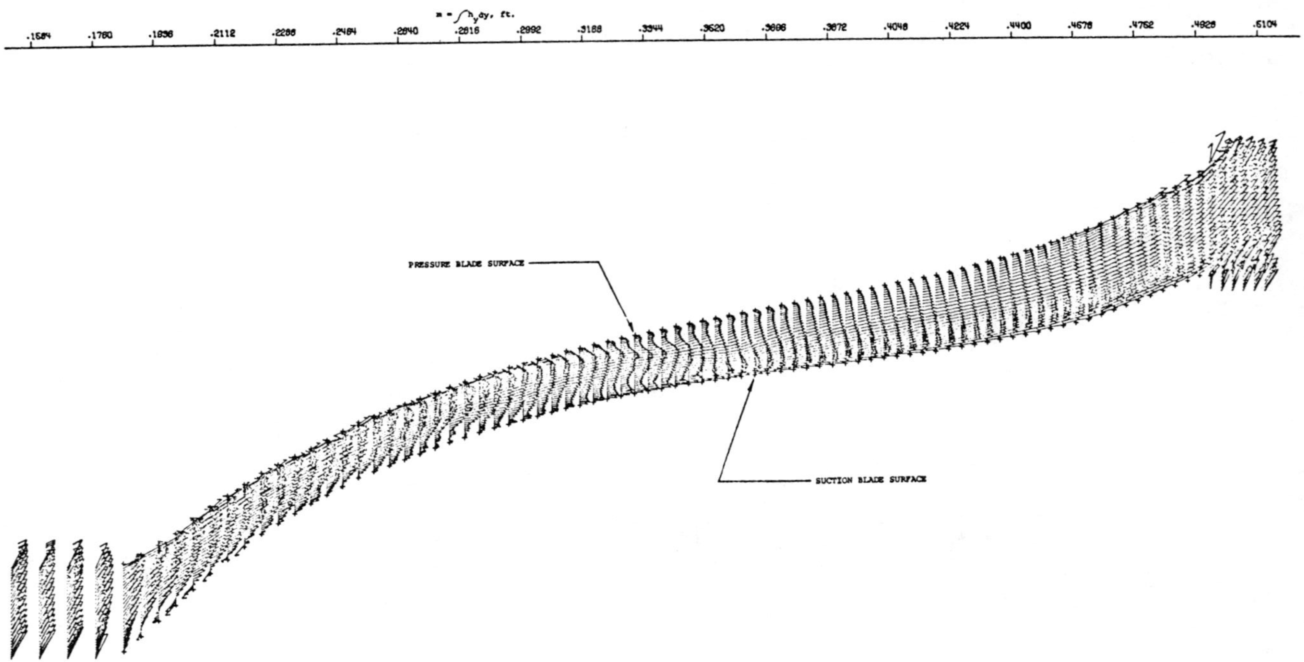

Figure 7. Relative Velocity Vector Plot of the Impeller Flow Field on the Blade-to-Blade Surface 19% of the Distance between the Hub and Shroud; $z = 0.04360$; + Symbols Indicate the Pressure and Suction Blade Surfaces.

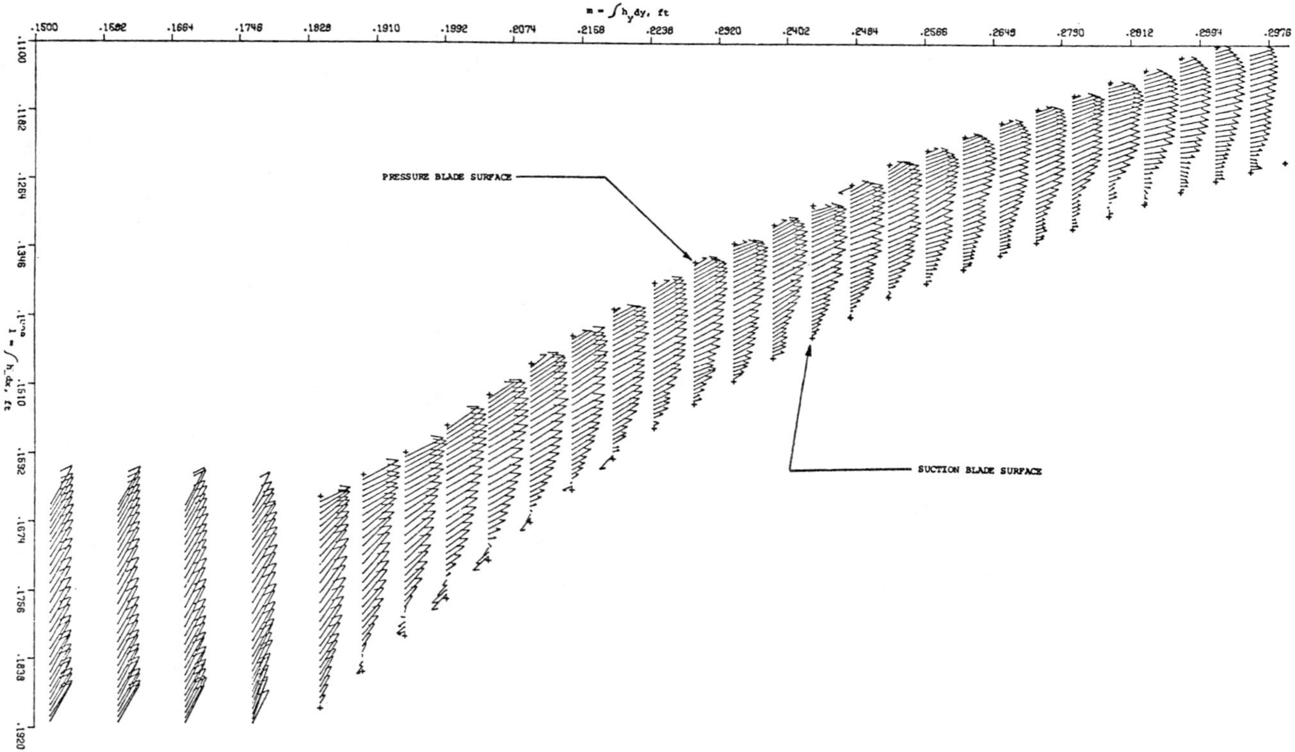

Figure 8. Relative Velocity Vector Plot of the Inducer Flow Field on the Blade-to-Blade Surface 19% of the Distance between the Hub and Shroud; z=0.04360; + Symbols Indicate the Pressure and Suction Blade Surfaces.

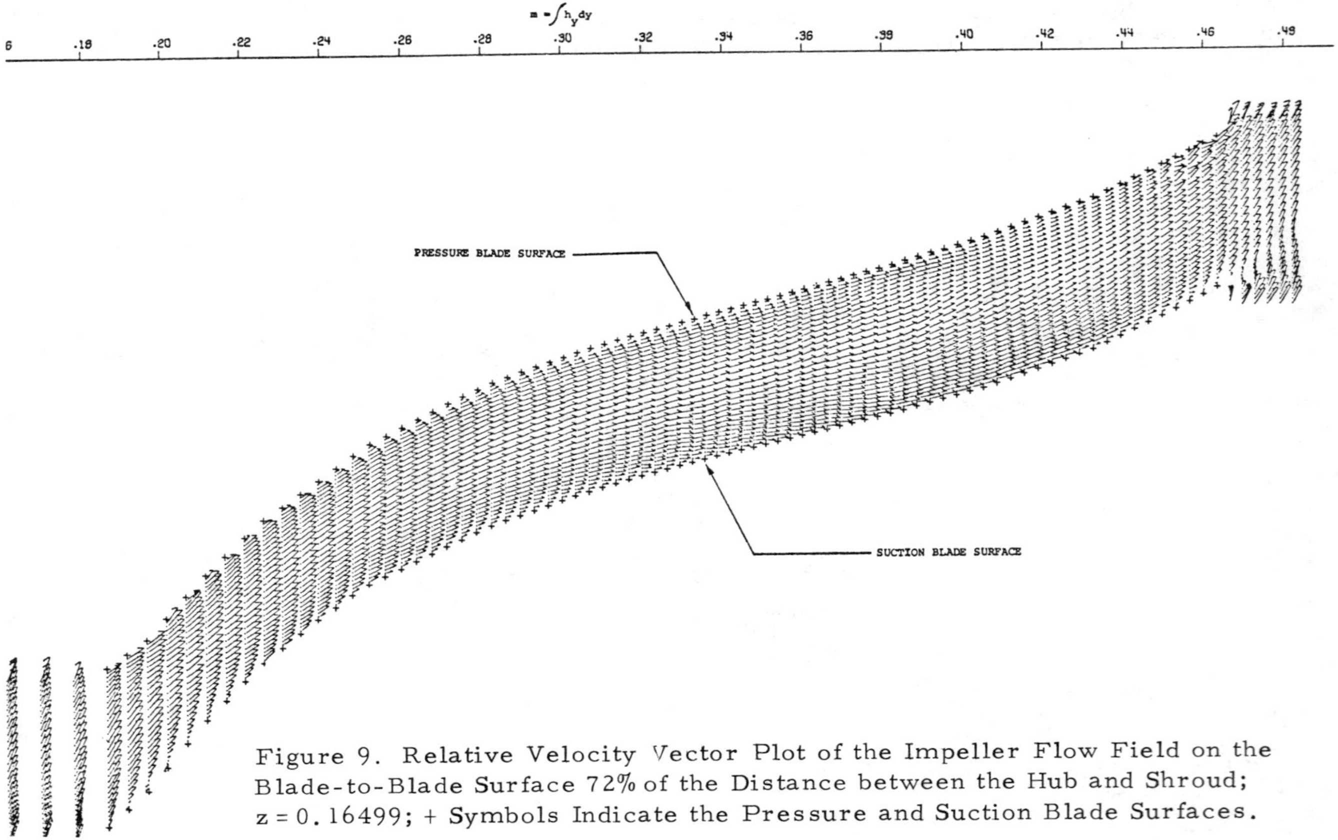

Figure 9. Relative Velocity Vector Plot of the Impeller Flow Field on the Blade-to-Blade Surface 72% of the Distance between the Hub and Shroud; z = 0.16499; + Symbols Indicate the Pressure and Suction Blade Surfaces.

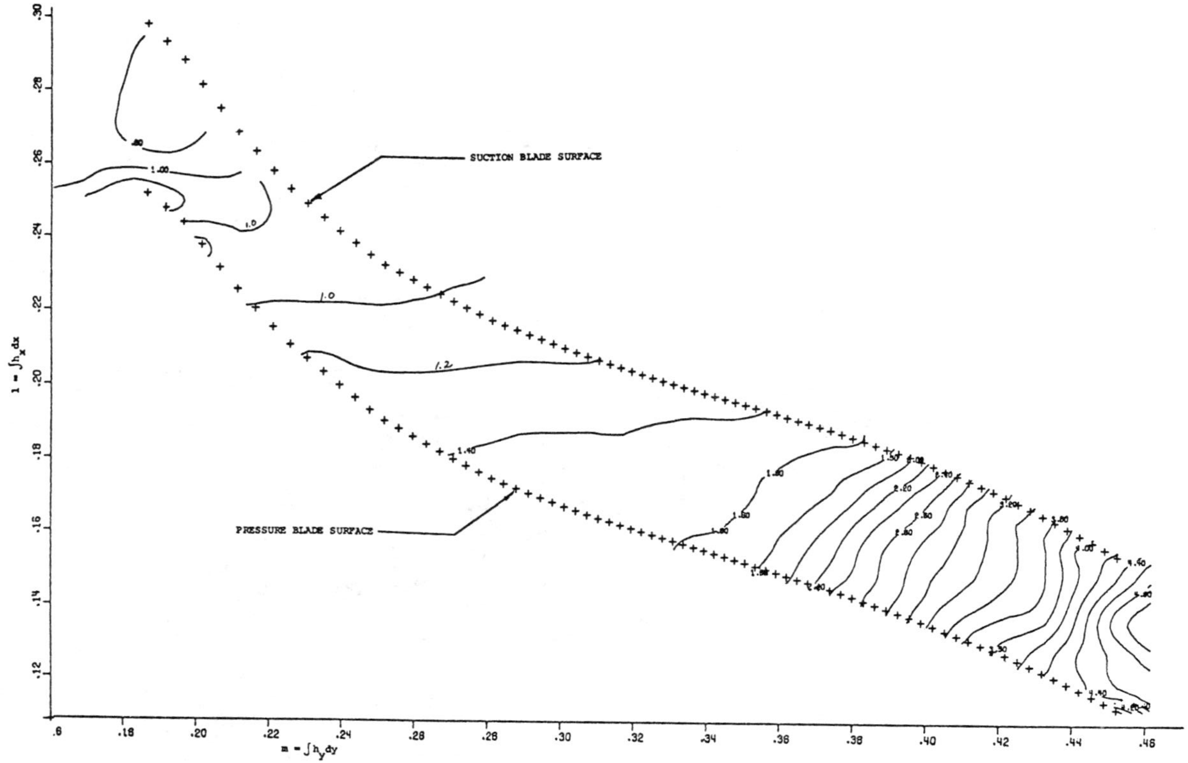

Figure 10. Pressure Contour Plot on the Blade-to-Blade Surface 98% of the Distance between the Hub and Shroud; z=0.22451; + Symbols Indicate Suction and Pressure Blade Surfaces.

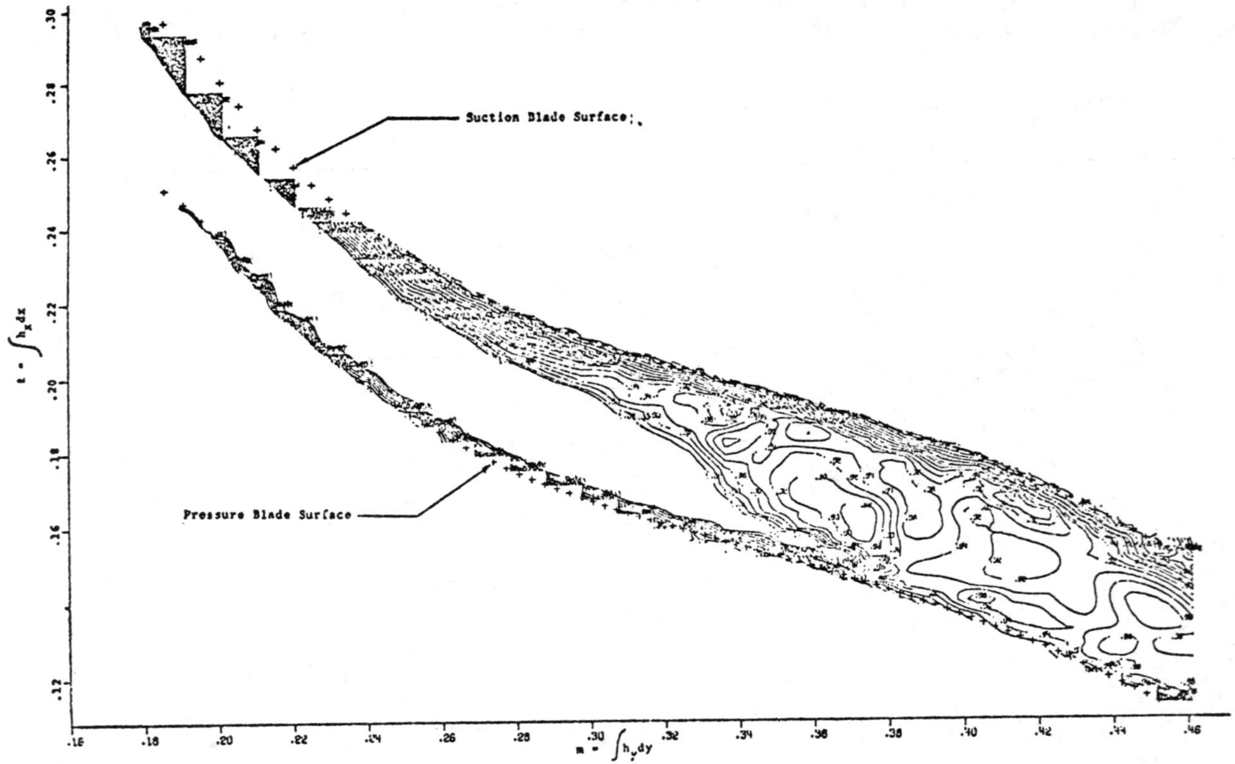

Figure 11. Relative Total Pressure Ratio Contour Plot on the Blade-to-Blade Surface 98% of the Distance Between the Hub and Shroud; $z = 0.23846$; + Symbols Indicate Pressure and Suction Blade Surfaces.

CONCLUDING REMARKS

It is believed that only with the full three-dimensional viscous Navier-Stokes solution can the true flow phenomenon in centrifugal compressors be determined. Perhaps eventually this work may permit major improvements in compressor performance.

REFERENCES

[1] Trulio, J. G. (1966): "Theory and Structure of the AFTON Codes," Air Force Weapons Laboratory Technical Report No. AFWL-TR-66-19.

[2] Walitt, L. (1969): "Numerical Studies of Supersonic Near Wakes," Ph.D. Dissertation., U.C.L.A.

[3] Katsanis, T. and McNally, W. D. (1973): "Fortran Program for Calculating Velocities and Streamlines on the Hub-Shroud Mid-Channel Flow Surface of an Axial or Mid-Flow Turbomachine," NASA TN D-7343.

DISCUSSION OF THE HARP PAPER

DODGE: Could you briefly explain the details of your method of solution?

HARP: It's an explicit time-marching method - the numerical technique originally developed by Dr. John Trulio. The thermodynamic zones overlap the momentum zones.

DODGE: What do you mean when you march from layer to layer? Do you do the 2-D time-dependent solution for each layer and then go from one to another?

HARP: We transform a two-dimensional time-dependent code to a 3-D steady flow by marching in time from the hub to the shroud.

MORETTI: I don't understand. This is a steady flow, right? Therefore, you have three variables there, x, y and z. Now you take x and y and make a calculation there and then you march on in the third direction whatever that may be.

HARP: Yes, some time later. The criteria is that we have steady flow in the machine.

MORETTI: But this layer-like calculation - does that mean that in one layer you only have two variables and only the derivatives with respect to these two variables and the velocity components which belong to those two variables or you have everything and ... I don't know.

DODGE: Is this equivalent to the parabolic method that's been expounded by Spalding and others?

HARP: The elliptic equations are in the program.

DODGE: You have downstream influence on one blade but you only have forward influence as you go from one layer to the other. The shroud can see the hub but the hub doesn't see the shroud, right? In some way it must be equivalent to the parabolic method of Spalding and others.

HARP: This is one of the reasons we need to start up in the front and march downstream and we'll march downstream again in time but at a rate slower than the gas velocity.

DODGE: Are you sure that it converges to the steady-state solutions?

HARP: This first iterative solution isn't the complete solution. Obviously, it doesn't pick things up that we've got to pick up in the second viscous solution as we start from the beginning and march down. Now the marching from front to rear - from upstream to downstream - has already been done and was funded by a NASA-AMES program. But that was strictly all supersonic and none of the downstream effects were able to be transmitted upstream. So this is the only way we knew that we could get the downstream effects into the solution and that's by going twice - hub to shroud and then upstream to downstream.

MORETTI: But you were not iterating with this viscous scheme?

RUNSTADLER: I have a real interest in that last impeller. That machine is the one we designed. I have a couple of questions on that. Could you speak a little about the criteria that you used in terms of the boundary layer separation criteria in rotational coordinate systems. How can you predict where and how separation will occur in the passage?

HARP: Dr. Wallit was the one who wrote the program. It's my understanding that the boundary layer separation just falls out of the Navier-Stokes solution.

RUNSTADLER: As I understand it, what you got in terms of the turbulent model solution is an apparent eddy viscosity model.

HARP: Right.

RUNSTADLER: That eddy viscosity model must have some criteria in it when you are going to have separation occur - relative to the rotational effects that you will get through the normal force acceleration in the viscous layer - predicting when you will get separation. I just wonder if you could speak at all as to what these criteria are.

HARP: I can't answer that question now. One of the things that Drs. Johnston and Eide of Stanford are studying is the fact that there's more turbulence on the pressure side and a tendency for stabilization on the suction side. This is not in the program but we do hope some time in the future to get it in.

RUNSTADLER: I presume that the downstream conditions you had at the discharge of the impeller were probably axisymmetric. The machine is actually run and will be run when tested with a main diffuser configuration. What impact would you expect the main diffuser configuration, i.e., the pressure field time dependent characteristics of the flow being felt upstream of the impeller to have on the solution scheme you have and what impact would that have on trying to incorporate such an unsteady time dependent behavior in your type of solution?

HARP: It gets tough for us to do 3-D unsteady solutions, and you've asked a question that we're still trying to figure out; we haven't got that far yet but we do know that there is an influence on the diffuser blades up into the impeller. My own feelings are that those unsteady effects are secondary compared to some of the more important things that are going to happen between the impeller and the diffuser. Other people feel the unsteady effects are pretty important. But when we get to that point we're going to have to figure out how we go from the impeller into the diffuser and whether or not we let the influences of the diffuser extend themselves up into the impeller.

RUNSTADLER: I presume that was a design point flow calculation that you showed in the slides for that impeller. I'm interested in trying to understand what happened to the whole stage and hence the impact of the impeller on the diffuser flow as you go down in flow from the design point because surge in this machine is a really critical item. With regard to that, how long does it take to run that type of solution?

HARP: To run this solution or just to find out if you're going to get separation?

RUNSTADLER: The full solution.

HARP: The full solution took just under ten hours on the CDC 7600. But on the first impeller it started to separate very quickly.

RUNSTADLER: Relative to the point that you made in discussing the kind of relevant diffusion you get in the wheel, a lot of performance of these centrifical stages is dependent upon the pressure recovery capability of the diffuser that follows the impeller; in fact that is the key element you've got to work on in these machines. Has anybody considered taking this technique and applying it to a radial diffuser?

HARP: To the radial diffuser? Yes, but we've been told we've got to crawl before we walk. That's been talked about on downstream. The downstream boundary condition was to establish the downstream pressure that would provide the design mass flow.

ERDOS: I'm still confused about how this scheme operates and if it takes ten hours on a 7600, it's worth asking a couple of more questions. You start with an inviscid solution on the hub ...

HARP: Right.

ERDOS: ... and you move up towards the shroud ...

HARP: Right.

ERDOS: ... When you get to the shroud you still have your initial inviscid solution on the hub.

HARP: No, we throw that away right at the beginning ... one effect that we don't get by starting on the hub is the fact that ... we have to assume when we start right out that that was a slip whereas the blades are no slip. So the inviscid solution ... we start with that but as soon as we get our first viscous solution we throw the inviscid solution away.

MORETTI: How do you get the inviscid solution?

HARP: By the method of Katsanis and McNally.

MORETTI: John made the point that ten hours of 7600 computer time ... I would have never scheduled that.

DEIWERT: I'm somewhat familiar with what Wallit has done - maybe I can shed a little bit of light on it - on this method. I think it's a first-order explicit finite different method to solve the Navier-Stokes equation but he gets by, I think, with really relatively short times because he uses a fairly coarse mesh and it's my feeling that he's not getting into sublayer scale at all. So what he is in effect really doing is lowering the Reynolds number by putting in a very large viscosity. So I think he's really in effect getting a laminar solution for what you think are turbulent flows. On the laminar flow you're probably getting laminar solutions.

MC CROSKEY: On that last point, how many points do you typically have across the turbulent boundary layer?

HARP: Now this is compressive. Did you say across the boundary layer?

MC CROSKEY: Yes.

HARP: One to three. Not many. But one has to weigh all factors - you get a lot of mesh points in the boundary layer and then you're talking about tens of hours on the CDC 7600 so it's a tradeoff between trying to open the mesh and, by the way, that was the reason we ran the laminar case with the viscous fluid in the first case to be able to have a viscous fluid and open the mesh up and have short machine time. We did that in about half an hour. Less than half an hour, I believe it was.

MORETTI: I think that those running times are too long.

SHOCK-FITTING IN TRANSONIC FLOW COMPUTATION

M. M. HAFEZ
Flow Research, Inc.

H. K. CHENG
University of Southern California

SUMMARY

Weak shocks are the main feature of many transonic external, as well as internal, flows of practical interest. For flow field calculations, shocks are either captured or fitted.

In both unsteady and iterative (relaxation) calculations, the shock-capturing technique depends upon an implicit (or explicit) artificial viscosity. When added to the inviscid model, this additional viscosity renders the discontinuous solution smooth, and results in a shock region or a transition layer in which the gradient is large but finite. As a result, a small grid size is needed in this region. The accuracy of the captured shock solution is improved only by grid refinement, which leads to lengthy and costly calculations. The advantage of this technique lies in its capability to capture shocks automatically (i.e., without special treatment of the discontinuity). The viscous term is active only in the shock region and "accurate" inviscid solutions can be obtained elsewhere.

On the other hand, the shock-fitting technique treats the shock as a discontinuity (infinite gradient) across which jump conditions (derived from conservation of mass, momentum and energy) are satisfied while the differential equation is valid in the upstream and downstream neighborhoods; hence, the grid size is controlled only by the accuracy requirement of the inviscid solution (where the gradient is not large), leading to much more economical calculations. The disadvantage of the shock-fitting approach is the additional program complexity which must be introduced in order to impose the proper jump conditions at those locations where shocks occur in the flow field. In unsteady calculations (to obtain the steady state) the first technique (shock-capturing) was used by Magnus and Yoshihara, while the second (shock-fitting) was used by Grossman and Moretti to solve the Euler equations in the transonic regime.

In 1970, Murman and Cole solved the potential equation within the transonic small-disturbance approximation by a line relaxation method which used type-dependent finite-difference schemes. This approach effectively introduced enough viscosity to guarantee stability and smoothness of the solution; however, the proper jump conditions were not satisfied. Later, Murman introduced a fully conservative scheme (shock-point operator) which successfully produced solutions with "correct" jumps only in the limit of vanishing grid size. This scheme gives good results if the flow downstream of the shock is subsonic; if the downstream flow is supersonic, the shock is smeared out.

Cheng and Hafez used line relaxation with the Murman and Cole finite-difference scheme but with a shock-fitting procedure based on the transonic small-disturbance shock polar. The finite-difference equation at the shock point is replaced by a nonlinear algebraic relation derived from the discretized shock relations. The use of these approximate shock jump relations results in a much simpler shock-fitting procedure than do the techniques which are based on the characteristic relations (such as have been used for the unsteady Euler equations). Consequently, solution accuracy which is comparable to that which is obtained in shock-capturing techniques can be obtained with much coarser grid spacings, and with only minimal increases in program complexity.

In passing, we notice that near the root of an embedded shock the transonic small-disturbance equation admits a weak singularity (Zierep-Oswatitsch) which has to be fitted in order to relieve the requirement for grid refinement at this location.

For the full potential equation, Jameson extended Murman's fully conservative schemes, while extension of Cheng and Hafez' shock-fitting procedure is currently underway.

In this paper, the previous work of Cheng and Hafez will be reviewed briefly and some new preliminary results, based on Gaussian Elimination rather than iterative line relaxation technique, will be presented.

TECHNICAL DISCUSSION

For two-dimensional flows, the transonic small-disturbance equation, in terms of a perturbation velocity potential, can be written in the following form (see Hafez and Cheng);

$$(K - \phi_x)\phi_{xx} + \phi_{yy} = 0, \tag{1}$$

where K is the similarity parameter. The irrotationality condition is simply

$$\phi_{yx} - \phi_{xy} = 0 . \tag{2}$$

Equations (1) and (2) admit a weak solution which is consistent with the Rankine-Hugoniot relations under the transonic small-disturbance assumptions, namely,

$$< K - \phi_x > = - \left(\frac{dx_D}{dy}\right)^2 \tag{1'}$$

$$\left(\frac{dx_D}{dy}\right) = - [\![\phi_y]\!] / [\![\phi_x]\!] , \tag{2'}$$

where $< >$ and $[\![\;]\!]$ denote the average and the jump across the shock $x_D = x_D(y)$. Equation (2') signifies the continuity of the tangential velocity component across the shock and can be replaced by

$$[\![\phi]\!] = 0 . \tag{2''}$$

Combining Eqs. (1) and (2), we have the shock polar,

$$< K - \phi_x > [\![\phi_x]\!]^2 + [\![\phi_y]\!]^2 = 0 \tag{3}$$

or

$$[K - \frac{1}{2}(\phi_{x_u} + \phi_{x_d})][\phi_{x_d} - \phi_{x_u}] + \beta\,[\phi_{y_d} - \phi_{y_u}] = 0 \qquad (3')$$

where

$$\beta = \frac{\phi_{y_d} - \phi_{y_u}}{\phi_{x_d} - \phi_{y_u}} = - \frac{dx^D}{dy} \qquad \begin{array}{l} d \equiv \text{downstream} \\ u \equiv \text{upstream} \end{array}$$

A shock-fitting procedure based on Equations (2'') and (3') is used. Equation (2'') assures the existence of an intersection of the surfaces belonging to two sides of the shock. The ridge where the two surfaces meet, therefore, locates the shock boundary. The first grid point downstream of the shock is identified as S. P. The finite-difference equation at this point is replaced by a discretized version of Eq. (3') using one-sided finite-difference approximation of the quantities ϕ_x, ϕ_y, as shown in the sketch below.

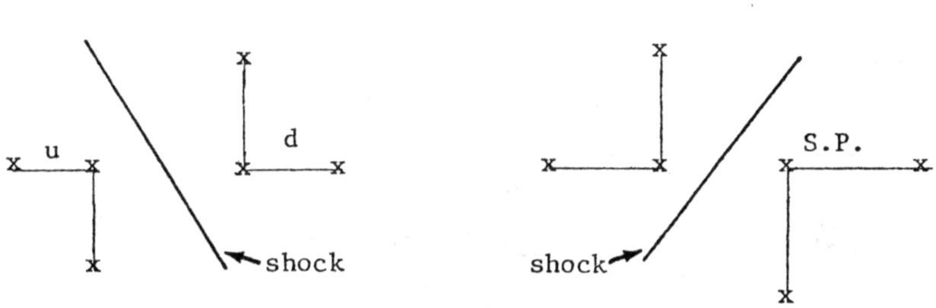

(A) (B)

Different algebraic equations are derived for different shock inclinations.

Some numerical results for transonic flows over a parabolic airfoil have been obtained using this shock-fitting procedure with Murman's line relaxation method for both embedded and bow shocks, as discussed in Hafez and Cheng and as shown in Figures 1 and 2.

A Direct Inversion Method

The nonlinear equation (Eq. 1) $F(\phi) = 0$, is solved iteratively by a variant of Newton's Method (False Position),

$$J(\delta\phi) = -F(\phi^n), \tag{4}$$

where

$$\delta\phi = \phi^{n+1} - \phi^n$$

$$J \equiv (K - \phi_x^n)\frac{\partial^2}{\partial x^2} + \frac{\partial^2}{\partial y^2} - \phi_{xx}^n \frac{\partial}{\partial x} .$$

Notice if ϕ_{xx}^n in the Jacobian J is replaced by a constant α, Eq. (4) can be written in the form of the unsteady transonic small-disturbance equation; namely,

$$\alpha\, \phi_{x_t} = (K - \phi_x)\phi_{xx} + \phi_{yy} . \tag{4'}$$

A fully implicit version of Murman's scheme is used with Gaussian Elimination to invert the super big matrix. A shock-fitting procedure is incorporated as follows: at each iteration,

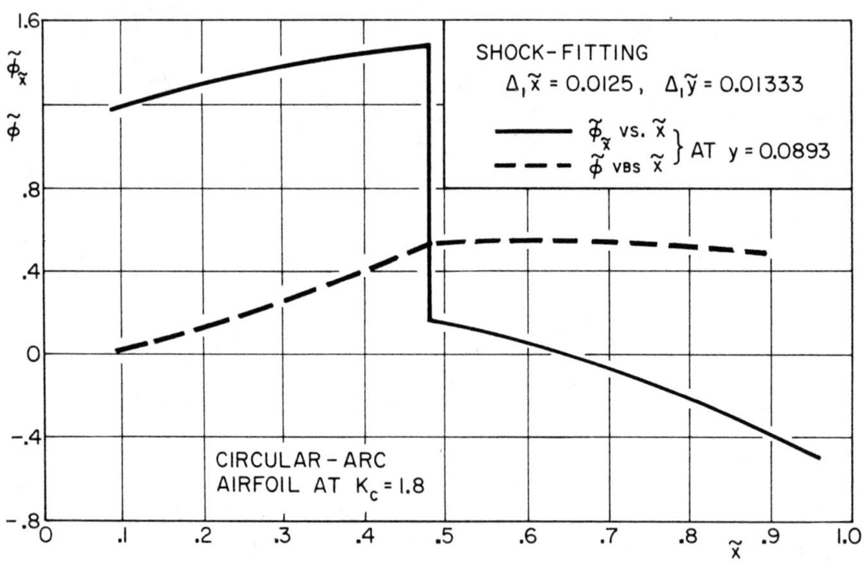

Figure 1

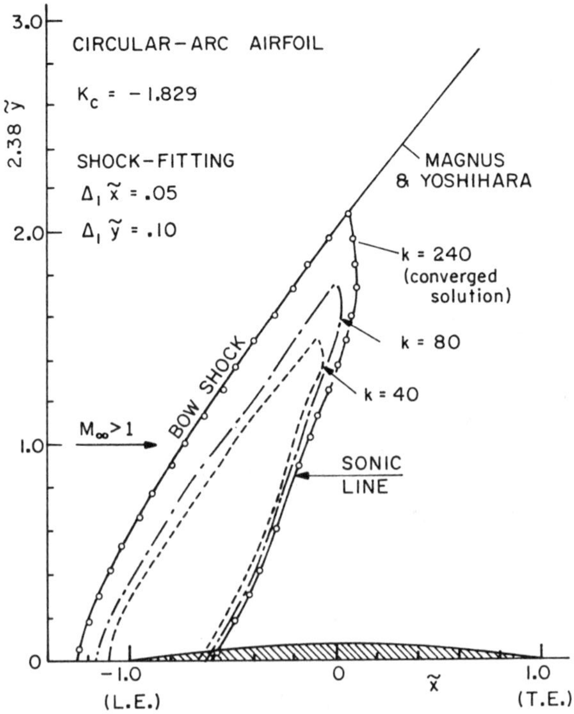

Figure 2a

shock points are located using Eq. (2''), as shown in the sketch below.

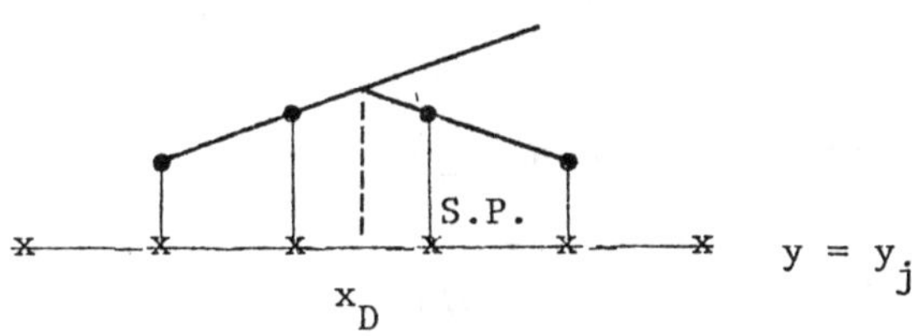

SHOCK-FITTING FOR UNSTEADY TRANSONIC FLOW

- SMALL-DISTURBANCE EQUATION:

$$\alpha \phi_{xt} = (K - \phi_x)\phi_{xx} + \phi_{yy}$$

- FULLY IMPLICIT SCHEME

— INVERTING MATRIX BY GAUSSIAN ELIMINATION

— INITIAL CONDITION: INCOMPRESSIBLE FLOW

— STEADY STATE: 10 ITERATIONS

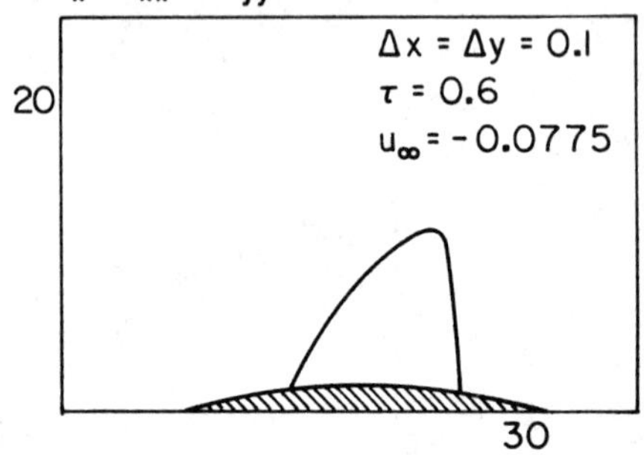

— COMPUTATIONAL REQUIREMENTS: 43 SEC & 200K

(MURMAN PROGRAM: 9 SEC & 60K)

Figure 2b

The finite-difference approximation of the differential equation at S. P. is replaced by a finite-difference version of Eq. (1') where the right-hand side is evaluated from previous iteration, namely

$$\frac{dx_D}{dy} = \frac{x_{D_{j+1}} - x_{D_j}}{\Delta y} \quad \text{and} \quad \left.\frac{dx}{dy}\right|_{y=0} = 0 \, .$$

(More accurate formulae can also be used.)

The matrix at each iteration is invertible even with the inclusion of shock relations at S. P. The steady state is reached after 10 iterations for 20 x 30 unknowns ($\Delta x = \Delta y = 0.1$, $M = 0.06$ and $u_\infty = -0.0775$) using 200,000 machine storage locations and 43 seconds of CPU time on a CDC 6600. (Using line relaxation with the previous shock-fitting sub-routine added to Murman's program takes only 60K and 10 seconds to get the same results.) In practice, Gaussian Elimination is not recommended for large systems. It is used for finite-element calculation (fewer unknowns) and it has been shown that the introduction of the shock relations will not make the matrix singular or disturb the convergence of iterations.

REFERENCES

Hafez, M. and Cheng, H. K. (1975): "Convergence Acceleration and Shock-Fitting for Transonic Aerodynamic Computations," AIAA Paper 75-51, 13th Aerospace Sciences Meeting; Pasadena, California. Also available as USC Report 132.

DISCUSSION OF THE HAFEZ-CHENG PAPER

DODGE: In terms of shock fitting - in the light of looking at turbomachinery problems - would you like to make a comment on what you feel the feasibility of that sort of approach is when you're talking about a system of very nonregular shocks where typically in Schlieren pictures of turbomachines in interesting cases you may find as many as four or five shocks in a lot of intersections and a lot of complex structure.

HAFEZ: For a multiple shock system the logic of the programming may not be as easy as for that of a single shock. In the former case, shock capturing (with shock-point operator) may be easier. However, the problem with internal flow, I believe, is that involving the velocity potential - if you have a good model and you supply the boundary conditions with the velocity potential, the shock fitting procedure may, in fact, be very attractive. I have not worked on internal flows, but I understand that Dr. William Rae of Calspan, Buffalo tried shock fitting for internal flow and he has not, as yet, obtained results. The problem is of interest when the number of grid points is very small; in this case a shock smeared on two or three grids would badly spoil the accuracy of the flow description.

EFFICIENCY GAINS OF SECOND ORDER ACCURATE METHODS WITH SHOCK FITTING

CZESLAW P. KENTZER
Purdue University

ABSTRACT

At least eight-fold gains in computing efficiency of transonic flow methods may be realized using second order accurate differencing throughout. These gains are lost if explicit shock fitting is not used. It is suggested how to combine second order schemes with shock fitting techniques to retain the highest possible computational efficiency while treating transonic flows as approximately irrotational.

INTRODUCTION

The aim of this paper is to discuss the formulation of a transonic boundary value problem with regard to the computational efficiency. Transonic flows, whether internal or external, are characterized by the presence of shock waves as downstream boundaries of imbedded supersonic zones. We shall first argue the advantages and efficiency gains of approximating shocked transonic flows by potential (irrotational) flows and then limit the discussion to the advantages and limitations of solving the potential equation in the presence of shock waves.

POTENTIAL FLOW APPROXIMATION

Whenever the assumption of irrotational flow is appropriate, the introduction of the velocity potential reduces the number of primitive physical variables to one scalar function. From the point of view of computational efficiency, it is to one's advantage to reduce the number of scalar variables because the computing time, the storage requirements, and the cost of computations are all approximately proportional to the number of variables. Thus the use of the velocity potential as a single scalar unknown will, for example, increase the computational efficiency five-fold as compared to a formulation of three-dimensional problems in terms of density, pressure and three velocity components. Consequently, the advantage of the potential flow approximation lies primarily in the gains in computational efficiency.

The occurrence of shock waves introduces discontinuous jumps in entropy. If the entropy jumps are small (weak shocks) and if the shock waves are long and have small curvature, then the entropy gradients downstream of the shock will be small. By Crocco's theorem, the vorticity will be likewise small and could be neglected. It appears, therefore, that such flows may be approximated by potential flows without undue loss of accuracy.

ORDER OF ACCURACY

The relative efficiency of computational schemes may be measured in terms of computing times (or, what is equivalent, computing cost) required to attain a given accuracy of the numerical results. For continuous solutions a convenient measure of the numerical accuracy is the order of the leading term in the truncation error which decreases as h, the mesh spacing, for first order methods, as h^2 for the second order methods, etc. Thus, in order to increase the accuracy by a factor of four, h has to be halved in the case of second order methods and has to be reduced to 1/4 h for a first order method. The number of mesh points along each coordinate must be doubled in the first case and quadrupled in the second. In three-dimensional calculations the total number of mesh points must be increased by 8 and 64 for the two cases, respectively. As a consequence, second order accurate methods are at least eight times more efficient in terms of computing time in regions where solutions are continuous. However, if the governing equation is elliptic everywhere with the exception of a finite imbedded hyperbolic zone, the overall accuracy of a given computational scheme will be only as good as that at the least accurate mesh point. Thus it is imperative that the desired order of accuracy be scrupulously maintained over the entire computational domain.

DISCONTINUOUS SOLUTIONS

The argument in favor of more accurate methods fails when the solution ceases to be continuous due to the spontaneous appearance of shock waves as is the case in transonic flows.

First we observe that the so called "shock capturing techniques" rely on the convergence of the finite-difference solutions to the weak solutions of the partial differential equation being solved. The weak solutions manifest themselves numerically as "smeared-out shocks" usually of thickness of 3h or greater. In such a shock layer of finite thickness all higher order derivatives of the solution are large and tend to infinity as $h \to 0$. The finite thickness of the shock is a result of a balance between the tendency of the discontinuity to become steeper, resulting in larger numerical values of the higher derivatives, and the numerical dissipation proportional to these derivatives tending to diffuse the discontinuity. Higher order methods have "numerical viscosities" of higher order in h and admit smaller dissipation effects thus producing thinner shock waves accompanied by larger numerical errors. Consequently, the advantage of using higher order methods is seriously compromised by allowing a discontinuity to be represented by a transition zone of a finite thickness. If the shock points are not handled carefully, large oscillations in the form of "over-shooting," "under-shooting" and "wiggles" destroy the usefulness of numerical solutions not only locally, but often in the entire computational domain. Very fine meshes must be used, at least locally, in order to maintain a reasonable accuracy. The cost of calculations increases and the efficiency suffers. One must stress the fact that before shock waves may actually form in the computational domain, the characteristic surfaces may tend to form an envelope. The occurrence of such envelopes signals a formation of a shock wave. Even if a shock formed in such a fashion is infinitesimally weak, second order methods are expected to react unfavorably to the fact that the higher derivatives are large. First order methods will not only ignore the singularity, but will obscure it to the point that the incipient formation of a shock wave may escape detection altogether.

Another aspect of the loss of accuracy of the "shock capturing" techniques is the error connected with the position and shape of the shock. On thin airfoils or wings a slight change in the conditions upstream or downstream of the shock may cause a disproportionally larger change in the shock position. Thus the lift, and in particular the drag and moment, may change by amounts exceeding the tolerated limits of accuracy. Numerical solutions become of little value when they respond strongly to the change in the numerical technique. Figure 11 of J. D. Cole's article [1], illustrates this phenomenon. A change from a non-conservative to a conservative differencing of the shock points moves the shock location downstream by approximately 10 percent chord. A non-conservative differencing does not satisfy the shock jump conditions which are the property of the potential equation (to be referred to as the "isentropic shock jump conditions"). Thus a non-conservative differencing corresponds to different jump conditions. Cole's example gives, therefore, an indication of the magnitude of errors in the shock position connected with changes in shock jump conditions. On the other hand, Steger and Baldwin, [2], calculated an identical supercritical flow by a similar relaxation technique and then evaluated the wave drag using exact Rankine-Hugoniot conditions and the isentropic shock jump conditions. The exact conditions resulted in a 30 percent increase in the drag coefficient. We must conclude that errors connected with shock position and shock jump conditions exceed the margins of engineering calculations.

It is possible, as ably demonstrated by Cole, to execute numerical calculations of a given system of equations with a surprising accuracy. The accuracy was checked by comparing drag as calculated from the pressure distribution and from the momentum integrals in the Trefftz plane. Since this problem was calculated under the assumption of irrotational flow and isentropic shock, the demonstrated accuracy is no indication of errors resulting from these assumptions.

STATE OF THE ART

With J. D. Cole's paper as an example of the most refined calculations of transonic potential flows, we may conclude that we have available methods that calculate accurately and efficiently solutions to the partial differential equations of potential flows in regions of continuous solutions and that such methods are successful because they apply numerical analogs of the isentropic jump conditions carefully along a line of points which line represents the location of the shock wave. In the sense that the line of shock points is automatically determined by the program, Cole's method may be viewed as being akin to the "shock fitting techniques." The objections to Cole's method may be summarized as follows.

1) Since the position of the shock is not known and the isentropic shock conditions are applied only along a single line of points, the computational mesh must be rather fine so that the shock may move freely over the mesh until it settles to a sufficiently accurately defined steady position.

2) Since the shock must be determined as a part of the solution of the boundary value problem, the shock and its position, as well as the solution everywhere else, depend uniquely and continuously on the boundary conditions for the velocity potential. The shock conditions should be viewed as boundary conditions given on an unknown boundary. To every set of shock conditions there must correspond a shock wave of certain shape and location. The isentropic and the exact shock conditions must produce shocks of different shape, extent and location thus influencing the rest of the flow field at transonic speeds.

3) The isentropic shock conditions agree with the exact Rankine-Hugoniot conditions only at Mach number of unity (see, e.g., Figure 1 of Steger and Baldwin, [2]) and underestimate M_2, the Mach number after the shock, e.g., at $M_1 = 1.3$ we have $M_2 \simeq .70$ for the isentropic relations instead of 0.75 for the exact shock relations. Even though shock waves at $M_1 = 1.3$ are relatively weak, their position on thin profiles may be quite sensitive to the influence of the downstream Mach number.

4) The computational scheme at supersonic points (equation 4.10 of [1]) is of first order accuracy due to noncentered (or "upstream") differencing. While it is true that such differencing preserves the basic local domain of dependence, the schemes that preserve the hyperbolic domain of dependence need not have to be noncentered. Interpreting relaxation procedures as integration with respect to some pseudo-time, we observe that it is necessary for stability that the domain of dependence of the partial differential equation be contained in the domain of dependence of the numerical solution, while stability and consistency imply convergence. Consequently, the use of noncentered schemes is unnecessary and it reduces the local, and possibly the global accuracy thus affecting the efficiency of the calculations.

We will discuss now some known techniques which would improve the efficiency of Cole's and similar methods. The objection 1) has to do with the conflicting requirements for the resolution of the details of the locally continuous and locally discontinuous solutions. Treating the shock as an explicit discontinuity removes the demand for high resolution of the shock wave requiring but a moderate mesh point density in the entire flow field. This alone will increase the computational efficiency of each iteration step manyfold. Explicit shock fitting has been used with success by, e.g., M. Salas, [3], and Marconi and Salas, [4], while the state of the art is discussed by G. Moretti, [5].

Once it is decided to treat the shock explicitly, it becomes a trivial matter to impose the exact rather than the isentropic shock conditions. This takes care of objections 2) and 3).

A second order accurate centered finite-difference scheme for hyperbolic points was proposed and used by the author in [6]. The scheme preserves the domain of dependence and is unconditionally stable. When applied to the full potential equation the scheme is implicit, but when the mixed derivatives are absent (e.g., in the small perturbation transonic equation) the scheme is explicit and remains unconditionally stable. Thus also the objection 4) may be removed.

PRESENT RESEARCH EFFORTS

As a continuation of the work on second order accurate relaxation schemes reported by the author in [6], research is in progress with the objectives of developing suitable shock fitting routines. When type-dependent difference schemes are used, the finite-difference equations applied on the opposite sides of a shock must be interpreted as analogs of two different partial differential equations rather than of a single equation of mixed type. The technique used by the author in [6] was to model a relaxation procedure after a suitably chosen time-dependent process. Thus the exact potential equation for time-dependent compressible flows,

$$\left(\frac{\partial}{\partial t} + \bar{q}\cdot\nabla\right)\left[\frac{\partial\phi}{\partial t} + \frac{1}{2}\bar{q}\cdot\bar{q}\right] = a^2\nabla^2\phi, \qquad \bar{q} = \nabla\phi,$$

when written as

$$\left(\frac{\partial}{\partial t} + \bar{q}\cdot\nabla\right)\frac{\partial\phi}{\partial t} + \bar{q}\cdot\frac{\partial\bar{q}}{\partial t} = a^2\nabla^2\phi - \bar{q}\cdot(\bar{q}\cdot\nabla)\nabla\phi$$

is approximated subject to the following rules:

(1) The right-hand side is always approximated by centered differences in order to maintain a compact computational module and to maintain a second order accuracy in space while being consistent with the steady state differential operator;

(2) The approximation to the temporal operator on the left-hand side must render the relaxation scheme type-dependent, that is, the numerical information must propagate in all the directions at subsonic points, and it must not propagate upstream at supersonic points.

The natural choices satisfying the above rules are

$$\frac{\partial \phi}{\partial t} = a^2 \nabla^2 \phi - \bar{q} \cdot (\bar{q} \cdot \nabla) \nabla \phi \tag{1}$$

at subsonic points, and

$$u_o \frac{\partial^2 \phi}{\partial x \partial t} = a^2 \nabla^2 \phi - \bar{q} \cdot (\bar{q} \cdot \nabla) \nabla \phi \tag{2}$$

at supersonic points. Equation (2) admits a single characteristic and, with u_o an arbitrary positive constant, leads to unconditional stability and carries numerical information only downstream in the direction of the stream velocity $U_\infty > 0$. Equation (1) is modeled after a diffusion process while Equation (2) resembles a telegrapher's equation or the time-dependent small perturbation transonic equation. The hyperbolic equation (2) determines the potential on the upstream side of the shock while the parabolic equation (1) allows for the upstream influence of the solution on the subsonic side of the shock. The fact that not both of the equations are hyperbolic with respect to time requires a modification of the exact procedures of shock fitting proposed by the author in [7].

CONCLUDING REMARKS

The relaxation procedures currently being investigated by the author offer the possibility of solving efficiently complicated transonic flows, e.g., three-dimensional shocked flows. The formulation combines the efficiency of second order methods with those of the shock fitting techniques. The formulation is exact except for the assumption of irrotationality downstream from the nonisentropic shock (Rankine-Hugoniot shock). It is felt that the assumption of isentropic shock is much more serious than that of isentropic flow downstream of long and almost straight shocks. The latter assumption permits the introduction of the velocity potential throughout the flow accompanied by a five-fold reduction in the number of equations (in 3-D flows) and a five-fold increase in computing efficiency.

REFERENCES

[1] Cole, J. D. (1975), SIAM J. Appl. Math., 29, pp. 763-787.

[2] Steger, J. L. and Baldwin, B.S. (1973), AIAA J., 11, p. 903.

[3] Salas, M. (1974), AIAA Paper No. 74-523.

[4] Marconi, F. and Salas, M. (1973), Computers and Fluids, 1, p. 185.

[5] Moretti, G. (1975), Lecture Notes in Physics, 35, pp. 287-292, Springer-Verlag.

[6] Kentzer, C. (1973), Proceedings of the AIAA Computational Fluid Dynamics Conference, Palm Springs, California, pp. 41-50.

[7] Kentzer, C. (1971), Lecture Notes in Physics, 8, pp. 108-113, Springer-Verlag.

DISCUSSION OF THE KENTZER PAPER

MORETTI: Could you convince us by showing us more examples than just telling us that it works ?

KENTZER: Well, the example that I had was one without fitting the shock wave, therefore the accuracy suffered. In case of shock capturing, you might get oscillations back - you call them wiggles. They shake the solutions. The shock oscillates and this is not a nice thing.

MORETTI: This happens in YOUR method; it doesn't happen in, e.g., Murman's or Jameson's method.

KENTZER: Well, it depends on how much time one spends in refining the program. My point of today's talk was that one has to combine the second order method with shock fitting; otherwise you lose the advantages of it. It's worse than the first order method, unless the shock is fitted explicitly.

MORETTI: I'd like to have a quick reaction from either Dr. Cheng or Dr. Hafez on the matter.

CHENG: We are afraid to go into the second order terms because when you go to the second order, your truncation error will be very large, if you have a singularity like that behind the shock. Now, unless you are not interested in what's going on near the body - but you are interested in what's going on on the body and if we don't do anything - do not fit the singularity - the Oswatitsch-Zierep singularity, ϕ_x behaves like x log x, x being the distance from the root of the shock - so because this singularity is rather weak the first order scheme suffices, but you can never claim accuracy for the second order derivative of your solutions. For the first derivative, say ϕ_x, you might say that to the leading order it is all right, but now if you go to the second order, then it is hard to convince me that your truncation error is not large. But you have a shock; if you want to fit the shock you had better fit the singularity too, - if you talk about the second order; that is my feeling.

MURMAN: There has been a considerable amount of work done on second order schemes for the transonic flow equation: Garabedian and Korn proposed a scheme which they have in their program. There is no shock fitting involved, but the hyperbolic scheme is second order. But even in regions of smooth recompression, isentropic recompression shock-free airfoils where shock fitting is not an issue, the second order methods have stability problems and you can show this by doing a stability analysis; and in decelerating flows there's a stability problem. Seebass and Yu have done considerable looking into second order methods also with shock fitting and they have proposed some schemes which are stable in decelerating regions but not conservative. There are people looking into those improvements in the basic method. Today, most of the proposed improvements don't seem to give a sufficient improvement - except for problems where the flow is changing so rapidly and there are discontinuities and so on - so that these instabilities develop anyway. On the second point on the analysis - the convergence analysis using the time dependent problem analogy -

I think the work of Tony Jameson - in papers in the last couple of years along the same lines you're talking about - has been used extensively by people doing transonic analysis, for deriving more efficient iteration algorithms.

KENTZER: I'm very glad there's work done along these lines advocating the use of higher order methods combined with shock fitting.

NOTE ADDED IN PROOF: [Reply by the speaker to E. M. Murman's comments]

In decelerating flows we expect space-time focusing of characteristics to occur. Multiple-valued solutions lead to shock formation. Even if the steady state solution is shock free, the transient might contain shocks. The presence of shock waves will destroy the usefulness of higher order methods when shock fitting is not used. This may explain your bad experience with second order methods.

It may be particularly important to fit shocks explicitly in the design of supercritical shockless airfoils in order to determine time histories of shocks and their eventual disappearance in the limit of steady state flow. First order methods, combined with "shock capturing," will smear any weak shocks to the point where the question of the existence of shock waves could not be answered with any degree of certainty.

SICHEL: There is also an analytical analog of the shock fitting vs. the shock-capturing that has been done in the case of nozzle flows. I've done some work on that; Tom Adamson has done some work on that. The idea is that you start with a small disturbance equation but include a compressive viscous term and then you can find analytical solutions in which the shock is captured or automatically appears in the solution. And then Adamson, and Adamson and Richey, have extended this idea in looking at nozzle flows, where instead of putting in the viscous term you, as it were, introduce a shock at the appropriate place - you find that there is one place where the upstream analytical solution and downstream analytical solution can be connected with an appropriate shock wave. I think it might be interesting at some point to do some sample problem where you compare the numerical solution of this type with what I think is an analytical analog of exactly the same problem.

SOLUTIONS TO INTERNAL TRANSONIC FLOWS VIA PARAMETRIC DIFFERENTIATION

W. WHITLOW, JR., AND W. L. HARRIS
Massachusetts Institute of Technology

INTRODUCTION

The recent advances made in our understanding of unsteady, internal transonic flows may be attributed to research efforts that are experimental, analytical, and numerical in their basic approach or emphasis. With the numerical effort somewhat in the forefront, a need for more analysis and experimentation is present. This paper addresses the need for more analysis and physical modeling of the phenomena. Within this research program, it is our objective to generate a mathematically self-consistent and a physically plausible predictive algorithm for unsteady, internal transonic flows. Such a predictive algorithm would be of optimal utility if it were useful over a range of a selected parameter which characterizes unsteady, internal transonic flows. Using parametric differentiation [1, 2, 3, 4], we have successfully initiated our efforts to generate the desired predictive algorithm.

Parametric differentiation provides a means by which both the small perturbation transonic potential equation and the full transonic potential equation and approximate boundary/initial conditions may be linearized in phase space. The transformed linear system of equations is obtained in such a way that the physics of the problem remains untransformed. Hence, the difficulty of defining boundary conditions associated with the linear, complex hodograph method is not a limitation or undesirable feature of the method of parametric differentiation. The linear equation in phase space also has the advantage of being solvable by a variety of techniques and methods not applicable to the original, untransformed, governing equation. For example, the physically motivated method of superposition of elementary flows may be exploited as a means of solving the linear equation in phase space.

Following a brief review of the concepts upon which the method of parametric differentiation is based, the results of applying the method to simple Meyer flows are presented.

THE METHOD OF PARAMETRIC DIFFERENTIATION

Essentially, the method of parametric differentiation is a procedure by which nonlinear problems involving parameters may be solved by transforming the nonlinear problem to an equivalent linear problem in some parameter space. It is noted that not all nonlinear equations, with appropriate boundary and initial conditions, are transformable to an equivalent system of linear equations by this method. The procedure is outlined below.

Suppose the nonlinear governing equation and boundary conditions are given by

$$N[\psi(\bar{x}, t; \lambda)] = 0 \qquad (1)$$

$$\psi(\bar{x}_B, t; \lambda) = \psi_B \qquad (2)$$

and/or

$$\psi'(\bar{x}_B, t; \lambda) = \psi'_B \qquad (3)$$

where N is a nonlinear operator, ψ is the dependent variable, $\bar{x}$ and t are the vector space coordinate and time, respectively, λ is a parameter and the subscript B denotes the boundary. A solution is assumed to be of the form $\psi_o = \psi_o(\bar{x}, t; \lambda_o)$ where λ_o is a limiting value of λ. The equivalent linear problem in $\bar{x}$, t, λ space is obtained, assuming the transformation is possible, by differentiating Eqs. (1), (2) and/or (3) with respect to the parameter. One obtains the following results:

$$L[g(\bar{x}, t, \lambda; \psi_o)] = 0 \qquad (4)$$

$$g(\bar{x}_B, t, \lambda) = g_B \qquad (5)$$

and/or

$$g'(\bar{x}_B, t, \lambda) = g'_B \tag{6}$$

where

$$g(\bar{x}, t; \lambda) \equiv \frac{\partial\psi(\bar{x}, t, \lambda)}{\partial\lambda} \tag{7}$$

Whereas N is a nonlinear partial or ordinary differential operator, L is a linear differential operator. Once Eqs. (4) - (7) are solved to obtain $g(\bar{x}, t, \lambda)$, the value of the dependent variable at a different value of λ, $\psi(\bar{x}, t; \lambda_o + \Delta\lambda)$, may be obtained by quadrature without restriction on the range of λ other than those imposed by the basic formulation. The equivalent linear system of equations has been obtained at the expense of an increase in the number of independent variables. Any "non-linearity" of the equivalent linear system of equations is confined to Eq. (7). Obviously, the nature of the computational problems depends upon L, $\psi_o(\bar{x}, t; \lambda_o)$ and the quadrature to obtain $\psi(\bar{x}, t; \lambda_o + \Delta\lambda)$.

Note that Eq. (7) yields

$$\psi(\bar{x}, t; \lambda_o + \Delta\lambda) = \psi_0(\bar{x}, t; \lambda_o) + \int_{\lambda_o}^{\lambda_o + \Delta\lambda} g d\lambda \tag{8}$$

The integration indicated by Eq. (8) requires that the base solution $\psi_o(\bar{x}, t; \lambda_o)$ to be known. See Figure 1.

A successful application of the MPD requires an answer in the affirmative to the following questions:

(a) Is parametric differentiation applicable to the proposed problem?

(b) Are the physical implications of the results obtained via the MPD plausible in the proposed problem?

(c) Is parametric differentiation cost-effective in the proposed problem area?

(d) Are solutions obtained a weak function of the base solution?

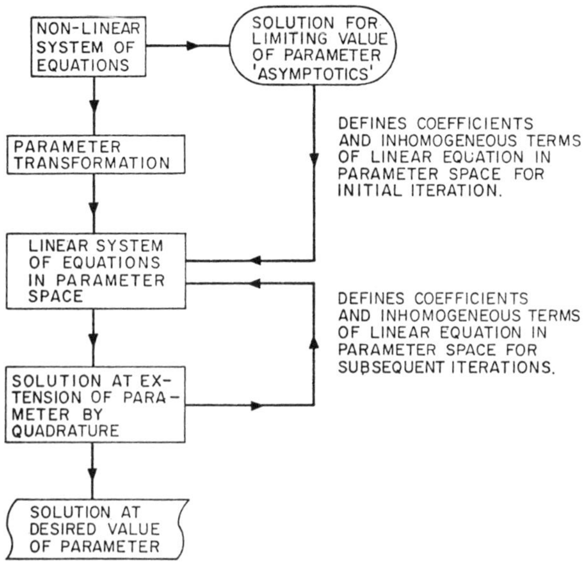

Figure 1. Iteration Procedure.

ANALYSIS

The problem to be investigated is that of perturbation from an inviscid, steady, sonic, irrotational, two-dimensional stream moving in the X-direction (see Figure 2). These perturbations have been studied for a simple acceleration from subsonic to supersonic flow (Meyer flow) by Adamson, et al. [5, 6, 7, 8] and Whitlow [9].

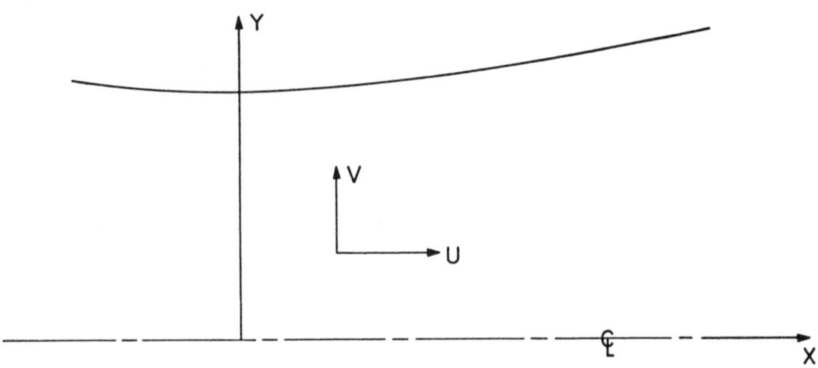

Figure 2. Coordinate System.

The equation governing the perturbations, as derived by Adamson and Richey [8], for viscous flow, is

$$-\phi_x\phi_{xx} + \phi_{yy} - \frac{2}{\tau\delta}\phi_{xt} - \frac{1}{\tau^2}\phi_{tt} + K_s(\phi_{xxx} + \frac{\delta}{\tau}\Gamma\phi_{xxt}) = 0 \qquad (9)$$

where $\phi(x, y, t)$ is the perturbation velocity potential, τ is a characteristic time associated with the disturbance, δ is a measure of the extent of the transonic region in the transverse direction, and K_s is a measure of the shock thickness. For inviscid flow, $K_s \equiv 0$, and our problem reduces to solving the nonlinear equation

$$-\phi_x\phi_{xx} + \phi_{yy} - \frac{2}{\tau\delta}\phi_{xt} - \frac{1}{\tau^2}\phi_{tt} = 0 \qquad (10)$$

Choosing our characteristic parameter as

$$\lambda = \frac{1}{\tau} \qquad (11)$$

and applying parametric differentiation, the equation to be solved is

$$-\phi_x g_{xx} - \phi_{xx} g_x + g_{yy} - \frac{2\lambda}{\delta} g_{xt} - \lambda^2 g_{tt} = 2\lambda\phi_{tt} + \frac{2}{\delta}\phi_{xt} \qquad (12)$$

The problem now becomes to solve a linear partial differential equation with variable coefficients.

As noted earlier, in order to employ the method of parametric differentiation, a known base solution, $\phi(x, y, t; \lambda_o)$ must be available to initiate the calculations. This solution was taken from work conducted by Adamson, et al. [5, 6, 7, 8]. Those investigators used the similarity method to obtain expressions for $\partial\phi/\partial x\ (x, y, t; \lambda_o)$ and $\partial\phi/\partial y\ (x, y, t; \lambda_o)$. Those investigators have obtained results which lead to a base solution of

$$\phi(x, y, t; \lambda_o) = 2bx^2 + 8b^2xy^2 + 8b^3y^4 + x(4b\beta - 2K\beta') + y^2(8b^2\beta - 2K^2\beta'') \qquad (13)$$

where $\beta = \beta(t)$.

The shape of the nozzle wall was chosen to be

$$y_{wall} = 1 + .02[\sqrt{1 + x^2} - 1] \qquad (14)$$

This corresponds to a gently varying nozzle, and, for convenience of computation, the throat is located at $x = 0$. The chosen nozzle is also rigid, imposing the condition that $(d/dt)(\partial\phi/\partial y) = 0$. This condition leads to an expression for β

$$\beta = c_1 e^{\frac{2bt}{k}} + c_2 e^{-\frac{2bt}{k}} - c_3 \qquad (15)$$

c_1 was set to zero, and c_2, the intersection of the sonic line with the nozzle centerline in the limit $t \to \infty$, was calculated to be .1155. For simplicity, c_2 was set equal to unity, and β now reads as

$$\beta = e^{-\frac{2bt}{k}} - .1155 \qquad (16)$$

The boundary condition at the nozzle wall is

$$-\frac{w}{\tau}\frac{\partial f_w}{\partial t} - \frac{w}{\delta}\frac{\partial f_w}{\partial x} + \frac{\delta^3}{(\gamma+1)}\left(\frac{\partial\phi}{\partial y}\right)_w \qquad (17)$$

where $w \ll 1$, γ is the ratio of specific heats (1.4 for this case), and f_w is the deviation of the nozzle contour from some constant value. For the case treated here

$$f_w = .02\left[\sqrt{1 + x^2} - 1\right] \qquad (18)$$

In phase space, the boundary condition becomes

$$\left(\frac{\partial g}{\partial y}\right)_w = - c_4(\lambda)\frac{\partial f_w}{\partial x} \qquad (19)$$

The transformed equation for g, with $\lambda = \lambda_o$, has been solved numerically by Whitlow [9] using finite differences. All derivatives were computed using a combination of forward and central differences, switching to backwards differences in y at the nozzle wall. Choosing n, m, p to represent an x, y, t nodal point, the derivatives are approximated as

$$g_x = \frac{g_{n+1,m,p} - g_{n-1,m,p}}{2\delta x} \qquad (20a)$$

$$g_{xx} = \frac{g_{n+1,m,p} - 2g_{n,m,p} + g_{n-1,m,p}}{(\delta x)^2} \tag{20b}$$

$$g_{y} = \frac{g_{n,m+1,p} - g_{n,m-1,p}}{2\delta y} \tag{20c}$$

$$g_{yy} = \frac{g_{n,m+1,p} - 2g_{n,m,p} + g_{n,m-1,p}}{(\delta y^2)} \tag{20d}$$

$$g_{tt} = \frac{g_{n,m,p+1} - 2g_{n,m,p} + g_{n,m,p-1}}{(\delta t)^2} \tag{20e}$$

$$g_{xt} = \frac{g_{n+1,m,p+1} - g_{n-1,m,p+1} - g_{n+1,m,p-1} + g_{n-1,m,p-1}}{4(\delta x)(\delta t)} \tag{20f}$$

where δx, δy, δt represent mesh spacings in the x, y, and t directions, respectively. At the nozzle wall

$$g_{y} = \frac{g_{n,m,p} - g_{n,m-1,p}}{\delta y} \tag{21a}$$

$$g_{yy} = \frac{g_{n,m,p} - 2g_{n,m-1,p} + g_{n,m-2,p}}{(\delta y)^2} \tag{21b}$$

For this study, δx, δy and δt were chosen as .01, .04, and .01, respectively.

The procedure used to calculate the flow field was to fix p, fix n, and calculate $g_{n+1,m,p+1}$ for all m; n is then incremented and $g_{n+1,m,p+1}$ is again computed for all m. This procedure is repeated until the entire flow field is computed for a fixed value of p. The value of p is then incremented and the procedure repeated until the flow field is computed over the desired time range. This procedure is flow charted in Figure 3.

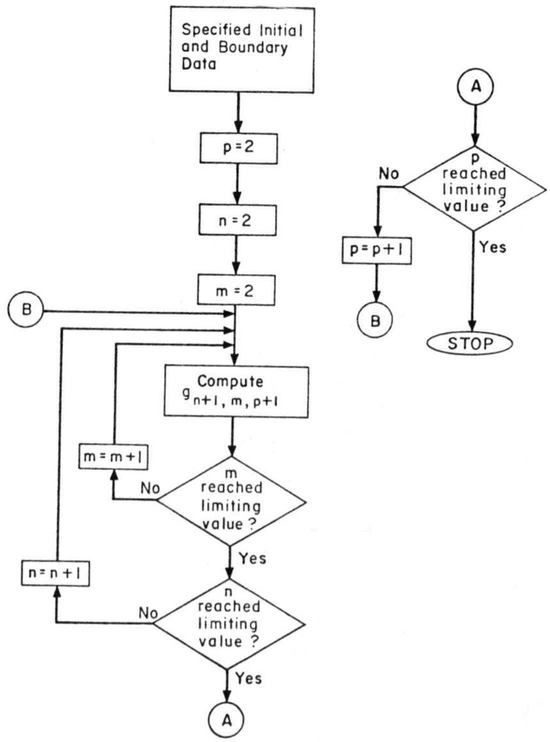

Figure 3. Procedure to Compute the Flow Field.

The computation of $g_{n+1, m, p+1}$ requires some initial values of g to be specified. The initial data required to initiate the calculations is shown in Figure 4. To specify the initial data $g(n, m, p, \lambda_o)$ was related to $\phi(n, m, p; \lambda_o)$ as

$$g(n, m, p, \lambda_o) = -\frac{\phi(n, m, p; \lambda_o)}{(1 + \lambda)^2} \tag{22}$$

The mesh point (n, m, p) = (2, 2, 2) was selected to correspond to the spatial point (x, y, t) = (0, 0, 0). Hence, the computation of $g_{n+1, m, 3}$ requires the specification of initial data for $t < 0$. Assuming that g grows with time $g_{n, m, 1}$ was related to $g_{n, m, 2}$ by

$$g_{n, m, 1} = .01\, g_{n, m, 2} \tag{23}$$

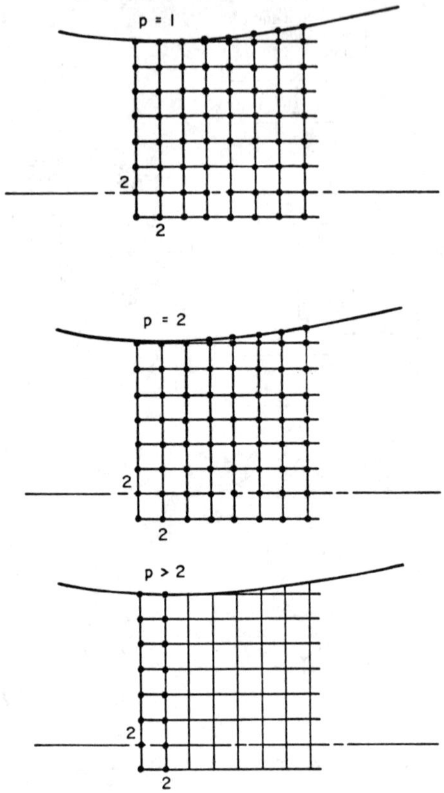

Figure 4. Required Initial and Boundary Data to Compute the Flow Field.

It should be noted that the relationship between $g_{n,m,1}$ and $g_{n,m,2}$ as well as that between $g(n, m, p, \lambda_o)$ and $\phi(x, y, t; \lambda_o)$ are not the only possible relationships, and others may be and should be tried.

The above procedure will compute the flow field for $\lambda = \lambda_o$. A procedure which will compute the flow field for various values of λ has also been worked out. In order to compute the flow field for the various values of λ, λ is incremented by $\Delta\lambda$. $\phi(n, m, p; \lambda_o + \Delta\lambda)$ is approximated by

$$\phi(n, m, p; \lambda_o + \Delta\lambda) = \phi(n, m, p; \lambda_o) + \Delta\lambda\, g(n, m, p, \lambda_o) \tag{24}$$

and the required initial data is computed as before. The flow field is then calculated as mentioned above, and the entire process is repeated until the flow field is computed over the desired range of λ. This procedure has been carried out for a portion of a flow field. To compute the portion of the flow field required 20,000 bytes of storage and .142 minutes on an IBM 370. The results are summarized in Table 1.

Having calculated g for various λ's, the potential at any value of λ can be computed by

$$\phi(x, y, t; \lambda) = \phi(x, y, t; \lambda_o) + \int_{\lambda_o}^{\lambda} g d\lambda \tag{25}$$

A sample calculation of the potential and velocity is also listed in Table 1.

TABLE I

SAMPLE RESULTS

λ	$\phi_{5,2,4}$	$g_{5,2,4}$	$\phi_{7,2,4}$	$g_{7,2,4}$	$u_{6,2,4}$
.01 ($\lambda = \lambda_o$)	.037	-3.21	.1116	-4.9	3.73
.02		2.4		3.59	
.03	.059	.215	.144	.259	4.25
.04		.0954		.0904	
.05	.061	.0472	.1461	.0211	4.225
.06		.0233		-.013	
.07	.0617	.0099	.1459	-.0315	4.21
.08		.0017		-.0424	
.09	.0617	-.0036	.1453	-.049	4.18
.10		-.0071		-.0531	
.11	.06158	-.00954	.144	-.0557	4.12

CONCLUSIONS

During the course of this study, it has been demonstrated that the method of parametric differentiation is successful in removing the nonlinearity from the governing perturbation potential equation. A rigid nozzle of specified shape has been chosen, and the boundary conditions, in phase space, have been formulated. A base solution corresponding to a rigid nozzle has also been found, using similarity results obtained by Adamson, et al. [5, 6, 7, 8].

Finite difference approximations to the derivatives of g have been developed, and the governing equation in phase space has been written using those approximations. The nozzle wall conditions in phase space have also been approximated in finite difference form. The initial and boundary data necessary to compute the flow field has been identified, and an expression connecting the base solution, $\phi(x, y, t; \lambda_o)$, and the chosen characteristic parameter has been written and differentiated to yield the necessary initial and boundary data. A computer routine has been written, with the specified initial and boundary data as input, that computes $(\partial\phi/\partial\lambda)$ and, eventually, ϕ and ϕ_x at all points in space and time over a range of characteristic parameters.

In this study, the method of parametric differentiation has been extended to a new level of complexity. Before this investigation was undertaken, the method had been used to study problems in two space dimensions with time fixed or one space dimension with time varying. Here, the method has been extended to two space dimensions with time varying.

One may ask what are the advantages of using parametric differentiation to linearize the problem over solving the potential equation numerically. The primary reason is that parametric differentiation removes the nonlinearity from the problem, and in doing so, reduces the amount of computing needed to obtain solutions of the governing equation. For the sample calculations presented here, the computation time and required storage were very reasonable. When using parametric differentiation, one can obtain solutions for all characteristic times and observe how they vary as the time is changed. It should be noted that in trying to solve a nonlinear equation, one is not guaranteed a unique solution; this is no problem with a linear equation. Hence, it is seen that parametric differentiation has great potential for applications to the study of transonic flows.

REFERENCES

[1] Harris, W. L. (1974), "Farfield Viscous Effects in Nonlinear Noise Propagation," Journal of the Acoustical Society of America, 56:2, pp. 313-322.

[2] Jischke, M. C. and Baron, J. R. (1969), "Application of the Method of Parametric Differentiation to Radiative Gasdynamics," AIAA Journal, 7:7, pp. 1326-1335.

[3] Rubbert, P. E. and Landahl, M. T. (1967), "Solution of Nonlinear Flow Problems through Parametric Differentiation," Physics of Fluids, 10:4, pp. 831-835.

[4] Rubbert, P. E. and Landahl, M. T. (1967), "Solution of the Transonic Airfoil Problem through Parametric Differentiation," AIAA Journal, 5:3, pp. 470-479.

[5] Adamson, T. C., Jr. (1972), "Unsteady Transonic Flows in Two-Dimensional Channels," Journal of Fluid Mechanics, 52:3, pp. 437-449.

[6] Adamson, T. C., Jr. (1971), "Unsteady Transonic Flows in Two-Dimensional Channels," Project SQUID Technical Report MICH-9-PU.

[7] Adamson, T. C., Jr. and Richey, G. K. (1973), "Unsteady Transonic Flows with Shock Waves in Two-Dimensional Channels," Journal of Fluid Mechanics, 60:2, pp. 363-382.

[8] Adamson, T. C., Jr. and Richey, G. K. (1972), "A Study of Unsteady Transonic Flows with Shock Waves in Two-Dimensional Channels," Project SQUID Technical Report MICH-10-PU.

[9] Whitlow, W., Jr. (1975), "Inviscid, Unsteady, Non-Linear, Internal Transonic Flow," Massachusetts Institute of Technology, S. M. Thesis.

DISCUSSION OF THE WHITLOW-HARRIS PAPER

ADAMSON: You're using the similarity solutions which are only valid for one wall shape, so I imagine that in your calculations this holds true, does it not?

HARRIS: Yes. My understanding of the Meyer flow solutions which you developed - the stream lines in fact do correspond to the wall for the rigid wall-condition.

ADAMSON: The point is that you can't take an arbitrary wall with a similarity solution.

HARRIS: Oh, sure.

ADAMSON: Yes. That's true also for this method?

HARRIS: Yes. That's certainly the case.

HAFEZ: How is this different from Golub and Bellman for a method to convert the boundary value problem to an initial value problem? The point here is that you embed your problem in a higher dimension, like solving the steady problem by an unsteady approach. So you increase the computation time of the number of unknowns because you have one more dimension.

HARRIS: No, it does not really increase the number of unknowns; it does increase the number of independent variables. To derive the unknown you want - like the stream function - is a direct quadrature of g.

HAFEZ: But if we have a boundary value problem in 2-D, you can make it easily an initial value problem in 3-D.

HARRIS: Yes. The method is basically one of extending the solution by analytic continuation, if you will; it's quite similar to the work that Van Dyke is now developing in which he generates his higher order series on a computer. We do not do that; we simply get ours by formalized quadrature.

PERTURBATION SOLUTIONS FOR BLADE-TO-BLADE SURFACES OF A TRANSONIC COMPRESSOR*

S. S. STAHARA AND D. S. CHAUSSEE
Nielsen Engineering and Research, Inc.

J. R. SPREITER
Stanford University

INTRODUCTION

Within the last several years, the development of numerical techniques to solve transonic mixed-flow problems has led to a number of significant advances in the calculation of such mixed flows in turbomachinery. A result of these developments, some of which have occurred quite recently and are reported at this workshop, is that the capability presently exists for calculating accurate inviscid solutions for an important class of transonic compressors and fans. At the present time, apparently, the most general and successful of these techniques are based on time-marching and finite-difference procedures for solving the Euler equations, such as developed for example by Erdos, et al. [1]. Steady state, or converged periodic solutions, are obtained asymptotically in time, almost inevitably with substantial computational requirements. These developments, however, form the basis of the present work which is directed primarily toward supplementing these calculations, with a view toward reducing the computational requirements now necessary. Although the current work has been

---

* Research supported by NASA Lewis Research Center under Contract NAS 3-19738.

underway for only a few months and our specific results are still modest, we are confident that the procedures we are developing will lead rapidly to significant progress in the practical solution of these problems.

DESCRIPTION OF METHOD

The objective of the present work, which is outlined in the first figure, is to provide a means for turbomachinery calculations, particularly where it is necessary to carry out a number of calculations for closely-related flows such as are needed in a parametric design study. We have initially chosen to study the transonic compressor for reasons which will become apparent later.

The basic idea underlying the method is to make use of a previously calculated base solution to determine first-order changes in the flow field due to variations in one or more of a variety of geometrical or flow field parameters. The assumptions which we have adopted are that the basic flow is inviscid, irrotational, and steady. While these assumptions are not absolutely essential and could be relaxed, we feel that this framework is sufficiently accurate to provide results useful in a practical sense and still not too complex to preclude adequate checking of the technique. The fundamental assumption associated with the perturbation solution is that the magnitude of the excursions from the base solution lie within the range of a linear perturbation analysis.

GOAL: To economize transonic compressor calculations, particularly parametric evaluations

METHOD: Utilize a known or base solution to determine first-order flow changes due to variations in geometrical or flow-field parameters

ASSUMPTIONS:
1. Inviscid, irrotational, steady flow
2. Excursions from base solution within scope of linear perturbation analysis

Figure 1. Basic Concepts

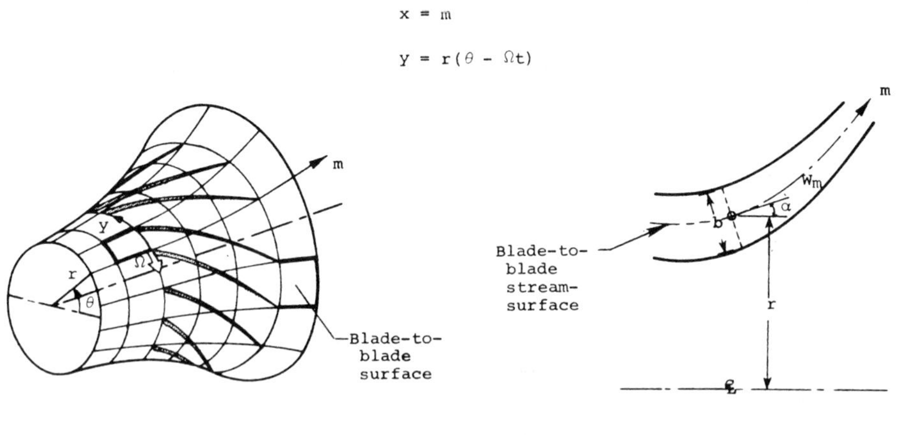

BLADE-TO-BLADE SURFACE OR REVOLUTION

STREAM CHANNEL IN MERIDIONAL FLOW PLANE

Figure 2. Model Problem. Transition Flow on the Blade-to-Blade Surface of a Single Row Compressor Rotor

The model problem which we have chosen to study is that of transonic flow on the blade-to-blade surface of a single row compressor rotor as indicated in Figure 2. Included in the formulation, which is carried out in the rotating coordinate system defined in that figure, are variations in the stream channel thickness b and radial divergence r in the meridional flow plane as shown in the sketch on the right. The basic reasons for this particular choice are (a) that this problem provides an ideal test of our perturbation technique as it is rich in a variety of parameters which can be varied, and (b) that solutions for the more accurate representations of such flows are substantial in cost in terms of computational time. Also, and quite significant, is the fact that for special subcases of the general problem, several different methods are available for solving the basic flow. These include small-disturbance transonic [2] and full potential methods [3], in addition to the Euler equation solutions mentioned previously. This is important from the point of view of potential applications of the method since it provides another basis for measuring the relative efficiency of the technique.

The level of accuracy of the base solution we have initially chosen to formulate the method on is the full potential

VELOCITY POTENTIAL: $u = \phi_X$, $v = \phi_Y$

DIFFERENTIAL EQUATION: $$\left[\phi_X^2 - A^2\right]\phi_{XX} + \left[\phi_Y^2 - A^2\right]\phi_{YY} + 2\phi_X\phi_Y\phi_{XY} = A^2\phi_X\left[\frac{1}{R}\frac{DR}{DX} + \frac{1}{B}\frac{DB}{DX}\right] - \phi_X\phi_Y\,\Omega\,\frac{DR}{DX}$$

WHERE $$A^2 = A_\infty^2 + \frac{1}{2}(\gamma - 1)\left[U_\infty^2 + V_\infty^2 - \phi_X^2 - \phi_Y^2\right]$$

BOUNDARY CONDITIONS:

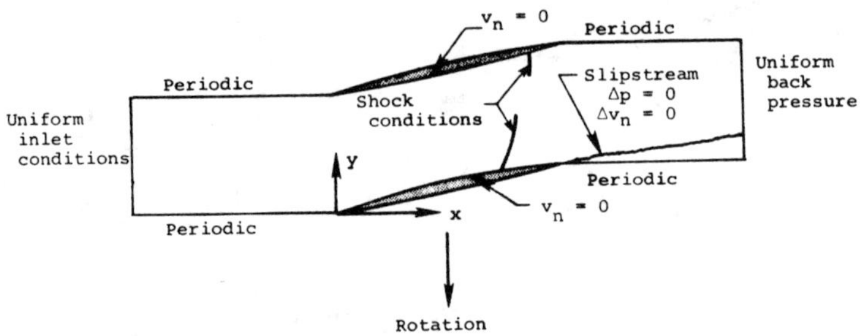

Figure 3. Equations of Motion

representation, for which the differential equation and boundary conditions are indicated in Figure 3 in a coordinate system rotating with the blades. We note that the variations in stream channel thickness b and radial divergence r enter as terms on the right-hand side of the familiar two-dimensional potential equation.

The perturbation analysis associated with this formulation of the base problem is outlined in Figure 4 which indicates the decomposition of the potential into a base plus perturbation component, and then the form of the governing linearized equation which then results for the perturbation component ϕ'. In the general case, all of the coefficients in the differential equation for ϕ' are functions of the base solution ϕ_o. However, for simpler cases, such as when the base flow is taken as a small-disturbance potential solution, a number of the coefficients will simplify or vanish.

Some of the typical perturbations that will be considered are indicated, and include changes in geometry, flow parameters and operating conditions. Of course, under certain flow conditions, such as choking, it is not possible to vary some of the parameters

DECOMPOSE: $\phi = \phi_0 + \phi'$

WHERE ϕ_0 = BASIC-FLOW SOLUTION

ϕ' = PERTURBATION COMPONENT RESULTING FROM CHANGES IN GEOMETRICAL OR FLOW CONDITIONS

LINEARIZED EQUATION FOR ϕ':

$$A\phi'_{XX} + B\phi'_{XY} + C\phi'_{YY} + D\phi'_X + E\phi'_Y = F$$

WHERE $[A, B, \ldots, F]$ ARE FUNCTIONS OF $\phi_0(X, Y)$

TYPICAL PERTURBATIONS:

GEOMETRICAL	FLOW	OPERATING CONDITIONS
• THICKNESS DISTRIBUTION	• INLET MACH NO.	• PRESSURE RATIO
• CAMBER DISTRIBUTION	• INFLOW, OUTFLOW ANGLES	• TOTAL PRESSURE, TEMPERATURE
• SOLIDITY		
⋮	⋮	⋮

Figure 4. Perturbation Analysis

independently of others. It is important to recognize and account for this in determining the perturbation solutions. In order to solve the linearized equation for the perturbation potential ϕ', we are employing a finite-difference procedure using a successive-line overrelaxation algorithm. The method is based on the Murman-Cole differencing scheme and uses the equations in appropriate conservation form to properly account for elliptic, hyperbolic, sonic, and shock points. Consequently, shock perturbations will be automatically accounted for in the method.

RESULTS AND DISCUSSION

Some preliminary results obtained from the perturbation procedures are shown in Figure 5. Here comparisons are made for results obtained by varying the thickness ratio of an unstaggered nonlifting cascade composed of biconvex profiles in a flow with oncoming free-stream Mach number $M_\infty = 0.60$. The plot on the left displays the results for a pitch-to-chord ratio, $H/C = 0.75$. The solid lines indicate the results obtained by using the basic solution method for the three thickness ratios shown. Those basic flow

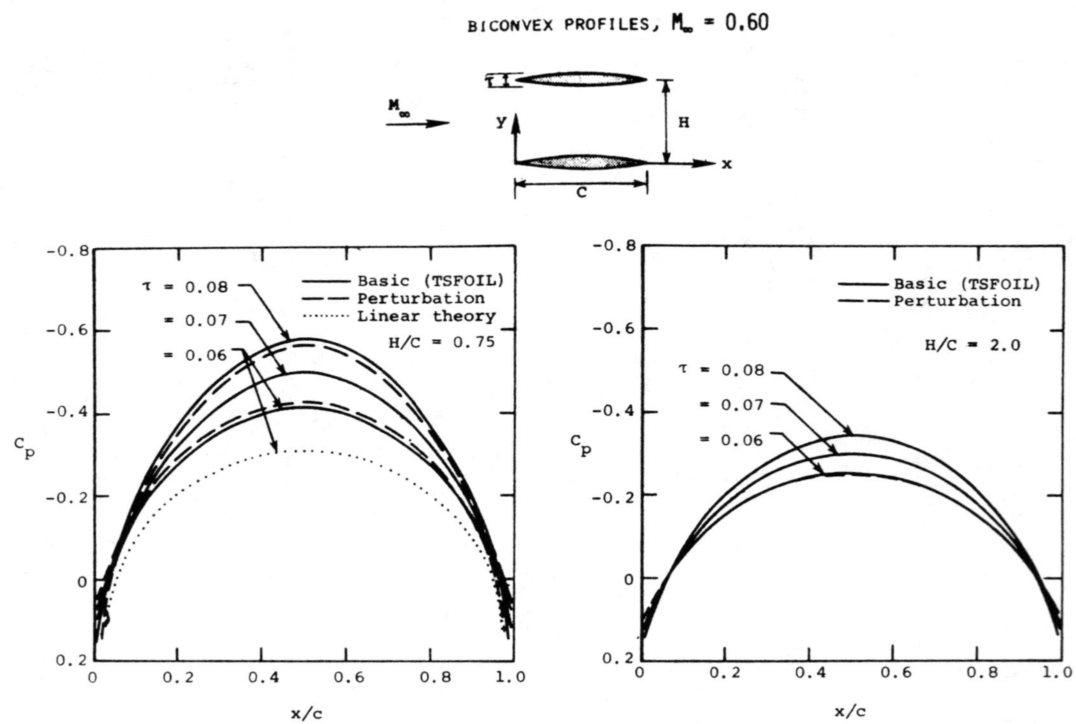

Figure 5. Perturbation Results for Thickness Ratio Change of an Unstaggered Cascade.

solutions were obtained by solving the transonic small-disturbance potential equation and were calculated by using the code TSFOIL [2]. The solid line results for thickness ratios τ equal to 0.06 and 0.08 are meant to be compared with those indicated by the dashed lines which were obtained by using the perturbation method together with the basic flow solution for thickness ratio equal to 0.07. The good agreement between the two results is impressive considering that at this spacing the flow is relatively sensitive to changes in geometry -- our numerical experiments show that this flow chokes at M_∞ just under 0.7. Also, a change in thickness ratio of 1 in 7 is 14%, which is not a small perturbation. Finally, to obtain some idea of the magnitude of the nonlinear effects present in this example, we have indicated with dotted lines the results of linear theory for $\tau = 0.06$. Comparison clearly indicates that nonlinear effects are sizable for this geometry at even this low a Mach number, thereby illustrating that a nonlinear calculation is required.

Analogous results are presented in the plot on the right for a pitch-to-chord ratio of 2.0. In this case, where the flow is not confined, the perturbation results are essentially identical with those predicted by using the basic flow method.

With regard to future work, the final figure indicates the three essential tasks on which we plan to concentrate. These include determination of the range of validity of the various perturbation solutions when utilizing representative base flow solutions, extension to supercritical, choked, and supersonic flows, and examination of the effect on accuracy and computational time of using various base flow solution methods.

- RANGE OF VALIDITY OF VARIOUS PERTURBATION SOLUTIONS

- SUPERCRITICAL, CHOKED, AND SUPERSONIC FLOWS

- EFFECT OF VARIOUS BASIC FLOW METHODS

Figure 6. Future Work

REFERENCES

[1] Erdos, J., Alzner, E., Kalben, P., McNally, W., and Slutsky, J. (1975): Time-Dependent Transonic Flow Solutions for Axial Turbomachinery. NASA SP-347, Part I, pp. 587-622.

[2] Murman, E. M., Bailey, F. R., and Johnson, M. L. (1975): TSFOIL-A Computer Code for Two-Dimensional Transonic Calculations, Including Wind-Tunnel Wall Effects and Wave-Drag Evaluation. NASA SP-347, Part II, pp. 769-788.

[3] Jameson, A. (1975): Transonic Potential Flow Calculations Using Conservation Form. Proceedings of AIAA 2nd Computational Fluid Dynamics Conference, Hartford, Connecticut, June 19-20, pp. 148-161.

DISCUSSION OF THE STAHARA, et al. PAPER

OLIVER: I think the idea of trying to find an alternative to executing an entire nonlinear calculation when you change parameters for design studies is a very admirable one, but it seems to me that there is a fundamental limitation here which is crucial for studying supercritical transonic flows which are the ones of major interest; that is, I think the linear perturbations are just not appropriate for, in fact, even small perturbations. The variations are inherently nonlinear and the simplest example of that is that the shock will not be transported properly - you cannot relocate shocks - by linear analysis. At least I don't believe you can. Do you have any comments about that fundamental limitation?

STAHARA: I'm aware of it. Until we really get into it, I don't have any results, actually, to bolster what I feel but I think there's going to be some range that we're going to be able to consider. I think there's an important point here. The reason we want to go with a linear analysis is that obviously we're going to do the calculation once for a unit change in whatever parameter we're considering. So that calculation once it's done, is done once and for all and it is valid for a range. Now what you're saying is that range is obviously going to be too small to be meaningful in a practical sense. I don't know.

OLIVER: What I'm saying is that I think it's incorrect. I don't think it's correct even to the smallest range because of its inherent nonlinearity.

ADAMSON: We have found, in our studies of nozzle flow, for example, that it is possible - using asymptotic analyses - to locate the shock using linearized equations, and then in the region very near the shock, presumably the nonlinear equation would hold. But there were certain cases even there, where the solution to the nonlinear equation is so simple, that in reality the shock shape, etc. - everything we calculated - was found using linear equations. So there are some range of parameters where this is possible.

STAHARA: [NOTE ADDED IN PROOF]: I think that this point can be clarified a great deal by considering the manner in which the linear perturbation problem is formulated and the solution actually carried out. Consider the simple example of transonic flow past a wedge profile with $M_\infty > 1$. Now let's say that the wedge angle is perturbed slightly, all else remaining the same. The bow shock would move to a new location slightly removed from the original or base flow location, and the rest of the flow field would also be perturbed slightly. Now, at first glance, it would seem that one could offer the argument that over the region that the shock has moved the perturbation quantities cannot be considered small or linear since they, in effect, are the cause of the shock movement, and thus must, in this region at least, be as large as the base flow quantities. However, this reasoning is misleading in the sense that it does not properly represent the mechanism by which the perturbation problem actually moves the shock. Consider the manner in which the perturbation problem is actually posed. We split the solution into the nonlinear base component plus a small linear perturbation. This obviously leads to a linear differential equation for the perturbation. Regard now the boundary conditions for the perturbation. These conditions are written for the new flow situation and involve specifications at the new wedge surface and bow shock which are at different locations than those of the base flow. In particular, the bow shock jump condition is to be satisfied at the new shock location. To implement this condition, however, a Taylor series expansion is used to transfer that condition back to the old base flow shock location. With the flow solution split into the base plus linear perturbation, the Taylor series expansion serves to provide the shock boundary conditions that must be satisfied by the linear perturbation at the old shock wave location. This condition involves, of course, the unknown displacement of the shock, but it turns out that this displacement can be eliminated, by appropriate manipulation, in favor of base flow properties. Thus, the linear perturbation problem is now formulated on the skeleton of the old base flow problem. One then proceeds to solve for the linear perturbation. With the solution for the linear perturbation in hand, the motion of the shock wave is solved for by a separate calculation. This is the fundamental mechanism by which the linear perturbation can move the shock, consistent with the total nonlinear problem.

SESSION III

Chairman:
ROBERT E. MELNIK
Grumman Aerospace Corporation

VISCOUS EFFECTS IN TRANSONIC FLOWS

ON THE PREDICTION OF VISCOUS PHENOMENA IN TRANSONIC FLOWS

GEORGE S. DEIWERT

NASA Ames Research Center

ABSTRACT

Viscous phenomena in transonic flows, such as viscous-inviscid interations, flow separation and buffeting, can be numerically simulated on modern computers. The equations appropriate to describe such flows are the Navier-Stokes equations for compressible flows. Efficient and accurate computing algorithms exist, and are being further developed, which solve these equations for flows about configurations of arbitrary geometry. For low Reynolds number, wholly laminar flows, the system is closed and directly amenable to solution. For higher Reynolds numbers, where the flow is turbulent or transitional, it is necessary to close the system of equations by means of turbulent transport models. Unfortunately, while many transport models exist for relatively simple flows, they remain the pacing item in the simulation of viscous effects in transonic flows.

Using algebraic eddy viscosity models, several flows over airfoil configurations are computed and compared with experiment. Where the viscous effects are weak (e.g., attached boundary layers, weak viscous-inviscid interaction, or small scale separation) the computations are quantitatively adequate. Where viscous effects are strong (e.g., large-scale separation, buffeting, or stronger viscous-inviscid interaction), the computations are only qualitatively correct. It is suggested that the key to predicting viscous effects in transonic flows lies in the understanding and prediction of turbulence phenomena.

INTRODUCTION

The prediction of viscous phenomena in transonic flows involves descriptions of both boundary layer and inviscid flow regions and their interaction with one another. For flows where the boundary layer remains attached, the two flow regions can usually be analyzed separately and their interaction determined iteratively. This generally requires solving the compressible Euler equations for the inviscid field and the boundary-layer equations for the viscous region near solid surfaces. When the viscous-inviscid interactions are strong, and there is flow separation or even buffeting, it is more reasonable to solve the Navier-Stokes equations for compressible flows. These equations describe the coupling between the viscous and inviscid regions and describe the elliptic behavior in regions of flow separation. They can be solved on modern computers by finite difference algorithms.

A particularly attractive algorithm has been developed by MacCormack [1-2]; it is explicit and is second-order accurate in both time and space. Application of this method to transonic flows is described by Deiwert [3-6] and is the method used in the present paper. Briefly, the time dependent Navier-Stokes equations are written in integro-differential form for an arbitrary fluid element volume.

$$\frac{\partial}{\partial t} \int_{vol} U \, d\,vol + \int_{S} \vec{H} \cdot \vec{n} \, ds = 0$$

where U is the vector of conservable quantities in the volume element, $\vec{H}$ is the vector of forces and fluxes acting on the surface of the volume, and $\vec{n}$ is a unit normal vector to the surface element ds about the volume. For two-dimensional flow the vectors U and $\vec{H}$ can be written in Cartesian coordinates as

$$U \equiv \begin{bmatrix} \rho \\ \rho u \\ \rho v \\ e \end{bmatrix} \qquad \vec{H} \equiv \begin{bmatrix} \rho \vec{q} \\ \rho u \vec{q} + \vec{\vec{\tau}} \cdot \vec{e}_x \\ \rho v \vec{q} + \vec{\vec{\tau}} \cdot \vec{e}_y \\ e\vec{q} + \vec{\vec{\tau}} \cdot \vec{q} - k \nabla T \end{bmatrix}$$

where

$$\vec{q} \equiv u\vec{e}_x + v\vec{e}_y$$

$$\vec{\vec{\tau}} \equiv \sigma_x \vec{e}_x \vec{e}_x + \tau_{xy} \vec{e}_x \vec{e}_y + \tau_{yx} \vec{e}_y \vec{e}_x + \sigma_y \vec{e}_y \vec{e}_y$$

$\vec{e}_x$ and $\vec{e}_y$ are unit vectors in orthogonal x, y space, ρ is the mass density, u and v are the velocity components, e is the internal energy, T is the temperature, and $\vec{\vec{\tau}}$ is the complete stress tensor. These equations are solved in rectangular Cartesian space for flow through volume elements of arbitrary geometry. This precludes the necessity of coordinate transformations and lends itself to simple extension to three-dimensional flows.

In the present examples the volume elements are defined by a body oriented mesh such as illustrated in Figure 1. The mesh elements are small in regions where viscous effects are expected to be important, and large in inviscid regions. The transition from small to large mesh regions is achieved both by mesh stretching and by mesh patching. Typically, mesh patching is implemented approximately at the sublayer - boundary layer interface and at the boundary layer - inviscid flow interface. At these mesh interfaces the algorithm is locally first-order accurate. The mesh stretching

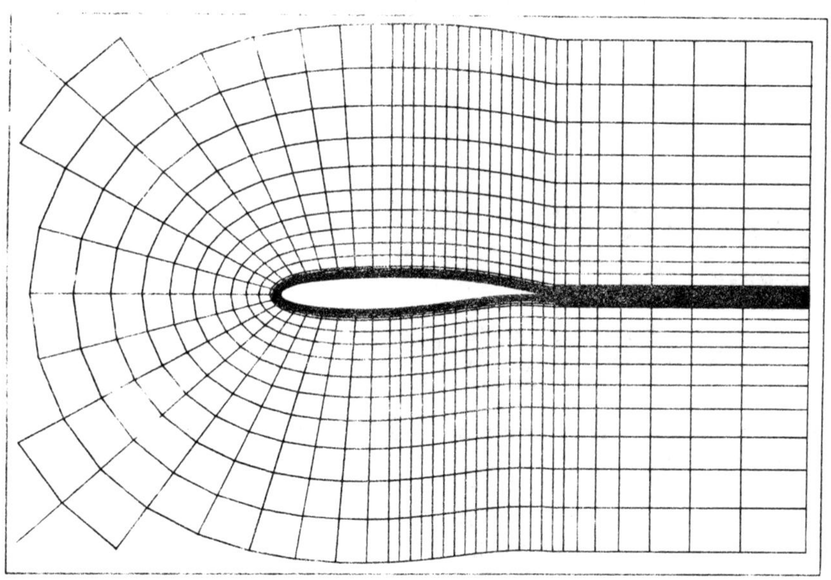

Figure 1. Mesh Configuration for Garabedian and Korn Shockless Lifting Airfoil.

is geometric and is small in regions of large gradients and large in regions of small gradients, thus preserving the over-all second-order accuracy of the method. Typically, the mesh spacing is constant in the sublayer region and highly stretched in the inviscid region. This permits the use of free-stream and Neumann boundary conditions with relatively few mesh points, thus achieving the desirable effect of mapping transformations.

It is the purpose of this paper to discuss the prediction of viscous effects in transonic flows of aerodynamic interest using the above method. Many of the concepts apply equally to flows in turbomachinery and the transition of the one field of interest to the other should be straightforward.

LAMINAR FLOWS

For wholly laminar flows the Navier-Stokes equations are exact and the components of the stress tensor can be written as

$$\sigma_x = p + \frac{2}{3}\mu\left(\frac{\partial u}{\partial x} + \frac{\partial v}{\partial y}\right) - 2\mu\frac{\partial u}{\partial x}$$

$$\sigma_y = p + \frac{2}{3}\mu\left(\frac{\partial u}{\partial x} + \frac{\partial v}{\partial y}\right) - 2\mu\frac{\partial v}{\partial y}$$

$$\tau_{xy} = \tau_{yx} = -\mu\left(\frac{\partial u}{\partial y} + \frac{\partial v}{\partial x}\right)$$

and the heat flux can be written as

$$k\nabla T = -\frac{\gamma\mu}{P_r}\left(\frac{\partial e}{\partial x}\vec{e}_x + \frac{\partial e}{\partial y}\vec{e}_y\right)$$

where γ is the isentropic exponent (C_p/C_v), P_r is the molecular Prandtl number, and μ is the coefficient of molecular viscosity.

For the low Reynolds number associated with wholly laminar flow the viscous regions can be thick and especially susceptible to separation. For airfoil configurations of reasonable thickness this can lead to stall or even buffeting at quite moderate angles of attack. While these laminar flows pose many interesting problems, their importance in transonic flows of aerodynamic interest is primary academic. Many of the important features of these flows can be and have been treated by incompressible analyses. The larger class of transonic flows is turbulent and is discussed in the following section.

TURBULENT FLOWS

Most transonic flows of interest are at high enough Reynolds number such that the flow is either turbulent or transitional either near the solid boundaries or in the near wake. While, in principle, the Navier-Stokes equations describe these flows, the length scale of the turbulent motion varies so widely that it is not practical to attempt to resolve it by finite difference procedures. Instead, the Reynolds averaged form of the equations are used and the consequent Reynolds stress terms are described by empirical model equations. In this case the form of the governing equations remains unchanged and, by using the eddy viscosity effect, the components of the stress tensor are approximated by

$$\sigma_x = p + \frac{2}{3}(\mu + \epsilon)\left(\frac{\partial u}{\partial x} + \frac{\partial v}{\partial y}\right) - 2(\mu + \epsilon)\frac{\partial u}{\partial x}$$

$$\sigma_y = p + \frac{2}{3}(\mu + \epsilon)\left(\frac{\partial u}{\partial x} + \frac{\partial v}{\partial y}\right) - 2(\mu + \epsilon)\frac{\partial v}{\partial y}$$

$$\tau_{xy} = \tau_{yx} = -(\mu + \epsilon)\left(\frac{\partial u}{\partial y} + \frac{\partial v}{\partial x}\right)$$

and the heat flux as

$$k\nabla T = -\frac{\gamma(\mu + \epsilon)}{P_{r_t}}\left[\frac{\partial e}{\partial x}\vec{e}_x + \frac{\partial e}{\partial y}\vec{e}_y\right]$$

where ϵ is the dynamic eddy viscosity coefficient and P_{r_t} is the total (molecular and turbulent) Prandtl number. The key to successful flow field simulation in this case is in adequately describing the turbulent transport by eddy viscosity model equations.

The simplest forms of these equations are algebraic, depending only on the local mean flow properties. More elaborate models include partial differential equations of the same degree of complexity as the Navier-Stokes equations being solved. Unfortunately, it is not intuitively clear that for the transonic flows of interest here the more elaborate differential models proposed to date will perform significantly better than algebraic formulations (e.g., see [6].

To illustrate the influence of including turbulent transport in the solution of the Navier-Stokes equations, flow fields have been computed for two different airfoil configurations for which experimental data are available for comparisons. An algebraic eddy viscosity model was assumed and flow conditions chosen where both large and small viscous-inviscid interactions are expected.

Biconvex Circular Arc

The first configuration is an 18 percent thick biconvex circular arc at zero angle of attack. This configuration is selected because even in a symmetric configuration at zero angle of attack it can exhibit both trailing edge and shock induced type separation. This configuration was also investigated experimentally by McDevitt, et al. [7] who found that at zero angle of attack for free-stream Mach numbers below 0.76 the flow field was steady, exhibiting only small scale trailing edge separation; and for free-stream Mach numbers greater than 0.78 the flow field was steady, exhibiting a massive merged shock induced - trailing edge separation. Oil flow photographs illustrating these two flow configurations are shown in Figure 2. For mach numbers between 0.76 and 0.78, unsteady buffeting is likely whereby the flow rapidly oscillates between the two steady flow configurations, alternately between the upper and lower surface. While it is beyond the scope of the present work to investigate unsteady transonic turbulent flows, the two steady flow configurations have been considered and comparisons with experiment are shown in Figures 3 and 4. The eddy viscosity model used was the "model 2" reported in [5] and [6].

Figure 3 shows experimental shadowgraphs of the flow field over the aft portion of the airfoil for a Mach number of 0.74 and 0.79 and a chord Reynolds number of four million. Included for comparison are computed Mach contours and velocity profiles. The shadowgraphs at the top of Figure 3 are of the aft region of the experimental flow field corresponding to the low and high Mach numbers. The shadowgraph at the left, for the lower Mach number, shows a nearly normal shock at about 63 percent of chord. The boundary layer behind the shock remains attached, yet shows substantial thickening. Near the trailing edge, it separates, denoted by the sudden change in curvature of the boundary-layer edge and

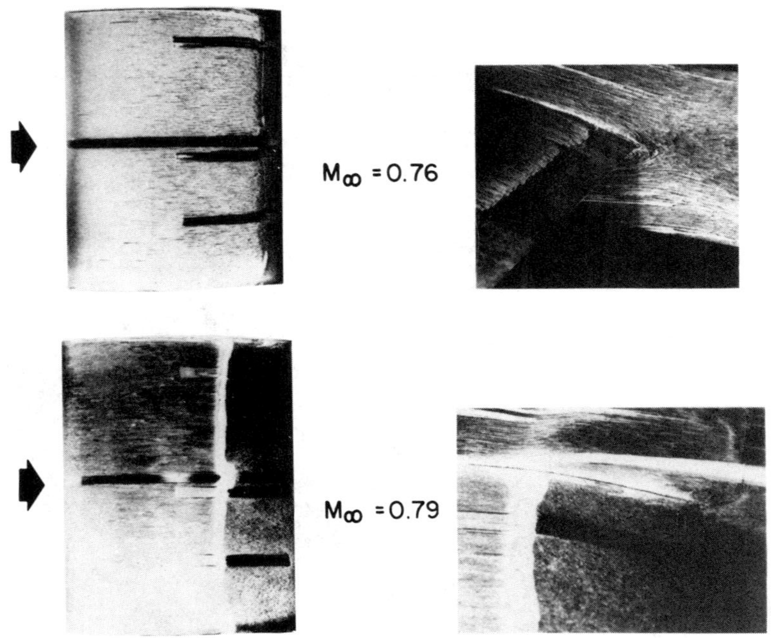

Figure 2. Oil-flow Patterns of Separation Regions on 18 Percent Thick Biconvex Circular Arc, $Re_c = 10 \times 10^6$.

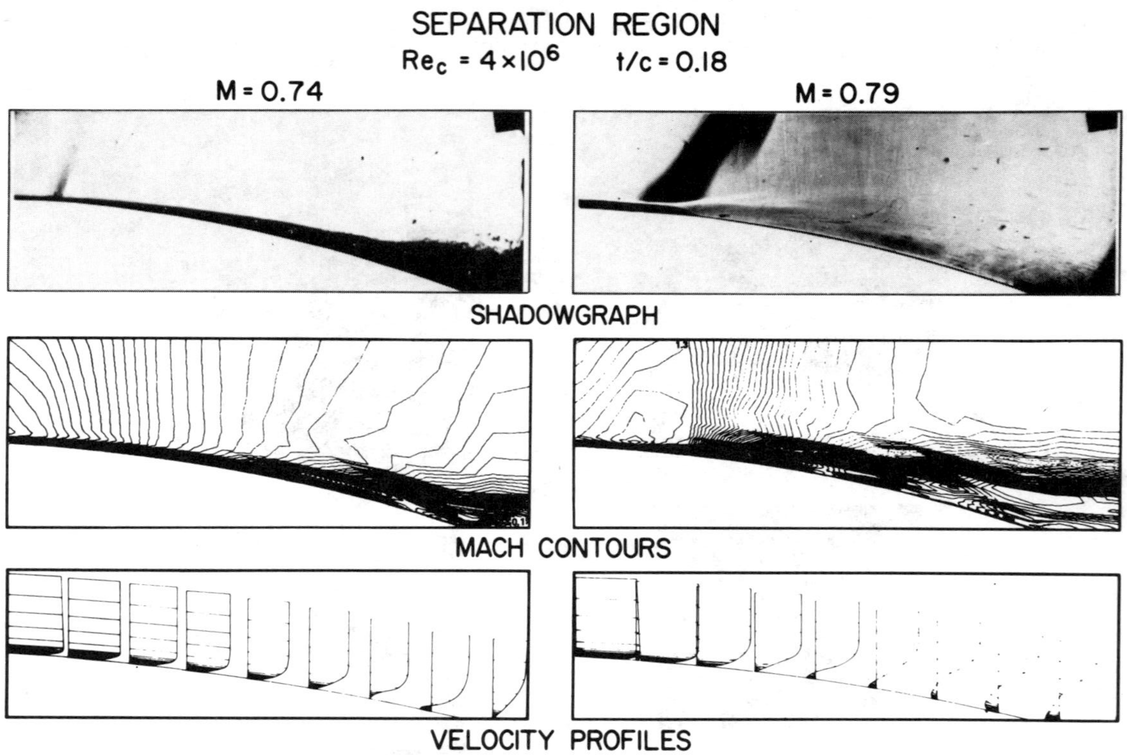

Figure 3. Flow-field Details over Aft Portion of an 18 Percent Thick Biconvex Circular Arc.

the massive thickening of the viscous region itself. The shadowgraph at the right shows an oblique shock at about 65 percent chord. The flow behind the shock is separated, as evidenced by the flow pattern behind the shock, and reattached downstream in the wake. The separation bubble is quite thick, as evidenced both by the existence of the oblique shock necessary to turn the flow and by the thickness of the viscous region itself.

Computed Mach contours for the same flow conditions are shown immediately beneath the shadowgraphs. Lines of constant Mach number are shown in increments of 0.02. For the lower Mach number case on the left the shock is centered about the 73 percent chord, in agreement with the computed surface pressure distribution. The boundary layer behind the shock thickens but remains attached until near the trailing edge where it separates. The flow field details indicated in both the shadowgraph and contour plot are remarkably similar in character although they differ in quantitative detail. For the higher Mach number flow on the right the shock is normal away from the airfoil surface and becomes oblique in the interaction region. The shock is centered near the 70 percent chord near the surface. The flow field behind the shock is separated and reattached in the wake. Except for some differences in shock position and orientation, the computed flow field is remarkably similar to that given by experiment in a qualitative sense.

Finally, at the bottom of Figure 3, computed velocity vector plots near the airfoil surface indicate the magnitude and direction of the flow. For the lower Mach number the separation region is small and confined to the vicinity of the trailing edge. For the higher Mach number the separation region extends from the shock impingement area into the wake and is noticeably thick.

Surface pressure comparisons for the two flow configurations are shown in Figure 4. The inviscid solutions agree with experiment over the forward half of the airfoil but are inaccurate in predicting shock strength and location, and in predicting the pressure level near the trailing edge. When the aft pressure recovery is strong, indicating that the shock-boundary layer interaction is weak, Figure 4(a), the viscous solution agrees well with experiment, predicting reasonably well the shock strength and location. The only significant difference appears to be near the airfoil trailing edge where the experimental pressure distributions do not indicate as extensive a separation region as does the numerical computation. Good agreement between computation and experiment is

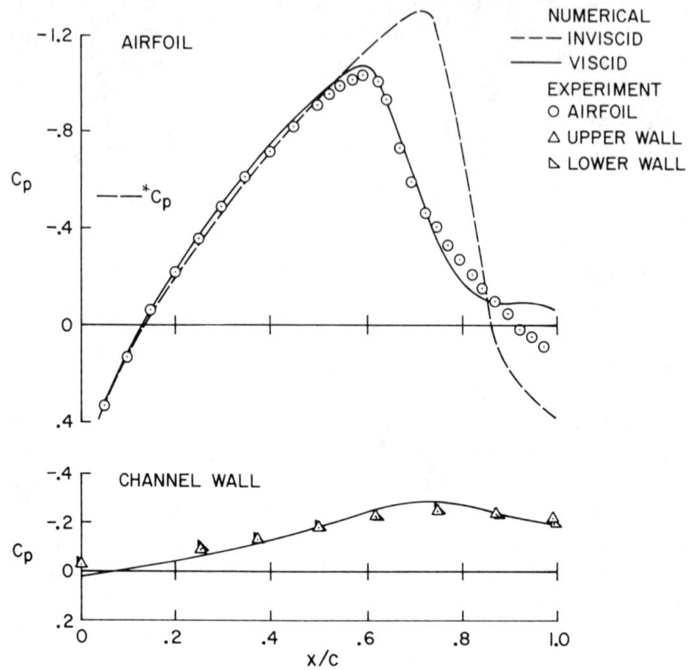

Figure 4(a). Surface Pressure Comparisons for 18 Percent Thick Biconvex Circular Arc. $M_\infty = 0.775$, $Re_c = 2 \times 10^6$.

also obtained at the channel walls, indicating that the streamline contouring used was reasonably accurate and was effective in this case for minimizing wall interference effects.

At the higher Mach number (Figure 4(b)), where separation is near the foot of the shock wave, the viscous solution does not predict the shock location and strength correctly, and greatly overpredicts the pressure recovery over the aft portion of the airfoil. At the channel walls the agreement between computation and experiment parallels that at the airfoil surface, - upstream of the airfoil shock the agreement is excellent but downstream the numerical method overestimated the pressure recovery and disagrees with experiment. This is probably due, in part, to inadequate turbulent modeling in the vicinity of the separation point and throughout the separated region. Also contributing to the disagreement downstream of the shock location was the fact that the channel upper

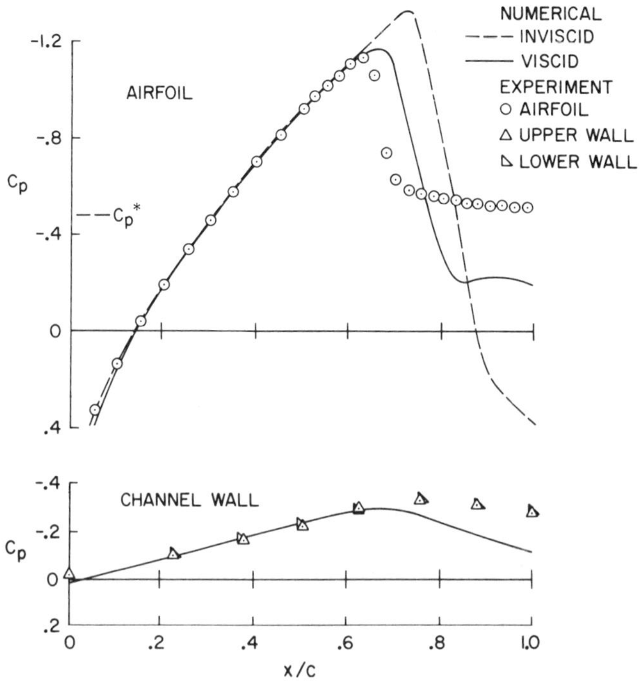

Figure 4(b). Surface Pressure Comparisons for 18 Percent Thick Biconvex Circular Arc. $M_\infty = 0.786$, $Re_c = 10 \times 10^6$.

and lower walls were contoured to match an inviscid flow solution at a slightly lower Mach number. The channel wall boundaries should, of course, be the same in both the numerical simulation and experiment for a proper comparison.

Shockless Lifting Airfoil

The second configuration considered is the shockless lifting airfoil developed by Garabedian and Korn [8]. This airfoil is nearly 12 percent thick and is designed for shockless flow at $M_\infty = 0.75$. It was investigated experimentally by Kacprzynski, et al. [9], for both the design condition and a variety of off design conditions.

Computed results are presented in Figures 5 through 7 for a free-stream Mach number of 0.755, a chord Reynolds number of 21×10^6 and an angle of attack of 0.12 degrees. The eddy viscosity model used to describe the turbulent transport is an algebraic mixing length model with relaxation in the flow direction to partially account for turbulence history. Based on ideas discussed in [6], the model is given by

$$\epsilon_i = \rho \ell_i^2 \left[\left(\frac{\partial u}{\partial y}\right)^2 + \left(\frac{\partial v}{\partial x}\right)^2 \right]^{1/2}$$

$$\ell_i = 0.41\, y\, [1 - \exp(-y/A)]$$

$$A = 26 \left[\frac{\mu_w}{\rho_w} \frac{1}{2\left(\dfrac{\partial (u^2 + v^2)^{1/2}}{\partial \eta}\right)_w} \right]$$

where ϵ_i is the eddy viscosity coefficient in the wall sublayer, ℓ_i is the mixing length in the wall sublayer, A is the van Driest damping parameter, and η is the outward normal to the airfoil surface. For the outer boundary layer region and the trailing edge

$$\epsilon_o = \rho \ell_o^2 \left[\left(\frac{\partial u}{\partial y}\right)^2 + \left(\frac{\partial v}{\partial x}\right)^2 \right]^{1/2}$$

$$\ell_o = 0.09\, |\delta - y_s|$$

where ϵ_o and ℓ_o are the eddy viscosity coefficient and mixing length respectively in the outer and wake regions, δ is the boundary layer on wake thickness, and y_s is the location of the dividing streamline which corresponds to the airfoil surface and wake centerline when the flow is unseparated). Inside the separation bubbles the viscosity is given by $\epsilon_s = \epsilon_o y/y_s$ over solid surfaces and $\epsilon_s = \epsilon_o$ in the wake. Turbulence history is considered by a simple relaxation expression given by

$$\epsilon(x, y) = (1-r)\epsilon_{eq}(x, y) + r\, \epsilon_{eq}(x - \Delta x, y)$$

where r is a relaxation parameter less than unity (0.3 was used in the present solution). Transition was assumed at 10 percent chord.

This simple model should yield reasonable results when the flow remains attached but is suspect in regions of reverse flow.

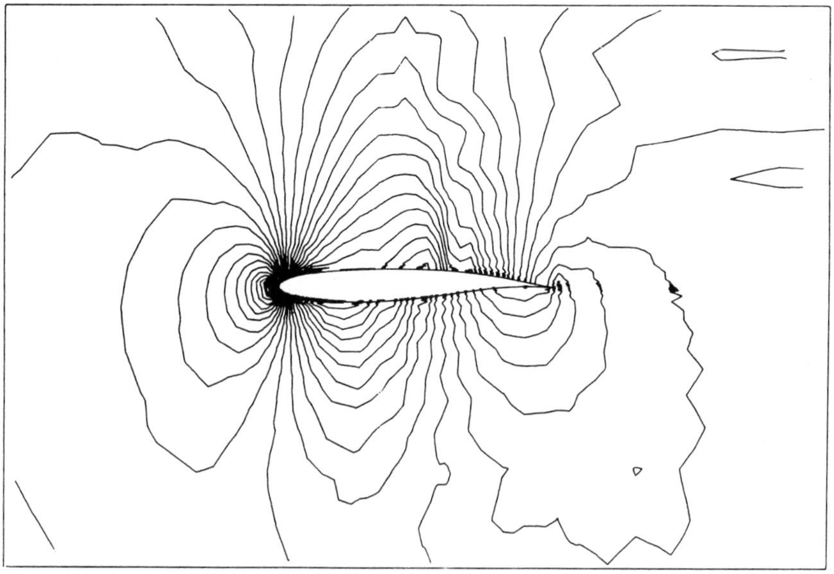

Figure 5. Computed Isobars for Garabedian-Korn Shockless Airfoil, $M = 0.755$, $Re = 21 \times 10^6$, $\alpha = 0.12^\circ$, $U_\infty t/c = 7.3$.

Further discussions of algebraic turbulence models can be found in [6].

The computational mesh contained 78 points around the airfoil and wake and 34 points from the airfoil surface to the far inviscid field. The innermost mesh contained 4 points distributed over a dimensionless length, η/c, of 0.0001, and the boundary layer mesh contained 12 points distributed over a dimensionless length of 0.0199. The mesh is extended radially a distance of 5 chord lengths and downstream of the trailing edge a distance of 6 chord lengths. Steady state was assumed after a dimensionless time of $U_\infty t/c = 9$.

Figure 5 shows computed pressure contours at a dimensionless time of 7.3 for intervals of $\Delta p/p_\infty = 0.02$. The flow over the upper surface is supercritical and essentially shock free. The flow over the lower surface is subcritical. Corresponding computed Mach contours at intervals of 0.02 are shown in Figure 6. Both the boundary layer and the location of the wake are discernible in this figure. Both figures indicate a shock free flow for which the configuration was designed and tend to confirm the validity of the

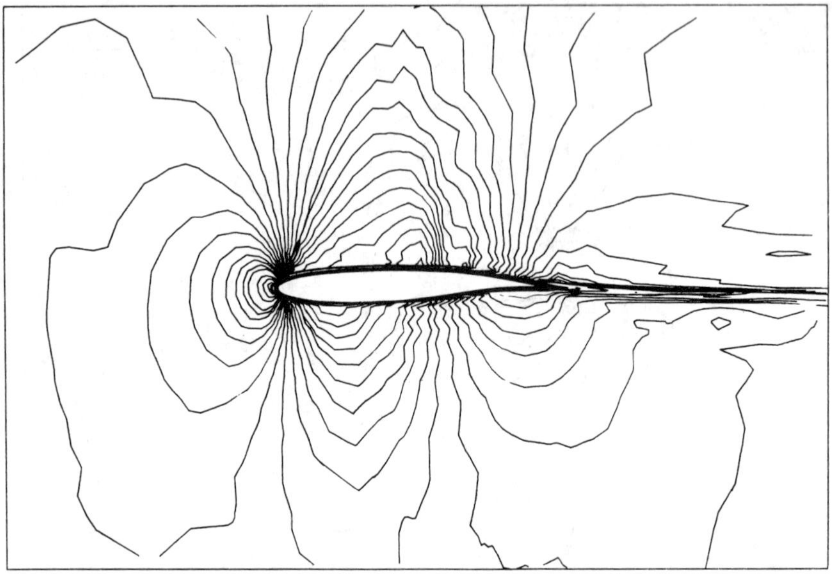

Figure 6. Computed Mach Contours for Garabedian-Korn Shockless Airfoil, $M = 0.755$, $Re = 21 \times 10^6$, $\alpha = 0.12^o$, $U_\infty t/c = 7.3$.

design.

Experimental and computed surface pressure distributions (at a dimensionless time of 9) are shown in Figure 7. For comparison, the inviscid results of Garabedian and Korn are included. The improved agreement of the viscous results with experiment as compared with the inviscid results is dramatic. In fact, the viscous solution shows excellent agreement with experiment. There is a small pressure plateau in the computed distribution on the upper surface at the trailing edge of the airfoil which is not discernible in the experimental results. This is due to increased displacement effects of a boundary layer that is near separation. This may be the first hint of a breakdown of the turbulence model or may be due to slight differences in the details of the airfoil geometry at the trailing edge. (For example, the experimental airfoil model has a finite trailing edge thickness whereas the analytic airfoil has zero thickness.) The experimentally measured shock at the upper nose region is not resolved by the coarse computational mesh used in this region. To further complicate matters the shock occurs just where transition was assumed and is probably near where transition

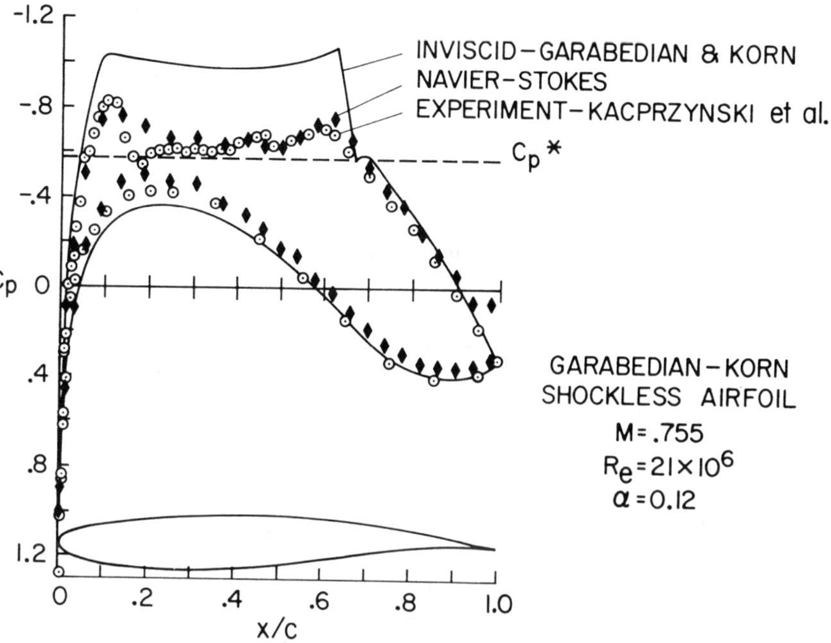

Figure 7. Surface Pressure Comparisons for Garabedian-Korn Shockless Airfoil, $M = 0.755$, $Re = 21 \times 10^6$, $\alpha = 0.12^{\circ}$, $U_\infty t/c = 9.0$.

actually occurs. The slight disagreement between computation and experiment on the lower surface can be attributed to two causes: first, the circulation is different because of lack of adequate resolution of the forward shock on the upper surface and secondly, the numerical solution isn't really fully converged at a dimensionless time of 9. In fact, the computed pressure distribution is slowly approaching the experimental values as time is advanced. Due to the nearness of the far field boundaries reflected waves begin to reach the surface at a dimensionless time of 10 and impose a second order disturbance on the solution. Hence, the solution was stopped at $U_\infty t/c = 9$.

CONCLUDING REMARKS

Using a simple algebraic eddy viscosity model to describe the turbulent transport process the compressible Navier-Stokes equations have been used to simulate three transonic flows. In the first, flow over an 18 percent thick biconvex circular arc at zero angle of attack, the viscous-inviscid interaction was weak and separation was confined to a small region near the trailing edge. Except for the details in the separation region the agreement between computation and experiment was excellent.

In the second, flow over the same geometry but at higher Mach number, the viscous-inviscid interaction was strong with a massive separation region extending from the base of the shock to the near wake. The agreement between computation and experiment was excellent ahead of the interaction but only qualitative in the interaction region. The primary reason for the lack of good quantitative agreement in the interaction region is the use of an inadequate turbulence transport model.

In the third, flow over a 12 percent lifting configuration designed for shock free flow, the viscous-inviscid interaction was weak with no flow separation. Except for a small region near the nose on the upper surface where the computational mesh was too coarse, and a small region on the upper surface at the trailing edge where analytical and experimental airfoil configurations differ, the agreement between computation and experiment is excellent.

In each of the three sample cases the viscous solutions show considerable improvement over corresponding inviscid results - even in the third case where there were no shocks or separation. Thus, the importance of including the viscous terms is clear. Equally clear, and probably the most critical, is the importance of turbulence modeling. When the flow remains attached the simple algebraic type models, such as used here, seem to be adequate. Even for flows with small scale separation reasonable results may be obtained. For flows where viscous-inviscid interactions are strong and where there is large scale separation, the simple models are not adequate and computed results are only qualitative.

It is suggested that the key to successful prediction of viscous effects in transonic flows is in the turbulence modeling. Without adequate descriptions of turbulent transport any prediction method is incomplete. In the words of Bradshaw [10], "a numerical procedure without a turbulence model stands in the same relation to a complete calculation method as an ox does to a bull."

REFERENCES

[1] MacCormack, R. W. (1971), "Numerical Solution of the Interaction of a Shock Wave with a Laminar Boundary Layer," Lecture Notes in Physics, vol. 8, Springer-Verlag, pp. 151-163.

[2] MacCormack, R. W. and Paullay, A. J. (1972), "Computational Efficiency Achieved by Time Splitting of Finite Difference Operators," AIAA Paper 72-154.

[3] Deiwert, G. S. (1974), "Numerical Simulation of High Reynolds Number Transonic Flows," AIAA Paper 74-603.

[4] Deiwert, G. S. (1975), "High Reynolds Number Transonic Flow Simulation," Lecture Notes in Physics, vol. 35, Springer-Verlag, p. 132.

[5] Deiwert, G. S., McDevitt, J. B. and Levy, L. L., Jr. (1975), "Simulation of Turbulent Transonic Separated Flow," Proceedings NASA Conference on Aerodynamic Analyses Requiring Advanced Computers, Langley, Va., NASA AP-347.

[6] Deiwert, G. S. (1975), "Computation of Separated Transonic Turbulent Flows," AIAA Paper 75-829.

[7] McDevitt, J. B., Levy, L. L., Jr. and Deiwert, G. S. (1975), "Transonic Flow about a Thick Circular-Arc Airfoil," AIAA Paper 75-878.

[8] Garabedian, P. R. and Korn, D. G. (1971), "Numerical Design of Transonic Airfoils," Numerical Solution of Partial Differential Equations - II, Academic Press, New York, pp. 253-271.

[9] Kacprzynski, J. J., Ohman, L. H., Garabedian, P. R. and Korn, D. G. (1971), "Analysis of the Flow Past a Shockless Lifting Airfoil in Design and Off-design Conditions," Aeronautics Report LR-554, National Research Council of Canada, Ottawa.

[10] Bradshaw, P. (1972), "The Understanding and Predition of Turbulent Flow," The Aeronautical Journal, 76, pp. 403-418.

DISCUSSION OF THE DEIWERT PAPER

KORN: Could you explain on that last picture what you've done about angle of attack? It isn't clear what you would take for the tunnel incidence, what you would have run for the inviscid solution, or what the design run would be - since there wasn't a design anyway because Mach 0.75 was the design. How would you know what the wall correction could be? How did you decide what lift or angle of attack to run at?

DEIWERT: I decided the lift and angle of attack because the paper that you had published co-authored with Garabedian, Kacprzynski and Ohman had a lot of experimental results in it and a lot of inviscid solutions. So I picked a case where there were both experimental data and your inviscid solutions at the same conditions. I didn't do anything about the tunnel effects, but I did get hold of a more recent code of Bauer and Korn and ran it for the same conditions and put those results on here [extra slide], too. Those are shown by the dots and you can see that this is just simple boundary layer correction and it makes considerable difference over the straight inviscid solution. I think that probably the tunnel effects are slight because both viscous solutions show pretty good agreement with the experiment. I think that the Navier-Stokes solutions are probably slightly better; they aren't fully converged here and they are slightly dropping down, but I think this shows that viscous effects are important.

KORN: I think your results are quite good. Did you run lift and angle of attack at the geometric incidence in the tunnel or did you say let's try to match up the lift, or what?

DEIWERT: No, I don't try to match up the lift; I just picked the Mach number, Reynolds number, and angle of attack that were published in the paper.

KORN: I see. That's the angle of attack that was run in the tunnel, that 0.12?

DEIWERT: That's right.

OLIVER: I have two questions. The first one is how the various zones for the eddy viscosity are distinguished in the calculations. I. e., how do you decide at each point what value of eddy viscosity to use?

DEIWERT: Over the boundary layer, I used the Clauser formulation to compute eddy viscosity all the way across the viscous region. Then I start from the wall and use the Prandtl mixing length theory with the van Driest damping factor and just start computing eddy viscosity as I go away from the wall. As soon as this exceeds the local value of the Clauser formulation, I switch to the Clauser formulation. In the separation bubble I have used different formulations. These have all been published in a recent AIAA paper given at Hartford and this probably explains it better than I can now because just about every run I make I change the turbulence model trying to get something that works better. I haven't had a whole lot of success, of course, but in these kinds of flows it's really kind of an art and you don't really know where the boundary layer is.

OLIVER: The reason I ask is that it sort of strikes directly at what we mean by prediction. As you pointed out the viscosity model or the turbulence stresses - how in fact those are going to be predicted - is precisely the question, and it really controls the whole nature of the solution.

DEIWERT: That's right.

OLIVER: If we use the experiment as a guide to detect where all the zones are, we come closer and closer basically to describing an experiment as opposed to predicting something that we don't know. It seems to me that the more the calculation relies on *a priori* information from the experiment about where to structure the zones, the less it is a predictive tool and the more it is a description of something that happens. Real prediction involves getting to things that are unknown - predicting ahead of time - predicting without knowing ahead of time.

DEIWERT: That's right. All my computations were total predictions; I didn't use any experimental input but I think at this stage, since turbulence is the problem, in my opinion, that we need to use these experiments to guide us in developing turbulence models. I think that's the crux of the whole problem here. That's the unknown: How do we model the turbulence?

OLIVER: The other question I had was my suspicion that it's probably the inviscid solution that is time-controlling in the calculation. I assume you're using a time-dependent numerical method (yes) and that it's the Courant condition (hopefully, that's so). Essentially, it is the Courant condition because of the fine mesh, that's controlling the computational progress.

DEIWERT: Yes, but that's been removed now.

OLIVER: Could you just comment briefly on that?

DEIWERT: All these solutions you've seen today - what you've said is true - that the inviscid stability condition controls the time-step. It's the inviscid stability condition in a fine mesh. MacCormack's new method - which is being incorporated into this code now, but hasn't been debugged - removes this limiting criteria and we aren't limited by the inviscid time-step in the fine mesh, but rather the inviscid time step in the outer mesh - so it's a much more happy situation. It's an unpublished method, by the way, but it should be published this June, I think.

ROBERTS: I wonder about the physical phenomena shown in the tunnel when you were showing your movie and you had a normal shock at the lower Mach number that, it seemed, did not separate the boundary layer. The boundary layer separated downstream near the trailing edge. Is that correct?

DEIWERT: That's right.

ROBERTS: And then at the higher Mach number you observed an oblique shock (yes). Is the oblique shock something that occurs in free flight or is that a tunnel phenomenon?

DEIWERT: That's free flight. I think the flow wants to have the smallest pressure jump that it can, so an oblique shock will give a smaller pressure jump. Also you have to have an oblique shock to go along with the turning of the flow at the point of separation.

ROBERTS: Is the oblique shock oblique to the chord of the airfoil or is it oblique to the flow also.

DEIWERT: It's oblique to the flow.

RUNSTADLER: I would agree with the statement that was made just a moment ago that the whole turbulence modeling is certainly the heart of how accurately one can predict what's going on. Also, being realistic, knowing that turbulence is a very difficult subject and being interested in the design applications of trying to get as good a tool available as possible, could you quickly tell us what you have in mind relative to trying different turbulence models; because it is basically having experimental data available and trying to dial all the parameters, that you need to come up with the kind of solutions that really represent reality.

DEIWERT: I don't really have any good ideas so next month I'm going to Imperial College and try to get Bradshaw to help me out. I'm going to stay six months with him on this very problem so maybe he can lend some insight.

NORMAL SHOCK WAVE–TURBULENT BOUNDARY LAYER INTERACTIONS IN TRANSONIC FLOW NEAR SEPARATION

T. C. ADAMSON, JR., AND A. F. MESSITER

University of Michigan

ABSTRACT

The problem of a normal shock wave impinging on a flat plate, turbulent boundary layer is considered for the case where the external flow is transonic. Asymptotic methods are employed. It is shown that there are two outer regions, including the outer part of the boundary layer and the external flow, in which inviscid flow governing equations hold, and two regions near the wall, in which Reynolds and/or viscous stresses need be taken into account. The solutions in the outer regions lead to the pressure distribution on the wall, for which an analytical expression is presented, valid under those conditions when separation is imminent but has not yet occurred. The solutions valid in the inner regions lead to a separation criterion.

INTRODUCTION

This paper is concerned with solutions found for the problem of a normal shock wave impinging upon a turbulent flat plate boundary layer in transonic flow, when separation is imminent. The method of matched asymptotic expansions is used in obtaining the

*Support for this work, by NASA-Langley Research Center, Grant Number NGR 23-005-523, is gratefully acknowledged.

solutions. It is hoped that this presentation will illustrate the utility of asymptotic methods in gaining an understanding of the structure of the interaction region. It is believed that it is only by careful application of these asymptotic methods that one can find the detailed description of the interaction region that is needed to understand the interplay between the various physical mechanisms near and at separation. Moreover, this work is not considered as an alternative but as an aid to computational studies. That is, the solutions illustrate the scale of the interaction region as well as the analytical form of the solution near the singularities, and in some cases provide the actual solution in the interaction region.

The first work on shock wave-turbulent boundary layer interactions in transonic flow, using asymptotic methods, was done by Adamson and Feo [1] and Melnik and Grossman [2] . Other analytical studies of the general problem by Bohning and Zierep [3] , and Inger and Mason [4] have involved approximations which are not asymptotic in nature. Adamson and Feo [1] considered the case where the shock is weak and oblique; thus, if $M_e - 1 = O(\epsilon)$, where M_e is the Mach number of the flow external to the boundary layer, and u_τ is the friction velocity made dimensionless with respect to the critical sonic velocity in the external flow, the case they considered was $u_\tau^2 \ll \epsilon \ll u_\tau$. In work finished soon after, Melnik and Grossman [2] presented solutions for the case where the shock wave is normal, and stronger, i.e., $\epsilon = O(u_\tau)$. In both papers, it was indicated that it appeared that $\epsilon = O(1)$ was a condition for separation. Here, we consider the case where the shock is normal and strong enough that separation is imminent, i.e., $\epsilon \gg u_\tau$, and show the resulting structure of the interaction region and the wall pressure distribution found under this condition. A discussion for the strong oblique wave case was given by Adamson [5] .

In the next section, the problem is formulated and the asymptotic structure of both the undisturbed boundary layer and interaction region are discussed and compared. In Section 3, the solutions valid in the inviscid flow regions are presented, as is the wall pressure distribution. In section 4, the inner regions near the wall, where Reynolds and viscous stresses are important, are discussed.

The flow is assumed to be two-dimensional and compressible,

and to consist of a gas which satisfies the perfect gas law and has constant specific heats.

PROBLEM FORMULATION AND STRUCTURE OF THE INTERACTION REGION

The problem considered, and the notation used is shown in Figure 1. We consider a normal shock wave in the external transonic flow, impinging on a turbulent boundary layer on a semi-infinite flat plate, at a distance $\bar{L}$ downstream of the leading edge. Overbars indicate dimensional quantities. The dimensionless, with respect to $\bar{L}$, time averaged thickness of the boundary layer at the intersection of the shock wave and boundary layer is of order δ, and the extent of the interaction region in the flow direction is of order Δ. The coordinates X and Y are made dimensionless with respect to $\bar{L}$, with origin located at the intersection of the shock wave in the external flow and the flat plate. The velocity components, U and V, are made dimensionless with respect to $\bar{a}_e^*$, the critical sonic velocity in the external flow, and the pressure, P, density, ρ, and temperature, T, are referred to their critical values in the external flow also.

The incoming external flow velocity and Mach number are defined as follows.

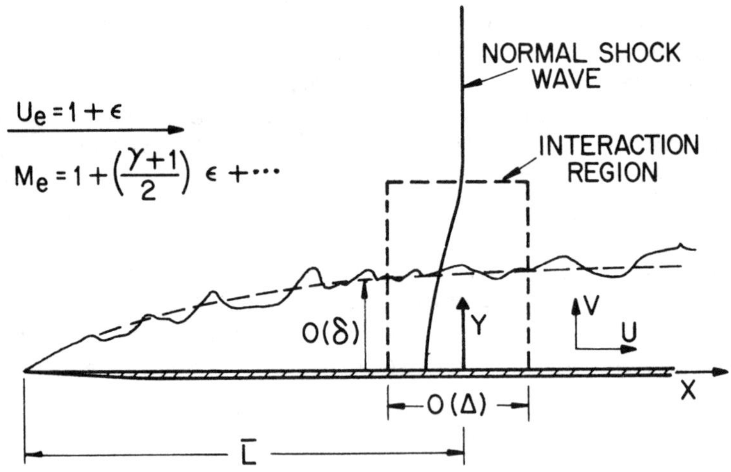

Figure 1. Sketch Showing Flat Plate, Boundary Layer, Shock Wave, Interaction Region, and Notation Used.

$$U_e = 1 + \epsilon \tag{1a}$$

$$M_e = 1 + \frac{\gamma+1}{2}\,\epsilon + \ldots, \tag{1b}$$

where, because the flow is transonic, $\epsilon \ll 1$, and Equation (1b) for the Mach number follows. The flow parameters of primary importance are, then, ϵ and the Reynolds number, defined as,

$$Re = \bar{\rho}^*_e \bar{a}^*_e \bar{L} / \bar{\mu}^*_e \,. \tag{2}$$

Rather than use the Reynolds number directly, however, we introduce u_τ, the undisturbed flow friction velocity evaluated at the intersection point, defined as follows,

$$u_\tau = \sqrt{\frac{\bar{\tau}_{wi}}{\bar{\rho}_{wi}}}\ \frac{1}{\bar{a}^*_e}\ \sqrt{\frac{T_e}{T_{wi}}} \tag{3}$$

where the subscript w refers to conditions at the wall, and the additional subscript i indicates initial or undisturbed conditions at the point $X = 0$. Solutions, then, are valid in the limit as $Re \to \infty$ or $u_\tau \to 0$, and as $\epsilon \to 0$, such that the limit $\epsilon \to 0$ is taken after the limit $u_\tau \to 0$; $\epsilon \gg u_\tau$.

Before considering the interaction region, we describe first the structure of the undisturbed compressible turbulent, flat plate boundary layer in asymptotic terms; this description of the undisturbed boundary layer is essentially the same as that given by Melnik and Grossman [2]. The typical two layer structure for the undisturbed turbulent boundary layer is illustrated in Figure 2, which contains a sketch of the velocity profile. In the outer, or velocity defect layer, which has a thickness of the order of δ because the wall layer is so thin compared to it, the stretched Y coordinates and the asymptotic expansion for the U velocity component are,

$$Y = \delta y \tag{4a}$$

$$U = U_e + u_\tau u_{01} + \ldots \,. \tag{4b}$$

Thus, $y = O(1)$ in the velocity defect region. Now, in general, $u_\tau =$

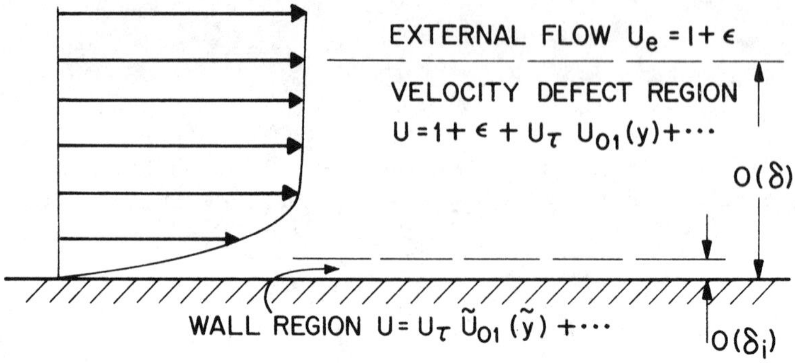

Figure 2. Sketch Showing Typical Velocity Distribution and Two Layer Structure in the Undisturbed Turbulent Boundary Layer, at $X = 0$.

u_τ (x) and $u_{01} = u_{01}(x, y)$ are found by substituting expansions similar to that shown in Equation (4b) into the boundary layer equation. For example, u_{01} is found from the equation of motion in the x direction, in which inertia and Reynolds stress terms are the most important terms. Here, however, it is possible to use the results of Maise and MacDonald [6] to write the expansion for U for a compressible boundary layer in terms of a corresponding incompressible boundary layer. Furthermore, because the interaction region is so thin in the flow direction, we need only consider the first terms in a Taylor expansion for U in the x direction [1] . Thus, Equation (4b) is an expansion for $U = U(0, Y)$, written therefore in terms of u_τ = constant, as defined in Equation (3), and in terms of $u_{01}(y)$. As it turns out, the variations due to compressibility may be accounted for by defining u_τ as in Equation (3), and u_{01} is the variable part of the corresponding incompressible flow velocity distribution. Therefore [7] , as $y \to \infty$, $u_{01} \to 0$ and $U \to U_e$, the external flow velocity, and as $y \to 0$

$$u_{01} \sim \alpha \ln y + \beta_0 \tag{5}$$

where α and β_0 are constants, α being the inverse of the von Karman constant, κ. The expansion for U is essentially the same as that used by Melnik and Grossman [2] .

In the wall layer, (see Figure 2), the important terms in the equation of motion in the flow direction are the viscous and

Reynolds stress terms. It is easy to show that in this case the order of the thickness of this region is $\delta_i = O[(\text{Re}\, u_\tau)^{-1}]$. The stretched Y coordinate and expansion for the U velocity component are, then, written at X = 0,

$$Y = \delta_i \tilde{y}; \qquad \delta_i = (\text{Re}\, u_\tau)^{-1} \tag{6a,b}$$

$$U = u_\tau \tilde{u}_{01} + \dots \tag{6c}$$

where $\tilde{y} = O(1)$, and $\tilde{u}_{01} = \tilde{u}_{01}(\tilde{y})$. As $\tilde{y} \to 0$, $u_{01} \to 0$, so the no slip condition is satisfied. Following Van Driest's [8] analysis, it can be shown that with a constant stagnation enthalpy, and a mixing length (or modified mixing length) model for the eddy viscosity, as $\tilde{y} \to \infty$

$$U \sim \sqrt{\frac{\gamma+1}{\gamma-1}} \sin\left[\sqrt{\frac{\gamma-1}{\gamma+1}} \sqrt{\frac{T_w}{T_e}}\, u_\tau (\alpha \ln \tilde{y} + \beta_i)\right] + \dots \tag{7}$$

If Equations (5) and (7) are matched, and if we choose parameters such that

$$\delta = u_\tau \tag{8}$$

then one finds that

$$u_\tau \alpha \ln \frac{\delta}{\delta_i} = b_0 + u_\tau (\beta_0 - \beta_i) + \dots \tag{9a}$$

$$b_0 = \sqrt{\frac{\gamma+1}{\gamma-1}} \sqrt{\frac{T_e}{T_w}} \sin^{-1}\left(\sqrt{\frac{\gamma-1}{\gamma+1}}\, U_e\right) \tag{9b}$$

Finally, if Equations (6b) and (8) are substituted into Equation (9a), it is seen that the asymptotic relationship between u_τ and Re is,

$$u_\tau = \frac{b_0}{\alpha \ln \text{Re}} \tag{10}$$

Anticipating the result that $T_e/T_w = 1 - \left(\frac{\gamma-1}{\gamma+1}\right) U_e^2$, one can show that for $\epsilon \ll 1$, $b_0 = \sqrt{\frac{2}{\gamma-1}} \sin^{-1} \sqrt{\frac{\gamma-1}{\gamma+1}}$.

The changes in the flow field brought about when a normal shock impinges upon a turbulent boundary layer are illustrated in Figures 3 and 4, where the structure of the interaction region, found using the asymptotic methods to be discussed, is summarized. It should be emphasized that the sketches in Figures 3 and 4 are not drawn to the proper relative scale, i.e. they are not asymptotically correct. Thus, $\Delta \gg \Delta_*$ and $\delta \gg \delta_*$, so to the scale of Δ and δ, one would not see the region ordered by Δ_* and δ_*; the sketches are drawn simply to illustrate the regions which must be considered and their relative positions. Starting with the external flow and working toward the wall, we see in Figure 3, that one must consider two inviscid flow regions. The so-called outer inviscid region is scaled by the order of the the undisturbed boundary layer, and a corresponding Δ (to be derived) in the x direction. In this region, which extends far into the external (to the boundary layer) flow, one sees a normal shock entering the velocity defect part of the boundary layer. The so-called inner inviscid region is scaled by the distance from the wall to the undisturbed flow sonic line, δ_*, and a corresponding $\Delta_* \ll \Delta$. Now, δ_* is a very important parameter because the extent of the upstream influence of the interaction is measured by the order of the extent of the subsonic flow in the undisturbed boundary layer. In this case, because $\epsilon \gg u_\tau$, the sonic line must be located in the logarithmic part of the undisturbed belocity profile. Thus, from Equations (4b) and (5), if $Y = \delta y = \delta_* y^*$ and the sonic line is defined to be located at $y^* = 1$,

$$\delta_*/\delta = e^{-\beta_0/\alpha}\, e^{-\epsilon/\alpha u_\tau + \dots} \tag{11}$$

Thus, since $\epsilon \gg u_\tau$, it is seen that in this case, δ_* is exponentially small compared to δ. The extent of the upstream influence, $O(\Delta_*)$, will be shown later, to be exponentially small compared to Δ. Finally, then, in the inner inviscid region, the thickening of the subsonic flow caused by the upstream influence, causes compressive characteristics emanating from the sonic line, to coalesce and form the shock wave which becomes more and more nearly normal as it passes through the boundary layer and becomes the shock seen in the outer inviscid layer.

Two important points concerning the solutions in the inviscid

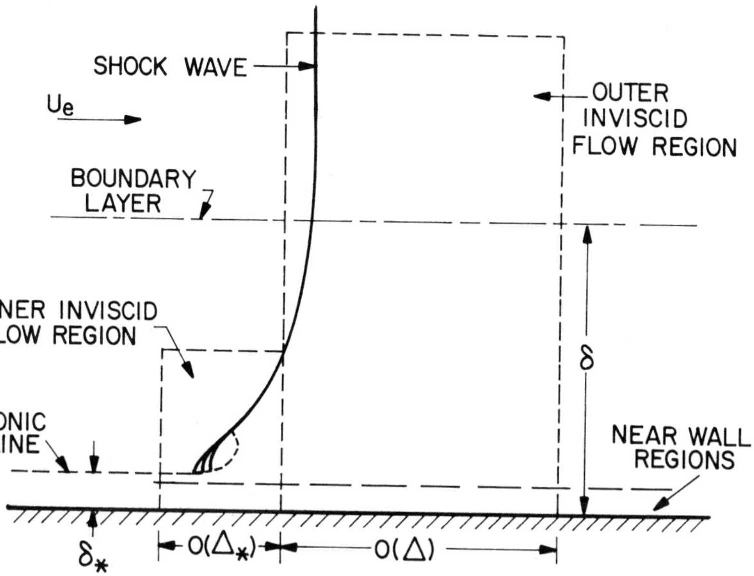

Figure 3. Structure of the Inviscid Flow Part of the Interaction Region for the Case Where the Incoming Shock Wave is Normal, Showing the Inner and Outer Inviscid Flow Regions.

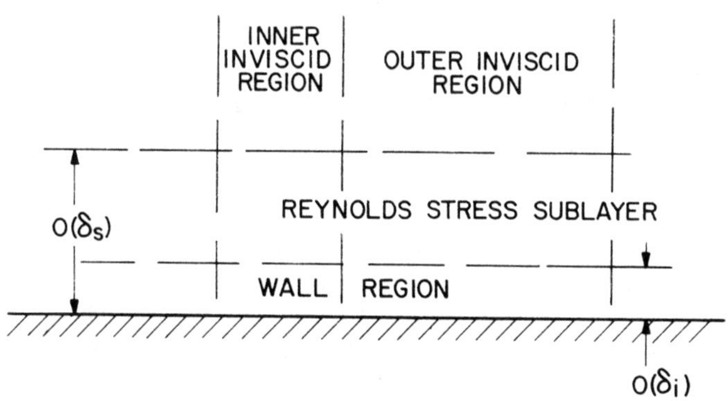

Figure 4. Structure of the Interaction Region Near the Wall, Showing those Regions in Which Reynolds Stresses (Reynold Stress Sublayer) and Both Reynolds and Viscous Stresses (Wall Layer) are Important.

flow regions should be kept in mind. First, although there are other regions between these inviscid flow regions and the wall, regions in which Reynolds and viscous stresses must be accounted for, these other regions will have the same characteristic X-dimension, but vanishingly small Y-characteristic dimensions, compared to the X and Y-characteristic dimensions in the corresponding inviscid flow region. Hence, to the order desired, $\partial P/\partial Y = 0$ in these other regions, and the limit form of the inviscid flow pressure as $y^* \to 0$ (inner inviscid region) or $y \to 0$, (outer inviscid region), is the pressure at the wall. Second, because the flow is not separated, i.e. the case of incipient separation is considered, the matching condition for the V velocity component as $y^* \to 0$, or $y \to 0$ is, as found previously [1, 2] $V^*(x^*, 0) = V(x, 0) = 0$, to the order considered. Hence, since no functional form for V need be found by matching, the inviscid flow region solutions may be found independently of the near wall region solutions. The resulting wall pressures are then used in the wall layer solutions to find the relation for the wall shear stress.

Two analytical solutions will be presented for the wall pressure. The first is valid in that part of the inner inviscid flow region where the shock wave is fully formed; the second is valid in the outer inviscid flow region. The result is a wall pressure distribution valid everywhere in the interaction region except at the beginning where the pressure first begins to diverge from the upstream, undisturbed pressure.

The so-called near wall regions, not shown in Figure 3 for reasons of clarity, are located between the inviscid flow regions and the wall, and are illustrated in Figure 4. Besides the wall region, which is located in the undisturbed flow, it is found necessary [1, 2] to add an extra region, the Reynolds stress sublayer. Thus, if one considers the form of the equation of motion in the X direction to the desired order of approximation, it is seen that in the wall region only Reynolds and viscous terms arise, while in the inviscid flow regions, inertia and pressure gradient terms are found. No terms are common to both regions; physically this means that there is no mechanism by which momentum can be transferred laterally from one region to the other. Therefore, another region, in which inertia, pressure gradient, and Reynolds stress terms are of equal order, must exist between the wall and inviscid flow regions. The mechanism by which momentum is transferred laterally is then, of course, turbulent transfer as given by the Reynolds stress terms, so this additional region is called the Reynolds stress sublayer.

In summary, then, the inviscid flow region solutions lead to a wall pressure distribution and the near wall region solutions lead to the shear stress along the wall, which, hopefully, leads to a separation criterion.

The general governing equations from which the particular equations valid in each of the above mentioned regions are derived, are, of course, the Reynolds equations, the energy equation, the equation of state, and, where necessary, an equation stating the closure condition employed. Here, adiabatic flow with both the Prandtl number and turbulent Prandtl number equal to one, is considered, so that the stagnation temperature is a constant.

Finally, it should be noted that U is defined in terms of the time averaged velocity component, $\overline{U}$, but

$$V = \frac{\overline{V}}{\overline{a}^*_e} + u_\tau^2 \frac{\langle \rho' V' \rangle}{\rho}$$

where fluctuating velocity components are nondimensionalized by $\overline{u}_\tau \sqrt{T_e/T_w}$ and $\rho' = \overline{\rho}'/\overline{\rho}^*_e u_\tau$; $\overline{u}_\tau = (\overline{\tau}_{wi}/\overline{\rho}_{wi})^{1/2}$.

INVISCID FLOW REGION SOLUTIONS

Inner Inviscid Flow Region

The inner inviscid flow region, sketched in Figure 3, is that region in which the flow field changes from its undisturbed form as a result of the upstream influence of the interaction. This upstream influence manifests itself by the adverse pressure gradient causing the flow to decelerate, thus thickening the subsonic region, and finally causing compressive waves emanating from the sonic line to coalesce and form a shock. Thus, the characteristic length associated with this region in the Y direction is δ_* , the distance from the wall to the undisturbed sonic line. If the corresponding characteristic length in the X direction is denoted by Δ_*, then in this region, the stretched independent variables are written as

$$X = \Delta_* x^* \qquad Y = \delta_* y^* \tag{12}$$

From Equations (4b), (5) and (11), one can show that the undisturbed flow velocity component, U, is, $U = 1 + u_\tau \alpha \ln y^* + \ldots$, that is the flow entering the shock wave has a velocity differing from sonic velocity by a term of $O(u_\tau)$. Hence, the flow velocity leaving the shock must have variations from sonic value of the same order and one can therefore write the following asymptotic expansions for the velocity components and pressure.

$$U = 1 + u_\tau u_1^*(x^*, y^*) + \ldots \tag{13a}$$

$$V = \nu_1(u_\tau) v_1^*(x^*, y^*) + \ldots \tag{13b}$$

$$P = P_e (1 + u_\tau P_1^*(x^*, y^*) + \ldots) \tag{13c}$$

$$\rho = P_e (1 + u_\tau \rho_1^*(x^*, y^*) + \ldots) \tag{13d}$$

$$T = 1 + u_\tau T_1^*(x^*, y^*) + \ldots \tag{13e}$$

where $\nu_1(u_\tau)$ is to be determined and we have anticipated that $\rho_0 = P_e$ and $T_0 = 1$.

From a consideration of the Reynolds stress transport equations (e.g., Ref. [9] , or following the steps shown in Ref. [10] for compressible flow), one can show that in this region, inviscid flow equations hold at least up to and including terms of second order (u_τ^2) in U, P, etc. Thus, up to second order, the governing equations reduce to the following:

$$\rho U \frac{\partial U}{\partial X} + \rho V \frac{\partial U}{\partial Y} = -\frac{1}{\gamma} \frac{\partial P}{\partial X} \tag{14a}$$

$$\rho U \frac{\partial V}{\partial X} + \rho V \frac{\partial V}{\partial Y} = -\frac{1}{\gamma} \frac{\partial P}{\partial Y} \tag{14b}$$

$$T + \frac{\gamma-1}{2} U^2 = \frac{\gamma+1}{2} \tag{14c}$$

$$P = \rho T \tag{14d}$$

$$(U^2-a^2)\frac{\partial U}{\partial X} + (V^2-a^2)\frac{\partial V}{\partial Y} + UV\left(\frac{\partial U}{\partial Y} + \frac{\partial V}{\partial X}\right) = 0 \tag{14e}$$

$$a^2 = T \tag{14f}$$

where the so-called gas dynamic equation, Equation (14e) derived from the continuity equation, replaces the continuity equation, and where a is the dimensionless (with respect to $\overline{a}_e^*$) local speed of sound.

If the expansions, Equations (13), and the stretched variables, Equations (12), are substituted into Equations (14), one finds from the energy equation, (14c), that since the first term in the expansion for U is one, that the first term in the expansion for T is one, i.e. $T_0 = 1$. Likewise, from the equation of state, then $\rho_0 = P_e$. Also, one finds that

$$\Delta_* = u_\tau^{1/2}\,\delta_* \tag{15a}$$

$$\nu_1 = u_\tau^{3/2} \tag{15b}$$

and that

$$P_1^* = -\gamma\, u_1^* + f_1^*(y^*) \tag{16a}$$

$$\frac{\partial v_1^*}{\partial x^*} = \frac{-1}{\gamma}\frac{\partial P_1^*}{\partial y^*} \tag{16b}$$

$$T_1^* = -(\gamma-1)u_1^* \tag{16c}$$

$$P_1^* = T_1^* + \rho_1^* \tag{16d}$$

$$(\gamma+1)\, u_1^*\frac{\partial u_1^*}{\partial x^*} = \frac{\partial v_1^*}{\partial y^*} \tag{16e}$$

The results shown in Equations (15) were obtained by requiring the balance of terms shown in Equations (16b) and (16e). These orders of magnitude are consistent with the usual transonic small-disturbance theory.

Since $P_1^* + \gamma u_1^*$ = constant across a shock wave in transonic flow, and far upstream of the shock wave $P_1^* = 0$ and $u_1^* = \alpha \ln y^*$, it is seen that

$$f_1^*(y^*) = \gamma\, \alpha \ln y^* \tag{17}$$

so that Equation (16b) becomes

$$\frac{\partial v_1^*}{\partial x^*} - \frac{\partial u_1^*}{\partial y^*} = -\frac{\alpha}{y^*}$$

Hence, Equations (16e) and (17) are finally the governing equations to be solved; they may be used to derive one nonlinear equation for u_1^*. It is seen that the flow in this region is rotational due to the rotation of the incoming flow.

Although a general analytical solution to the nonlinear equation for u_1^* is not available, a similarity solution has been found for $x^* \gg 1$ and $y^* \gg 1$. Thus, the solution is valid in a region intermediate to the inner and outer inviscid flow regions, where the shock wave has been formed. It is not valid for $x^* = O(1)$, $y^* = O(1)$, where the compression waves are beginning to coalesce. The solutions found for u_1^*, v_1^* and P_1^*, and the similarity variable are as follows:

$$u_1^* = -\alpha \ln y^* - \alpha \ln\left(\frac{S^2}{\gamma+1} + 1\right) \tag{18a}$$

$$v_1^* = -2\alpha\,(\gamma+1)^{1/2}(\alpha \ln y^*)^{1/2}\left[\tan^{-1}\frac{S}{\sqrt{\gamma+1}} - \frac{\pi}{2}\right] \tag{18b}$$

$$P_1^* = \gamma\left[2\alpha \ln y^* + \alpha \ln\left(\frac{S^2}{\gamma+1} + 1\right)\right] \tag{18c}$$

$$S = \frac{x^*}{y^*(\alpha \ln y^*)^{1/2}} \tag{18d}$$

The shock wave is located at S = 0, and the wall pressure can be found by writing P_1^* in the limit as $S \to \infty$. Thus, since $P_e = 1 - \gamma\epsilon$

$+ \ldots,$

$$P_w = 1 - \gamma\epsilon + u_\tau 2\gamma\alpha \ln x^* + \ldots \qquad (19)$$

It is this form, valid in the limit as $x^* \to \infty$, which must match with the wall pressure found in the outer inviscid region, written in a form valid in the limit as $x \to 0$.

Outer Inviscid Flow Region

In the outer inviscid flow region, which is scaled by δ, the order of the thickness of the boundary layer, one sees a normal shock entering the boundary layer, which is represented by the velocity defect layer; this region is sketched in Figure 5. To order ϵ, the incoming velocity is uniform and the shock is therefore planar; because there are variations from this uniform incoming velocity, in the boundary layer, there are corrections to the shock shape, as will be seen.

As shown in Figure 5, the incoming velocity outside the boundary layer is $U = 1 + \epsilon$, while in the boundary layer, it is $U = 1 + \epsilon + u_\tau u_{01} + \ldots$. Because the upstream influence is confined to the inner inviscid region, which is vanishingly small to the scale of the outer inviscid region under consideration, one need consider only the undisturbed flow passing through the shock wave. Thus,

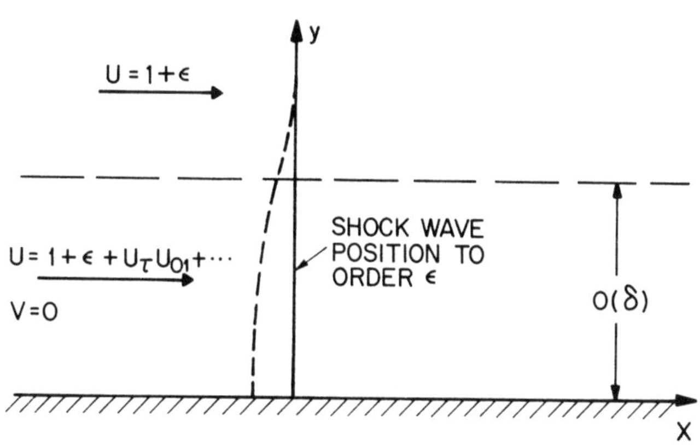

Figure 5. Sketch of Flow Structure in the Outer Inviscid Flow Region, for the Case Where the Incoming Shock Wave is Normal. The Dotted Line Shows the Corrected Shock Position to Order u_τ.

since the incoming flow is $1 + \epsilon$ plus corrections of order u_T, the velocity downstream of the shock is $1 - \epsilon$ plus terms of order u_T, and we choose to write the expansions for U, V and P as follows, for reasons which will become apparent later.

$$U = 1 - \epsilon + u_T u_{01}(y) + u_T u_1(x, y) + \ldots \tag{20a}$$

$$V = \epsilon^{1/2} u_T (\gamma+1)^{1/2} v_1(x, y) + \ldots \tag{20b}$$

$$P = 1 + \gamma\epsilon + u_T P_1 + \ldots \tag{20c}$$

with similar expansions for ρ and T. The stretched variables y and x are defined as

$$Y = \delta\, y, \quad X = \Delta\, x \tag{21a, b}$$

Again, using the Reynolds stress transport equations [9,10] , one can show that inviscid flow equations hold in this region at least up to terms of second order. Hence, the governing equations are, again, Equations (14). If the expansions given in Equations (20) and similar expansions for ρ and T and the stretched variables, Equations (21), are substituted into Equations (14), one can show that $\Delta = O(u_T \epsilon^{1/2})$ and, in fact, we choose to define Δ as

$$\Delta = (\gamma+1)^{1/2} \epsilon^{1/2} u_T \tag{22}$$

Also, one finds that

$$P_1 = -\gamma\, u_1 \tag{23a}$$

$$\frac{\partial v_1}{\partial x} = -\frac{1}{\gamma}\frac{\partial P_1}{\partial y} = \frac{\partial u_1}{\partial y} \tag{23b}$$

so that a potential exists for the perturbation velocity components. From Equation (14e), the governing equation for this potential function is found to be

$$\frac{\partial^2 \phi_1}{\partial x^2} + \frac{\partial^2 \phi_1}{\partial y^2} = 0 \tag{24}$$

i.e., Laplace's equation holds downstream of the wave, with the following boundary conditions:

$$\phi_{1y}(x,0) = 0 \qquad 0 < x < \infty \tag{25a}$$

$$\lim_{y \to \infty} \phi_{1y}(x,y) = 0 \qquad 0 < x < \infty \tag{25b}$$

$$\phi_{1x}(0,y) = -2u_{01}(y) \qquad 0 < y < \infty \tag{25c}$$

$$\lim_{x \to \infty} \phi_{1x}(x,y) = 0 \qquad 0 < y < \infty \tag{25d}$$

A solution for ϕ_1 may be found by employing a source distribution along $x = 0$ (i.e., along the shock wave) to give the proper value of ϕ_{1x}, and symmetric in y so that $v_1 = 0$ at $y = 0$. That is, one considers a half plane problem symmetric about $y = 0$. The solution may be written as follows,

$$\phi_1(x,y) = -\frac{2}{\pi}\int_{-\infty}^{\infty} u_{01}(\eta)\ln\sqrt{x^2 + (y-\eta)^2}\,d\eta \tag{26}$$

Therefore,

$$u_1 = \phi_{1x} = -\frac{2x}{\pi}\int_{-\infty}^{\infty}\frac{u_{01}(\eta)\,d\eta}{x^2 + (y-\eta)^2} \tag{27a}$$

$$v_1 = \phi_{1y} = -\frac{2}{\pi}\int_{-\infty}^{\infty}\frac{(y-\eta)u_{01}(\eta)\,d\eta}{x^2 + (y-\eta)^2} \tag{27b}$$

and from Equations (20c), (23a) and (27a), it is seen that the wall pressure distribution is, then

$$P_w = 1 + \gamma\epsilon - u_\tau\gamma\,\phi_{1x}(x,0) + \ldots \tag{28}$$

Now, as $y \to 0$, since $u_{01}(y)$ is considered to be an even function of y,

$$\phi_{1x}(x,0) \quad -\frac{4x}{\pi}\int_{0}^{\infty}\frac{u_{01}(\eta)}{x^2 + \eta^2}\,d\eta$$

and as $x \to 0$, the main contribution to this integral occurs near $\eta = 0$, where $u_{01}(\eta) \propto \ln \eta$, from Equation (5). In this case, then, as $x \to 0$,

$$\phi_{1x}(x, 0) = -2\alpha \ln x + \ldots \tag{29}$$

Finally, from Equations (12), (21b), (15, (22) and (11), one can show that,

$$x = \Delta_* x^* / \Delta \tag{30a}$$

$$\Delta_*/\Delta = \left((\gamma+1)\,\frac{u_T}{\epsilon}\right)^{-1/2} e^{-\beta_0/\alpha}\, e^{-\epsilon/\alpha u_T} \tag{30b}$$

Now, using Equations (30), (29) and (28), one can write the limiting form (as $x \to 0$) of the outer solution for P_w, in inner variables. The resulting equation should match term by term with the limiting form (as $x^* \to \infty$) of the inner solution for P_w, Equation (19). Upon performing the indicated substitutions, it is seen that the outer solution not only matches with the inner solution, but contains it. Hence, Equation (28) gives the solution for P_w valid throughout the interaction region except at its initiation, where the pressure first begins to increase from its undisturbed flow value.

If the shock position at any value of y is denoted by $x_s(y)$, and if $[U] = U_d - U_u$ where the subscripts u and d denote conditions immediately upstream and downstream of the shock respectively, then

$$\frac{dX_s}{dY} = -\frac{[V]}{[U]}$$

and one can write, in stretched variables,

$$\frac{dx_s}{dy} = \frac{u_T}{\epsilon}\,\phi_{1y}(0, y) + \ldots \tag{31}$$

where ϕ_{1y} is given by Equation (27b).

Numerical calculations for P_w/P_0, the ratio of wall pressure to upstream stagnation pressure, and x_s, the shock shape, have been performed using Coles' [11] form for u_{01},

$$u_{01} = \alpha \ln y - \frac{\alpha}{2} (1 + \cos \pi y) \qquad y < 1 \qquad (32a)$$

$$= 0 \qquad y > 1 \qquad (32b)$$

and for ϵ = 0.167 (M_e = 1.20), and u_τ = 0.028 (Re = 0.954 x 10^6). The Mach number chosen is, according to Gadd [12], one very close to the limit for unseparated flow over a flat plate, i.e. conditions are those for incipient separation. The results are shown in Figures 6 and 7. Available experimental data at the given Mach numbers and Reynolds numbers (e.g., Gadd's data [12] for axisymmetric flow) appear to require correction for effects of geometry, and so comparison with such data must await extension of the present results to include these effects. However, the available experimental data corroborate the general pressure distribution

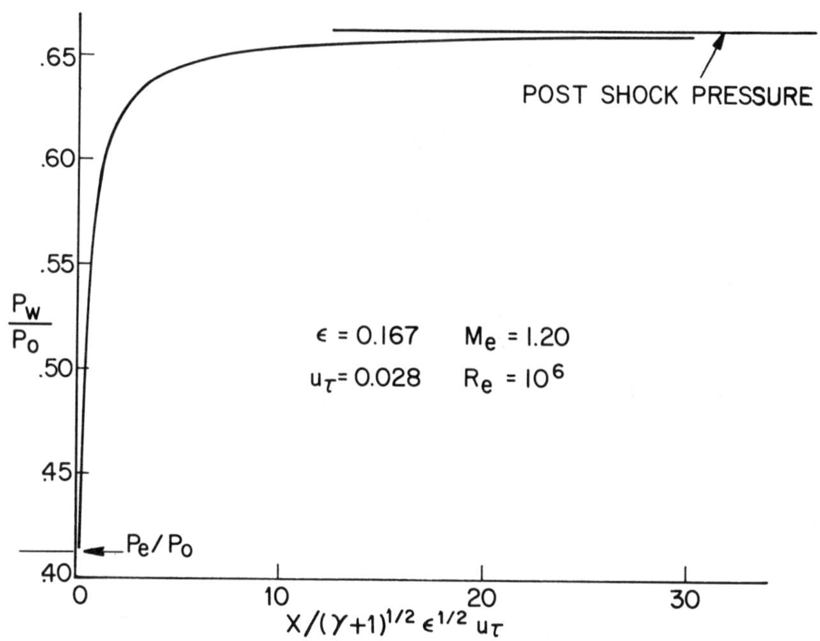

Figure 6. Calculated Wall Pressure Distribution for ϵ = 0.167 (M_e = 1.20) and u_τ = 0.028 (Re = 10^6), For Normal Shock Wave Interacting With a Turbulent Flat Plate Boundary Layer.

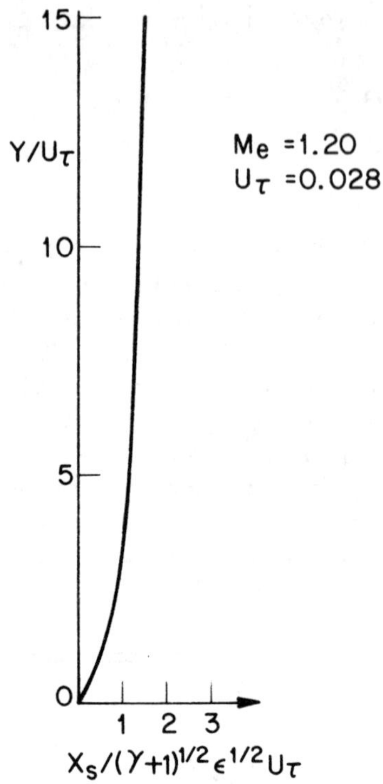

Figure 7. Shock Shape Corresponding to Conditions Given in Figure 6.

found, and more importantly, corroborate the shock displacement shown in Figure 7. Thus, as a result of the interaction between the shock wave and the boundary layer, the incoming shock is no longer planar, but is displaced forward, by a distance of several boundary layer thicknesses for the chosen parameters.

NEAR WALL REGIONS

The near wall regions consist of the Reynolds stress sublayer region and the wall region, in each of which Reynolds shear stress terms arise. Thus, a closure condition is necessary, that is, a model for the Reynolds shear stress must be chosen. Although a modified mixing length model is being used by the present authors, the general form of the solution is not dependent upon the specific model used, except that it must be such that as the undisturbed flow is approached ($x^* \rightarrow -\infty$), the velocity has a logarithmic variation in the overlap region between the wall and velocity defect regions.

Only a very brief discussion of the solutions in the near wall layers is given here because the details of the calculations necessary to derive a separation criterion are beyond the scope of this paper and a description of the calculations of only the first order terms in the wall shear stress merely recapitulates work described elsewhere [1, 2]. Suffice it to say that if the inviscid region solutions shown here are matched with the Reynolds stress sublayer solutions, and the Reynolds stress sublayer solutions are matched with the wall layer solutions, a relation for the wall shear stress can be derived which is of the same form, but different in detail from the solutions shown in References [1] and [2]. The differences come about because of the functional differences in P_1^* and P_1 as ϵ increases in order, relative to u_τ. The details of the calculations involving the higher order terms necessary to derive a separation criterion will be given in a subsequent paper.

An important point to be noted here is that in the calculations referred to above, the terms retained in the equation of motion in the flow direction are such that the equation reduces to

$$\frac{\partial}{\partial Y}\left\{-u_\tau^2 \;\rho < U'V' > + \frac{1}{Re}\,\mu\,\frac{\partial U}{\partial Y}\right\} = 0 \qquad (33)$$

in the wall region. Now, this equation cannot be valid when separation occurs. That is, the only terms are those due to Reynolds and viscous shear stresses. However, when separation occurs, and a reverse flow exists, it must be that pressure and shear forces are of equal order in the wall region. Thus, opposing forces must be of the same order, if the flow is to be turned, and in fact, reversed. Therefore, at the most, Equation (33) may hold only up to incipient separation, and any theory which holds for the separated case must contain an explanation of how the Reynolds and viscous shear stress terms decrease in magnitude until at separation they are of the same magnitude as the pressure gradient term.

REFERENCES

[1] Adamson, Jr., T.C. and Feo, A.: "Interaction Between a Shock Wave and a Turbulent Boundary Layer in Transonic Flow," SIAM J. Appl. Math., 29 (1975), pp. 121-145.

[2] Melnik, R.E. and Grossman, B.: "Analysis of the Interaction of a Weak Normal Shock Wave with a Turbulent Boundary Layer," AIAA Paper No. 74-598 (1974).

[3] Bohning, R. and Zierep, J.: "The Normal Shock at a Curved Wall in the Viscous Case," Symposium Transsonicum II (Eds. Oswatitsch, K., and Rues, D.), Springer-Verlag (1976), pp. 236-243.

[4] Inger, G.K. and Mason, W.H.: "Analytical Theory of Transonic Normal Shock Boundary Layer Interaction," Symposium Transsonicum II (Eds. Oswatitsch, K. and Rues, D.), Springer-Verlag (1976), pp. 227-235.

[5] Adamson, Jr., T.C.: "The Structure of Shock Wave-Boundary Layer Interactions in Transonic Flow," Symposium Transsonicum II (Eds. Oswatitsch, K. and Rues, D.), Springer-Verlag (1976), pp. 244-251.

[6] Maise, G. and McDonald, H.: "Mixing Length and Kinematic Eddy Viscosity in a Compressible Boundary Layer," AIAA J., 6 (1968), pp. 73-79.

[7] Bush, W.B. and Fendell, F.E.: "Asymptotic Analysis of Turbulent Channel and Boundary Layer Flow," J. Fluid Mech., 56 (1972), pp. 657-681.

[8] Van Driest, E.G.: "Turbulent Boundary Layer in Compressible Fluids," J. Aero. Sci., 18 (1951), pp. 145-160, p. 216.

[9] Cebeci, T. and Smith, A.M.O.: Analysis of Turbulent Boundary Layers, Academic Press, New York (1974).

[10] Hinze, J.O.: Turbulence, An Introduction to Its Mechanism and Theory, McGraw-Hill, New York (1959).

[11] Coles, D.E.: "The Young Person's Guide to the Data," Pro-

ceedings of the AFOSR-IFP-Stanford Conference, Vol. II (Eds., Coles, D.E. and Hirst, E.A.), (1968), pp. 1-45.

[12] Gadd, G.E.: "Interactions Between Normal Shock Waves and Turbulent Boundary Layers," A.R.C. 22,559, F.M. 3051 (1961).

DISCUSSION OF THE ADAMSON-MESSITER PAPER

OSWATITSCH: I am concerned about a little detail at the beginning of the shock which you have sketched. According to our considerations, I think that if the sonic line goes round as you sketched and here you have converging Mach lines in this case and the shock starts, the sonic line goes at an angle and not tangential to the shock.

ADAMSON: I think you are right. I can't even blame it on the artist because I drew it! Our solution was valid beyond where this shock has started so I really wasn't very careful with the drawing. Thank you.

MURMAN: 2 questions: (1) In this initial region in which you can't compute the initial pressure rise from the similarity solution, do you anticipate that this region will be solved numerically in sort of the same way Melnik and Grossman presented their solutions?

ADAMSON: It could very well be, but we have an idea that we can get an approximation to this solution also. The trouble is that it's rotational flow. The problem is, I guess, how rotational.

MURMAN: It also seems that it would be nonlinear.

ADAMSON: Yes, that is true.

MURMAN: (2) I don't understand what the difference is between the inner inviscid region and outer inviscid region physically. The inner inviscid region contains the boundary layer, too?

ADAMSON: Physically, the difference is that one region is ordered by the thickness of the boundary layer and one is ordered by the distance to the sonic line. E.g., in Melnik and Grossman's case - they're of the same order. But in this case, the distance to the sonic line is very small compared to the boundary layer thickness. So therefore you have to look at another region. The point is that the thickness of the subsonic region orders the upstream influence so that this inner inviscid region is an important region.

MURMAN: But in the outer inviscid region the shock is specified; that's the initial shock location, is that right? Then in the inner inviscid region the shock moves?

ADAMSON: Actually, the shock position and shape (i.e., it is normal) is set to order epsilon, but there are corrections to the shape of order u_T, in both inviscid regions.

MC CROSKY: This looks very interesting. I don't understand the significance, in the beginning, of setting u_T equal to δ and then what that might do later to your ordering for the flat plate case, as you've done here. Is that consistent later on with having ϵ large compared to u_T which means also compared to δ? Would this in fact limit you if you tried to look at the more complicated case of a boundary layer that's exposed to an adverse pressure gradient without the shock wave and then a weak shock on top of that? That is to say, a supercritical airfoil with an adverse pressure gradient and a very weak shock wave, so that this order ϵ much, much greater than u_T, if u_T is equal to δ, wouldn't really be appropriate any more?

ADAMSON: I think insofar as the flat plate boundary layer is concerned u_T and δ are of the same order and making them equal is just simply a way of getting rid of a constant. I think that that remark would hold also for the case you imagine but it would just be a different constant, but I'm not sure; I'll have to look into that.

HAFEZ: The change of the order of the wall layer - how would this feed into the inviscid solution? I mean, suppose we have a separation bubble, a very smooth separation bubble downstream. The streamline would be deflected and the shock would be changed in the inviscid solution. How would this interaction show up in the formulation?

ADAMSON: In the inner regions? (yes) Well, the way we picture this is that the existence of a separation bubble will cause a displacement in the flow which will cause a different shock structure. It will cause a lambda shock to form, and you will no longer have just a straight normal shock.

INTERACTIONS OF NORMAL SHOCK WAVES WITH TURBULENT BOUNDARY LAYERS AT TRANSONIC SPEEDS

ROBERT E. MELNIK AND B. GROSSMAN

Grumman Aerospace Corporation

ABSTRACT

Inviscid and boundary layer techniques have proven to be accurate and useful methods for predicting flow fields over solid surfaces at high Reynolds numbers. It is generally recognized that the boundary layer approach breaks down in strong interaction regions at shock waves and trailing edges on lifting airfoils at transonic speeds. In this note we show that boundary layer methods break down because of the appearance of singularities in the second-order inviscid solutions at shock waves and trailing edges. We discuss the nature of the singularities that arise in interactions with fully developed turbulent boundary layers. We will also briefly describe the structure of local "inner solutions" developed by the author and his colleagues at Grumman for interactions without flow separation using formal asymptotic techniques. We show that the solution in each of the strong interaction regions has a three layer structure. The flow in the outer most layer is described by a "Lighthill type model" of inviscid rotational flow. Normal pressure gradients across the boundary layer are important, while Reynolds stresses have no direct influence in the outer layer of each of the interaction zones. The trailing edge studies point to the importance of the pressure drop produced by the momentum deficit in the highly curved wake behind an airfoil. Results obtained in the shock interaction studies explain why static pressures behind normal shock waves do not approach the full Rankine-Hugoniot values in experiments in channels and pipes.

1. INTRODUCTION

Inviscid and boundary layer techniques have proven to be accurate and useful methods for predicting flow fields over solid surfaces. It is generally recognized, however, that the boundary layer approximations break down in strong interaction regions. Although the interactions of shock waves with turbulent boundary layers are known to produce significant effects on the flow field and have been studied for many years, they remain poorly understood. Recently, progress has been made on the development of rational theories for analyzing turbulent shock wave-boundary layer interaction problems [1-3]. These works considered problems involving weak transonic shock waves interacting with fully developed turbulent flows at high Reynolds numbers.

In [1], Adamson and Feo considered very weak, oblique shock waves which did not penetrate into the boundary layer. The flow remained supersonic throughout, and the pressure rises were small and did not separate the boundary layer. In [2] and [3], the present authors considered cases of normal shock waves with somewhat larger shock strengths. Under these conditions the shock waves penetrated into the boundary layer, and the flow field was essentially transonic, involving mixed regions of subsonic and supersonic flow outside the boundary layer. Although the shock strengths considered were an order of magnitude larger than studied in [1], they were still relatively weak and did not involve boundary layer separation. The two theories were entirely consistent, with the Adamson-Feo theory [1] appearing as a special limiting case of [2] and [3] in a weak shock limit.

In the present discussion, we will review the theory of [2] and [3] for interactions of normal shock waves of moderate strength with turbulent boundary layers on flat surfaces. As in [2], the boundary layer is assumed to be attached throughout the interaction and the flow field is considered two dimensional. The theory is based on a formal asymptotic expansion of the Navier-Stokes equations in the limit of Mach number $M \to 1$ and Reynolds number $R \to \infty$. The analysis is consistent with the physical ideas developed by Lighthill [4], where it was shown that supersonic interactions could be treated mainly as an inviscid rotational flow, linearized about the velocity profile in the upstream boundary layer. Our analysis ([2] and [3]) formalized Lighthill's ideas and generalized his model to include nonlinear mixed flow effects that are important in transonic interactions. In addition, the new theory (and also [1]) gave, for the first time, a treatment of the inner, dissipative layers that properly accounted for the logarithmic behavior of turbulent velocity profiles near walls.

The interaction equations developed in [2] were solved by a finite difference technique to obtain the shock shape and pressure distribution. The results for the surface pressure were compared with several sets of experimental data including those of Gadd [5]. Although the comparison showed good agreement in the initial stages of the pressure rise, the agreement on the downstream side was poor. The large differences were due to the failure of the experimental data to approach the downstream static pressure given by the shock wave relations.

In the later study [3], we determined the reasons for this discrepancy and modified the theory of [2] accordingly. The comparison with Gadd's data [5] in a circular pipe was repeated using modified interaction equations and a more accurate finite difference method. The new theoretical results are shown to be in good agreement with the experimental data throughout the entire pressure rise.

In Section 2, we consider the behavior of second order boundary layer solutions near points where shock waves impinge on turbulent boundary layers. We discuss the nature of singularities that arise in interactions with fully developed turbulent profiles. In Section 3, we briefly sketch the main features of the strong interaction theory [2] and [3]. In Section 4, we discuss some results obtained from numerical solution of the interaction equations. We also present a comparison of our results with Gadd's data and with results of the weak shock theory of Adamson and Feo. In Section 5 we list some general conclusions regarding the strong interaction theory.

2. SECOND ORDER BOUNDARY LAYER THEORY

It is well known that the asymptotic developments underlying inviscid/boundary layer theory provide a rational approximation to the solution of the full Navier-Stokes equations in the limit $R \to \infty$. However, these approximations completely fail in strong interaction problems. When shock waves are involved, the failure of the standard theory is signaled by the appearance of singularities in the second order outer solution, as illustrated in Figure 1. The lowest order inviscid solution imposes a jump in pressure on the boundary layer which, from simple momentum considerations [6],

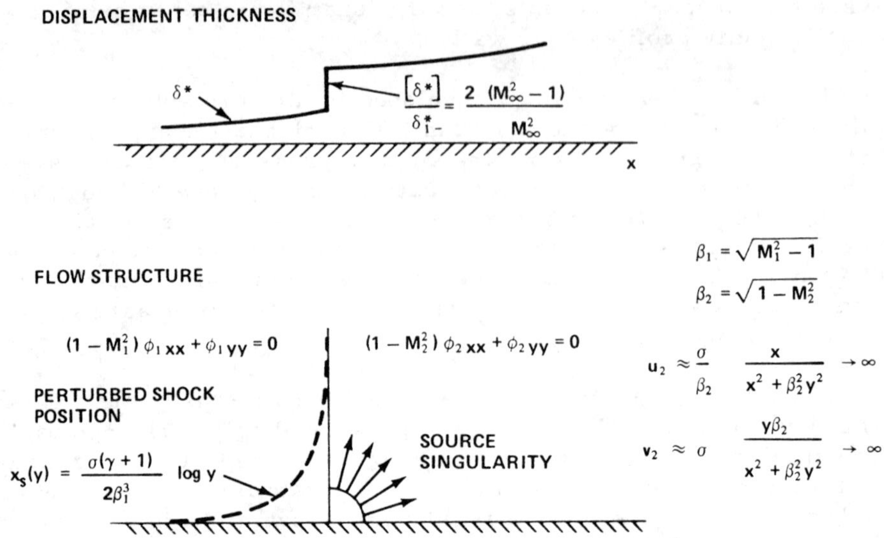

Figure 1. Second Order Boundary Layer Theory

leads to jumps in the boundary layer integral parameters. The magnitude of the discontinuity in displacement thickness

$$[\delta_1^*] = \delta_{1+}^* - \delta_{1-}^*$$

through a normal shock wave is given in the figure for the limiting conditions $R \to \infty$, $M \to 1$.

The second order inviscid solution is obtained by solving the linearized perturbation equations written in Figure 1, where x, y are the nondimensional coordinates along and normal to the plate; ϕ_1 and ϕ_2 are the perturbation potentials; δ_{1-}^*, δ_{1+}^* are the displacement thicknesses just upstream and downstream of the shock wave; and M_1, M_2 are the Mach numbers from the first order inviscid solution. The solution of this system behaves as the flow over a small step, and this leads to a source singularity in the velocity components u_2, v_2. The source strength, σ, is determined by the jump in the displacement thickness and is given by

$$\sigma = \frac{2\beta_1^2 \delta_{1-}^*}{\pi M_1^2}$$

It can be shown from the shock wave relations that the source singularity leads to a logarithmic singularity in shock position. The second order solution represents an "outer solution," in the sense of the method of matched asymptotic expansions. The strong interaction theory developed in [2] and briefly described in the following section is an "inner solution," which is valid in a small, local region near the shock impingement point.

3. STRONG INTERACTIONS THEORY

The main features of the interaction problem are sketched in Figure 2. The flow is controlled by two basic parameters, the free stream Mach number and the Reynolds number, R, based on the distance, L^*, between the leading edge and the shock impingement point. For convenience, a small parameter, ϵ, equal to the friction velocity, is introduced in place of the Reynolds number according to the relation

$$\epsilon = \sqrt{C_{f_o}/2} \approx 1/\log R \qquad (1)$$

where $C_{f_o}(R)$ is the skin friction coefficient of the undisturbed boundary layer just upstream of the shock wave. The theory is then cast in terms of the double limit $\epsilon \to 0$, $M_\infty \to 1$. A distinguished limit is shown to exist for the case where the velocity

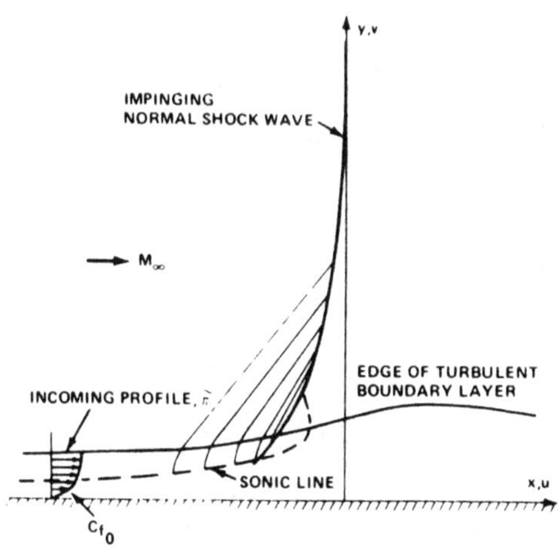

Figure 2. Schematic of Flow Field

change across the shock wave is the order of the velocity variation across the outer or wake part of the turbulent flow. This leads to the basic interaction parameter

$$K_t = (M_\infty^2 - 1)/\epsilon M_\infty^2,$$

which is held fixed in the double limit process. This limit implies weak shock waves and fully developed turbulent boundary layers.

The theory leads to the multilayer structure indicated in Figure 3, where we indicate the ratio of the length scale to the reference length, L, appropriate to each region. In the regions upstream and downstream of the interaction zone, the flow has a conventional inviscid/boundary layer structure, with the boundary layer exhibiting the well known law of the wall/law of the wake two layer structure (see discussion in [2]. The additional parameter, $\hat{\epsilon}$, which scales the wall layer thickness, is given by

$$\hat{\epsilon} = 1/(\epsilon^2 R).$$

The interaction zone divides into three regions: (1) an outer region made up of the main part of the boundary layer and the

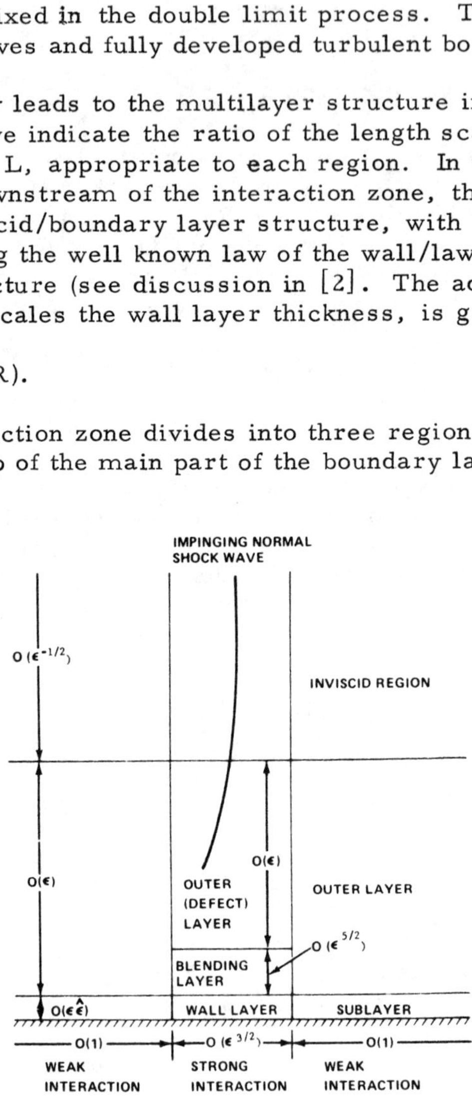

Figure 3. Asymptotic Structure

external inviscid flow, (2) a wall layer which is a continuation of the wall layer in the upstream boundary layer, and (3) a blending layer separating the outer and inner layers. The outer layer is basically an inviscid, rotational stream, while the inner layer is a conventional equilibrium layer with the total shear stress, laminar plus turbulent, constant across the layer. The Reynolds stresses are frozen to their upstream values in the outer layer and are determined by the local value of skin friction in the inner layer. This leads to a discontinuity and mismatch of the Reynolds stresses in the outer and inner regions. This discontinuity is resolved by the blending layer solution. The derivation of the governing equations, boundary and matching conditions for each of the regions is given in [2] along with further details and discussion.

In the present study, we will be concerned only with the solution in the outer region. This will lead to a theory for the determination of the shock shape and pressure distribution. The outer solution is formulated in terms of stretched coordinates, x, y, defined in terms of physical coordinates, x^*, y^* by

$$x^* = \epsilon^{1/2}\delta_o^* x, \qquad y^* = \delta_o^* y, \qquad \delta_o^* = \epsilon L^* \delta_o \tag{2}$$

where δ_o^* is the boundary layer thickness in the undisturbed boundary layer at the shock impingement point. The velocity components, u, v, are normalized with respect to free stream velocity and are expanded in the form

$$u = 1 + \epsilon[u_o(y) + \phi_x(x, y)] + \ldots \tag{3a}$$

$$v = \epsilon^{3/2}\phi_y(x, y) + \ldots \tag{3b}$$

where ϕ is the perturbation velocity potential and $u_o(y)$ is the defect part of the upstream velocity profile. We represent it as a Coles' law of the wall/law of the wake form

$$u_o(y) = \frac{1}{\kappa}\log y - \frac{\tilde{\pi}}{\kappa}[1 + \cos \pi y] \tag{4}$$

where $\kappa = 0.41$ is the von Karman constant, and $\tilde{\pi}$ is Coles' wake parameter. Substitution of these expansions into the Navier-Stokes equations leads to the following equation for the perturbation potential in the outer region

$$[\chi_t + M_\infty^2(\gamma + 1)(u_o + \phi_x)]\phi_{xx} - \phi_{yy} = 0 \tag{5a}$$

where

$$\chi_t = \frac{M_\infty^2 - 1}{\epsilon} \equiv M_\infty^2 K_t \tag{5b}$$

and γ is the ratio of specific heats. Equation (5) is the nonlinear transonic small perturbation equation generalized to account for weak upstream nonuniformities generated by the turbulent flow near the wall. A linearized version was proposed by Robinson [7] for shock wave boundary interaction problems. Robinson's linear theory was appropriate for the study of the reflection of very weak disturbances from the boundary layer.

In [2] we obtained solutions to Eq. (5) for normal shock waves using a (nonconservative) mixed flow differencing scheme of Murman and Cole [8]. The results were compared with data of Gadd [5] for flow in circular pipe at $M_\infty = 1.12$ and with data of Vidal, et al. [9] for flow in rectangular channel at a nominal Mach number of $M_\infty = 1.4$. The comparisons showed reasonably good agreement only in the initial stages of the interaction. The comparison showed very poor agreement on the downstream side, due mainly to the failure of the experimental data to approach the downstream static pressure given by the shock wave relations.

In [3], it was found that the reason the data did not approach the appropriate shock values downstream was due to the three dimensional nature of the experimental configuration. The boundary layer thickening through the shock wave reduced the equivalent area outside the boundary layer, which resulted in a higher velocity in the downstream region. In [3], we showed that although this effect was formally of higher order, it was numerically significant. It was shown that good agreement could be obtained with Gadd's results if the flow geometry was more accurately represented. Thus in [3], the basic formulation leading to Eq. (5) was extended to axisymmetric flow in order to represent Gadd's pipe flow experiment better.

For axisymmetric flow, Eq. (5) is replaced by

$$[\chi_t + M_\infty^2(\gamma+1)(u_o+\phi_x)]\phi_{xx} - \phi_{yy} = \frac{\epsilon\phi_y}{y_c - \epsilon y} \tag{6}$$

where

$$y_c = \epsilon D^*/2\delta_o^* = \delta_o(D^*/2L^*) = 0(1) \qquad (D^* = \text{pipe diameter})$$

where x and ($\epsilon y - y_c$) are cylindrical coordinates, with y measured from the wall of the pipe. The right side of Eq. (6) is formally negligible and the flow is locally two dimensional near the wall in the limit $\epsilon \rightarrow 0$. With this term retained, Eq. (6) is a composite equation valid both in the strong interaction region near the wall and in the weak interaction region in the interior part of the flow near the axis, $y = \epsilon^{-1} y_c$.

The numerical solution of Eq. (6) with a symmetry condition on the axis, $y = y_c \epsilon^{-1}$ is far easier to accomplish than the solution of the equations with the far field matching conditions involving a logarithmic singularity in shock position (see Fig. 1). In addition, retaining the right side of Eq. (6) leads to an important correction accounting for the equivalent area reduction in the far field, downstream of the shock waves. For inviscid flow downstream of a normal shock in a circular pipe, the streamwise velocity should approach a value given by the shock relations appropriate to Eq. (6). However, many experiments suggest that the velocity does not approach the value given by the shock relations. As sketched in Figure 4, we show that this is due to the thickening of the boundary layer through the shock wave, which acts to reduce the area of the flow outside the equivalent body formed by the boundary layer displacement surface on the walls of the pipe. This reduction of equivalent pipe area acts to speed up the flow in the downstream region. The magnitude of the velocity correction can be computed from Eq. (6) by carrying out a mass flow balance far upstream and downstream of the pipe. If u_2 is the magnitude of the velocity far downstream

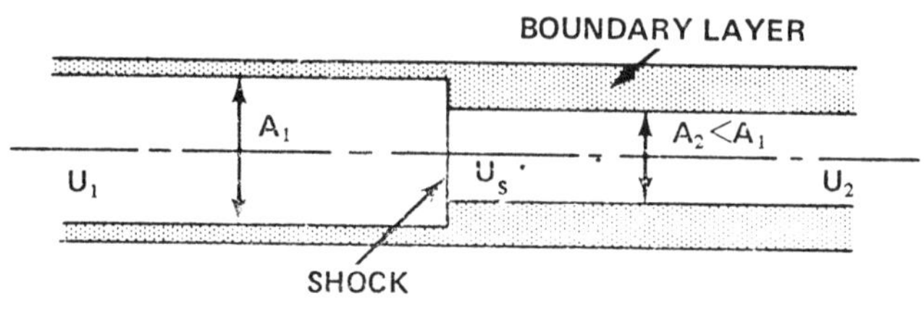

Figure 4. Downstream Boundary Condition

$$u_2 = u_s + \left[\frac{1+\tilde{\pi}}{\kappa}\right]\frac{\epsilon}{y_c} - \left[\frac{1 + 2\pi(1 - 4/\tilde{\pi}^2)}{4\kappa}\right]\frac{\epsilon^2}{y_c^2} \tag{7}$$

where

$$u_s = -2K_t/(\gamma+1)$$

is the value given by the shock relation. Equation (7) was evaluated using the initial profile given in Eq. (4).

Thus, in this study, the following boundary conditions are employed in the solution of the composite equation given in Eq. (5)

$$\phi = \phi_x = 0 \qquad x \to -\infty \qquad 0 \leq y \leq \epsilon^{-1} y_c \tag{8}$$

$$\phi_x = u_2 \qquad x \to +\infty \qquad 0 \leq y \leq \epsilon^{-1} y_c \tag{9}$$

$$\phi_y = 0 \qquad -\infty < x < \infty \qquad y = 0, \quad \epsilon^{-1} y_c \tag{10}$$

where u_2 is given by Eq. (7). Equations (8) and (9) require the solution to approach uniform flow far upstream and downstream of the interaction, while Eq. (10) requires the normal component of velocity ϕ_y to vanish on the wall and at the axis of symmetry.

In [2], we solved a two dimensional version of Eq. (5) and Eqs. (8) - (10) without the downstream boundary condition correction, using a nonconservative difference scheme. In [3], we solved the full system of equations given in Eqs. (6) - (9) using a fully conservative differencing technique([10] and [11]). We employed a nonuniform grid obtained by mapping the upper half plane to a finite rectangle with a concentration of mesh points in the boundary layer and near the shock wave. The x, y mesh contained 48 x 80 points with 16 points in the boundary layer. A typical solution required about 4 minutes CPU time on an IBM 370/168 computer.

4. RESULTS

In this section we discuss the results of applying the theory of the previous section to the experiment of Gadd [5] for a normal shock wave in circular pipe with an upstream Mach number M_∞ = 1.12. Fitting the velocity profile form given in Eq. (4) to Gadd's measured profile upstream of the interaction yielded $\epsilon = 0.034$, $D^*/2\delta_o^* = 18.66$, $\tilde{\pi} = 0.5$, $K_t = 5.98$.

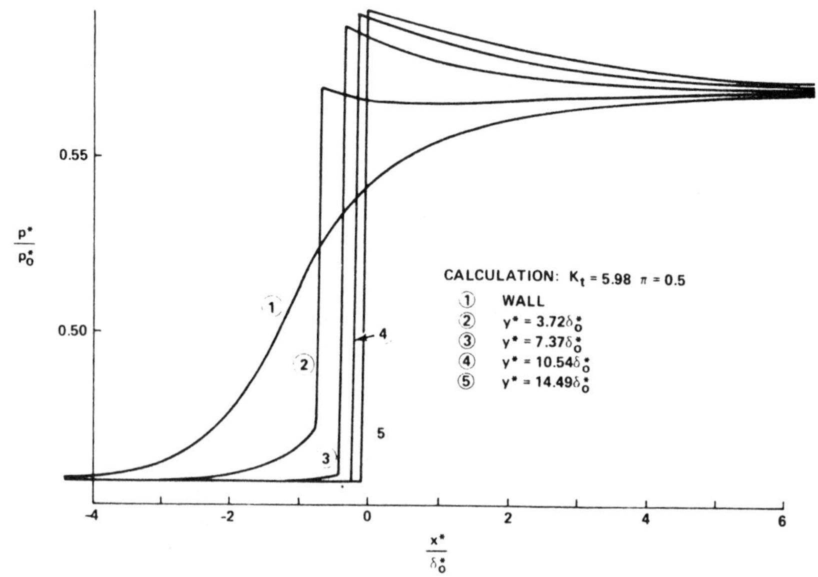

Figure 5. Computed Pressure Distribution within a Pipe

The solution for the ratio of static pressure, p^*, to stagnation pressure p_o^* is given as a function of x^*/δ_o^* for several values of y^* across the pipe in Figure 5. This figure illustrates the transition from discontinuous pressure distributions outside the boundary layer to a smooth, continuous distribution at the surface. This result also shows evidence of the rapid, exponential decay of pressure disturbance upstream, and the slow, algebraic decay ($\Delta p \sim x^{-1}$) downstream. The pressure distribution in the flow field clearly shows a smooth precompression, followed by a discontinuous jump to the post shock value, and then by a more gradual expansion to the asymptotic far field value, which is lower than the value behind a $M_\infty = 1.12$ normal shock wave. The post shock expansion is consistent with a ring source behavior of the far field solution induced by a nearly discontinuous displacement surface. The results in the figure also show large normal pressure variations through the boundary layer near the shock wave, which decays upstream and downstream of the interaction zone.

The theoretical solution for the pressure distribution on the wall is compared with the experimental data of Gadd in Figure 6. Also included in the figure is the asymptotic level (denoted by ▭) that the solution approaches far downstream, along with far field

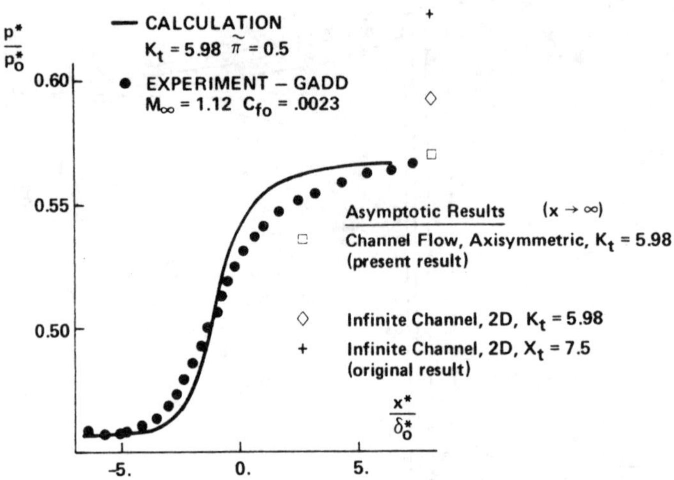

Figure 6. Surface Pressure along a Pipe

values obtained with several other approximations. The agreement of the new results with experiment is substantially improved over the results presented in [2]. The improvement is due primarily to the use of the mass corrected far field boundary condition retention of the M_∞^2 factor in the coefficient of ϕ_{xx} in Eq. (6). The magnitude of the latter effect in the far field is given by the difference between the $K_t = 5.98$ (symbol ◇) result and the $\chi_t = 7.5$ (symbol +) result, where the M_∞^2 factor was set equal to one. The parameter

$$\chi_t = (M_\infty^2 - 1)/\epsilon$$

is the limit form of interaction parameter. The $K_t = 5.98$ point (◇) also corresponds to the values of static pressure given by the exact Rankine-Hugoniot shock wave relations. The difference between the symbols ◇ and □ shows the influence of the mass flow correction on the far field pressure level. This result clearly demonstrates the large effect of the wall boundary layers in reducing the effective area of the pipe and in accelerating the flow in the far field.

The small pressure gradient evident in the data in the downstream region of the pipe is likely due to weak interaction effects generated by boundary layer growth in the far field that have been ignored in the present theory.

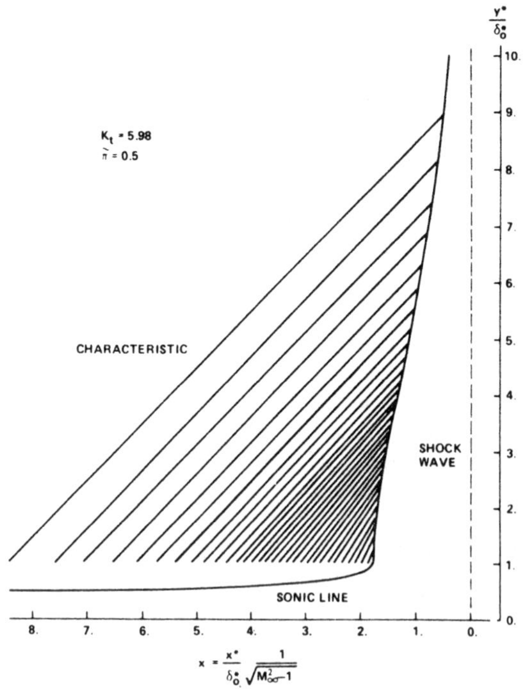

Figure 7. Computed Shock Wave, Sonic Line, and Characteristics

In Figure 7 we show the computed shock shape, sonic line, and compression waves generated by the curved sonic line upstream of the shock wave. The straight compression waves were determined after the solution was completed from the computed velocities at the edge of the boundary layer. The shock wave crosses the x-axis at the pipe centerline $y^*/\delta_o^* = 18.66$. This result clearly shows the long range effect of the source-like flow in the subsonic region behind the shock wave. The flow behind the shock wave pushes the shock wave forward until it runs into the upstream compression waves generated by the interactions. The compressions oppose the downstream flow and deflect the shock wave downward, resulting in smooth continuation of the shock wave into the sonic line within the boundary layer. Although the characteristics tend to merge together near the shock, there is no indication of wave focusing to form another leg of the shock pattern. Thus, in this case, there is no evidence of a Mach reflection or of the appearance of a supersonic tongue.

As a final point, we consider the behavior of solution to Eq. (5) in the limit $\chi_t \to 0$ corresponding to the weak shock limit considered by Adamson and Feo [1]. In [2], it was shown that this limit leads to a singular perturbation problem with the outer region of the present study splitting into two layers, consisting of the original rotational layer and an outer much taller irrotational region. The pressure is constant across the inner rotational layer, and the flow here behaves like an inviscid boundary layer. The solution in the outer layer is governed by the usual nonlinear, irrotational transonic small disturbance equation. The solutions in the two regions are coupled, with the inner layer generating flow deflections in response to pressure gradients imposed by the outer solution. This results in a free interaction description for turbulent flow, with the solution independent of downstream conditions. The solution upstream of the shock wave is governed by simple wave solutions of the small disturbance equation and can be found in closed form.

In the notation of the present paper the solution for the pressure can be written in the form

$$p_1(x, 0) = -\phi_x = -\frac{\chi_t}{\gamma + 1}(z - 1) \tag{11a}$$

$$x = \text{constants} + \frac{A}{3}\sqrt{\frac{\gamma + 1}{\chi_t}}\left[\log\left(\frac{1 - z^{\frac{1}{2}}}{1 + z + z^{\frac{1}{2}}}\right)\right.$$

$$\left. + 2\sqrt{3}\tan^{-1}\left(\frac{2z^{\frac{1}{2}} + 1}{\sqrt{3}}\right)\right] \tag{11b}$$

where

$$A = \frac{3\sqrt{\gamma + 1}\,(\tilde{\pi} + 1)}{2} \tag{11c}$$

The limit solution agrees with the solution first obtained by Adamson and Feo [1]. In Figure 8, we compare the above solution with the numerical solution of the full equation [Eq. (5)] for $\chi_t = 7.5$, with the two solutions matched at the sonic point. In the figure, we show the pressure computed from the full solution at the wall and at the edge of the boundary layer. Although there are relatively large pressure changes across the boundary layer for this case, the agreement is seen to be quite good if the free interaction solution

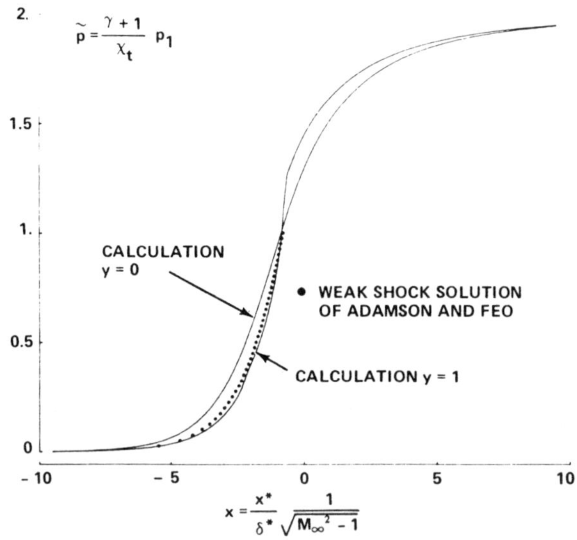

Figure 8. Comparison of Weak Shock Solutions of Eq. (11) with Numerical Solution of Full Equations for $\chi_t = 7.5$

is assumed to apply at the edge of the boundary layer.

5. CONCLUSIONS

Several general conclusions can be drawn from the present work:

- The relatively good agreement with Gadd's data tends to validate the theoretical model.
- The flow downstream of the shock exhibits a ring source type character. The results indicate that the reason the experimental pressures do not approach the Rankine-Hugoniot values downstream is due to the reduction of equivalent area of the pipe by boundary layer growth through the shock.
- The theoretical solution leads to a very simple shock pattern. There is no evidence of a Mach reflection or of the formation of a supersonic tongue for the case considered.

. The theory indicates that the streamwise length scale for an unseparated interaction is

$$\Delta x^* \approx \sqrt{M_\infty^2 - 1}\,\delta_o^*,$$

which is less than a boundary layer thickness for Mach numbers near one.

. The moderate shock strength theory of [2] and [3] is consistent with the weak shock solution of Adamson and Feo [1] in the limit $K_t \rightarrow 0$.

. The results point to the importance of sidewall boundary layers in wind tunnel experiments. This implies relatively large three dimensional effects in nominally two dimensional channel flow experiments. We suggest that basic experimental studies on shock wave-boundary layer interactions should be carried out with axisymmetric geometries in order to validate theoretical models.

6. REFERENCES

[1] Adamson, T. C., Jr. and Feo, A. (1975), "Interaction Between a Shock Wave and a Turbulent Boundary Layer in Transonic Flow," SIAM J. Appl. Math., 29.

[2] Melnik, R. E. and Grossman, B. (1974), "Analysis of the Interaction of a Weak Normal Shock Wave with a Turbulent Boundary Layer," AIAA Paper 74-598.

[3] Melnik, R. E. and Grossman, B. (1975), "Further Developments in an Analysis of the Interaction of the Interaction of a Weak Normal Shock Wave with a Turbulent Boundary Layer," presented at IUTAM Symposium Transsonicum II, Gottingen, Germany, Springer-Verlag, September 1975; also Grumman Research Department Report RE-511, October 1975.

[4] Lighthill, M. J. (1953), "On Boundary Layers and Upstream Influence. II. Supersonic Flows with Separation," Proc. Roy. Soc., Series A, 217.

[5] Gadd, G. E. (1962), "Interactions between Normal Shock Waves and Turbulent Boundary Layers," ARC 22559, R and M 3262.

[6] Green, J. E. (1970), "Interactions between Shock Waves and Turbulent Boundary Layers," article in Progress in Aeronautical Sciences, 11, Pergamon Press.

[7] Robinson, A. (1951), "Wave Reflection Near a Wall," Proc. Camb. Phil. Soc., 47.

[8] Murman, E. M. and Cole, J. D. (1971), "Calculation of Plane Steady Transonic Flows," AIAA J., 9.

[9] Vidal, R. J., Wittliff, C. E., Catlin, P. A., and Sheen, B. H., (1973), "Reynolds Number Effects on the Shock Wave Turbulent Boundary Layer Interaction at Transonic Speeds," AIAA Paper 73-661.

[10] Murman, E. M. (1974), "Analysis of Embedded Shock Waves by Relaxation Techniques," AIAA J., 12.

[11] Jameson, A. (1975), "Potential Flow Calculations Using Conservation Forms," Proceedings of AIAA 2nd Computational Fluid Dynamics Conference, Hartford, Connecticut.

DISCUSSION OF THE MELNIK-GROSSMAN PAPER

HAFEZ: Without allowing interaction between the inner layer and the inviscid layer, we don't expect hat the viscous effect is meaningful. Actually, what you took care of is the defect of the velocity in the common shear layer and this is a continuity matter. And you took care that the incoming flow profile has shear or vorticity and the vorticity sticks to the streamline and then the conclusion is that nothing spectacular happened. Yes, because what you actually solved is the small disturbance equation and you left the similarity parameter a function of what ? So we don't expect much out of it. Nevertheless, I admire the way you fixed the downstream condition so that you got good agreement with the experiment. But, I feel that unless we have the interaction between the viscous layer and the inviscid layer, we cannot have meaningful viscous effects - I mean separated flow, even with a small bubble and deflection of the streamline and these interactions coming in - especially if you use the computer and if you have a composite solution and solve this problem numerically - and especially if we have some results using boundary layer equations in the separated flow, which is simple to do.

MELNIK: When I said nothing spectacular happens, it didn't mean I expected anything. I felt all along that I understood these interactions and didn't expect anything to happen. What I'm saying is that if you take into account the Mach number profile in front of the shock wave, for these unseparated cases and take into account in the channel the fact that the boundary layer gets thicker through the shock wave, just those two very simple physical ideas explain almost all shock boundary layer interactions without separation. No complicated patterns; the next step is - (comment in background) - well, there's nothing viscous about it. My comment was: I don't believe many of these interactions are viscous. I believe that if we solve the inviscid Euler equations, taking into account the vorticity in the boundary layer, we'll explain most of the data. I believe secondly that for separated flows - for laminar flows - it is clear that we don't need the Navier-Stokes equations. Carter, M. Werle, and T. Davis have shown that you can do flows with separation bubbles using the boundary layer equations. The next step is: what happens to turbulent separated flows? I think you're going to get something simpler than Navier-Stokes equations but more complicated than the boundary layer equations. You need normal pressure gradients but I believe we need nothing more than the inviscid Euler equations with upstream vorticity. When we have a scheme to do that, then we'll be able to do flows with separation, I believe.

ADAMSON: You mention that you thought that your solution would include ours in the upper limit and I disagree because I think that there is one fundamental difference between the solutions. Your source strength at the wall is $O(M_e - 1)$ which, in my notation, would be of order ϵ. The source strength we get is of order u_T. Now, for the case you're looking at, where ϵ is of order u_T, that makes no difference but if you get to stronger and stronger shocks then this would make a difference. Our source strength isn't that large and, in fact, we do not predict a jump in displacement thickness - at least to the order that you mention. So there is a difference.

MELNIK: I wasn't saying it was the same. All I'm saying is that the equations we're solving numerically include the linearized equations that you have in your model and that if we could find a numerical solution to our system of equations, we have all the terms and all the boundary conditions in the stronger shock case, and we should be able to reproduce the stronger shock ...

ADAMSON: I'm talking about the boundary conditions; I really don't agree with that.

MELNIK: O.k. We can talk about it a little bit further.

MURMAN: On the last slide you showed where you talked about the airfoil problem and the observation that the pressure downstream of the shock doesn't bear much relation to the normal shock value and you wondered what you are going to do about that. It seems to me that both your paper and Tom Adamson's dealt with the flat plate case where there was no curvature. In the airfoil case, there is a curvature effect which does allow a mechanism for a normal pressure gradient, and for a lower pressure to exist on the surface than will exist somewhere out in the flow field behind the stronger shock. I think this is the mechanism that will allow you to compute a solution for an airfoil problem that does have a much milder pressure rise through the shock on the surface, but out in the flow field may have a stronger pressure rise. I can show you some things similar to that.

MELNIK: There are two speculations. One of them is that it's the curvature effect. I looked at the equations, the formal structure of those terms, and they don't seem to be large enough to explain it. So right now, I would say we just have to speculate. We have some ideas we're going to try. But I believe that is the central problem that has to be resolved before we can do an airfoil calculation.

COUPLED INVISCID/BOUNDARY-LAYER FLOW FIELD PREDICTIONS FOR TRANSONIC TURBOMACHINERY CASCADES

P. R. GLIEBE

General Electric Company

ABSTRACT

A prediction procedure has been developed for transonic flow through turbomachinery cascades which couples the inviscid flow and the blade surface boundary layers, resulting in a mutually compatible solution. This procedure utilizes the Time-Dependent Computation technique for the inviscid flow solution calculation, and the Stratford and Beavers integral method for the blade surface boundary layer calculation. The procedure has been applied to predictions for several cascade configurations, and the inviscid-flow/boundary-layer interaction calculation has proved to be stable and convergent, with no evidence of strong-interaction instability for supersonic flows. For most cascades, including those typical of high Mach number fan rotor tip sections, the computed results showed little effect of the boundary layers on the flow distribution. For one cascade with supersonic inflow and subsonic outflow, the inclusion of the blade boundary layer effects produced a large effect on the cascade passage flow conditions.

INTRODUCTION

The development of a computational procedure for predicting the blade-to-blade flow through a turbomachine blade section in the transonic regime has become of prime importance in recent years. The Time-Dependent Computation Method (TDC) has been studied as one approach to calculating transonic cascade flows. References

[1-3] report the application of TDC methods to cascade flows in the transonic regime. A General Electric Aircraft Engine Group computer program has been developed which employs the TDC method to calculate the inviscid transonic flow through turbomachine blade sections.

Considerable experience has been gained with the TDC method, through analysis of the flow through many cascade geometries for which either experimental or alternate theoretical pressure distributions were available. The results of these comparison analyses seemed to indicate that, if the blade surface boundary-layer effects had been included in the analytical prediction, better agreement between prediction and experiment would have been achieved. In particular, shock/boundary-layer interaction effects appeared to be very important in some cases.

A program was therefore undertaken to develop a method for correcting the inviscid flow calculation for blade surface boundary layer effects within the framework of the TDC procedure. A boundary-layer calculation method was selected on the basis of reasonable accuracy without unduly burdening the TDC computation in terms of program modification and computational processing time. The method selected was an integral-momentum technique developed by Stratford and Beavers [5].

The following sections briefly describe the combined inviscid/ boundary-layer transonic flow cascade analysis technique, followed by a presentation and discussion of some results of computations made with the combined procedure.

TDC INVISCID ANALYSIS

The approach used to apply the time-dependent calculation method to transonic, inviscid cascade flows is similar to those derived in [1-4]. The basic approach is similar to that of [2], but is more general in that variable radius and streamtube thickness effects are taken into account. A sketch of the cascade system is shown in Figure 1.

With the TDC approach, the steady-state solution is sought as the asymptotic limit for large times of an initial flow distribution which is allowed to vary with time. Using a finite-difference grid representation of the flow field, the solution is updated at small time intervals until changes in the solution are no longer significant,

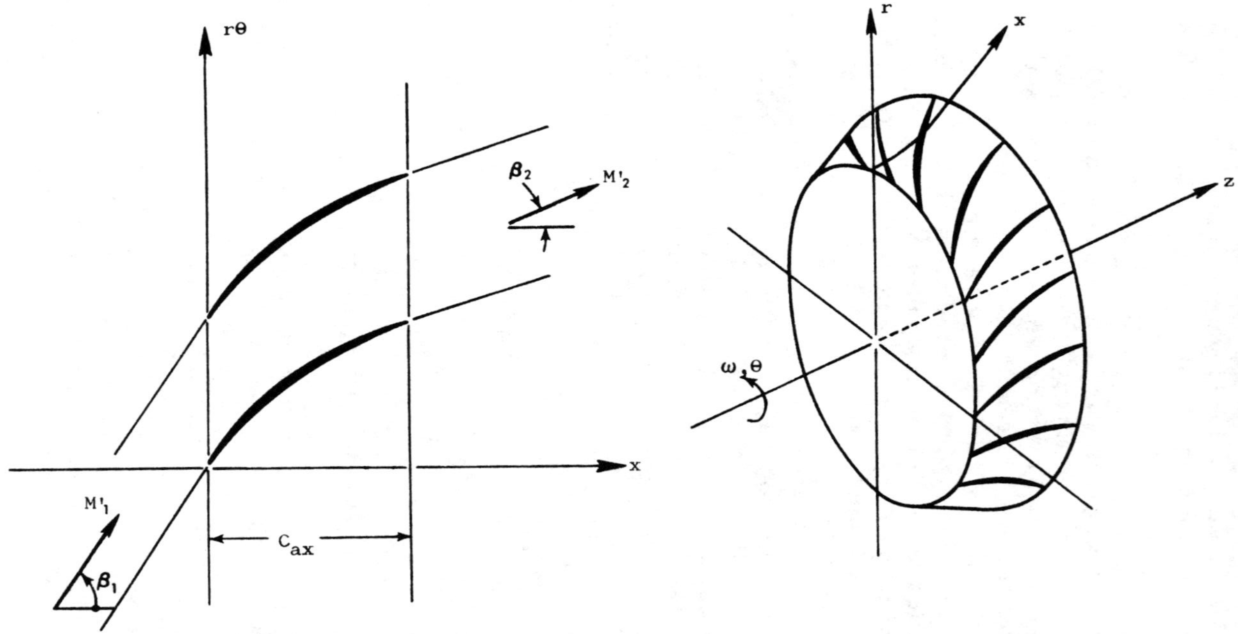

a) Cascade Plane

b) Streamsurface-of-Revolution

Figure 1. Cascade System Geometry.

at which time the flow field is considered to have reached its steady state condition.

When shock formations appear in the transient flow evolution, the numerical solution procedure becomes unstable unless specific steps are taken to accommodate the shock formations. The approach used in the TDC cascade program is to model all shock formations as rapid but continuous changes in flow properties through the use of the "artificial viscosity" technique [2]. A coordinate transformation is employed so that bounding grid lines will coincide with the blade surfaces, forming a channel between two adjacent blades in the cascade. The numerical algorithm for solving the time-dependent equations of motion is basically a two-step predictor-corrector procedure. The alternating-direction field-sweep technique is also employed, resulting in savings in both computational time and program storage. Special equations are used when computing blade surface boundary points which are based on unsteady method-of-characteristics theory. The procedure assumes that the axisymmetric streamsurface-of-revolution properties (radius, meridional flow angle, curvature, normal distance between stream surfaces) are fixed input conditions.

BLADE BOUNDARY LAYER ANALYSIS

As was outlined in the Introduction, the blade surface boundary layer effects on the inviscid flow were speculated to be important in some cases. To account for these effects, a method for computing the blade surface boundary layer properties was required. An integral method was deemed the most practical, from the standpoint of minimal computation time while providing reasonable estimates of boundary layer displacement thickness distributions.

The boundary layer method selected was that of Stratford and Beavers, proposed in [5], and discussed extensively in [6]. This method is essentially a correlation of several other integral methods, and employs a simple algebraic expression for momentum thickness and displacement thickness as a function of local Mach number, Reynolds number, and equivalent flat-plate length distance along the surface (defined as that distance over which a boundary layer growing on a flat plate would acquire the same thickness as the real boundary layer at a given location on the actual surface).

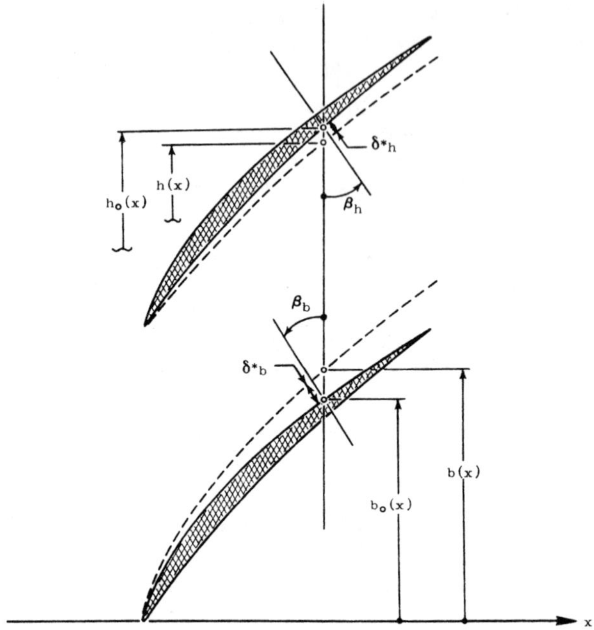

Figure 2. Blade Surface Geometry Boundary Layer Corrections.

BLADE SURFACE CORRECTIONS

Once the boundary layer calculations have been made, the blade surface geometry is corrected for boundary layer displacement effects. Referring to Figure 2, the effective coordinates of the lower and upper surfaces forming the blade passage are given by b(x) and h(x), respectively. The corresponding surface slopes are b'(x) and h'(x). Let the physical (hardware) coordinates and slopes be denoted by subscript "0". Given a displacement thickness distribution δ_b^* and δ_h^* along surface $b_o(x)$ and $h_o(x)$, respectively, the effective coordinates and slopes are then calculated by adjusting the surfaces "inward" and amount $\delta^* \sec\beta$; surface slopes are found by a least-squares error curve-fitting procedure. The new values of b(x), h(x), b'(x) and h'(x) are then used to continue the TDC inviscid solution for a pre-selected number of time steps.

THE "WAKE" REGION

Proper treatment of the wake behind the trailing edge of the

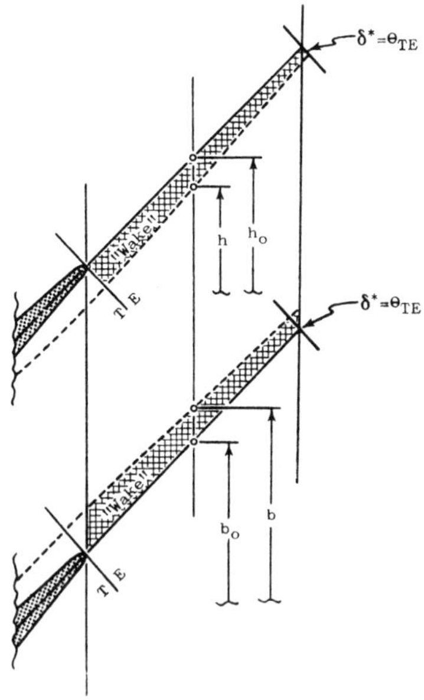

Figure 3. "Wake" Region Model.

blades is a difficult task when using a coupled inviscid/boundary-layer calculation system. This task is further complicated by the fact that the location of the trailing-edge stagnation streamline and its downstream trajectory are not explicitly known from the inviscid TDC solution.

For these reasons, a simple method for handling the boundary layer wake which accounts for effective flow area reduction in an elementary fashion was selected. Figure 3 shows qualitatively how the wake region is treated. The bounding upper and lower grid lines of the flow field downstream of the trailing edge are positioned to approximate a trailing-edge streamline. The displacement thickness is assumed to vary linearly from its trailing-edge value (at the last grid point on the blade) to a value equal to the trailing-edge momentum thickness at the downstream boundary. The upper and lower grid lines are then adjusted "inward" using the calculated displacement thickness.

THE COMBINED INVISCID/BOUNDARY-LAYER CALCULATION SEQUENCE

The calculation sequence begins as in the inviscid case, with an initial condition specified throughout the flow field. The inviscid TDC calculation is then commenced, for a pre-determined number of time steps; say 50, for example. From the blade surface distributions prevailing at this point in time, the blade surface boundary layer displacement thickness distributions are next computed, using the Stratford and Beavers procedure. From the resulting displacement thickness distributions so obtained, the blade surface and wake boundary grid line coordinates and slopes are adjusted to their "effective" values. The TDC inviscid calculation is then continued for another pre-determined number of time steps using the "effective" geometry.

The above procedure is repeated until the flow field solution has converged to a steady state. The boundary layer calculation is "quasi-steady," in that no time-dependent solution is sought, and the boundary-layer is assumed to respond immediately to whatever surface pressure distribution prevails. Also, the boundary layer is not (necessarily) updated every time step, but only at selected intervals, e.g., every 50 time steps.

EXAMPLE CALCULATION RESULTS

Utilizing the combined TDC/boundary-layer calculation procedure for transonic cascade flows, three example cases were tested. The following paragraphs discuss the results of these computations.

(a) <u>Transonic Fan Pitchline Selection Cascade:</u>

The first example of a combined inviscid TDC and boundary-layer calculation is shown in Figure 4. This example geometry is a pitchline cascade section for a transonic fan rotor, characterized by a low supersonic ($M_1 \approx 1.1$) upstream relative Mach number and appreciable camber in the blades. Figure 4 shows blade surface pressure distributions for three different conditions, as follows:

(1) Inviscid TDC solution with no boundary layer corrections;

(2) Inviscid/boundary-layer combined solution, boundary

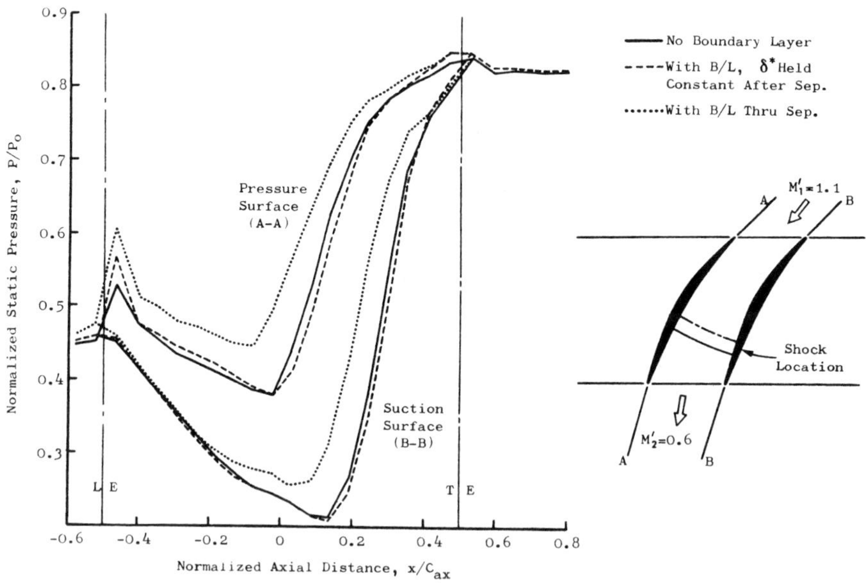

Figure 4. Transonic Fan Rotor Pitchline Section Cascade: Effect of B/L on Blade Static Pressures.

layer thickness held constant after indicated point of separation;

(3) Inviscid/boundary-layer combined solution, boundary layer thickness calculation continued through separated region.

The first combined TDC/boundary-layer calculation provided for maintaining displacement thickness constant beyond separation, should separation be encountered. As is seen from Figure 4, the boundary-layer corrections had little effect on the surface pressure distributions using this approach. When the same calculation was repeated with continuation of the boundary layer computation through the separation zone, however, a small but significant change in the inviscid flow pressure distribution resulted.

Figure 5 shows distributions of boundary layer displacement thickness, for both methods of handling separation. The distributions are similar up to the point of separation, after which the second approach computes a considerable rise in δ^* beyond the separation point. Also shown in Figure 5 is a boundary-layer calculation using an alternate method (method of Walz [8]), so that the

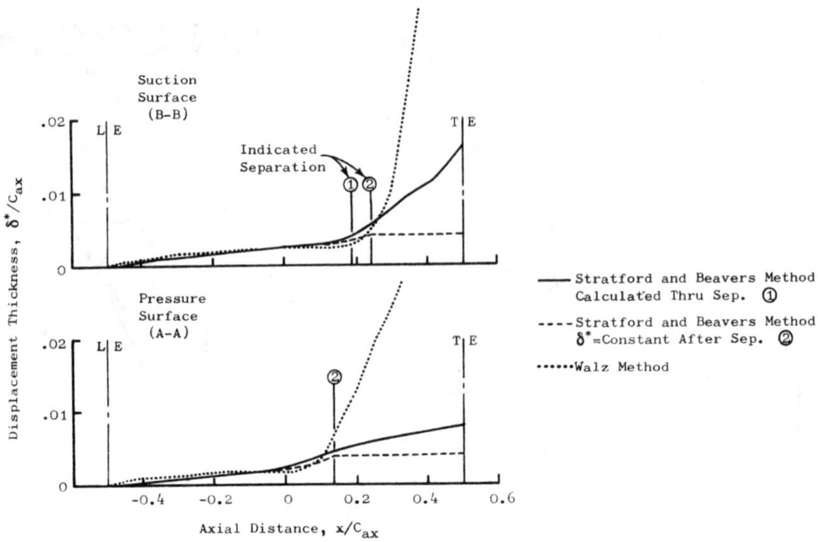

Figure 5. Transonic Fan Rotor Pitchline Section Cascade: Blade Surface Boundary Layer Displacement Thickness Distribution.

Stratford and Beavers prediction accuracy could be assessed. The Walz method is an integral momentum method with verified engineering accuracy. It is observed from Figure 5 that, up to the point of separation, the Stratford and Beavers method gives good results. The small deviations near the leading edge are due to the fact that the Stratford and Beavers method assumes a turbulent boundary layer starting at the leading edge and ignores the laminar region followed by transition to turbulent flow.

(b) Transonic Fan Tip Section Cascade:

A typical fan tip section was also analyzed. This cascade is characterized by a supersonic inlet relative Mach number, $M_1' = 1.5$, and a near-sonic exit relative Mach number. The airfoil section has relatively little camber, and the blades have high (61°) stagger. Figure 6 shows the surface pressure distributions obtained for the inviscid TDC solution with no boundary layer corrections; and for inviscid TDC solutions with boundary layer corrections, with the boundary layer calculation continued through the separated region.

The results shown in Figure 6 indicate that the boundary layer had little effect on the inviscid solution. Figure 6 also shows the boundary layer displacement thickness distributions obtained.

(c) <u>Supersonic Fan Pitchline Section Cascade</u>:

A pitchline section from an 1800 fps tip speed fan rotor was analyzed. This cascade has a relatively high inlet Mach number ($M_1' = 1.3$), but also considerable camber, and a subsonic exit relative Mach number ($M_1' = 0.7$). Figure 7 shows the surface pressure distributions obtained with and without boundary layer corrections. This particular case demonstrated a dramatic change in the inviscid flow field as a result of the boundary layer corrections.

The inviscid flow field without boundary layer corrections shows a two-shock system, with a weak oblique shock in the front of the blade passage and a strong normal shock standing near the exit of the blade passage. The weak oblique front shock actually is locally a normal shock where it stands off in front of the blade leading edge, due to the large leading-edge wedge angle and bluntness. Reacceleration of the flow around the leading edge, however, produces supersonic expansion waves which weaken the shock front and cause it to become oblique as it propagates across the passage. The flow after the shock reaccelerates to rather high velocity relative to inlet ($M = 1.4$) before forming a normal shock to meet the required back pressure.

The effect of the boundary layer corrections was, in this case, to move the rear normal shock farther forward in the passage. The shock is weaker because the flow has not reaccelerated as much before encountering the normal shock, and there is now substantial subsonic diffusion of the flow after the shock.

Shown in Figure 8 are static pressure contour plots for the solutions obtained (a) without boundary layer effects, and (b) with boundary layer effects, respectively. These figures dramatically illustrate the importance of including boundary layer effects for this particular configuration.

DISCUSSION

The results presented in Figures 4-8 demonstrate a successful inviscid/boundary-layer interaction calculation method for transonic

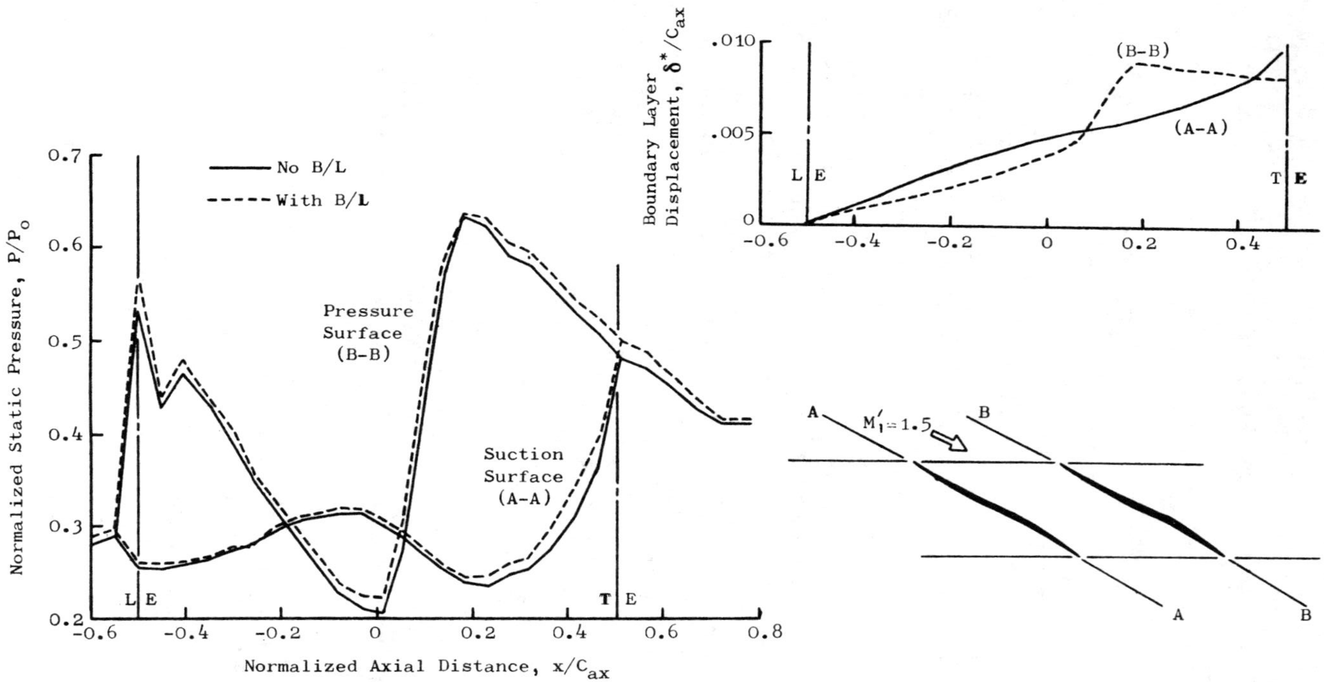

Figure 6. Transonic Fan Rotor Tip Section Cascade: Effect of Blade B/L on Surface Static Pressures.

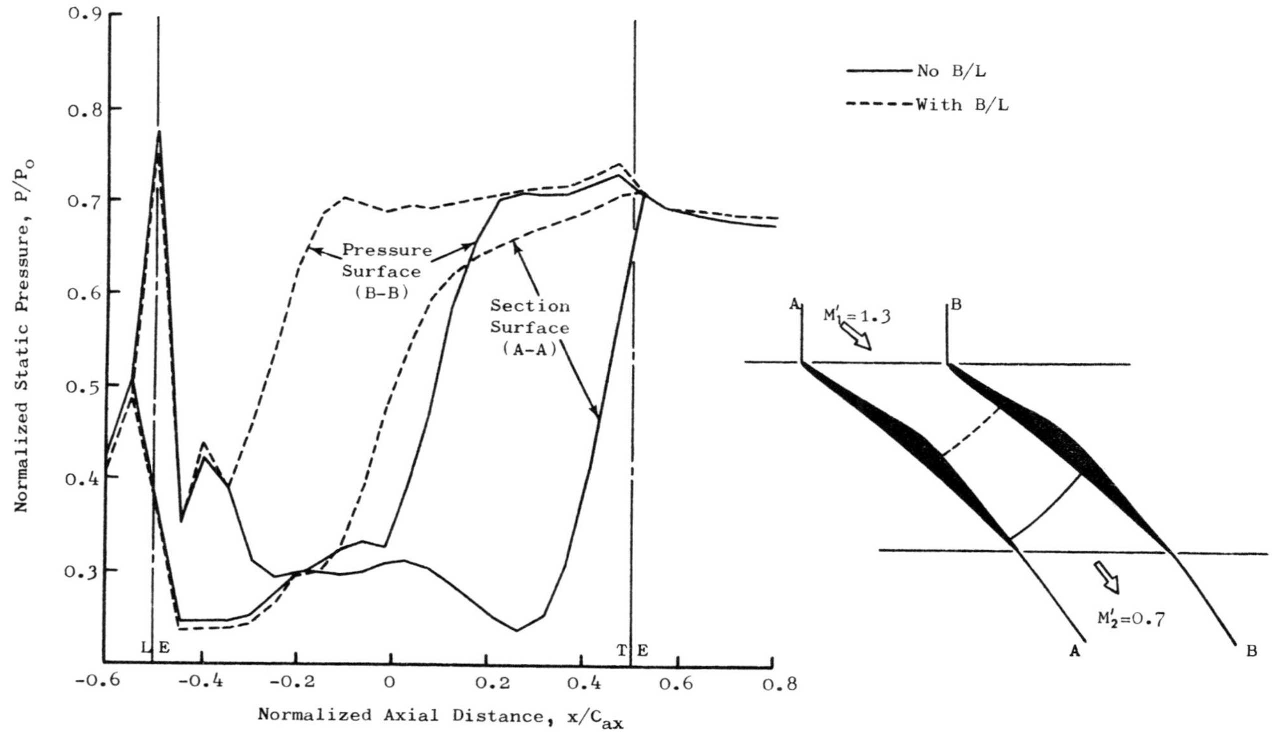

Figure 7. Supersonic Fan Rotor Pitchline Section Cascade: Effect of Blade B/L on Surface Static Pressures.

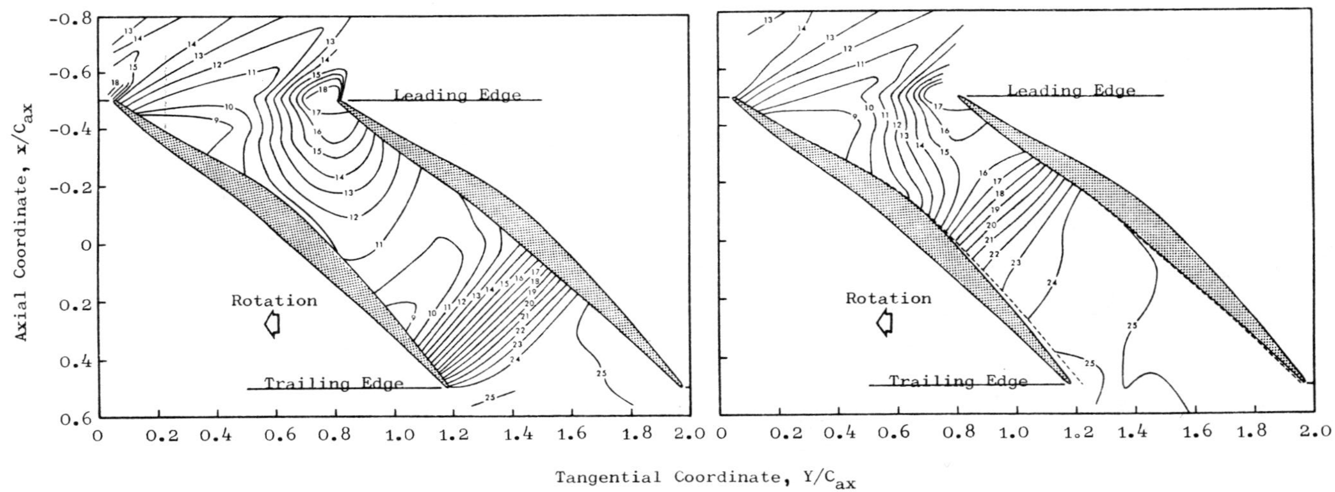

(a) Without Boundary Layer (b) With Boundary Layer

Figure 8. Supersonic Fan Rotor Pitchline Section Predicted Static Pressure Contours.

cascade flows. It had been speculated that the iterations between the inviscid calculation and the boundary-layer calculation would be divergent, especially in the supersonic flow regions. This speculation was based on previous experience with attempting inviscid-viscous interaction calculations, e.g., Reference [9], using conventional techniques.

Considering the supersonic flow over a concave surface, the boundary layer growth tends to increase the effective flow turning (compression), which in turn increases the adverse pressure gradient of the boundary layer, which causes a more rapid thickening of the boundary layer, which increases the effective flow turning, which in turn increases the adverse pressure gradient, etc., etc., to destruction of the calculation. This type of self-feeding calculation instability is at least partially a result of using a hyperbolic/parabolic computation method, where both external inviscid flow and boundary layer flow are calculated simultaneously by step-by-step marching downstream from an initial value datum line.

The combined calculation procedure does not exhibit this self-feeding instability for diffusing supersonic flows. The primary reason is that the inviscid TDC method permits feedback upstream through the propagation of unsteady waves. Also, the fact that the TDC method uses artificial viscosity for shock modeling provides some suppression of instability.

It is worth noting that, because the shocks are modelled as rapid but continuous changes in flow properties, the boundary layer calculation does not have to negotiate a discontinuity, thus enhancing the stability of the calculation procedure.

A physical explanation for the lack of apparent sensitivity of the inviscid flow to the boundary disturbances caused by the presence of the boundary layer is offered on two fronts. First, the flow in a cascade is (primarily) a channel type of flow, and hence the response of the inviscid "core" to a contraction in flow area due to boundary layer displacement is to accelerate, reducing adverse pressure gradients, and therefore thinning the boundary layer. Second, the inviscid flow is much more sensitive to change in blade surface slope ($\partial\delta^*/\partial S$) than to displacement ($\delta^*$), and these tend to be small except near separation. Some experimental evidence that boundary layer displacement effects on the inviscid flow are small is presented in Figures 9-11, taken from [10]. These figures show comparisons of equal-velocity contours from the prediction method described herein with laser-velocimeter

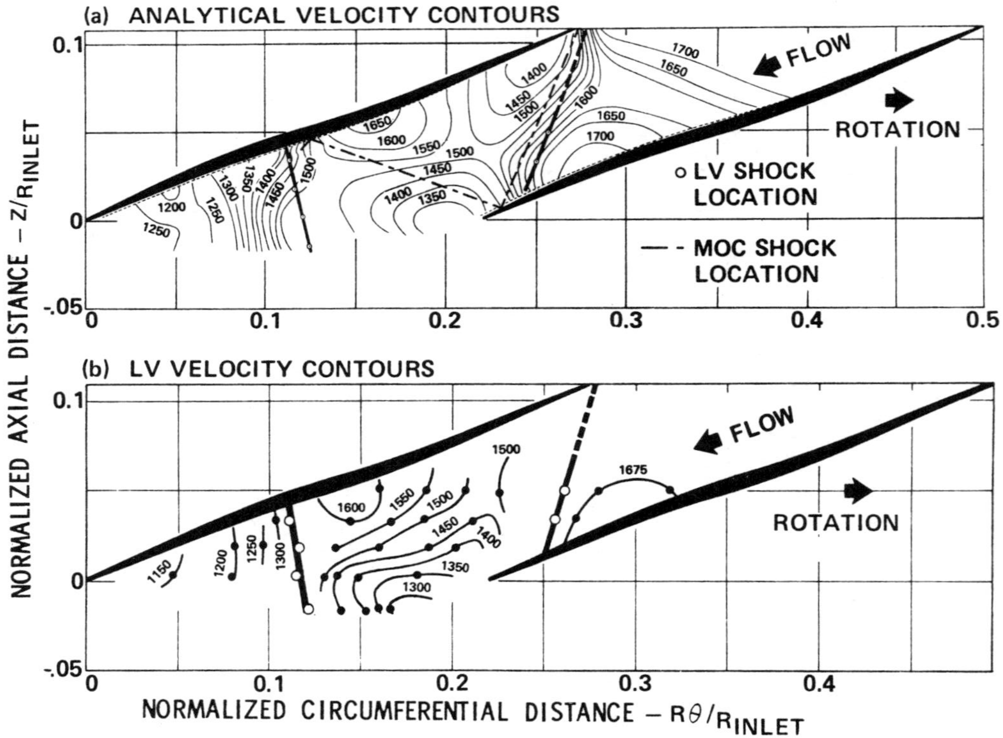

Figure 9. Comparison of Predicted and Measured Velocity Contours for a High Speed Fan Rotor Section, 10% Flow Stream Surface (Figure Taken from Ref. [10]).

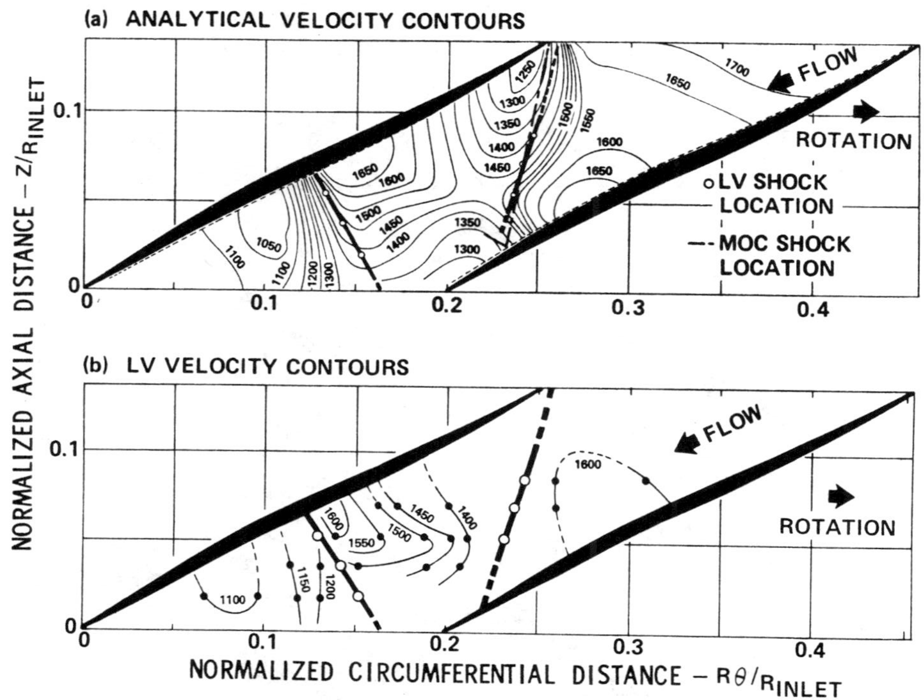

Figure 10. Comparison of Predicted and Measured Velocity Contours for a High Speed Fan Rotor Section, 30% Flow Stream Surface (Figure Taken from Ref. [10]).

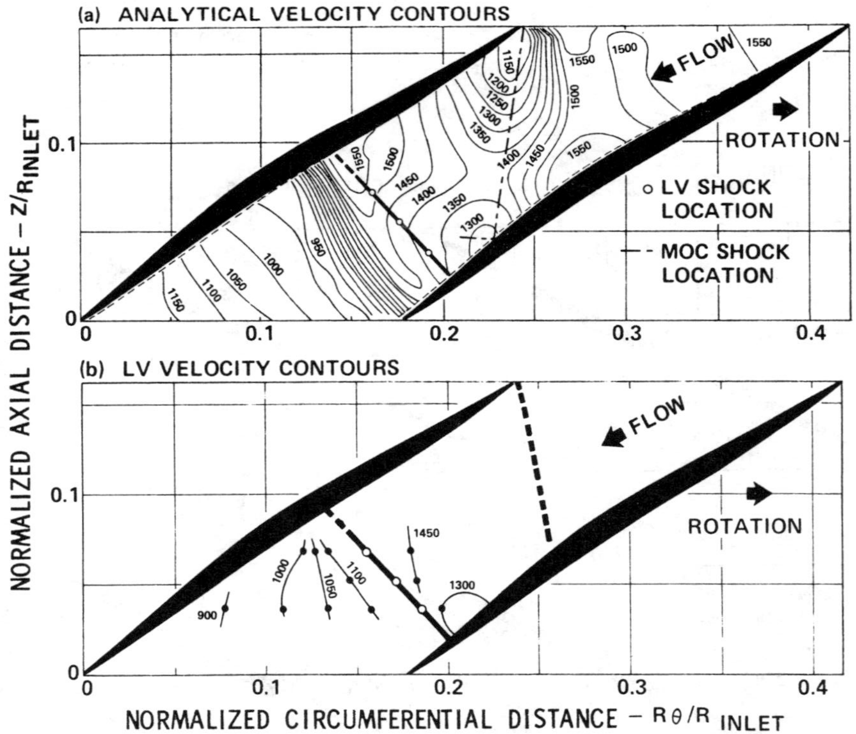

Figure 11. Comparison of Predicted and Measured Velocity Contours for a High Speed Fan Rotor Blade Section, 50% Flow Stream Surface (Figure Taken from Ref. [10]).

measurements. These measurements were taken on a high speed fan rotor at several spanwise locations. There seems to be generally good agreement with the measurements, and the solutions obtained with and without boundary layer corrections differed very little, on the order of that shown in Figure 6.

REFERENCES

[1] Gopalakrishnan, S. and Bozzola, R. (1971): "A Numerical Technique for the Calculation of Transonic Flows in Turbomachinery Cascades," ASME Paper No. 71-GT-42.

[2] Gopalakrishnan, S., and Bozzola, R. (1972): "Computation of Shocked Flows in Compressor Cascades," ASME Paper No. 72-GT-31.

[3] McDonald, P. W. (1971): "The Computation of Transonic Flow Through Two-Dimensional Gas Turbine Cascades," ASME Paper No. 71-GT-89.

[4] Gopalakrishnan, S., and Bozzola, R. (1973): "Numerical Representation of Inlet and Exit Boundary Conditions in Transient Cascade Flow," ASME Paper No. 73-GT-55.

[5] Stratford, B. S. and Beavers, G. S. (1959): "The Calculation of Compressible Turbulent Boundary Layers in an Arbitrary Pressure Gradient - A Correlation of Certain Previous Methods," ARC R&M No. 3207.

[6] Keith, J. S., et al. (1973): "Analytical Method for Predicting the Pressure Distribution about a Nacelle at Transonic Speeds," NASA CR-2217.

[7] Schlichting, H. (1960): Boundary Layer Theory, McGraw-Hill, New York, 4th edition.

[8] Felsch, K. O, Geropp, D., and Walz, A. (1968): "Method for Turbulent Boundary Layer Prediction," Proceedings of the Stanford Conference on Computation of Turbulent Boundary Layers, vol. 1.

[9] Werle, M. J., Dowyer, D. L., and Hankey, W. L. (1973): "Initial Conditions for the Hypersonic-Shock/Boundary-Layer Interaction Problem," AIAA Journal, 11:4, pp. 525-530.

[10] Wisler, D. C. (1976): "Shock Wave and Flow Velocity Measurements in a High Speed Fan Rotor Using the Laser Velocimeter," ASME Paper No. 76-GT-49.

DISCUSSION OF THE GLIEBE PAPER

OLIVER: In the second calculated case of the supersonic fan, which showed the most significant impact of boundary layer on the shock locations, do you have any comparisons with data for those cases? Also, I wonder if in the cases that you did show, the analytical velocity contours were indeed calculated with boundary layer corrections and compared more favorably than those without boundary layer corrections.

GLIEBE: For the case for which there was a dramatic change I do not have data. It happened to be a blade section, for the last three slides, which was farther down toward the hub and no measurements were made on that section. On the sections for which I showed Laser velocimeter measurements, the calculation was done with and without the boundary layer corrections, and in fact, the difference was very small. It was kind of like the difference you saw on the supersonic tip section cascade.

OLIVER: I assume in those calculations the Stratford-Beavers criterion did not indicate separation.

GLIEBE: I think in the last two calculations, in fact, the separation point was predicted. It happens close to the trailing edge so that it doesn't have a chance to affect the solution - the flow field - very much before exiting from the cascade.

FARN: Have you done any analysis for turbine blades? And, if yes, is the boundary layer, in fact, important?

GLIEBE: Yes, I have and no. There has been one case where it did make a difference. This is on a turbine rotor where the inlet Mach number was sufficiently high that a supersonic bubble occurred on the suction surface near to the leading edge, and the inviscid prediction of that supersonic bubble showed a much higher Mach number before the terminal shock than did the inviscid calculation; in that case there were surface pressure measurements from a cascade test which tended to verify the coupled boundary layer inviscid solution rather than the inviscid one alone.

FARN: When you put the boundary layer in, if you have shock waves impinging on the boundary layer and the displacement thickness is not smooth, do you have to numerically smooth it?

GLIEBE: It turns out that because of the artificial viscosity approach for modeling shocks, the pressure rises across the shock are not discontinuous and neither is the displacement thickness. It's a rapid but continuous change, so there's no need to smooth it.

FARN: Are you confident using combination boundary layer and inviscid solutions to predict the loss in turbine and compressor cascades?

GLIEBE: I don't think this procedure will predict the losses. I don't believe the shock losses that the inviscid calculation gives me because of the numerical errors that can be encountered in using the artificial viscosity approach. We have not tried to evaluate the boundary layer skin friction losses or the wake losses associated with the boundary layer, so I really can't give you an answer.

SESSION IV

Chairman:
ARTHUR J. WENNERSTROM
Wright Patterson Air Force Base

EXPERIMENT

REVIEW OF EXPERIMENTAL WORK ON TRANSONIC FLOW IN TURBOMACHINERY

WILLIAM D. McNALLY

NASA Lewis Research Center

In this paper I would like to review the status of experimental work on transonic flow in turbomachinery, which is being conducted throughout the United States and in some locations in Europe. If all the engine companies, government laboratories, smaller companies, and universities involved in turbomachinery research and development are considered, a tremendous amount of work is underway. However, in this Workshop we are primarily concerned with analytical methods for calculating internal flows in turbomachinery. Therefore, I will concentrate on the modern experimental techniques of high response and laser supported instrumentation. With these techniques, detailed steady and unsteady data can be obtained inside of and in the vicinity of transonic blade rows. Such data is sorely needed by the analyst for the verification of his computer codes.

A detailed picture of the general turbomachinery blade row problem is shown in Figure 1. The blade passage includes regions of relatively ideal flow, which are complicated by shocks off the blades and dampers, blade and end wall boundary layers, tip clearance flows, corner boundary layers and vortices, a variety of secondary flows, leading edge separation bubbles, trailing edge separation and flow deviation angle, slipstream and wake mixing regions, and unsteady effects due to blade row interaction, upstream distortion, and flutter. This is an extremely complicated flow situation, which presents many obstacles to both analysis and testing.

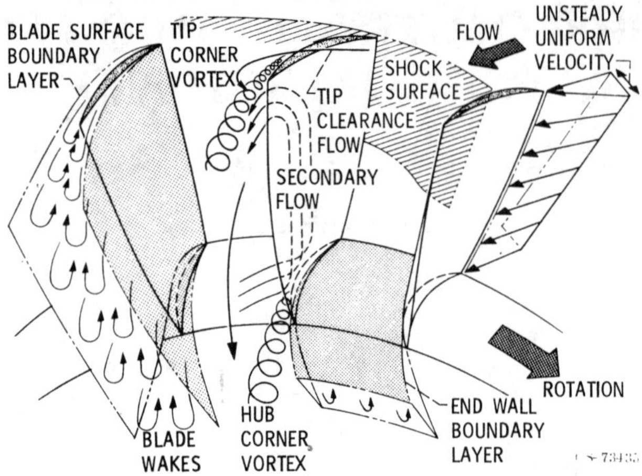

Figure 1. Rotor Blade Row Flow Phenomena

To design and analyze such machinery, a wide variety of techniques have been developed, many of which have been discussed at this workshop. These include velocity gradient and streamline curvature methods, subsonic 2-D stream function techniques, transonic time-marching or shock-capturing techniques, relaxation methods, floating shock-fitting approaches, 2-D and 3-D method of characteristics, perturbation methods on both linear and nonlinear equations, hodograph techniques, viscous marching solutions, and, recently, full Navier-Stokes solution. And these are only some of the approaches.

Faced with a variety of design and analysis techniques, the experimentalist should sense a very strong demand for good, thorough experimental data, both inside and in the vicinity of all types of turbomachinery blading. Such data is needed both to assist in the formulation of theoretical models, and also to verify the many analysis approaches which are being developed and applied.

Breakthroughs in modern experimental techniques have now made it possible to obtain some of the data which is required. I am speaking of such items as:

1. High Response Transducers
2. Hot Wire Probes
3. Hot Film Gages

4. Laser Velocimetry
5. Laser Fluorescence
6. Laser Holography

For what follows, I will report on how these techniques are being used to obtain transonic flow data in turbomachinery. There are many groups using these techniques, so the detail on any one of them will be minimal. However, in all cases, responsible individuals are listed and references given, and they can both be consulted for further details. Some individuals listed are not specifically doing transonic research, but are obtaining useful data for modeling flows or verifying analytical codes.

The first work discussed is that being conducted at the large engine companies: Pratt & Whitney (and United Technologies Research Center), General Electric, Allison, and AiResearch. This work is summarized in Table 1.

I will first discuss Pratt & Whitney. The overall picture on the use of high response instrumentation, hot wires, and hot films at Pratt & Whitney can be obtained from John T. Carroll, while detailed work on this instrumentation is being done by Howard Grant. Pratt & Whitney is making extensive use of Kulites in both component and multi-stage work. Kulites are used in probes downstream of components to determine the unsteady blade performance. They also have been used in the casing to do some mapping of shocks in the tip region. More recently, holograms have been used to determine the location of tip shocks and then LDV used to get quantitative values of velocity. Kulites are also mounted on the blade surfaces to measure both static and dynamic performance. Unsteady phenomena studied include rotating stall, the development of surge, and flutter. Surface-mounted Kulites have only been used in the front stage of machines because of the temperature sensitivity. In order to elevate the temperature range over which Kulites can be used, they are often water-cooled when used in the casing or in probes in the later stages.

Hot wires have been used for turbulence measurements in the front stages of machines and in cold combustion tests up to a Mach number of 0.5, and temperature of 533° K (500° F). Hot films are used on probes up to 0.5 Mach number and 644° K (700° F), and on blade surfaces to study boundary layer transition, separation, and shock location up to a Mach number of 1 and 755° K (900° F). Both the transducers and the hot films can function adequately up to 50,000 g's of centrifugal loading. Most of the details of this work,

TABLE I

High Response and LDV Testing by the Large Engine Companies

1. Pratt & Whitney -- High Response

J. T. Carroll H. Grant	Rig -- Component tests Tests -- Kulites in probes and rakes to determine unsteady performance Kulites in casings to map shock patterns and track rotating stall Kulites on blades for surge warning, and to study rotating stall and flutter Reports -- Some

2. Pratt & Whitney -- High Response

R. Mazzawy	Rig -- Component & single stage tests Tests -- Kulite and hot wire probes to determine unsteady performance with distortion Reports -- Some
R. Mazzawy	Rig -- Multistage tests Tests -- Kulites in probes & outer casings to study unsteady performance Hot films on walls & blades to study development of stall and surge Reports -- No

3. Pratt & Whitney -- High Response

H. Stargardter	Rig -- Fan flutter rig Tests -- Kulites on blades to study unsteady loading Hot films on blades to determine boundary layer separation Reports -- Some

TABLE I (Continued)

4. Pratt & Whitney -- Velocimeter

G. Alwang C. Williams	Rig -- Supersonic cascade and fan components Tests -- Two velocity component, within blade passages at several radial planes, to determine velocities and map shock structure Reports -- Some

5. United Technologies Research Center -- High Response

B. Johnson J. Bennett	Rig -- 5 Ft. subsonic rig Tests -- Kulite total pressure probes & hot wire probes to determine time dependent performance - pressure, velocity & flow angle Hot film gages to locate stall Reports -- Some

6. U.T.R.C. -- High Response

F. Carta	Rig -- 5 ft. subsonic rig Tests -- Kulites & hot films on blades, and Kulite & hot wire probes, to determine time dependent performance, both distorted and undistorted Reports -- No
F. Carta	Rig -- Flutter cascade, subsonic sections Tests -- Kulites & hot films on blades to study separation and high incidence stall flutter Reports -- Some

7. U.T.R.C. -- Velocimeter

J. Bennett B. Johnson K. Owen	Rig -- Supersonic cascade Tests -- Two velocity component, inside fan blading, to determine velocities and shocks Reports -- Some

TABLE I (Continued)

8. General Electric -- High Response

D. Prince	Rig -- Components, new fan stages and some inlet core stages Tests -- Kulites in casing over entire blade row, to map shock structure in tip region Reports -- Yes

9. General Electric -- High Response

M. Thomas	Rig -- Turbine components Tests -- Kulites in total and static probes behind turbine blades, to compare steady state and time-varying velocity diagrams Reports -- No

10. General Electric -- High Response

W. Steenken W. Tesch	Rig -- Full scale engines Tests -- Kulites in compressor face rakes to define inlet distortion in space and time Kulites in shroud above each stage, to study mismatch of stages as engine goes into surge Reports -- No

11. General Electric -- Velocimeter

D. Wisler M. Thomas	Rig -- Model compressor and turbine stages Tests -- Two velocity component, within blade passages, to determine velocities and map shock structure Reports -- Yes

TABLE I (Continued)

12. Allison -- High Response

R. Alverson	Rig -- Experimental high tip speed rotor Tests -- Kulites in shroud, to map shocks, detect rotating stall and surge, and identify flutter Reports -- No
R. Alverson	Rig -- Full scale engine Tests -- Kulites in casing & Kistler in probes, to study distortion & detect rotating stall & surge Reports -- Some

13. Allison -- Velocimeter

J. Fagan S. Fleeter	Rig -- Supersonic wind tunnel with transonic airfoil sections Tests -- Two velocity component, in interblade passages & blade exits, to define flow & measure turbulence intensities Reports -- Some

14. Allison - High Response

S. Fleeter	Rig -- 4 ft. subsonic rig Tests -- Kulites on blade surfaces to study blade loading Kulites in total pressure probes to study rotor wakes Reports -- No
S. Fleeter	Rig -- Flutter cascade Tests -- Kulites on blades to determine unsteady pressure with blades in flutter, or under forced vibrations Reports -- Yes

15. AiResearch -- High Response

E. Palmreuter	Rig -- Full scale engine Tests -- Kulites in casing above each blade row, to detect rotating stall and surge Reports -- No

particularly instrumentation development advances, are not reported in the open literature, although some information is given in contract reports. Some details on transducer and hot film development are contained in [1].

Robert Mazzawy has been responsible at Pratt & Whitney for using high response instrumentation to study the response of both single stage and multi-stage machines to inlet flow distortion. In a single stage rig with 0.7 hub-tip ratio and low camber blades, he has used both high response probes and hot wires, first with uniform inlet flow and then with screen generated distortion. He has obtained quantitative measurements of the non-steady losses and turning of blade rows under both undistorted and distorted conditions. Some details of this work were reported in [2].

Mazzawy is also studying responses to distortion in multistage rigs, with 7 to 12 stages. Blade and wall mounted hot films are used to obtain a qualitative measure of where rotating stall originates (usually in the front stages) and where surge begins (usually in the back stages). Attempts are then made to redesign the blading where rotating stall occurred, and consequently postpone surge. Kulites are used in this work in wall statics and in total pressure interstage rakes, and upstream and downstream of the engine. These give quantitative values used to evaluate unsteady performance. Generally, reports are not available on this multistage work.

Hans Stargardter is using high response instrumentation at Pratt in a 472 m/s (1550 ft/s) research fan rig to study flutter. Stall flutter is being investigated at about 70% speed with a tip speed of 305 m/s (1000 ft/s). Conventional strain gauges are used to measure the amplitude and frequency of vibration as well as the inter-blade phase angles. Blade mounted Kulites (6 on the suction side and 10 on the pressure side) are used to study the nonsteady surface pressure distribution both in and out of flutter. Blade surface hot films are also mounted at the same locations as the Kulites to determine nonsteady surface velocity and boundary layer separation point. Blade motion is measured by recording, with high speed motion pictures, the reflection of laser light from tiny surface-mounted mirrors and diffraction gratings. This blade motion work is reported in [3], and the other work will be reported in the future.

Laser doppler velocimeter work at Pratt & Whitney is under the direction of Gilbert Alwang. Carl Williams can also be contacted for details of the work. A 2-component LDV system was developed in a free jet, and later checked in a supersonic cascade with single-stage fan blading. Recently, the system has been applied to measure flows in a JT8D fan engine. The technique has also been used to study velocity levels and to measure turbulence intensity in experimental combusters. In fan blading the system is being used to determine the velocity field within the blade row and also to map the shock structure in the tip region. Tangential variation is obtained at each radial plane. Harry Griswold is comparing LDV measurements with results from analytical codes. Some aspects of the Pratt LDV system are reported in [4].

Alwang reports that although laser holography has been used at Pratt to obtain 3-D shock locations in fan blading, that at the present time LDV seems like a more fruitful technique. Therefore, the shift in emphasis is now strongly in that direction.

At United Technologies Research Center, a great deal of work is underway in addition to the program at Pratt & Whitney. In the area of high response instrumentation, both Bruce Johnson and John Bennett have been doing the development work. They have done much of this work on total pressure probes and hot wire probes in the 1.5 meter (5 ft) diameter, low-speed rotating rig with a single row of subsonic blading. Using Kulites and hot wires total pressures, velocities, and flow angles are obtained, and performance measurements are compared to those obtained with conventional instrumentation. Hot film gauges have been developed to check stall characteristics on both turbomachinery and helicopter blading. Some details of the hot film gage work are reported in [5], while the other development work has generally not been reported.

Frank Carta and his co-workers have been using high response instrumentation to study both distortion and flutter. In the 1.5 meter (5 ft) diameter, low speed rig with a rotor alone, he is using Kulite transducers and hot films on blade surfaces, and also hot wire probes in the interblade regions, to get a basic understanding of the blade surface and interblade flows under both distorted and undistorted conditions. Tests will be run in three stages: first, with an empty rig plus distortion screens in order to determine circumferential velocity variations in the incoming flow generated by the screens alone. Second, screens will be pulled and a rotor-alone test performed to obtain undistorted pressure-ratio weight-flow characteristics for several rotor speeds. Finally, screens will be added again to apply the distortion to the rotor and to take unsteady aerodynamic information in the passages and wake regions along several speed lines. An attempt will be made to get the blades into rotating stall.

A triaxial hot wire probe with total pressure Kulite directly above it can be traversed circumferencially and radially within one-half chord of the trailing edge. Six Kulites on both sides of a single blade will be used to calculate lift and moment. Neighboring blades will have conventional static taps in the same locations to obtain steady state time averaged values. The unsteady components from the Kulites will be superimposed on these steady state values.

Carta is also using Kulites and hot films on blade surfaces in the UTRC oscillating flutter cascade where he is studying subsonic compressor blade sections. Twenty Kulites are mounted on one blade at center span in order to study unsteady effects as the blades are driven in a prescribed pitching motion. Hot films are used to study transition and separation. Some of this work has been reported in [6].

Laser doppler velocimetry techniques are being developed and applied at UTRC by Bruce Johnson, John Bennett, and Kevin Owen. A two-component system is being used presently and a 3-component system is under development for use in the 1.5 meter (5 ft) rig. The 2-component system has been used extensively in the 1.2 meter x 1.8 meter (4 x 6 ft) subsonic tunnel to study flow fields about helicopter blading [7]. It has also been used heavily in combustion work [8, 9, 10]. The system has been used to study a subsonic fan, [11], and is now being applied in the supersonic cascade tunnel to obtain velocities and map shock patterns in transonic fan blading. This later work is not yet reported.

The General Electric effort is quite similar to that going on at Pratt & Whitney and likewise concentrates on the use of high response instrumentation and LDV. David Prince and Marshall Thomas are the people most responsible for the development and use of high response techniques at G.E. Prince has been using Kulites in the casings above all new fan stages (except where tip shrouds interfere) and in some inlet core stages to predict shock structure in the tip region [12]. Seven or eight Kulites are mounted directly above the blade tip, with 3 more upstream to determine the apparent inflow to the blade, and another pair downstream to study the relative outflow. Of the 8 Kulites above the blades, at least 3 are needed near the front of the blade to distinguish between normal and oblique shocks at the passage entrance. Some results of this work have been reported by G. E. [13] and also by NASA [14].

Curt Koch at G. E. is responsible for some testing in a 1.5 meter (5 ft) diameter low speed multi-stage rig where high response instrumentation is being used with 46 m/s (150 ft/s) tip speed blades. A rotating transducer, moving with the rotor, is used to study wake structure behind the blades. Surface mounted hot films are used to study boundary layer transition and separation.

Marchall Thomas of G. E. has been using Kulites in probes downstream of various turbine rotors and stages with high subsonic outlet flow. Included are a high pressure turbine stage for the Air Force, an intermediate pressure ratio NASA core stage, and a low pressure ratio but highly loaded research turbine stage. Time varying velocity diagrams and flow angles are being determined at various radial locations and circumferential positions relative to the vane trailing edges at a short distance downstream of the vanes. Some of this work will be reported soon for NASA, but the work for the Air Force is proprietary.

William Steenken and William Tesch of G. E. have been using high response instruments to study the development of stall and surge in both military and commercial full-scale machines. Kulites are used in the walls (2 per stage) and in compressor face rakes to study these transients in somewhat the same way as they are studied by Mazzawy at Pratt. All of this work is proprietary.

David Wisler and Marshall Thomas are developing and applying the laser velocimetry system at G. E. Wisler has been using LDV on a scale model of the rotor for a 550 m/s (1800 ft/s) tip speed fan stage. An improved LV optics system and signal processor have been developed.

Velocities are determined and shock structure is mapped at 10, 30 and 50% immersion from the casing both near stall and along the operating line. The instrumentation is set to record at 8 axial locations on each radial plane, and a continuous tangential variation is obtained as the blades sweep by. Shock jump conditions and change in angle across shocks can be determined with this 2-component system. The results are being used to verify analytical codes used in design. Verification of the LDV system is reported in [15], while LDV measurements on the scale model compressor and comparisons with theory are reported in [16]. Work is also underway at G. E. on a 3-component LDV system.

Marshall Thomas is using the G.E. LDV system on an Air Force high pressure turbine stage. Measurements are taken at 10 axial locations at the rotor midspan, and again a continuous tangential variation of velocity is obtained. Details of this work for the Air Force are proprietary.

Allison is likewise doing high response and LDV work, and applying these techniques in S. Fleeter's flutter cascade. Richard Alverson, working with a high-tip-speed experimental rotor, has been using Kulites mounted in the shroud primarily to detect rotating stall and surge. He is also using Kulites to some degree to map shock patterns and identify the onset of flutter. He has used Kulites in the shroud and Kistlers in probes to study distortion effects and to detect rotating stall in a full scale machine. Some of the full scale engine work is reported in [17], while the high speed rotor work for the Air Force is classified.

John Fagan and Sanford Fleeter have been working on the development of an LDV system at Allison. They are developing this 2-component system in their supersonic wind tunnel on a cascade of compressor airfoils which represent the 90% span section of some 550 m/s (1800 ft/s) tip speed blading. Only inter-blade passage and blade exit surveys have been done to date, with no in-blade work yet. They do not plan 3-component work in the near future. The system is also being used in the Allison heat transfer cascade facility to measure turbulence intensities in high temperature regions. The wind tunnel cascade work is reported in [18].

Fleeter at Allison has been using high response instrumentation in several ways. He will be using Kulites in dynamic total pressure probes to study rotor wakes in Allison's 1.2 m (4 ft) diameter subsonic rig. Twenty surface Kulites will be mounted on this low speed blading near mid-span to study blade loading. In his flutter cascade, Fleeter uses strain gauges to measure blade deflections. He uses Kulites on the blade surfaces (6 per surface, or 12 on a blade) to study the time dependent pressure distribution with the blades harmonically oscillated in a torsional mode or undergoing forced vibrations from an oscillating inlet flow. Hot films are also used on the blade surfaces to detect boundary layer separations and transition. This cascade work is well documented in [19, 20, 21].

AiResearch is using some high response instrumentation in their multi-stage axial compressor work. E. Palmreuter has mounted Kulites in the shroud near leading and trailing edges of each blade row to detect both rotating stall and surge. This work, however, is not reported.

The holography work at AiResearch, done for NASA on the 489 m/s (1600 ft/s) lightly loaded fan stage, has not been continued, but is thoroughly reported in [22, 23]. LDV work is just getting underway at AiResearch.

Many smaller companies are also involved in turbomachinery development of one form or another. The work being done by some of these companies is summarized in Table II.

Among the smaller companies, one is doing significant LDV development, and that is Creare Corporation. Peter Runstadler is in charge of this work, where 2-component LDV is being used to study the flow in the entrance tip region of the inducer and also in the vaneless space at impeller discharge of an 8:1 centrifugal compressor. LDV will be used to define the impeller exit flow and the flow about the diffuser leading edge and in the diffuser throat. This will be done for several settings of the impeller relative to the diffuser. The Creare LDV system is thoroughly developed and has been reported [24]. LDV data from the compressor will be reported in the near future.

Gary Ludwig at Calspan has been making extensive use of crossed (2-component) hot wire probes to study rotating stall phenomenon in the reworked front stages of a J79 engine, run at low speed. Blade tip speeds do not exceed 60 m/s (200 ft/s). Radial and circumferential distributions of axial and tangential velocity and flow angle are obtained both upstream and downstream of rotors and stators. Factors such as inception of rotating stall, number of stall cells, propagation speed, and the influence of distortion on stall are studied. Several piezoelectric transducers, mounted in the shroud, have also been used in this work to study the inception and propagation of rotating stall. Reference [25] is an older report on this work; several other reports will be published in 1976.

At Northern Research, Willem Jansen has used both hot wires and Kulites in work being done on centrifugal compressors in the 6:1 to 9:1 range. Kulites are used in a probe at the impeller exit to study velocity and angle fluctuations. Hot wires are used at the inlet and at vaned diffuser discharge to detect rotating stall and surge in flows with velocities up to 180 m/s (600 ft/s). Some of this work is reported in [26].

TABLE II

High Response and LDV Testing by Smaller Companies

1. Creare -- Velocimeter

P. Runstadler	Rig -- Centrifugal compressor Tests -- Two velocity component, in inducer inlet, impeller exit, & diffuser inlet, to study entrance flow, transition flow from impeller to diffuser, & diffuser flow Reports -- Yes

2. Calspan -- High Response

G. Ludwig	Rig -- Full scale engine Tests -- Transducers in shroud, to detect rotating stall Reports -- Yes

3. Northern Research -- High Response

W. Jansen	Rig -- Centrifugal compressor Tests -- Kulites in probe at impeller discharge, to determine velocities and angles Hot wires near vaned diffuser, to detect rotating stall & surge Reports -- Yes

Several universities are doing very good work using both high response instrumentation and LDV. These efforts are summarized in Table III. One of the universities, MIT, has pioneered an entirely new technique, laser fluorescence, for both qualitative and quantitative measurements within blade rows.

The blade-down compressor test facility at MIT, under the direction of Jack Kerrebrock, is now quite well known [27]. Kulite high response transducers are used in total and static pressure probes and in the outer casing at several locations in front of and behind a single test rotor. The propagation of shock waves and noise from the front end of the compressor has been extensively studied as well as the decay and merging of wakes downstream from the blade trailing edge [28]. The new laser fluorescence technique in which sheets of laser light are used to fluoresce the argon-freon test gas, seeded with butane-dione, permits the visualization of shock waves and boundary layer regions, as well as an accurate quantitative mapping of the density field within the blade row [29, 30].

At Virginia Polytechnic Institute, Walter O'Brien is using Kulite high response instrumentation on axial fan blade rows in a low speed rig, to study blade response to circumferential inlet flow distortion. Shortly this work will be extended in a high speed rig. Five Kulites have been used on the blade surfaces at mid-span, and the data transmitted from the rotating machinery using FM radio telemetry techniques. This work has been thoroughly reported [31 through 35].

At Penn State, Robert Henderson is likewise studying the response of a compressor axial flow fan blade row to inlet circumferential distortions. Henderson's rig permits testing with a full stage, with rotor-stator spacings varying up to 2 chord lengths, and rotor tip speed up to 100 m/s (325 ft/s). Only rotor alone tests are reported to date [36, 37]. The unsteady normal force and pitching moment on a chordwise element of the rotor blade is determined from strain gage readings on a torque tube and force cube built into the blade. Tests cover sinusoidally varying circumferential distortions with different numbers of distortion cycles.

Also at Penn State, B. Lakshminarayana has been using triaxial hot wire probes to study the wake development behind compressor rotor blade rows. The probes have been used in a stationary mode [38] and also rotating with the blades [39] in the region from the trailing edge to one chord downstream. Probe measurements

TABLE III

High Response and LDV Testing at the Universities

1. Massachusetts Institute of Technology -- High Response

J. Kerrebrock W. T. Thompkins	Rig -- Blowdown rig, axial rotor Tests -- Kulites in total and static probes, & Kulites in casing, to determine shocks & noise out front end, & study wake decay downstream Reports -- Yes

2. Massachusetts Institute of Technology -- Laser Fluorescence

J. Kerrebrock A. Epstein	Rig -- Blowdown rig, axial rotor Tests -- Photograph laser-illumunated fluoresced gas, to study shocks & boundary layers, & to map density through the blades Reports -- Yes

3. Virginia Polytechnic Institute -- High Response

W. O'Brien	Rig -- Compressor rotor Tests -- Kulites on blade surfaces, to determine unsteady response to circumferential distortion Reports -- Yes

4. Penn State -- High Response

R. Henderson	Rig -- Compressor rotor Tests -- Strain gages built into blades to determine unsteady response to circumferential distortion Reports -- Yes

TABLE III (Continued)

5.	Penn State -- High Response	
	B. Lakshminarayana	Rig -- Low speed compressor blades Tests -- Triaxial hot wire probes to study wake development and boundary layers Reports -- Yes
6.	Naval Postgraduate School -- High Reponse & LDV	
	R. Shreeve D. Collins	Rig -- Transonic compressor stage Tests -- Kulite & hot wire probes, & Kulites in casing, to determine performance and study shock structure Two velocity component LDV within blade rows, to determine velocity levels & study shocks Reports -- No
7.	Purdue University -- Velocimeter	
	H. D. Thompson	Rig -- Two-dimensional wind tunnel Tests -- Two velocity component, to define flow and compare to analysis Reports -- Yes

permit the determination of instantaneous velocities, turbulence intensity, and Reynolds stresses in all three directions. These measurements have been made in a low-speed single-stage compressor rig with heavily loaded blading and a tip speed of 76 m/s (250 ft/s), and also in an axial flow fan rig with a single row of low-loaded fan blades and a tip speed of 30 m/s (100 ft/s). In the single stage compressor rig, the rotating probe measurements will soon be made with the probe inside the blade row [40] to study the end wall boundary layer development.

At the Naval Postgraduate School, work has begun using both Kulites and LDV to study the flow through a transonic compressor stage. In work under the direction of Raymond Shreeve, 26 Kulites have been mounted in the casing above the entire stage in order to map shock structure and define the flow in the rotor. A Kulite probe and hot wires will also be used downstream of the rotor. There are no plans to use hot films or Kulites on the blade surfaces. Daniel Collins of the Postgraduate School also plans to make LDV measurements in this compressor stage. None of this work is reported yet.

At Purdue University, H. D. Thompson has been making LDV measurements in a 2-D laboratory wind tunnel using various nozzle configurations. These have compared favorably with other measurements and with analytical solutions [41 - 43].

Finally, Table IV summarizes the LDV and high response work being conducted at the various government laboratories here and abroad. The most thorough program using these techniques is the one underway in Germany at the DFVLR by Doctors Weyer, Eckardt, and Schodl. This work is being done on both axial and centrifugal compressor rigs. On the axial machines, Weyer and his co-workers have used Kulites in the casings to define shock patterns and in total pressure probes to study wave development [44].

In a centrifugal compressor with vaneless diffuser region, both high response instrumentation and laser velocimetry have been used. Eckardt has mounted transducers in the casing within the rotor region and on the front and back walls of the diffuser to measure static pressure. Transducers are also used in total pressure probes downstream of the impeller in the beginning of the vaneless diffuser region in order to study the jet-wake flow. This unsteady pressure is imposed on a time-averaged value obtained with a special pitot-probe attached to a standard pressure

TABLE IV

High Response and LDV Testing at Government Laboratories

1. DFVLR -- High Response

H. Weyer D. Eckardt	Rig -- Axial compressor rigs Tests -- Kulite type in walls above rotors, to define shock patterns Kulite type in probes behind rotors, to study wake development Reports -- Yes

2. DFVLR -- High Response

D. Eckardt	Rig -- Centrifugal compressor rig Tests -- Kulite type in casing to give static pressure Kulite type in probes downstream of impeller, to study jet-wake flow Shielded hot-wire probe, to study flow angle in jet-wake region Reports -- Yes

3. DFVLR -- L2F Velocimeter

R. Schodl D. Eckardt	Rig -- Centrifugal compressor rig Tests -- Two velocity component, at five cross-channel planes through the compressor, to define the flow, and the jet-wake region Reports -- Yes

4. AFAPL -- High Response

A. Wennerstrom	Rig -- Axial compressor rig Tests -- Kulites in casing, to map shock patterns Reports -- No

TABLE IV (Continued)

5. NAVY Propulsion Test Center -- High Response

T. Bogia	Rig -- Full scale engine test Tests -- Kulites in total pressure rakes in inlet to study circumferential and radial distortion Reports -- No

6. NASA Lewis -- High Response

W. Britsch	Rig -- Single stage compressor rig Tests -- Kulites in probes behind rotors, to define unsteady performance Hot wire probes behind rotors, to study unsteady velocity & flow angle Hot film probes before rotors to track stall patterns Reports -- No

7. NASA Lewis -- High Response

J. Adamczyk W. Stevans	Rig -- Flutter rigs Tests -- In-house and contract tests, to study various forms of flutter Reports -- No

8. NASA Lewis -- High Response

T. Moffitt	Rig -- Turbine test facility Tests -- Kulites in probes behind rotor, to study secondary flows & wake history Reports -- No

9. NASA Lewis -- Velocimeter

R. Seasholtz L. Goldman	Rig -- Turbine annular cascade Tests -- Two velocity component, behind vanes, to measure performance, and define wake regions Reports -- Yes

transducer. In order to study the strongly fluctuating flow angle at the impeller exit, a special shielded hot-wire probe is used which protects the wire from the destructive jet flow. This probe also permits location of the wakes relative to the blade tips at the impeller exit. This high-response work is reported in [45 - 46].

Also, at DFVLR, R. Schodl and D. Eckardt are applying Schodl's L2F laser velocimeter technique [47, 48] to study the flow in this same centrifugal compressor [49]. Two-component LDV measurements have been obtained at 5 different optical measuring cross-channel planes through the compressor from the inducer leading edge to the impeller exit. These planes are at 8, 43, 59, 87 and 101% along streamlines in the meridional plane. The complete cross-channel velocity profiles are given at these 5 locations and clearly show the wake region from the 59% profile on. Secondary flow patterns have been studied as well as the jet-wake interaction.

A small amount of high response work has gone on at both the Air Force Aero-Propulsion Lab and at the Navy Propulsion Test Center. At AFAPL, A. Wennerstrom was responsible for in-house tests where Kulites were placed above the rotor blades in a single stage compressor rig. These were used to map shock patterns and to detect the presence of stall. This work is not reported. T. Bogia at the Navy has been using Kulites in total pressure rakes to study time-varying radial and circumferential distortions coming into full stage engines. This data is all proprietary. No LDV work is being done on turbomachinery at either of these centers.

Finally, at NASA Lewis Research Center, both high response and LDV work are underway. In the single stage compressor rig at NASA, Werner Britsch is developing 3-Kulite probes to measure total pressure and angle behind the rotors. Hot wire probes are also being developed for use behind rotors to measure velocities, flow angle and temperature rise. Hot films are used on probes in front of the rotor to determine stall patterns and to track rotating stall. Kulites will be used on the blade surfaces in the future. None of this work is reported.

John Adamczyk and William Stevens of the Compressor Branch at Lewis are setting up a test program to study various types of flutter. This work will begin soon and will obviously make use of various forms of high response instrumentation.

Thomas Moffitt of the Turbine Branch at NASA will be using

Kulites in probes to survey on several axial planes in front of the stator and behind the rotor of a single stage high pressure core turbine to be tested soon in NASA's large turbine test facility. A study of the wake history behind this high-radius-ratio and low-aspect-ratio turbine with its large secondary flows will be made. This work will begin soon.

LDV work is underway at Lewis by Richard Seasholtz and Louis Goldman in NASA's turbine stator annular cascade facility. Two-component LDV is being used [50] to measure the performance behind the stator vane from a Pratt J75 engine against the performance determined by conventional instrumentation [51]. A discrete point method was used initially but has now been replaced by a much faster continuous scan method. The vane from the high pressure core turbine mentioned above will be tested this year in the annular cascade using LDV. Two-component, and eventually 3-component, LDV will be used to define the flow behind these vanes and later both types of LDV will be used within the blade rows. An LDV system is also under development at Lewis for the single stage compressor test facility.

That concludes my review, in which I am certain some groups were neglected; I hope none mentioned were misrepresented. From my background in analytical code development, I am most impressed with the need for experimental data to verify these codes. After compiling this review, I am likewise impressed with the fact that this data is now on its way. I am most impressed in this regard with the work being done at the major engine companies, Creare, DFVLR and MIT. As far as techniques are concerned, it seems obvious that LDV is the most promising technique for the near future for the quantitative mapping of internal flow and shock structure within blade rows.

REFERENCES

[1] Carroll, J. T. (1973): "Dynamic Flow Studies by Application of Pressure and Velocity Sensors on Axial Flow Fan Blades," Instrumentation for Air Breathing Propulsion. Progress in Aeronautics and Astronautics, vol. 34, MIT Press.

[2] Mazzawy, R. S. (1975): "Multiple Segment Parallel Compressor Model for Circumferential Flow Distortion," Proceedings of AGARD Conf. on Unsteady Phenomena in Turbomachinery, AGARD-CPP-177, September.

[3] Stargardter, H. (1976): "Optical Determination of Rotating Fan Blade Deflections," ASME 76-GT-48, ASME International Gas Turbine Conference, New Orleans.

[4] Walker, D. A., Williams, M. C. and House, R. D. (1975): "Intra-Blade Velocity Measurements in a Transonic Fan Utilizing a Laser Doppler Velocimeter," Proceedings of the Minnesota Symposium on Laser Anemometry, October.

[5] Bellinger, E. E., Patrick, W. P., Greenwald, L. E. and Landgrebe, A. J. (1974): Experimental Investigations of the Effects of Helicopter Rotor Design Parameters on Forward Flight Stall Characteristics. USAAMRDL-TR-74-1, April.

[6] Carta, F. O. (1975): Unsteady Surface Probe Behavior on a Cascade of Airfoils Oscillating Below Stall. Project SQUID Technical Report UTRC-1-PU, September.

[7] Johnson, B. V. (1974): "LDV Measurements in the Periodic Velocity Field Adjacent to a Helicopter Rotor," Proceedings of 2nd International Workshop on Laser Velocimetery, Purdue University, March.

[8] Owen, K. (1975): "Laser Velocimeter Measurements in Free and Confined Coaxial Jets with Recirculation." AIAA Paper 75-120, January.

[9] Owen, K. (1976): "Laser Velocimeter Measurements of a Confined Turbulent Diffusion Flame Burner." AIAA Paper 76-33, January.

[10] Owen, K. (1976): "Simultaneous Laser Velocimeter and Laser Induced Photo-Luminescence Measurements of Instantaneous Velocity and Concentration in Complex Mixing Flows." AIAA Paper 76-35, January.

[11] Magliozzi, B., Hanson, D. B., Johnson, B. V. and Metzgar, F. B. (1973): Noise and Wake Structure Measurements in a Subsonic Tip Speed Fan. NASA CR-2323, November.

[12] Prince, D. C. (1975): "Some Flow Pattern Features in Supersonic Rotor Cascades," Univ. of Tennessee Space Institute Short Course on Aero Propulsion, November.

[13] Bilwakesh, K. R., Kocl, C. C. and Prince, D. C. (1972): Evaluation of Range and Distortion Tolerance for High Mach Number Transonic Fan Stages. NASA CR-72880, June.

[14] Miller, G. R. and Bailey, E. E. (1971): Static Pressure Contours in the Blade Passage at the Tip of Several High Mach Number Rotors. NASA TM S-2170, February.

[15] Wisler, D. C. and Mossey, P. W. (1973): "Gas Velocity Measurements within a Compressor Rotor Passage using the Laser Doppler Velocimeter." Journal of Engineering for Power. Trans. of the ASME, Series A, vol. 95, April.

[16] Wisler, D. C. (1976): "Shock Wave and Flow Velocity Measurements in a High Speed Fan Rotor using the Laser Velocimeter," ASME 76-GT-49, ASME International Gas Turbine Conference, New Orleans.

[17] Korn, J. A. and Garzon, J. R. (1972): Circumferential and Radial Total Pressure Distortion Testing of the TF-41-LP-IP Compressor System. Allison EDR-7519.

[18] Fleeter, S., Holtman, R. L., McClure, R. B. and Sinnet, G. T. (1975): Experimental Investigation of a Supersonic Compressor Cascade. ARL-TR-75-0208, June.

[19] Fleeter, S., Holtman, R. L., McClure, R. B. and Sinnet, G. T. (1975): "Supersonic Inlet Torsional Cascade Flutter," AIAA Journal of Aircraft, 12:8, August.

[20] Fleeter, S., Novick, A. S. and Riffel, R. E. (1975): "The Unsteady Aerodynamic Response of an Airfoil Cascade to a Time Varying Supersonic Inlet Flow Field," Proceedings of AGARD Conference on Unsteady Phenomena in Turbomachinery, AGARD-CPP-177, September.

[21] Fleeter, S., Novick, A. S. and Riffel, R. E. (1976): "Experimental Determination of the Unsteady Aerodynamics in a Controlled Oscillating Cascade," ASME 76-GT-17, ASME International Gas Turbine Conference, New Orleans.

[22] Wuerker, R. F., Kobayashi, R. J., Heflinger, L. O., and Ware, T. C. (1974): Application of Holography to Flow Visualization within Rotating Compressor Blade Row. NASA CR-121264, February.

[23] Benser, W. A., Bailey, E. E. and Gelder, T. F. (1975): "Holographic Studies of Shock Waves within Transonic Fan Rotors," Journal of Engineering for Power, Trans. of ASME, January.

[24] Runstadler, P. W. and Dolan, F. X. (1975): "Design, Development and Test of a Laser Velocimeter for High Speed Turbomachinery," Proceedings of Laser Doppler Anemometer Symposium, Copenhagen, Denmark, August.

[25] Ludwig, G. R., Menni, J. P. and Arendt, R. H. (1973): "Investigation of Rotating Stall in Axial Flow Compressors and the Development of a Prototype Rotating Stall Control System." AFAPL-TR-73-45, May.

[26] Northern Research and Engineering Corporation (1974): Improvements in Surge Margin and Diffuser Performance. Report no. 950-350, March.

[27] Kerrebrock, J. L., Epstein, A. H., Haines, D. M. and Thompkins, W. T. (1974): "The M.I.T. Blowdown Compressor Facility," Journal of Engineering for Power, Transactions of ASME, October.

[28] Thompkins, W. T. and Kerrebrock, J. L. (1975): "Exit Flow from a Transonic Compressor Rotor," Proceedings of AGARD Conference on Unsteady Phenomena in Turbomachinery, AGARD-CPP-177, September.

[29] Epstein, A. H. (1975): "Quantitative Density Visualization in a Transonic Rotor," AIAA Paper 75-24, AIAA 13th Aerospace Sciences Meeting, Pasadena, California, January.

[30] Epstein, A. H. (1975): Quantitative Density Visualization in a Transonic Compressor Rotor." M.I.T. Gas Turbine Laboratory Report no. 124, September.

[31] Sexton, M. R., O'Brien, W. F. and Moses, H. L. (1973): "Pressure Measurements on the Rotating Blades of an Axial-flow Compressor," ASME Paper 73-GT-79.

[32] O'Brien, W. F., Moses, H. L., and Carter, H. R. (1974): "A Multi-Channel Telemetry System for Flow Research on Turbomachine Rotors," ASME 74-GT-112.

[33] Moses, H. L. and O'Brien, W. F. (1974): "The Chordwise Pressure Distribution on a Rotating Axial Flow Compressor Blade," Proceedings of the 2nd International Symposium on Air Breathing Engines, Sheffield, England, March.

[34] Sexton, M. R., O'Brien, W. F. and Moses, H. L. (1975): "An On-Rotor Investigation of Rotating Stall in an Axial-Flow Compressor," Proceedings of AGARD Conference on Unsteady Phenomena in Turbomachinery, AGARD-CPP-177, September.

[35] Neal, T. E. (1975): "The Lift Response of a Rotating Blade of an Axial Flow Compressor to Inlet Flow Distortion," M.S. Thesis, Virginia Polytechnic Institute and State University, August.

[36] Bruce, E. P. and Henderson, R. F. (1975): Axial Flow Rotor Unsteady Response to Circumferential Inflow Distortions. Project SQUID Technical Report, PSU-13-PU, September.

[37] Bruce, E. P. and Henderson, R. E. (1975): "Axial Flow Rotor Unsteady Response to Circumferential Inflow Distortions," Proceedings of AGARD Conference on Unsteady Phenomena in Turbomachinery, AGARD-CPP-177, September.

[38] Lakshminarayana, B. and Poncet, A. (1974): "Method of Measuring Three Dimensional Wakes in Turbomachinery," ASME Journal of Fluids Engineering, June.

[39] Raj, R. and Lakshminarayana, B. (1975): On the Investigation of Cascade and Turbomachinery Rotor Wake Characteristics. NASA CR-134680.

[40] Gorton, C. A. and Lakshminarayana, B. (1975): "A Method of Measuring the Three Dimensional Mean Flow and Turbulence Characteristics Inside a Rotating Turbomachinery Passage," ASME 75-GT-4.

[41] Flack, R. D. and Thompson, H. D. (1973): An Experimental and Analytical Investigation of Internal Transonic Flow. U.S. Army Missile Command Technical Report, RD-73-20, October.

[42] Flack, R. D. and Thompson, H. D. (1975): "Comparison of Pressure and LDV Velocity Measurements with Predictions in Transonic Flow," AIAA Journal, 13, January.

[43] Flack, R. D. (1975): "The Application of a Laser Doppler Velocimeter (LDV) in Interpreting Turbulent Structure," Ph. D. Thesis, Purdue University School of Mechanical Engineering, December.

[44] Weyer, H. and Schodl, R. (1973): "Methods for Measurement of Fluctuating Quantities in Turbomachinery," Presented at the Politecnico Milano Lecture Series on Experimental Techniques in Turbomachinery, October.

[45] Eckardt, E. (1974): "Applications of Dynamic Measurement Techniques for Unsteady Flow Investigations in Centrifugal Compressors," Von Kármán Institute Lecture Series 66, Advanced Radial Compressors, March.

[46] Eckardt, E. (1975): "Instantaneous Measurements in the Jet-Wake Discharge Flow of a Centrifugal Compressor Impeller," Journal of Engineering for Power, Trans. of ASME 97:3, July.

[47] Schodl, R. (1974): "Laser Dual-Beam Method for Flow Measurements in Turbomachines," ASME Paper no. 74-GT-157.

[48] Schodl, R. (1975): "A Laser Dual Focus Velocimeter for Turbomachine Applications," Von Kármán Institute Lecture Series 78, Advanced Testing Techniques in Turbomachines, April.

[49] Eckardt, D. (1976): "Detailed Flow Investigations within a High-Speed Centrifugal Compressor Impeller," ASME 76-FE-13, ASME International Gas Turbine Conference, New Orleans.

[50] Seasholtz, R. G. (1974): Laser Doppler Velocimeter Measurements in a Turbine Stator Cascade Facility. NASA TM X-71524, March.

[51] Goldman, L. J., Seasholtz, R. G. and McLallin, K. L. (1976): "Velocity Surveys in a Turbine Stator Annular Cascade Facility Using Laser Doppler Techniques." Proposed NASA TN (E-8637).

DISCUSSION OF THE MC NALLY PAPER

KORN: Do you know of any work on supercritical cascade blades, i.e. testing, and possibilities of doing that type of testing?

MC NALLY: No, I don't. Are there people from engine companies (here) who are doing that sort of thing? Or from Langley? I didn't contact anyone at Langley because I was particularly concerned with turbomachinery type blading, but I think they would also be using these techniques heavily.

KORN: I don't think that Langley does cascade testing.

MC NALLY: I don't know of any group using that sort of blade.

WOOD: I wanted to point out one area of work that is certainly not operational now but certainly promising in the future, which you really didn't touch on at all. And that's the area of work including laser Raman and laser absorption instrumentation which, hopefully, we will develop to the point of being able to determine temperature and density distributions, species concentration, that sort of thing in hot flows. Now that is not so applicable to compressors as it is to turbines and combustion, but I think it is something that ought to be pointed out because there is a lot of work going on in it, quite a bit being supported by Project SQUID. In fact, we had a workshop on it a while ago and the results of that are published and available. If you want any more information on the proceedings of that meeting, talk to Professor Murthy.

FLOW IN A TRANSONIC COMPRESSOR ROTOR

J. L. KERREBROCK
Massachusetts Institute of Technology

ABSTRACT

Some results of a comprehensive study of the flow through a transonic compressor rotor are presented. The rotor produces a pressure ratio of 1.6 at a tip Mach number of 1.2, has a cylindrical casing, sloped hub, and MCA blade sections. The gas density has been measured in the rotor by gas fluorescence, which reveals details of the shock and boundary layer structure as well as density in the passages. Time resolved measurements of the flow just downstream of the rotor give the blade wake structure, which is related to the features shown by the fluorescence. Of particular interest is a complex three-dimensional separation near the sonic radius. The flow through the rotor has also been computed in three dimensions by a time-marching technique, and the results of this computation are compared to those found by flow visualization.

INTRODUCTION

This presentation deals mainly with the experimental aspects of a comprehensive program of research on the aerodynamics of

*The research reported here was supported by the NASA Lewis Research Center under Grant NGL 22-009-383 and in part by the General Electric Company and Pratt & Whitney Division, United Technologies Corporation.

transonic compressors which is pursued in the MIT Gas Turbine Laboratory. In addition to the experimental work, which is built around the capabilities of the Blowdown Compressor Facility, the program includes development of a linearized theory for high work transonic rotors, described at this Workshop by Professor J. E. McCune, and development of a three-dimensional transonic computational capability, described by Dr. D. A. Oliver. Professor McCune discusses the correlation of the linearized theory with the experiments to be summarized here, and the three dimensional computation will be related to experiment in this discussion.

The Blowdown Compressor Facility has been described in detail elsewhere, so will not be discussed at length, however a summary of its unique characteristics will be given for background. It has made feasible two new diagonistic techniques, a quantitative density measurement by laser-induced gas fluorescence and simultaneous time-resolved local measurement by means of miniature pressure transducers of the stagnation pressure, static pressure, and three components of Mach number. These two techniques together yield information about the details of the flow in the rotor which has not been available in the past. Comparison of the experimental results to the linearized theory and to the results of the three dimensional computation yields a definition of the results of viscous shear.

A caveat concerning the transonic rotor is in order. It was designed by simple techniques to produce a pressure ratio of 1.6 at a tip Mach number of 1.2 and to be more or less typical of operational rotors. It was not intended to show an advance in the art of compressor design, and indeed it does not. Since the data to be discussed here was acquired, a stator has been added to complete the stage. The stage shows a pressure ratio of over 1.52 at the tangential Mach number of 1.2, with an efficiency of 0.87. That the stage performs this well indicates the rotor flow is probably not dominated by unusual loss mechanisms and can in this sense be considered typical.

THE BLOWDOWN COMPRESSOR

The concept of the Blowdown Compressor is diagrammed in Figure 1. A rotor (or stage) is positioned between two tanks, separated from one, the supply tank, by a diaphragm. The second, or dump tank, is evacuated and the rotor brought to speed in the vacuum. The diaphragm is ruptured allowing the gas from the

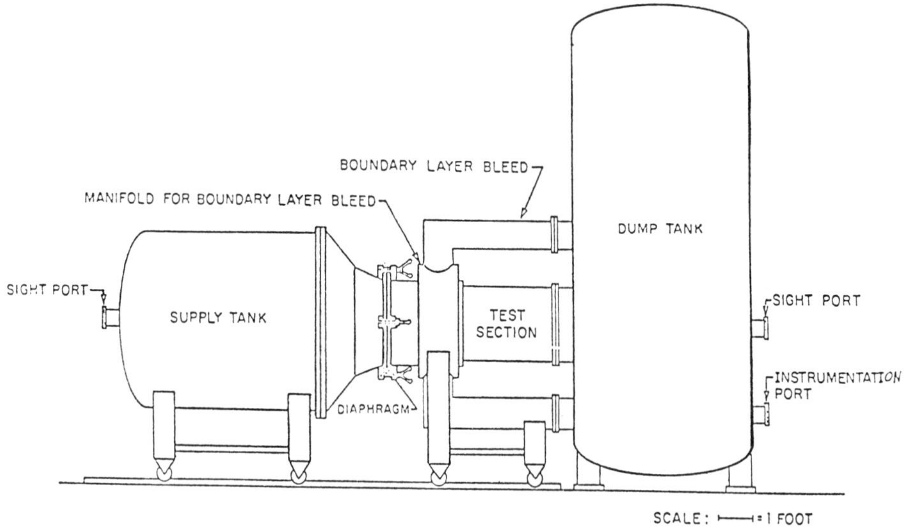

Figure 1. Scale Drawing of Blowdown Compressor Facility, Sized for 23.25-inch Diameter Rotor.

supply tank to flow through the rotor for a time of the order of 0.1 second, during which the rotor is driven by its inertia. By suitable matching of the rotor's moment of inertia and the supply tank pressure, the rotor slowing can be matched to the decrease of sound speed in the supply tank, so that the rotor operates at nearly constant tangential Mach number throughout the test time. The axial Mach number is set by a choked discharge into the dump tank.

The size of the MIT apparatus was dictated by the requirement that high response pressure transducers should resolve the unsteady flow to about the tenth harmonic of blade passing frequency, resulting in a rotor of about two-foot diameter and the facility shown to scale in Figure 2.

A typical time history for one test is shown in Figure 3, where the pressures in the supply tank (p_1), ahead of the rotor (p_2), behind the rotor at the casing (p_3) and in the dump tank (p_4) are shown. There is an initial period of very unsteady flow, followed by a test time of about 40 ms, which is terminated by unchoking of the throttling orifice. During the 40 ms the pressures all decay, but in constant ratio. The power input to the rotor is deduced from its rate of deceleration. During the 40 ms, a probe can be traversed

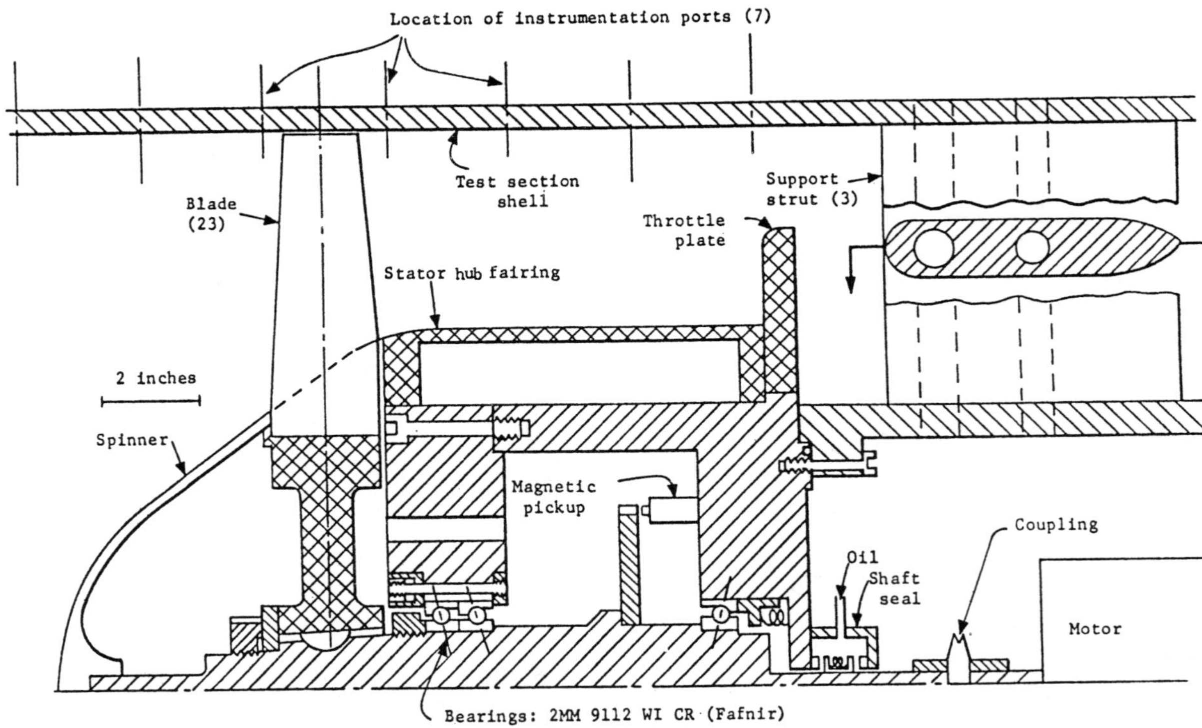

Figure 2. Scale Drawing of the Rotating Assembly and Test Section, Showing Locations of Rotor Instrumentation Ports, and Throttle Plate .

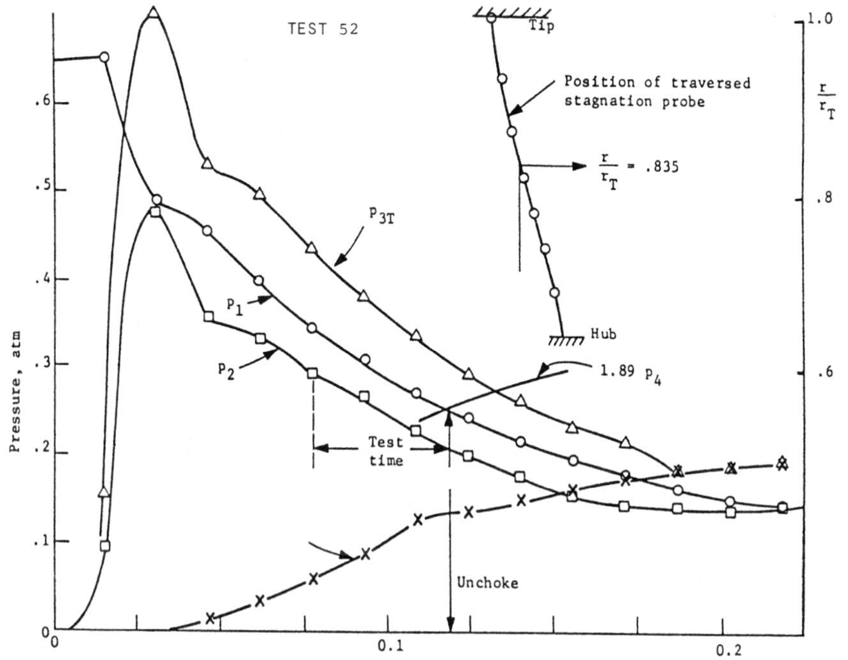

Figure 3. Variations of Supply Tank (p_1), Rotor Inlet (p_2), Rotor Outlet (p_{3T}) and Dump Tank Pressures for a Near-Design Full Speed Run without Stall. The Position of the Downstream Traversing Stagnation Probe as a Function of Time is Shown at Upper Right.

across the annulus to give a complete survey upstream and/or downstream of the rotor, and pressure transducers can also be installed in the casing.

As an indication of the capability of the facility for time resolution, the output of a pressure transducer (Kulite) mounted just upstream of the rotor is shown in Figure 4. At the top of the figure the complete test history is shown. Next, a small portion during the test time is expanded to reveal a well developed set of rotating stall cells and the individual blade passing signatures. Finally, the form of the shock-expansion pattern is displayed in detail in the oscilloscope picture. The point is that the pressure is obtained quantitatively on all time scales from D. C. to a high harmonic of blade passing from a single transducer. Drift problems usually

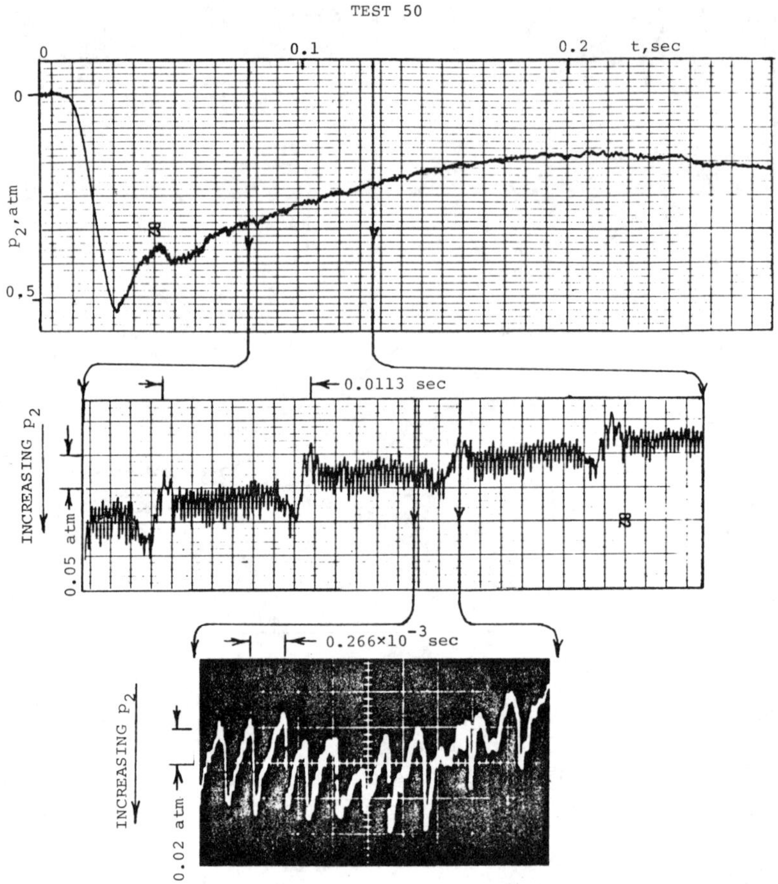

Figure 4. Static Pressure Upstream of Rotor on Three Time Scales, for Run with (Single Cell) Rotating Stall. At Top is Blowdown Time Scale. In the Center, Several Stall Periods are Displayed, and at the Bottom, Details of Pressure for a Few Blade Passing Periods.

associated with the semi-conductor type transducers are virtually eliminated by the short test time. Further, to provide a full calibration of the transducers, the dump tank is opened to the air shortly after the end of the test, sweeping the facility through the full pressure range of the experiment for calibration.

All data is recorded on a 14-channel FM tape recorder with

bandwidth from D. C. to 80 kHz and is then digitized for automatic data reduction.

DENSITY MEASUREMENT BY GAS FLUORESCENCE

The clean, hospitable environment of the Blowdown Compressor and the freedom to choose the test gas more or less arbitrarily stimulated an examination of optical diagnostics. A method was sought which would permit three-dimensional resolution of the flow with simpler interpretation than that required in holographic methods and preserving the time resolution.

A technique of density measurement by gas fluorescence which meets these criteria was developed by Epstein [2]. An organic gas, Biacetyl, added to the Argon-Freon test gas in small fractions, is excited by a flashlamp-pumped dye laser. It fluoresces, emitting radiation of slightly longer wavelength than that from the laser, with an intensity proportional to the exciting light intensity and proportional to gas density. The level scheme for the molecule is shown in Figure 5. The fluorescence is very fast (10^{-8} s), so the time resolution is set by the laser at about 0.5 μs. The intensity is independent of gas temperature. A phosphorescence also occurs on a longer time scale (~ 1 ms), at a longer wavelength, and is temperature sensitive, offering future potential for combined density and temperature measurements.

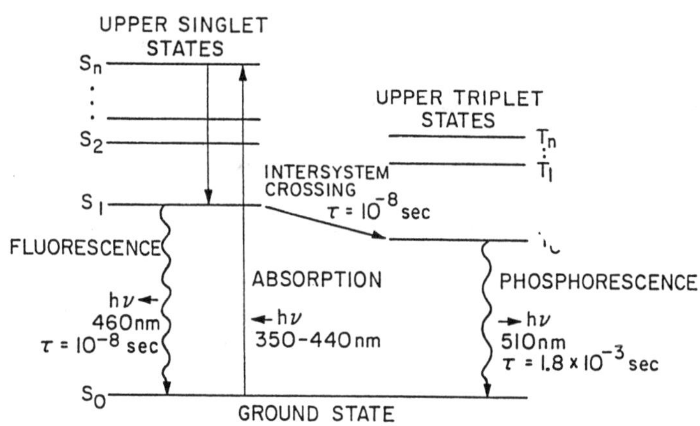

Figure 5. Internal Structure of 2, 3 Butanedione ($CH_3COCOCH_3$).

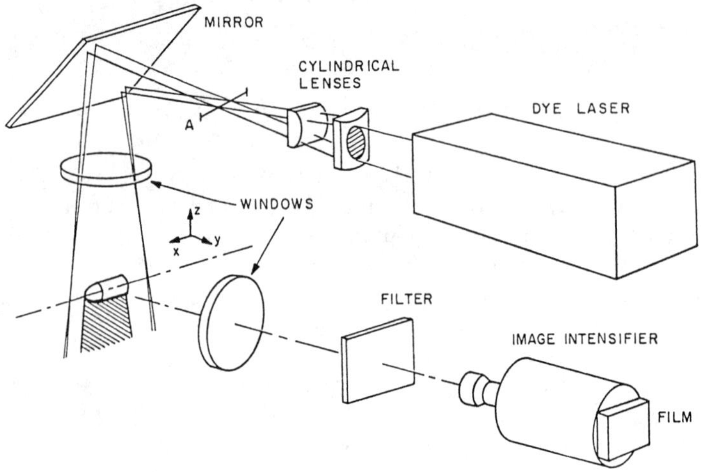

Figure 6. Schematic of Ballistic Range Optics.

The geometry of the optical setup is shown in Figure 6. A small mirror is used to project the laser light into the blade passages as a thin (1 mm) sheet roughly normal to the blade axis, thus defining a two dimensional cut of the flow through the passage, the radial location depending on the angle of the mirror.

Due to the low quantum efficiency of the process ($\simeq 10^{-7}$) the emitted light cannot be directly photographed, but must be amplified by an image intensifier. Financial limitations of the program necessitated the use of an intensifier of the type produced in quantity for dark vision applications. Such devices have high gain but substantial noise and image distortion. Most of the deficiencies of the photographs to be presented are traceable to this image intensifier. Better devices have been developed for astronomical and other scientific applications, and one is being procured for future experiments.

By careful calibration of the film, digital microdensitometry, correction for distortion in the image intensifier and nonuniformity in the laser beam, the photographic negative can be converted to a quantitative map of gas density [2]. Figure 7 indicates the degree of success achieved since it shows a photograph and corrected map obtained in quiescent gas. The uncertainty in density is about $\pm$ 5 % and this is largely due to the image intensifier. A precision of $\pm$ 2% is expected with the better quality intensifier.

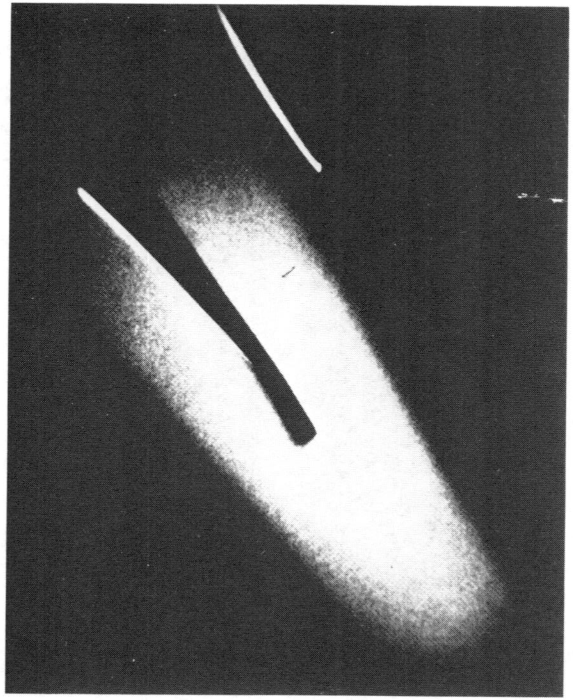

UNPROCESSED

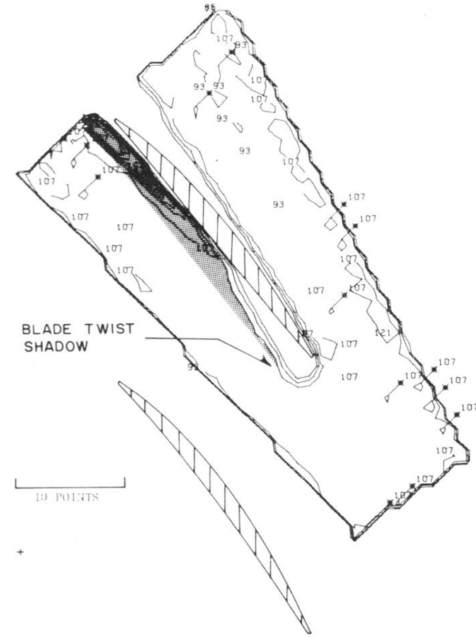

FULLY CORRECTED

Figure 7. Visualized Stationary Gas. The Density is Uniform; Therefore all Variations Result from Instrument Imperfections.

Typical photographs are shown with their density plots for several radii in Figures 8 to 11. A cut near the tip (Figure 8) shows a well defined bow shock and passage shock, the latter terminating in a lambda at the suction surface of the blade. A separation can be discerned behind the lambda shock if the negative is printed with higher contrast. The density map shows a ratio of about 1.3 across the bow shock, an overexpansion down to about 0.85 on the suction surface, a return to about 1.25 across the passage shock then diffusion to a density ratio of about 1.4 at the trailing edge.

At a smaller radius (r/r_T = 0.83) the lambda shock has disappeared, as shown in Figure 9, even though the passage shock is still quite strong. A dark patch, indicating low density, appears behind the shock surface intersection. Both in comparison to Fig. 8 indicate a strong three-dimensionality in the flow structure. At a still smaller radius, (Figure 10), (r/r_T = 0.80) the shock has been replaced by a rapid compression. Finally, very near the hub (r/r_T = 0.65) we see (Figure 11) a severe overexpansion on the suction surface followed by a rapid recompression, then a gradual diffusion to the outlet. These last features will be correlated with the three dimensional computation later.

The structure which seems to emerge from these maps has a strong passage shock near the tip, separating the boundary layer there and generating a blockage. At a smaller radius this blockage sheds a streamwise vortex off its inner end, resulting in the very low pressure patch. Inside this the boundary layer is well behaved, perhaps because of strong radial flow in it, but the pressure rise behind the tip shock is felt by the flow at the hub, resulting in the rapid recompression there. Again, the flow is manifestly three dimensional.

TIME-RESOLVED MEASUREMENTS

Time resolved measurements have been made both upstream and downstream of the rotor. Only the latter will be discussed here. They were made with the probe shown in Figure 12. By calibration it is possible to deduce the two flow angles, ϕ and θ, the static pressure and the stagnation pressure from the absolute pressures measured by the four semiconducting strain gage diaphragms. The frequency response of the probe is limited by its dimensions to about 40 kHz, which is some 10 times the blade passing frequency in the Blowdown Compressor.

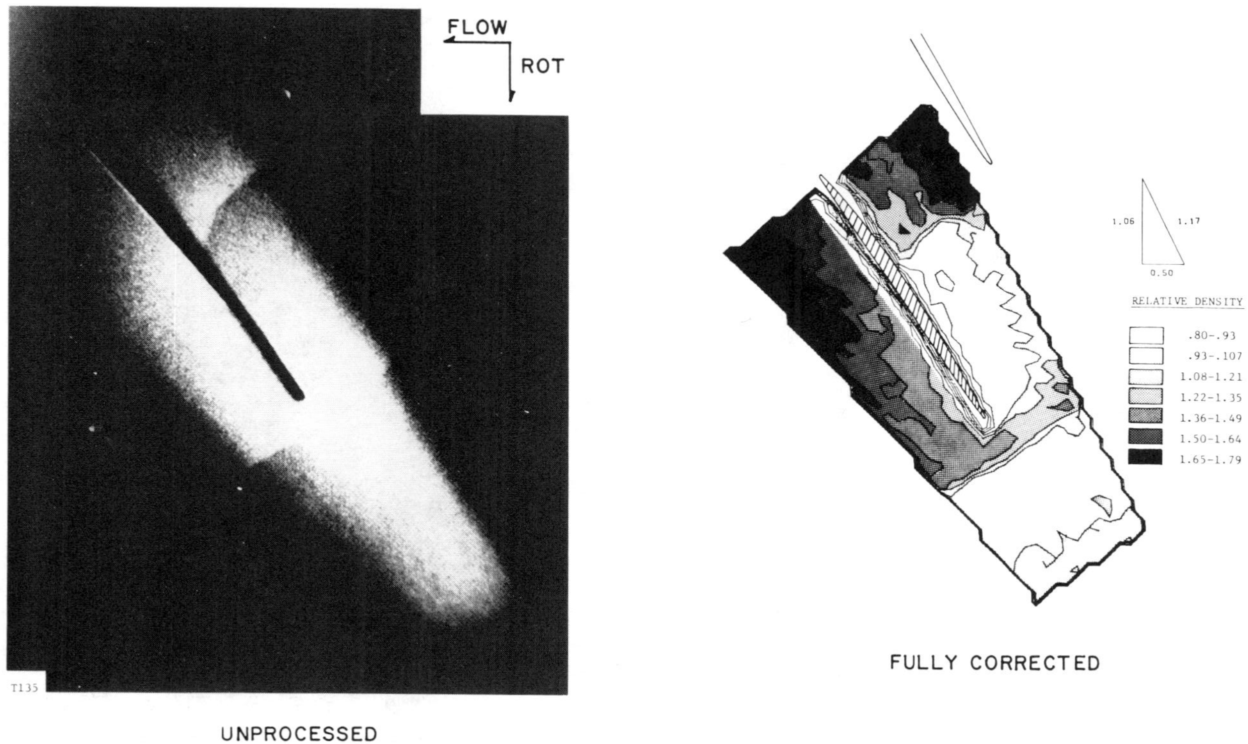

Figure 8. Visualized Flow at $R/R_T = .88$.

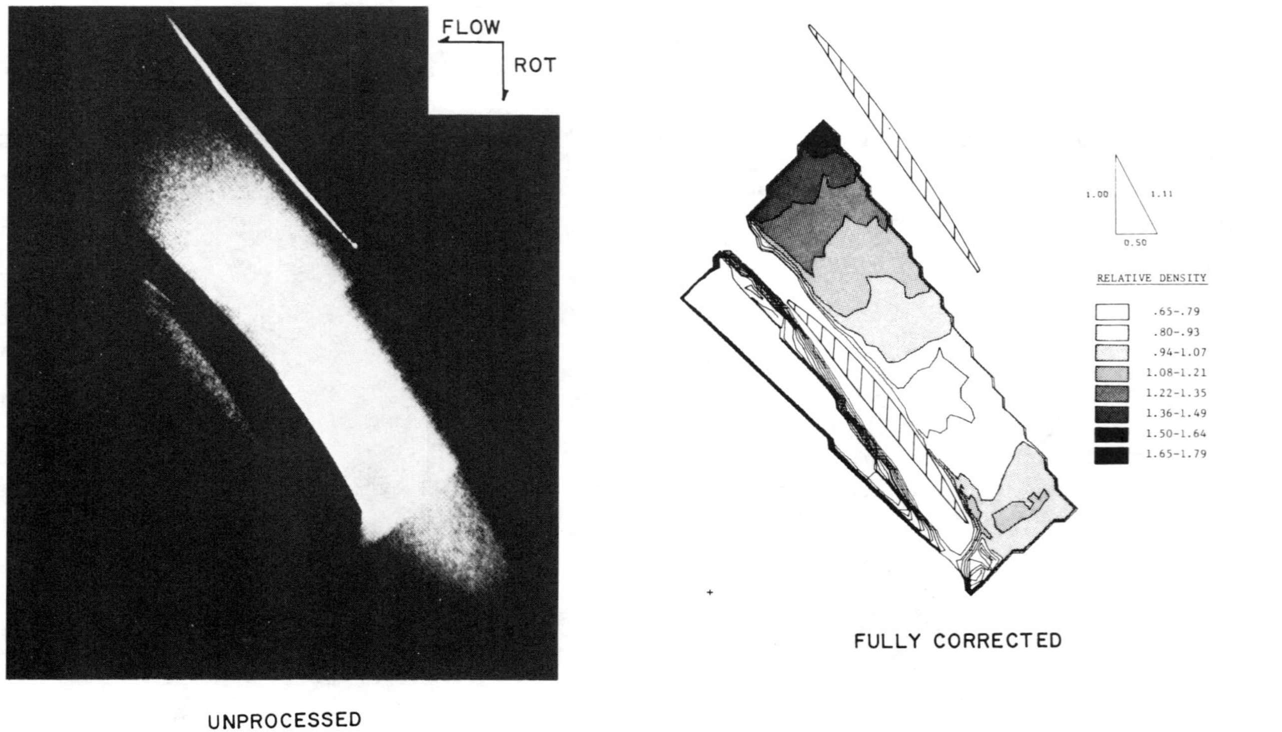

Figure 9. Visualized Flow at $R/R_T = .83$.

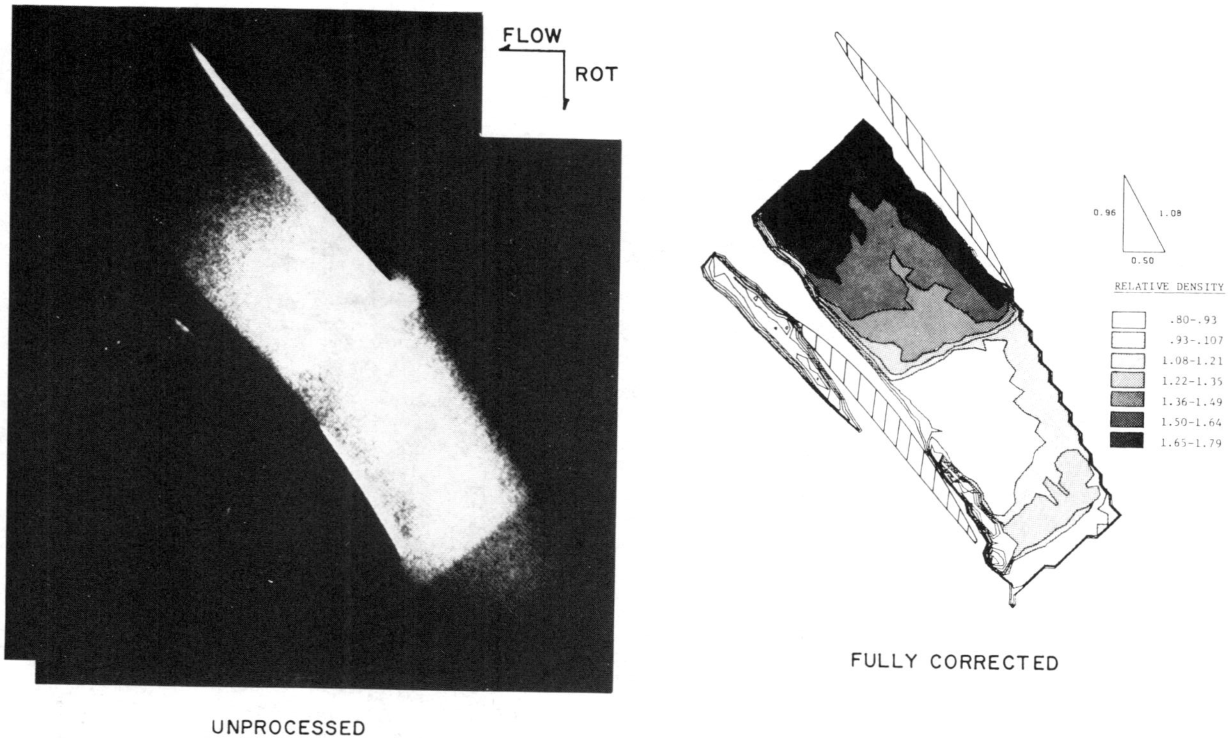

Figure 10. Visualized Flow at $R/R_T = .80$.

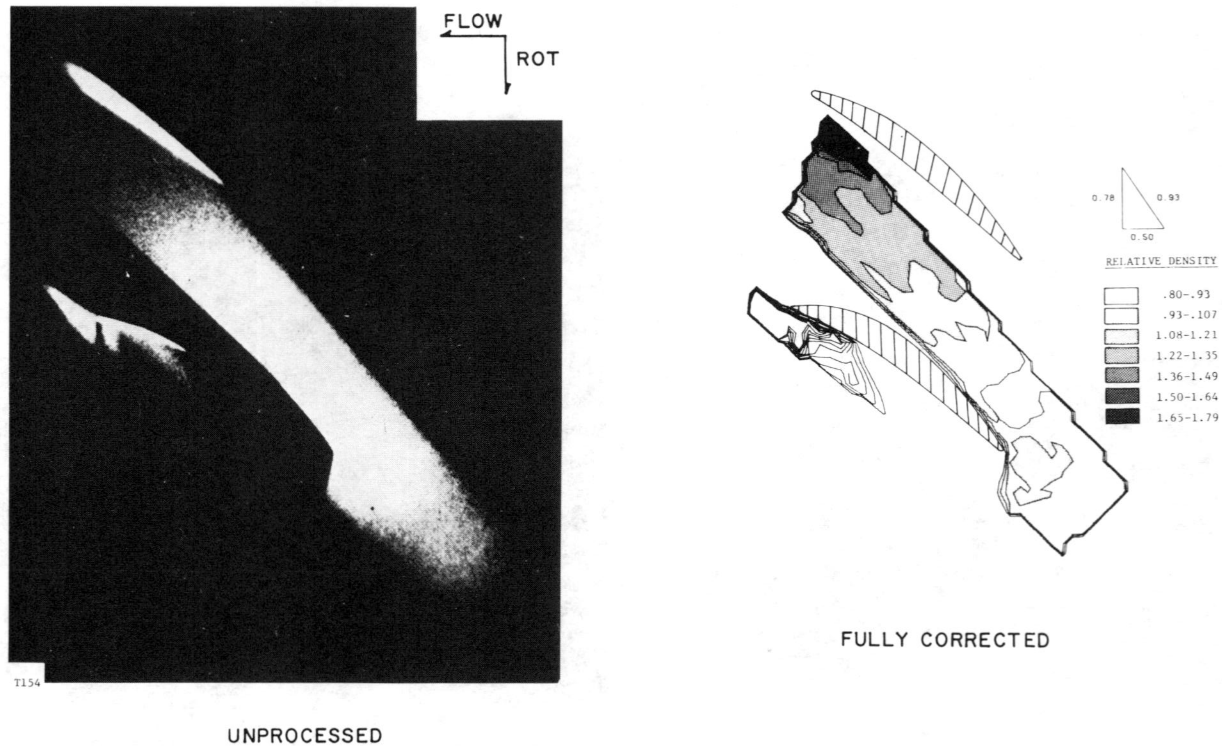

Figure 11. Visualized Flow at $R/R_T = .65$.

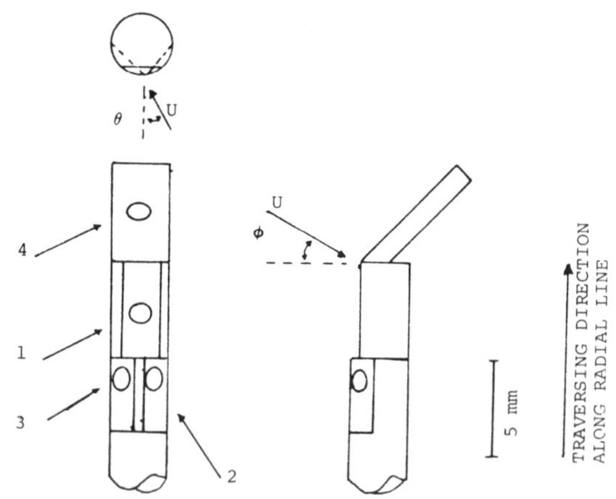

Figure 12. Sketch of Four-Diaphragm Probe Showing Air Angle Definitions.

Typical time histories over a few blade passing periods are shown in Figures 13 and 14. The axial and tangential Mach numbers shown in Figure 13 show a regular blade-to-blade periodicity typical of the behavior in the near wakes at small radii. The total and static pressure histories of Figure 14 on the other hand show an irregularity which is typical of the tip region. It is argued in [3] that this unsteadiness may be connected to the tendency of

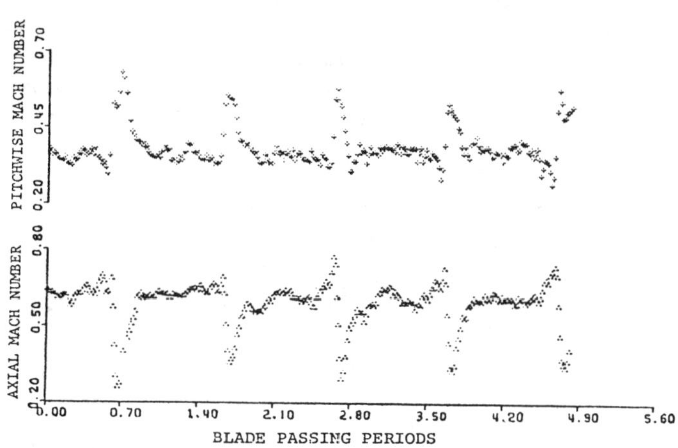

Figure 13. Mach Number Components. R/RT = 0.738. 0.1 Axial Chords Downstream of Rotor.

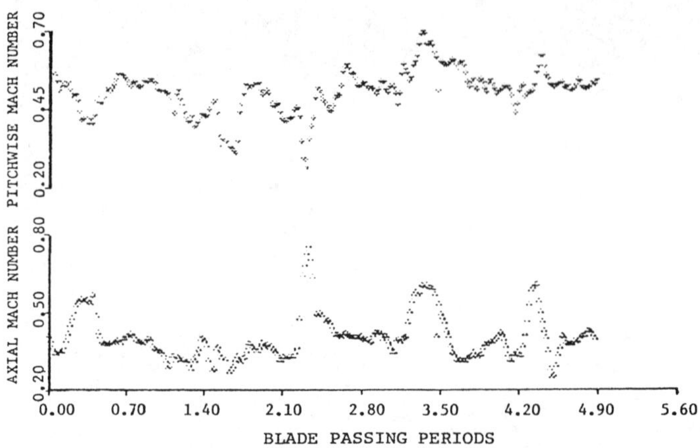

Figure 14. Total and Static Pressure Ratios. R/RT = 0.870.

the wake flow to evolve into a set of vortices parallel to the axis which represent one of the eigenmodes of the duct flow with strong rotation.

By combining samples of such data obtained at different radii, but all from the same test, it is possible to generate maps in r and $r\theta$ of each of the flow properties. It must be borne in mind that these are not instantaneous resolutions of the flow, but composites of a series of flows which are not periodic near the tip. Nevertheless, they reveal some interesting features of the flow. Figure 15 shows a map of the stagnation pressure derived in this way. It exhibits strong local peaks at the radius where the streamwise vortex was identified in the photographs, but is otherwise rather uniform. Figure 16 shows the pitchwise Mach number, M_θ, which is representative of the work distribution. The regions of high M_θ are also regions of high stagnation temperature and apparently result from excess viscous (and radial flow) work done on the wakes. This identification is supported by the entropy distribution shown in Figure 17, the local maxima of entropy coinciding with the regions of high M_θ. The peaks of static pressure shown in Figure 18, which coincide with those in the stagnation pressure shown in Figure 15, lie on the pressure side of the wakes. Their existence is confirmed by the density map of Figure 9.

Measurements at a distance of about one chord downstream of the rotor show that in the outer part of the annulus the wake flows have reorganized into a cellular vortex structure with a θ spacing of about 1.4 blade spacings. The radial Mach number distribution of this flow is shown in Figure 19; as noted above it is argued in

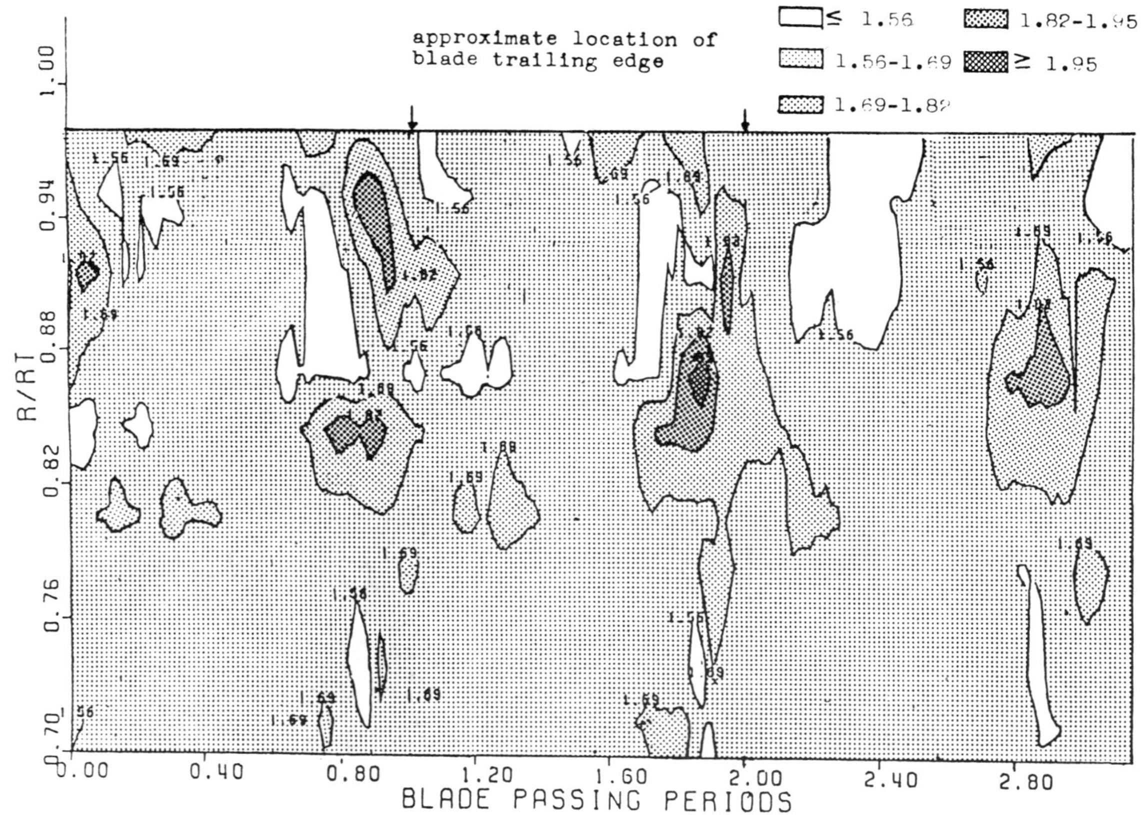

Figure 15. Total Pressure Ratio Map. 0.1 Chords Downstream of Rotor.

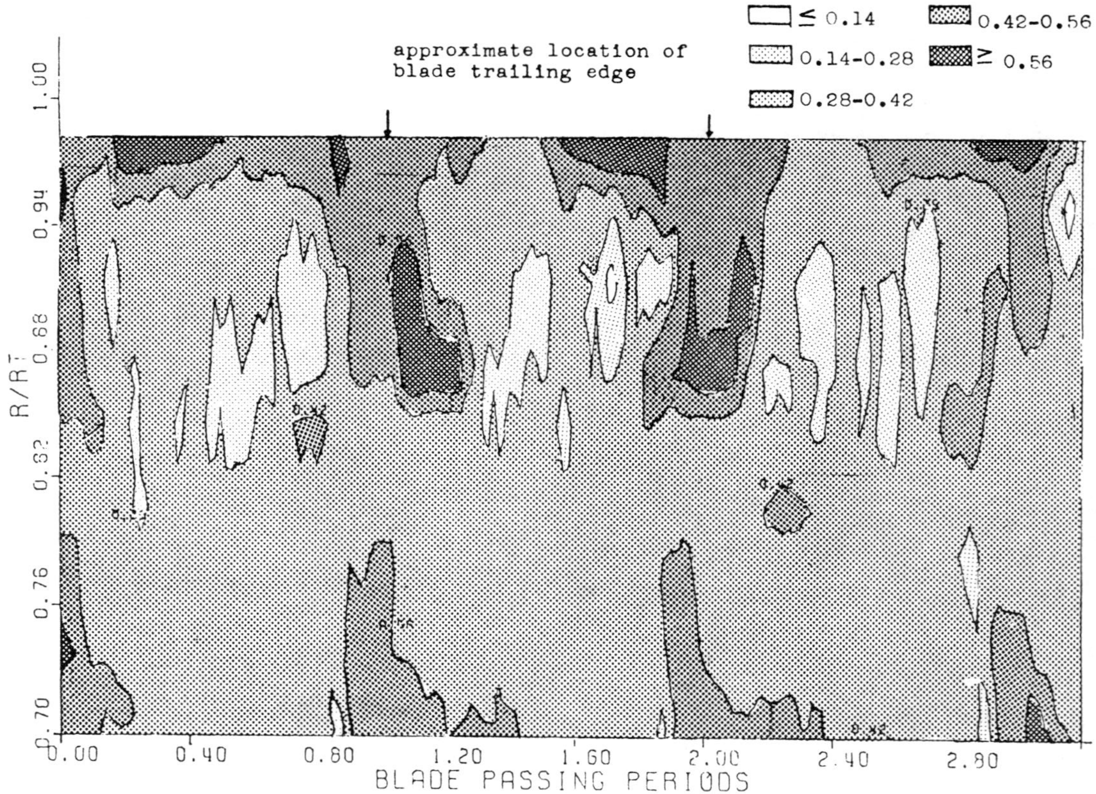

Figure 16. Pitchwise Mach Number Map. 0.1 Chords Downstream of Rotor.

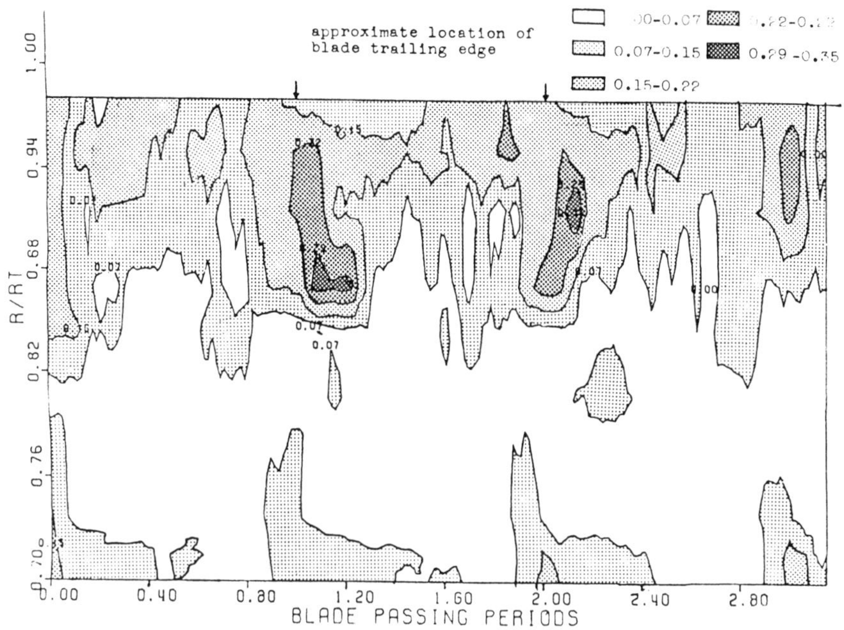

Figure 17. Entropy Rise Map. 0.1 Chords Downstream of Rotor.

[3] that the cellular vortex flow is an eigenmode of the flow in the annulus with a strong rotation.

From the time resolved measurements, θ or time averaged values at any radius can be computed. They will be compared in the next section to predictions of the three dimensional computation.

RESULTS OF THREE-DIMENSIONAL COMPUTATION

Except for inaccuracies resulting from finite grid spacing and lack of full convergence, the three-dimensional computation should represent an exact inviscid solution for the flow through the rotor. Comparison of its structure to that of the actual flow field should then reveal the results of viscous shear.

Figures 20, 21 and 22 show comparisons of the measured and computed exit air angles, axial Mach numbers,. and tangential Mach numbers. From Figure 20 we see that the actual angle

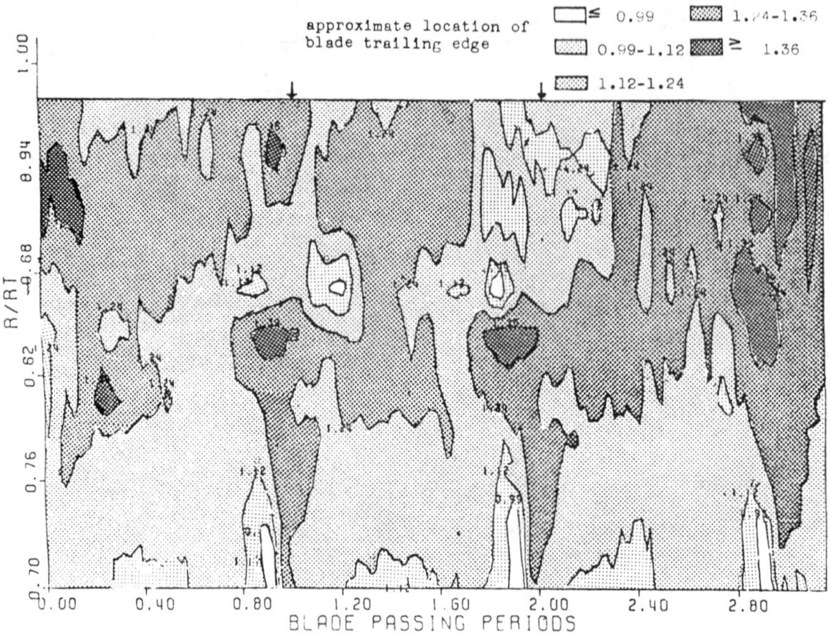

Figure 18. Static Pressure Ratio Map. 0.1 Chords Downstream of Rotor.

is very close to the inviscid value except near the tip, where a separation was noted in the photograph. From Figure 21 the measured axial Mach number is much lower near the tip than that computed, while from Fig. 22 there is an excess of tangential Mach number near the tip. Both support the idea that boundary layer fluid of high energy but low axial Mach number is accumulated near the casing due to radial flow in the boundary layers and wakes.

The three-dimensional computation also yields detailed distributions of all flow properties throughout the blade passages. Samples of the Mach number distributions near the tip and hub are shown in Figures 23 and 24. In Figure 23 we see a severe overexpansion on the suction side, as noted in Fig. 8. The overexpansion and rapid compression near the tip on the suction side is also evident on Fig. 24, in agreement with Fig. 11. This agreement supports the contention that the rapid recompression is due to the radial inflow induced by the shock near the tip, as it seems inviscid in origin.

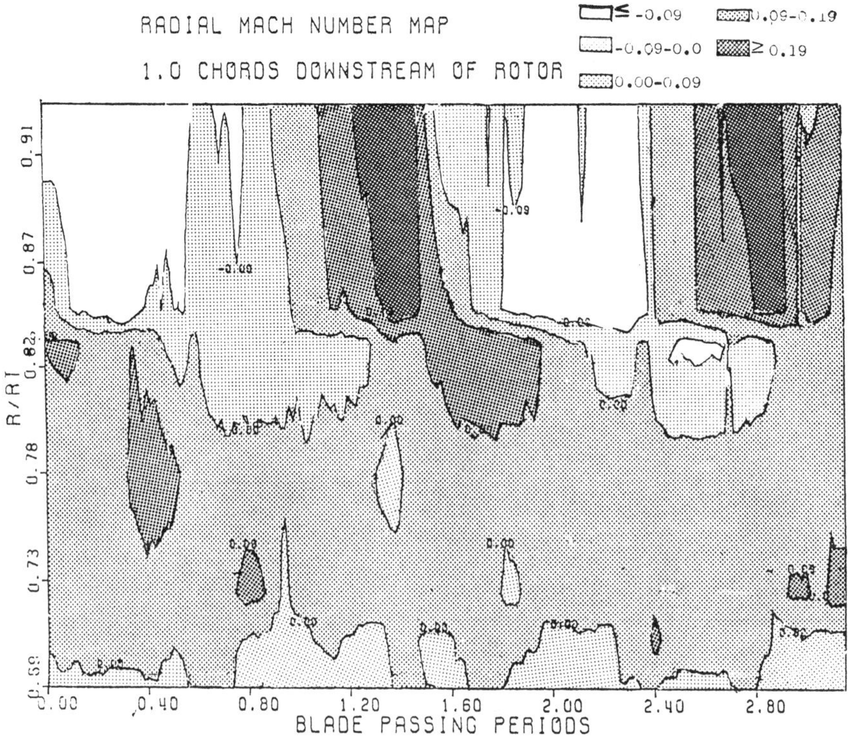

Figure 19. Radial Mach Number Map. 1.0 Chords Downstream of Rotor.

Finally, Figure 25 shows three stream surfaces derived from the computation. The surface is displaced radially inward on the suction surface at all radii shown, but most strongly near the hub. There seems to be no evidence of the strong radial shear which this would produce in the experimental maps of the radial Mach number, so perhaps viscous effects suppress the shift.

CONCLUDING REMARKS

The program of research on transonic compressors, of which a part has been reviewed here, and the remainder by Professor McCune and Dr. Oliver, is meant to provide sufficiently detailed knowledge of the actual three-dimensional flow so that rational approaches can be evolved for reducing the secondary flows and the associated losses. The viability of the Blowdown Compressor

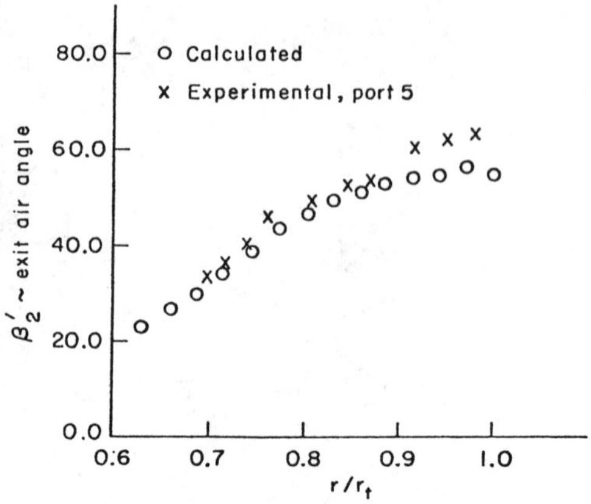

Figure 20. Rotor Exit Air Angles Relative Coordinate Frame Theta Averaged Values.

and its associated instrumentation for providing the required experimental data has, we believe, been demonstrated. Further, the feasibility of a full three-dimensional computation of the inviscid flow has been demonstrated. The linearized theory provides a valuable tool for understanding the effects of shed vorticity due to radial variation of the circulation.

From the data on hand it is clear that the actual flow deviates greatly from that implied by the design systems now in use, which

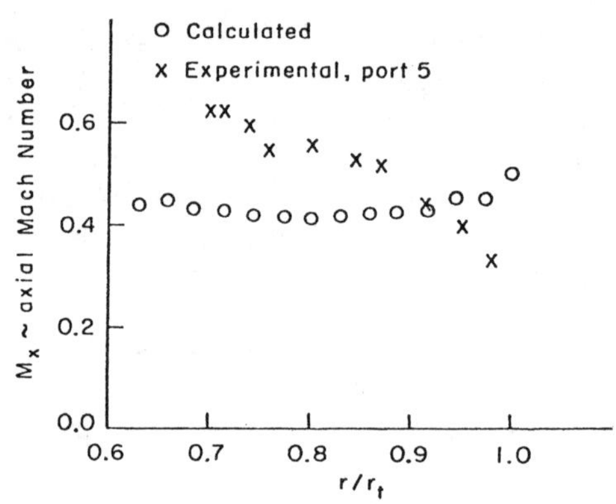

Figure 21. Rotor Exit Axial Mach Number Profile ~ Theta Averaged Values.

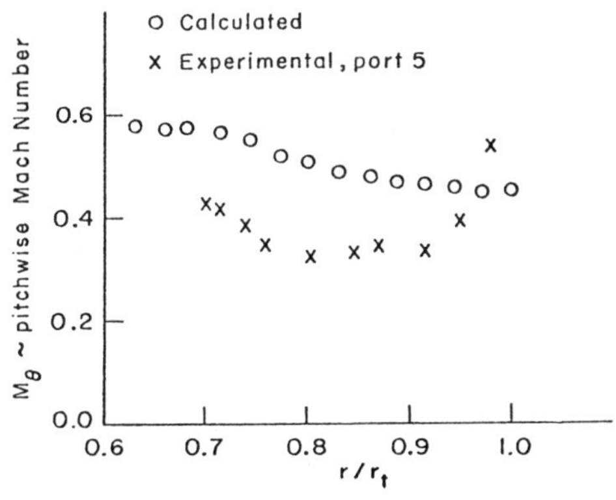

Figure 22. Rotor Exit Pitchwise Mach Number Profile ~ Theta Averaged Values.

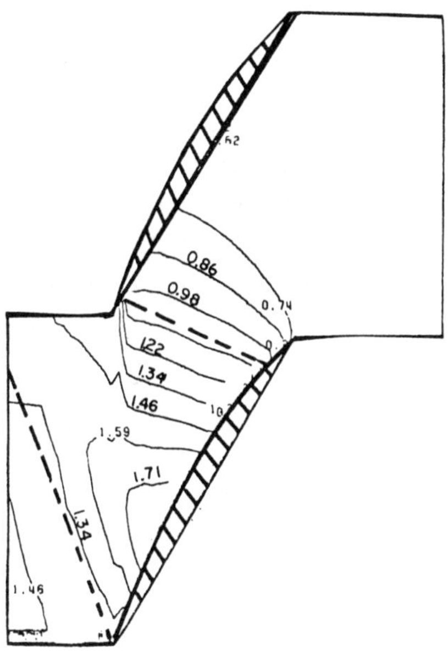

Figure 23. Mach Number Contours for Computational Blade to Blade Surface Number 15; Tip Surface.

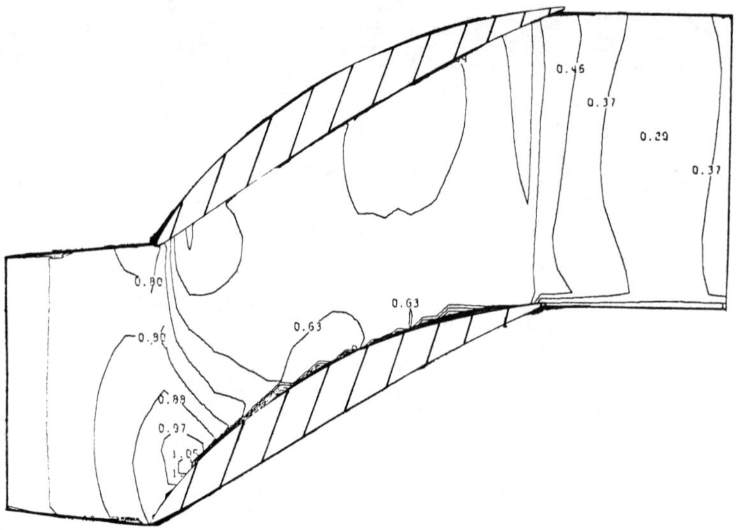

Figure 24. Mach Number Contours for Computational Blade to Blade Surface Number 2; Hub Surface.

do not account for the strong radial flows due to both viscous and inviscid gapwise variations. But clearly the utility of such detailed information can be demonstrated only by making use of it in the design of a better rotor. Hopefully, such an improved design and evaluation can be executed in a timely way.

ACKNOWLEDGEMENTS

The author is indebted to the several students who have contributed to the Blowdown Compressor development. In particular, Dr. A. H. Epstein and Mr. W. T. Thompkins, Jr. have been largely responsible for converting the concept into a useful experimental tool. The continued support of the NASA Lewis Research Center Compressor Research Branch has made the program possible.

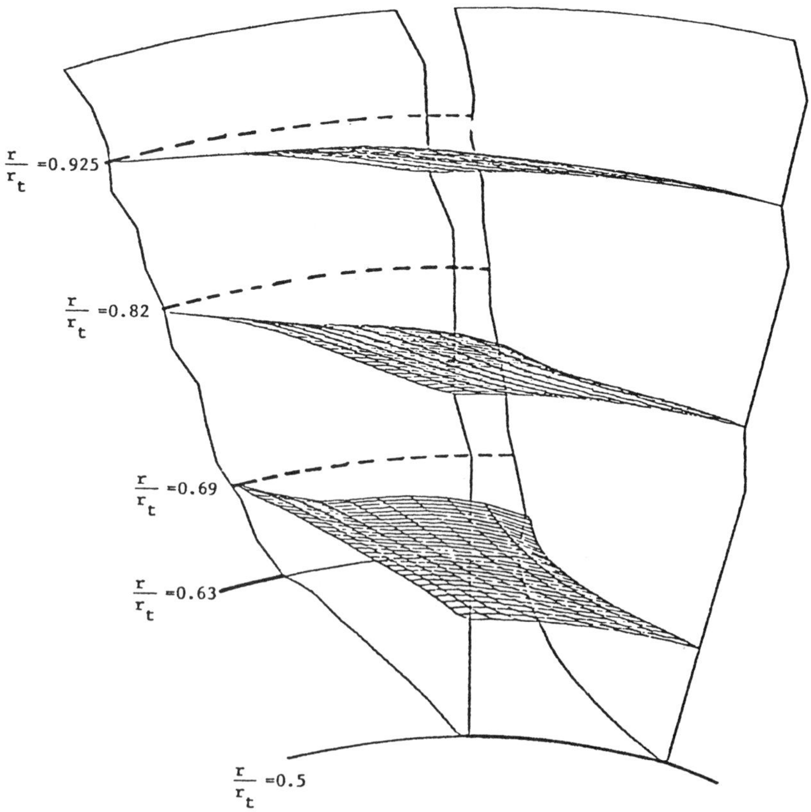

Figure 25. Computed Blade to Blade Stream Surfaces. View as Seen Looking Downstream.

REFERENCES

[1] Kerrebrock, J. L. et al., (1974), "The M.I.T. Blowdown Compressor Facility," J. Engr. for Power, 96, A, 4, October, pp. 394-405.

[2] Epstein, A. H. (1975), "Quantitative Density Visualization in a Transonic Rotor," AIAA Paper 75-24; also Ph.D. Thesis, Aero and Astro., Massachusetts Institute of Technology, September.

[3] Thompkins, W. T., Jr. and Kerrebrock, J. L. (1975), "Exit Flow from a Transonic Compressor Rotor," Proc. AGARD Propulsion & Energetics Panel 46th Meeting, Monterey, September, Paper No. 6; also MIT Gas Turbine Lab. Report 123, September 1975.

DISCUSSION OF THE KERREBROCK PAPER

MC DONALD: Is the laser fluorescence inhibited or quenched by air?

KERREBROCK: No. The phosphorescence is, but not the fluorescence.

MC NALLY: Would you comment a little about how difficult the fluorescence technique is to apply and how much of an effort it would be to train someone else to do now what Epstein does.

KERREBROCK: It's difficult to judge individual capabilities but Epstein is not a usual guy. He did it as a Ph. D. student in a period of about three years and he had to reinvent the dye laser in the process. I think that basically it's very easy to use; the only thing that was really difficult about extracting this quantitative data other than getting the dye laser to work was the series of corrections that had to be made. When you take data which has built in distortions of the factor of 3 and try to correct it back to within 5 percent you realize you have to be pretty fastidious. But with a better image intensifier, all of that would be unnecessary. I really believe that that technique can be used routinely at least in this facility to get density maps $\pm$ a couple percent. You just fire it; take a picture. That's it. There isn't a lot of complicated data reduction required.

RUNSTADLER: What's the concentration used?

KERREBROCK: It's biacetyl. Butanedione, which you eat everyday; they use it to flavor butter. Smells terrible. It's 20 mm out of about 500.

RUNSTADLER: What does that do to your speed of sound level?

KERREBROCK: We change the argon freon mixture to compensate for that.

RUNSTADLER: Give a measure of how much.

KERREBROCK: It doesn't make much difference.

RUNSTADLER: Two per cent?

KERREBROCK: Yes. The only thing difficult about the butanedione is that it is an inflammable chemical so you can't just dump it out in the air without some thought, and it smells. We did this for two or three years; just before we got finished they caught up with us and then we had to put in a liquid nitrogen trap to catch the stuff for the last two runs.

RUNSTADLER: What's the upper ... I was thinking about using it for the centrifugal compressor work which goes up to quite high temperature ratios, 8:1 or 9:1 pressure ratio. What's the upper temperature ratio for either flash point or chemical breakdown?

KERREBROCK: It's pretty high. At least 200° - 300° F. The real problem here is not that, but rather condensation of the stuff when you go through expansion.

RUNSTADLER: I see. Do you have any reports?

KERREBROCK: Yes. There's a gas turbine report by Epstein which I'll be happy to send you. It was also presented at an AIAA meeting last winter.

MC NALLY: Would you comment on the potential of this technique for obtaining velocity plots as opposed to density plots.

KERREBROCK: The possibility there is because the phosphorence has a long lifetime - something like a millisecond - you can think of firing the laser and then you essentially mark the fluid. Then if you watch the drift of the phosphorent fluid in time you can use it very much like a hydrogen bubble technique. That's a way of measuring velocity, potentially. There's also a potential for a direct temperature measurement because the phosphorence is quite temperature sensitive and the fluorescence is not. So the ratio of the two intensities is a direct measurement of temperature.

DODGE: If you're injecting this stuff in a mixture of gas, you would not get discrete traces like you could in a hydrogen bubble, isn't that right?

KERREBROCK: No, the point is with the hydrogen bubble you mark the fluid by making a little bubble. In this case you mark it by causing the stuff to fluoresce with a collimated laser beam.

DODGE: You mark a single spot, then watch the spot move?

KERREBROCK: You can mark a line. You can draw a line through the fluid and it'll fluoresce for a millisecond, then you can watch it move.

DODGE: Why did you choose an image intensifier rather than photomultiplier?

KERREBROCK: Because we wanted a picture. In a photomultiplier all you get is one number.

DODGE: You can scan but it makes the reduction of the data a great deal simpler.

KERREBROCK: That's true and you could use this technique to get local time-resolved measurements. Then we would be back to trying to make a picture out of a bunch of squiggly curves. I have big arguments with Alan Epstein on this; we finally settled on the picture taking because it's nice to have that qualitative information. But I think both are valuable techniques.

MIKOLAJCZAK: I am a little confused on the data you showed where you show the angle at the exit from the rotor and then you show the tangential velocity and axial velocity. If I remember right, the axial velocity at the hub is much higher measured than predicted and the tangential velocity is much lower measured than predicted. And, yet, you come out with the same angle at the hub. Are you using the same set of data to arrive at that conclusion?

KERREBROCK: Yes. The angle is the relative angle in the rotor and that resolves the problem. The angle that I showed is beta prime, the angle in rotor relative coordinates, and it's right on. But what that's saying is that the absolute magnitude of the velocity vector is considerably larger in the relative coordinate system than it should be. So that gives you a larger axial velocity but a smaller tangential velocity. Right? It still has a positive stagger at the hub.

MIKOLAJCZAK: What really saves you is the lock of the wheel speed to give you a good or reasonably close angle agreement, right?

KERREBROCK: No. I think they're consistent. You see, if you say that the angle is right on the blade angle, then if you say the magnitude of the velocity vector in the relative coordinate system is larger than it should be, that will give you a deficiency in absolute tangential velocity at the hub and an excess of axial velocity. Maybe that's the point that's missing, Alex, tangential Mach number is absolute.

COMPARISON OF PREDICTION OF TRANSONIC FLOW IN A FAN WITH FLOW MEASUREMENTS TAKEN USING A LASER DOPPLER VELOCIMETER

A. A. MIKOLAJCZAK

Pratt and Whitney Aircraft Company

INTRODUCTION

In recent years extensive effort has been directed toward prediction of flow in turbomachines. At P&WA we have an active program directed at developing analytical predictions and at the acquisition of accurate flow measurements to check and refine the analytical models where necessary. We believe that improvement in flow prediction in transonic stages will allow us to

- increase efficiency of fans and compressors
- increase tip speeds to achieve reduced number of stages
- identify blade geometries to increase blade durability so that peak compressor performance can be retained over long engine life.

In this presentation some velocity measurements made in a transonic fan will be shown and compared to flow predictions.

FLOW MEASUREMENTS

To provide a data base for checking transonic flow predictions velocity measurements were made in the intrablade flow field of a transonic fan using a dual beam, confocal, backscatter optical

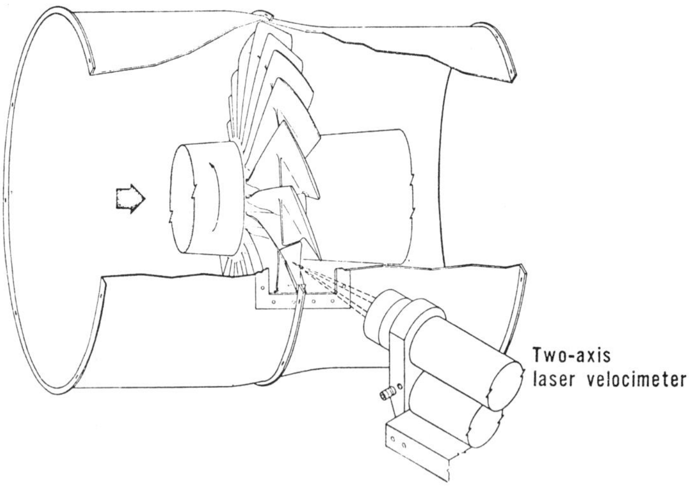

Figure 1. Velocity Measurement in the Interblade Flowfield

system looking through a 3-inch window placed in the compressor case (Figure 1). By traversing the optics and by varying the delay time between trigger signal and the acquisition of velocity samples, measurements were made in a large portion of the rotor intrablade flow field.

The optical system (Figure 2) was developed at P&WA and has been described in detail by Walker, _et al._ [1]. The two pairs of laser beams are of different color and are orthogonal to each other. At the focal point of the beams, referred to as the probe volume, two perpendicular interference fringe patterns are formed. One such fringe pattern is shown in Figure 3. When a particle traverses the probe volume bursts of light are emitted in the fringe pattern and translated into velocity by the optical system and the associated electronics. Two velocity components are determined simultaneously, from the two fringe patterns. A large number of readings are taken at every location in the flow and a histogram developed. A typical histogram is shown in Figure 3. It shows a distribution of the number of measurements obtained in different velocity intervals and allows the average velocity during the sampling time to be determined.

The flow was seeded with Di (2-ethylhexyl) phtalate (DOP). The seed particle size was estimated to be less than 2 microns in

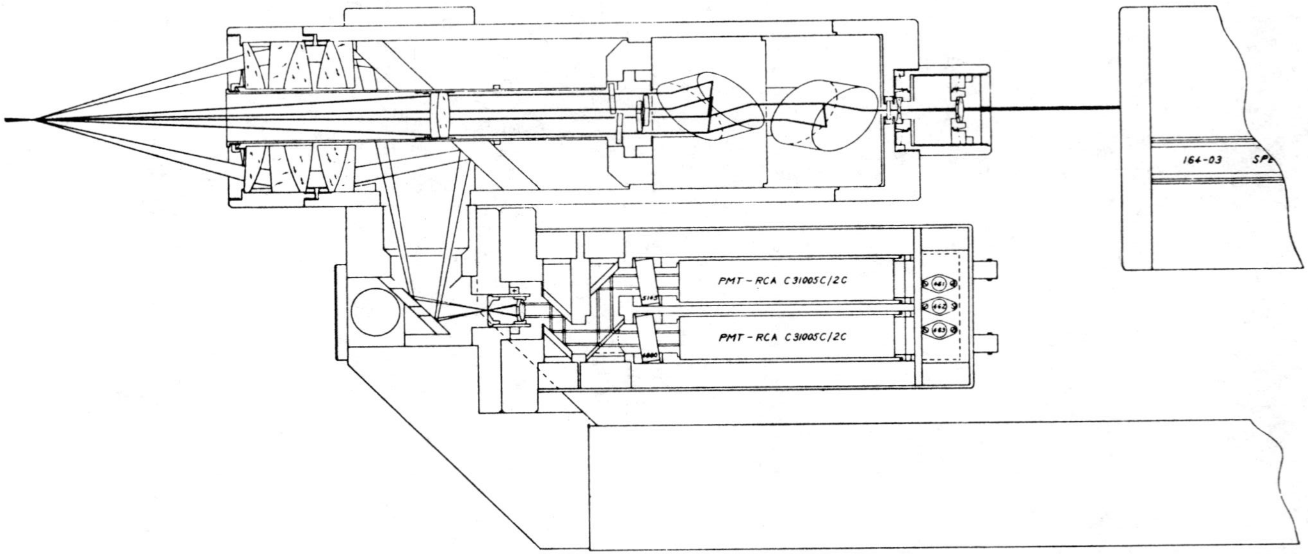

Figure 2. Optical Layout of Two-Color, Two-Component Confocal Backscatter Laser Doppler Velocimeter

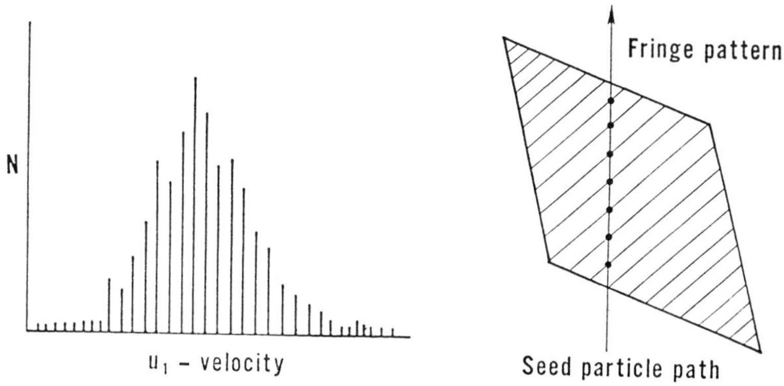

Figure 3. Histogram

diameter, which is sufficiently small to allow the particles to move essentially at the gas velocity. The velocity spread observed in the histograms was thus attributed directly to the unsteadiness which exists in the flow during the measurement time. This unsteadiness, expressed in terms of standard deviation normalized by mean velocity, was calculated to be approximately 9%. Using

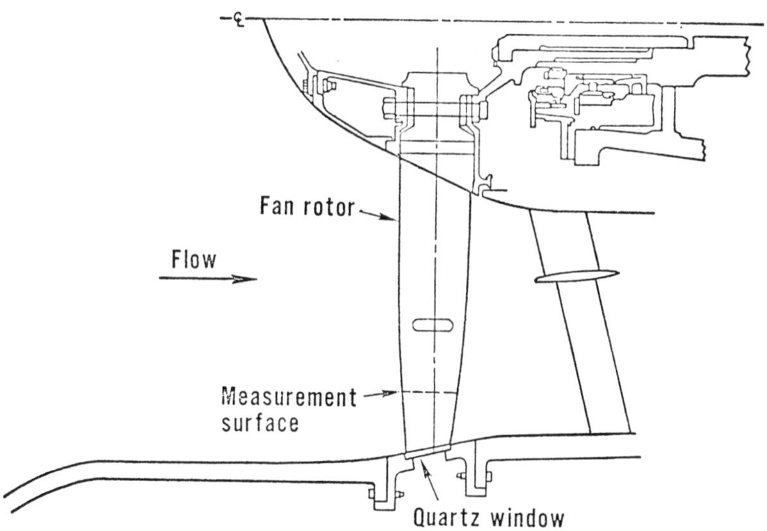

Figure 4. LV Measurement in Transonic Fan; Tip Speed 1550 F.P.S.

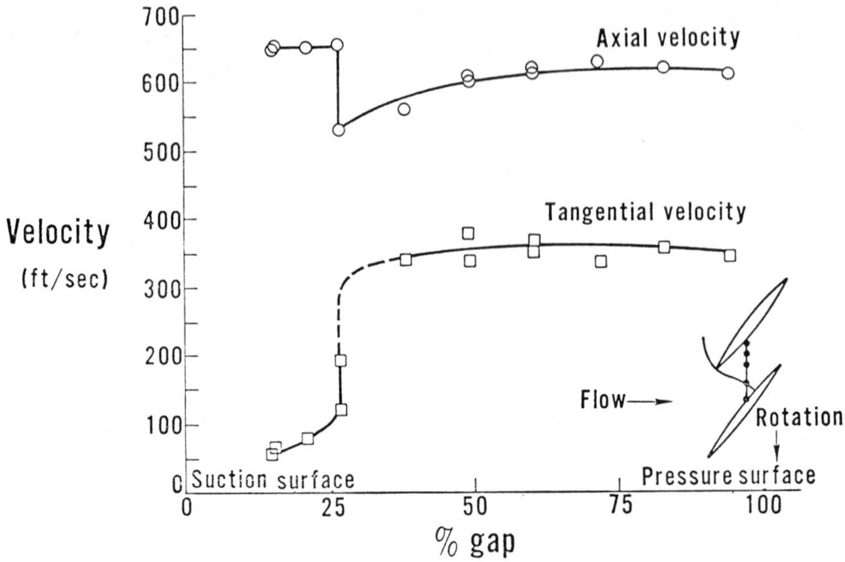

Figure 5. Measured Gapwise Velocity Distribution

a large number of measurements the accuracy of the average velocity was calculated to be within 12% over a 95% confidence level interval.

The velocity measurements shown in Figures 5, 6 and 7 were obtained in a transonic fan on a cylindrical surface at 80% of the span outboard of the hub as shown in Figure 4. The local blade chord was 3.8 inches, the gap to chord ratio was .72. At the design tip speed of 1550 feet per second the relative Mach number at the measuring station was 1.3 and the fan total pressure ratio was 1.67.

Velocity distribution was determined in approximately 80% of the blade passage. The twist and camber of the airfoil outboard of the measuring surface produced a shadowed region in the vicinity of the pressure surface from 0 to 50% chord and near the suction surface from 50% to 100% chord. Accurate measurements were possible to within 0.1 inches from blade surfaces. Gapwise absolute velocity distribution taken at 50% of axial chord is shown in Figure 5. Both the axial and the tangential velocity components gave an excellent definition of the passage shock. Velocity measurements were found to be very repeatable as evidenced by duplicate data points shown in Figure 5 which were obtained at different test times. Chordwise absolute velocity distribution measured at mid-gap is shown in Figure 6. Both velocity components gave a very sharp definition of the shock location. The chordwise absolute velocity distribution, shown in Figure 6, has been converted to a velocity relative to the rotor and is shown in Figure 7.

COMPARISON WITH ANALYTICAL PREDICTIONS

The LDV measurements are compared to a quasi-three-dimensional passage analysis in Figure 7. The blade-to-blade calculation procedure has been developed by Ni and is based on his thesis work [2]. This time-marching finite-difference approach has been developed to allow calculation of unsteady aerodynamic coefficients required for flutter predictions. Ni has extended his calculation procedure, which already accounts for shock losses, to include stream tube contraction and stream tube radius change through the blading.

For the predictions shown in Figure 7, a relatively coarse calculating grid (20 x 20 points) required approximately four minutes on IBM 370 computer. The stream tube contraction ratio,

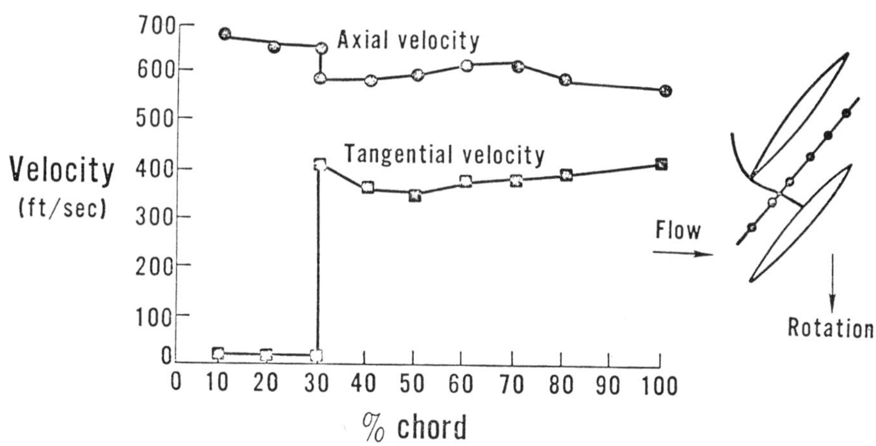

Figure 6. Measured Mid-Gap Velocity Distribution

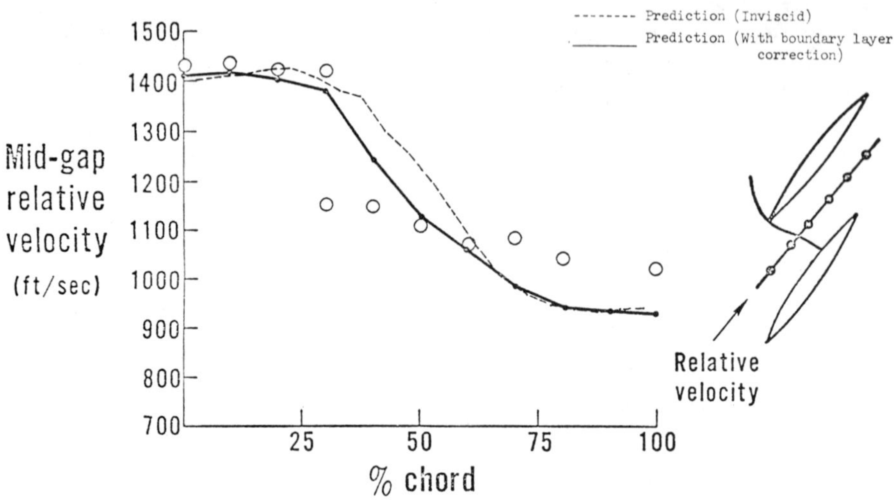

Figure 7. Comparison of Quasi Three-Dimensional Flow Prediction with Measurements

measured across the blade row to be 0.85, was assumed to be distributed linearly in the chordwise direction. It is apparent that the calculated shock is smeared over three grid points and that it is predicted to occur somewhat downstream of the measured shock. The predicted exit velocity is approximately 9% lower than measured.

As a first order correction to the inviscid analysis, the boundary layers on the blade surfaces were calculated using the Fish-McDonald [3] finite difference boundary layer procedure, and the flow field recalculated assuming that the blading thickness distribution has been increased by the boundary layer displacement thickness (Figure 7). The predicted shock location has moved forward, however the exit velocity has remained unaffected by the thin boundary layer calculated for this very efficient blade section. The agreement between measurements and predictions is very good considering that a very coarse calculating grid was used, a simple boundary layer correction was applied, and it was assumed that the flow in the calculation stream tube was unaffected by the adjacent stream tubes, that the blade leading edge was sharp, and that the stream tube convergence could be distributed linearly.

To eliminate some of the above restrictions and to assess

their importance we are currently developing a fully three-dimensional analysis, as described by Tony Ganz earlier at this workshop, and also a conformal mapping and relaxation approach [3] for quasi-three-dimensional calculations which will allow us to design finite leading edge airfoils.

CONCLUSIONS

Accurate intrablade velocity measurement techniques have been developed and are being applied successfully to measure velocities in transonic blading.

Acceptable quasi-three-dimensional analyses have been developed. However comparison between predictions and measurements indicate a need for:

- improved boundary layer analyses which take into account shock interaction and boundary layer separation
- extension of the passage analyses to include airfoils with blunt leading edges. This capability is needed to allow designers to identify high performance practical blading.
- additional detailed flow measurements.

REFERENCES

[1] D. A. Walker, M. C. Williams, R. D. House (1975): "Intrablade Velocity Measurements in a Transonic Fan Utilizing a Laser Doppler Velocimeter," Minnesota Symposium on Laser Anemometry Proceedings, October 22-24, University of Minnesota.

[2] H. McDonald, R. W. Fish (1973): "Practical Calculations of Transitional Boundary Layer," International Journal of Heat & Mass Transfer, 16:9, pp. 1729-1744, January.

[3] D. C. Ives and J. F. Liutermoza (1976): "Analysis of Transonic Cascade Flow Using Conformal Mapping and Relaxation Techniques," AIAA Paper No. 76-370, July.

DISCUSSION OF THE MIKOLAJCZAK PAPER

KERREBROCK: I was very much impressed a while back in reading the von Karman lectures by A. M. O. Smith in which he described the way that people have been able to actually use boundary layer theory to predict the points of separation and, in fact, the details of the performance of very high lift wing systems. It occurred to me that we just don't do that for compressors, even though they have very high lift systems. Do you think it is feasible if we had sufficiently detailed information about the flow field in a compressor to really design a compressor blade in such a way that it minimizes boundary layer loading and thereby minimize the losses?

MIKOLAJCZAK: Yes. In fact, they're trying to do that in some of the work done by Professor Korn and some of the work in house. The difficulty is that we have to have a reliable boundary layer procedure first and Harry MacDonald has been developing one for us that's been very successful. We hope to use that in conjunction with these techniques, to do just that - to give us very low loss blades. The other feature that people don't recognize is that these designs have to operate well off design and this is where a good boundary layer analysis is really useful to you. Because you must have enough stability margin and a very low loss profile and that's the combination you have to strive for.

PLATZER: I wonder if you could tell me a little more about Ni's 2-D oscillating blade computer program. This is a time marching technique, isn't it? (yes) Have you computed for purely transonic flow there?

MIKOLAJCZAK: No. This will work for any velocity. You can go from subsonic to supersonic.

PLATZER: How do you make it 3-D?

MIKOLAJCZAK: It's quasi 3-D because it puts in the stream tube convergence which is in essence like a 1-D analysis which you correct. You change, in fact, the velocity through the streamline.

PLATZER: Do you assume that as far as the inflow is concerned you have subsonic flow close to the hub and supersonic flow close to the tip?

MIKOLAJCZAK: No, I am sorry if I gave you that impression.

Ni's analysis is essentially a 2-D analysis. But it has a correction. If you take a streamline through a rotor, in a 2-D case the streamline height remains constant. If you do have a convergence of the streamline, the streamline shape changes as it goes through a rotor. What he does is make a correction for that change in the streamline height. But if you are asking me, if that streamline is subsonic here or supersonic there, I don't know what he does at that point.

PLATZER: The problem I think one is struggling with is what kind of initial conditions one has to prescribe in the case of a mixed subsonic/supersonic inflow or at least that is the problem I am struggling with, because in the supersonic part you would have to satisfy the unique incidence condition, wouldn't you (yes) and toward the subsonic part you would have to come up with some kind of transition into the subsonic part.

MIKOLAJCZAK: I don't know how that one very special streamline is handled. It's a good point, though.

FARN: You emphasize using an easier method to predict shock loss in one of the slides. (yes) Are you quite happy with the accuracy?

MIKOLAJCZAK: Well, he has checked this out. Certainly, in the one-dimensional flow he gets excellent accuracy. We still have to do some more checking in the multiple shock system. We're not yet quite satisfied.

FARN: When you say 1-D you mean he checked the 1-D nozzle ?

MIKOLAJCZAK: That's right. You get an excellent agreement that way.

FARN: My experience shows that for one-dimensional flow you almost get exact loss but for 2-D you cannot count on it.

MIKOLAJCZAK: We're looking at this right now. It's quite hard to separate out the various components of the loss when comparing predictions to measurements. And this is the part we're trying to unravel right now. But in some calculations we've done, we've found that the thing overpredicted a lot.

NONINTRUSIVE MEASUREMENTS OF THE FLOW VECTORS WITHIN THE BLADE PASSAGES OF A TRANSONIC COMPRESSOR ROTOR

R. SCHODL AND H. WEYER

Deutsche Forschungs- und Versuchsanstalt for Luft- und Raumfahrt

ABSTRACT

In this paper a nonintrusive flow velocity measuring method on an optical basis is described. This technique has been developed for applications in turbomachines, where extremely difficult test conditions are normally present. Some results of flow investigations within the blade passages of a transonic compressor rotor are discussed to demonstrate the new possibilities offered by this technique for future research work on turbomachines in order to increase the performance and efficiency as well as to develop the aerodynamic calculation method.

INTRODUCTION

For experimental investigations in turbomachines, especially for detailed flow studies within the rotor blade channels, optical measuring methods are required.

Using the well established Laser Doppler technique, flow velocity measurements become often more difficult in high speed turbomachinery because of the unfavorable test conditions. This is especially true, when in addition to the high speed problems, the flow channels are very narrow. There often a strong background-radiation is generated by laser light reflections at the hub and at the casing windows rendering any measurements more difficult.

Considering this situation the development of a new optical measuring technique was initiated, which should render such difficulties easier to be solved. Based on a proposal of Tanner [1], a new optic and a data processing system was developed, resulting in a compact measuring apparatus, which is very easy to handle and which allows to change the measuring point without any readjustments of the optical arrangement. This technique is already described in detail by R. Schodl [2] and is called "Laser-Dual-Focus" (L2F) method.

DESCRIPTION OF THE LASER-DUAL-FOCUS METHOD

Similar to the Laser Doppler technique the Laser-Dual-Focus (L2F) velocimeter measures also the velocity of small-light scattering-particles as they are normally contained in every fluid. The principal difference, however, lies in the fact that the fringe pattern in the probe volume is replaced by two discrete light beams. The result is a "light-gate" with - compared to LDV - a 100-times greater light concentration and a corresponding high signal/noise-ratio.

The details within the L2F-control volume are shown in Figure 1. As a result of focusing, both beams have a converging/ diverging cross-section. The minimum beam-diameter is about 5-10 μm, the distance between the beams is fixed to about 0.5 mm and along the optical axis the photo-optics detect particle-scattered light $\pm$ 1 mm off the focal plane. A particle passing both light beams in this range produces two successive pulses of scattered light. Given the distance between the two laser beams, the time elapsed between the pulses yields the velocity of the flow perpendicular to the optical axis.

In order to register this double-pulse, it is necessary, however, that the plane containing the laser beams is parallel to the flow direction; consequently, it is possible to determine the flow direction. The flow angle resolution of this technique depends on the beam diameter-to-distance ratio and is for the set-up described about $\pm$ 1 deg.

The optical L2F-arrangement is illustrated schematically in Figure 2. The center of a polarization (Rochon) prism, used to

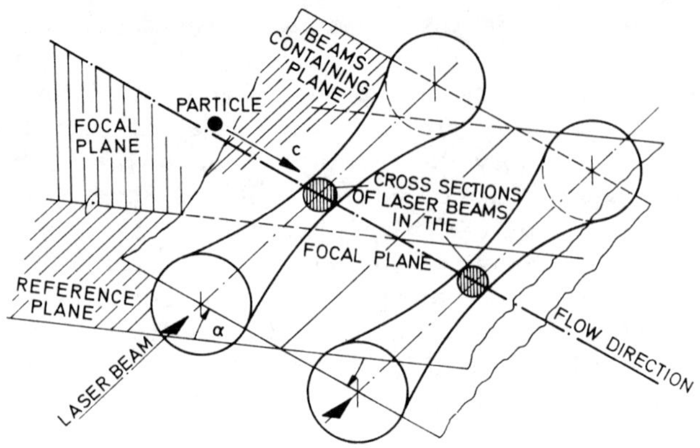

Figure 1. Principal Sketch of L2F-Measuring Volume.

split the initial laser beam (detail A), is located at the focal point of the immediately following lens L_1. As a result, the beam axes leave the lens parallel to each other.

As illustrated in the enlargement of Figure 2, two parallel beams are formed, each of which is highly focused in the focal point of lens L_1 such that the "light-gate" desired is created.

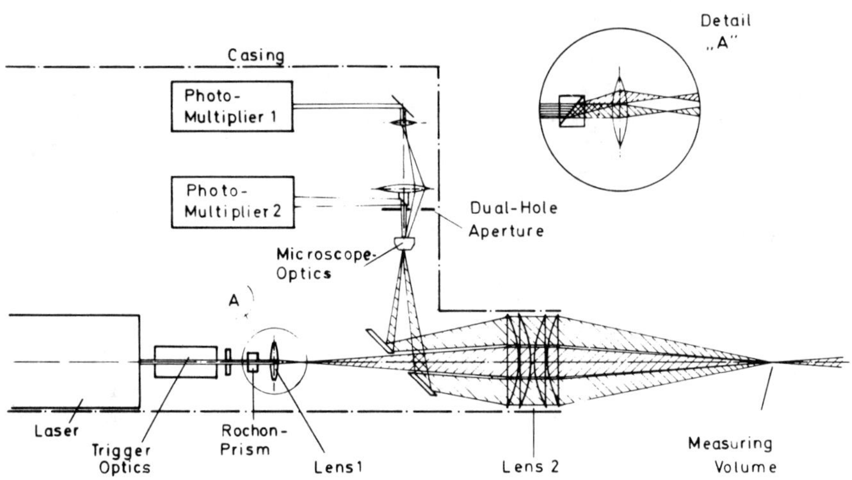

Figure 2. Optical Set-Up of the L2F-Measuring.

A second lens complex L_2 is installed to increase the distance between probe volume and measuring apparatus to the desired length; the same lens system L_2 picks up the backscattered light, which is then turned by a perforated mirror to a microscope. By means of the following optics each of the photomultipliers is adjusted to one of the focused laser beams. The microscope optics and the double-hole aperture minimize the noise level, due to background radiation, so that measurements down to 1.5 mm off the flow channel walls can be performed. The setting angle α is changed by twisting the beam splitter.

Signal processing is carried out with modern electronic equipment of sufficiently high time response, operating in a start-stop mode. In laminar flows only a few measurements are needed to evaluate the velocity vector. Turbulent flows, however, require a considerably greater number of individual measurements. In order to analyze such a flow adequately, it is expedient to adjust the setting angle α in 8-10 steps around the mean flow direction and to take up to 1,000 measurements at each step. The collected data are then classified - by means of a multichannel analyzer - to probability density distributions.

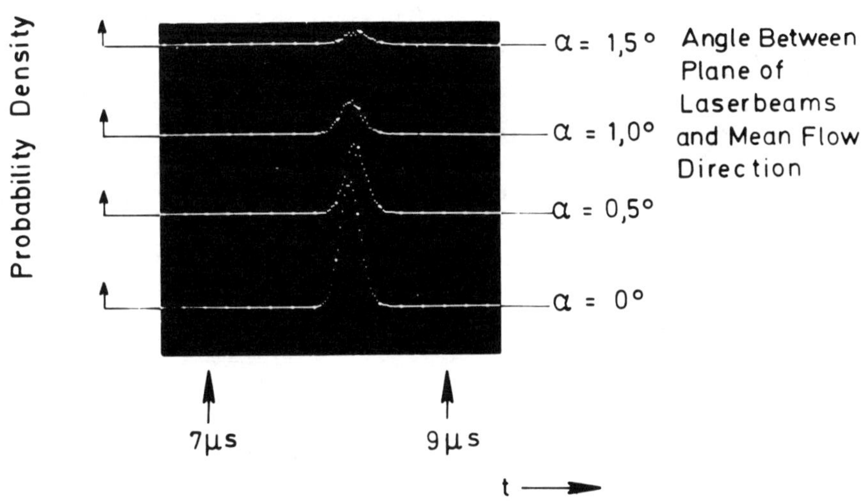

Figure 3. Probability Density Distributions for Different Positive Setting-Angles α-identical Neg. α-distributions Not Shown - Referred to the Same Number of Particles Entering the Probe Volume.

Figure 3 shows probability density distributions corresponding to different setting angles α. At the abcissa the time is plotted which the particles need to pass through both laser beams. The quantity of each time-measurement, corresponding to a distinct velocity in the turbulence spectrum, is arranged along the ordinate.

The probability that a particle, traveling along a streamline, will be irradiated by both laser beams is maximum when $\alpha = 0^o$, corresponding to the mean flow direction. The peak of that distribution curve indicates then the mean velocity whereas the width at the baseline illustrates the maximum velocity fluctuation.

Actually, the mean flow direction can only be detected by an estimation at the beginning of the measuring process; then magnitude and direction of the average velocity as well as the degree of turbulence, parallel and perpendicular to the mean flow vector, can be evaluated accurately by a complete statistical analysis of the probability distributions [3].

In order to carry out measurements within the blade channels of turbomachine rotors, a trigger unit, consisting mainly of a Pockels cell, has been provided. The laser beam is interrupted - controlled by a blade trigger signal from the rotor - and released to a discrete point in every blade channel to take the measurements until enough data for an adequate statistical evaluation are gathered. More detailed information on this technique is given in [4]; the paper deals with measurements in wind tunnels, in turbulent tube flows (L2F-results are compared with hot-wire measurements) and in an axial compressor stage; furthermore, additional equipment for compressor tests is described - a window cleaning device and an apparatus for air-flow seeding with 0.5 μm diam. SiO_2-particles.

FLOW INVESTIGATIONS OF A TRANSONIC AXIAL COMPRESSOR STAGE

During one and a half years, the Laser-Dual-Focus velocimeter is operating at our compressor test stands. Besides the preliminary tests in the subsonic axial compressor extensive investigations in a high loaded centrifugal compressor have been carried out [5]. Furthermore, this technique is used very successfully to analyze the rotor flow of a transonic axial compressor [6].

This transonic stage has been designed for a total pressure ratio of 1.5, a mass-flow of 17.3 kg/s and a rotational speed of 20,260 rpm [7]. The rotor inlet-diameter is 400 mm, the blade height 100 mm. The rotor blades consist of double-circular-arc profiles near the hub and of multiple-circular-arc-profiles from 50 to 100% blade height. The performance data of the stage and the results of detailed investigations of the flow field using conventional measuring methods are described in [8].

The investigations with the nonintrusive technique were concentrated on the rotor flow analysis at 55% design speed with a mass-flow of 8 kg/s. This is an operating point between maximum isentropic efficiency and surge line, where also detailed results of quasi-three-dimensional potential flow calculations (not including friction forces) were available for the rotor.

Figure 4 shows a part of the compressor flow path. The horizontal lines denoted by the numbers 1 to 11 designate the computed stream surfaces along which the blade-to-blade calculations were carried out, the vertical lines - numbers 1 to 38 - indicate the axial planes. The theoretical results are available in the intersection points of both the line types. In consequence, it was obvious to carry out the experiments with the L2F-velocimeter as far as possible in the same points. Over one blade spacing as far as 15 distinct measuring points were selected - especially just downstream of the rotor. Inside of the blade channels, however, it was not always possible - due to the strong blade twisting - to measure immediately in the vicinity of the blade pressure and suction surfaces, except near the tip.

The dashed lines in Figure 4 demonstrate the actual leading and trailing edge position of the rotor blades at 55% speed; the thick full lines indicate the blade position at rest. During operation the blades become considerably untwisted ($\simeq$ 2.5 deg. at tip) and - at the same time - tilted upstream. It is evident, that the difference in the blade position leads to discrepancies between the real flow and the corresponding calculated data, because - above all - the experimental incidence angles do not agree with those of the computation.

The Figures 5 to 7 represent the relative flow Mach number distribution M_{aw} in the planes 14, 19 and 38 (compare Fig. 4), shown in a perspective view against the flow direction. The computed data are always plotted on the left hand diagrams, the experimental results on the right hand graphs; thereby the circles mark

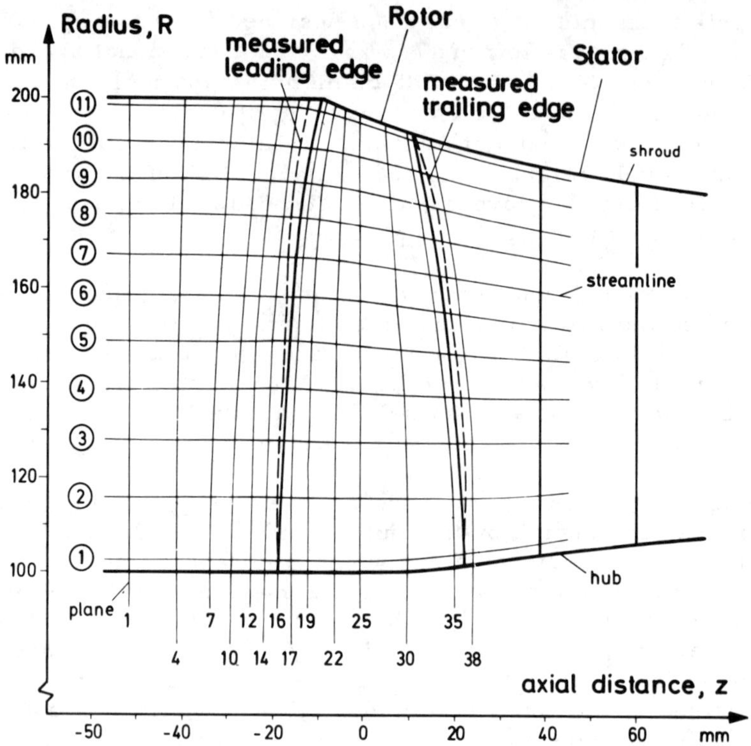

Figure 4. Compressor Flowpath with the Computational and Measuring Grid.

the measuring points.

Figure 5 illustrates the relative Mach number distribution just upstream of the rotor blade row (plane 14) and demonstrates the strong effect of the rotor on its upstream flow-field. Furthermore, it makes evident the increase of the relative velocity from hub to tip and also from blade pressure to suction side. Thereby, the same tendency of flow behavior is observed with the computed and the test results. In the right hand plot the small area of reduced relative velocity near the blade leading edge indicates roughly the position of the stagnation streamlines. Up to 70% blade height the calculated and measured relative Mach numbers agree quite well, whereas in the tip region (near the suction side) the calculated velocity decreases considerably below the experimental values,

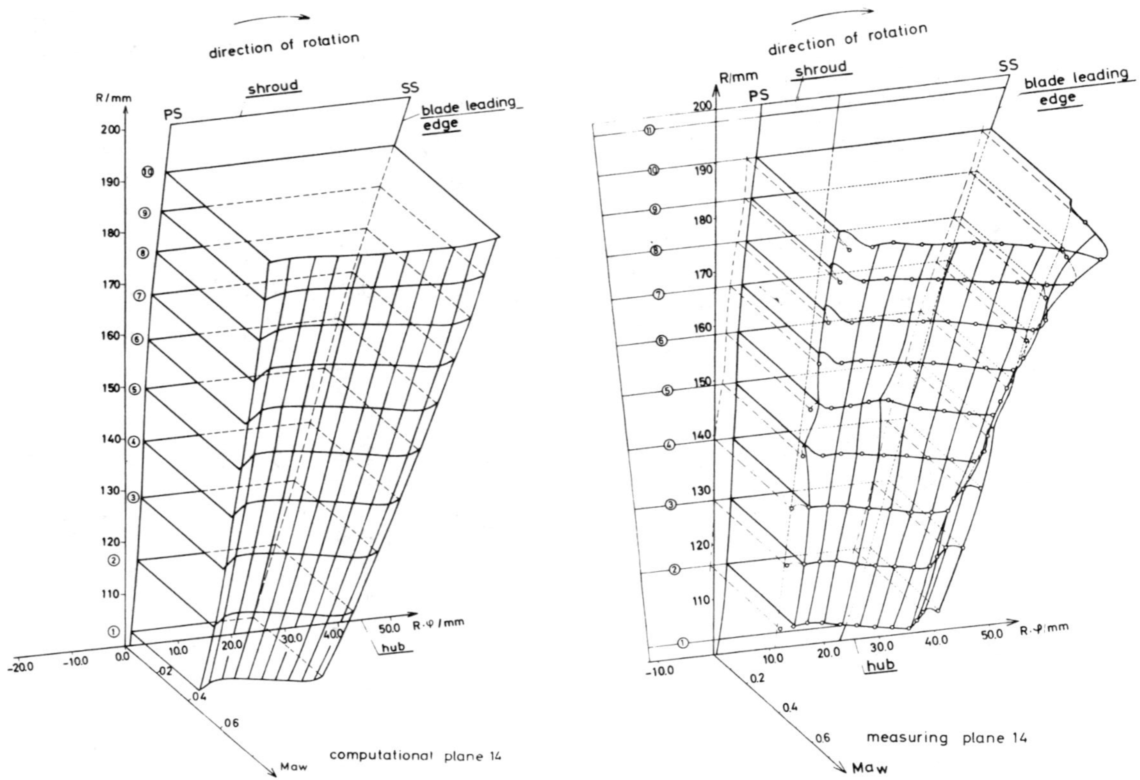

Figure 5. Perspective Plot of Computed and Measured Relative Mach Number, M_{aw}, Over One Blade Channel (View Against Flow Direction). Distribution Just Upstream of the Blade Row.

primarily due to numerical problems in the consequence of the high inflow Mach number. Another reason may be the flow separation in the sharp corner of the outer casing just in the leading edge plane (Figure 4).

Figure 6 shows the relative Mach number (M_{aw}) distribution short behind the blade leading edge (plane 19). Again, a tendentiously good agreement between theoretical and experimental data is achieved. Stronger differences occur only in the tip region.

Figure 7 represents the distribution of the relative Mach number M_{aw} in plane 38, just downstream of the blade trailing edge. Remarkable in these plots is the blade wake, clearly identified by the experiments. The extrapolation of the blade trailing edge to the measuring plane - under consideration of the circumferential velocity component - indicates the wake being in the extension of the blade suction side.

The tendency of calculated and measured distributions agrees quite well, however, the measured values always exceed the theoretical data due to boundary-layer and flow separation displacement which surely occur in this off-design point where the stage is operating near the surge line.

Figure 8 illustrates as a further noticeable result of the measurements behind the rotor that the relative flow angle varies strongly within the wake. In this case the variation amounts to about 20 deg.

Comparing the measured results with the calculated data great differences exist which have been expected, because the potential flow theory does not predict friction effects as boundary-layers, flow separations and secondary flows. These effects may be better described surely by 3-D or quasi-3-D computational methods including turbulent boundary layer calculations, which are just under development.

Not only the amount and the direction of mean flow vectors can be determined by these nonintrusive measuring techniques, but also the flow fluctuations. This quantity is, e.g., very helpful in interpreting the development and the position of flow separation and is also discussed in [6].

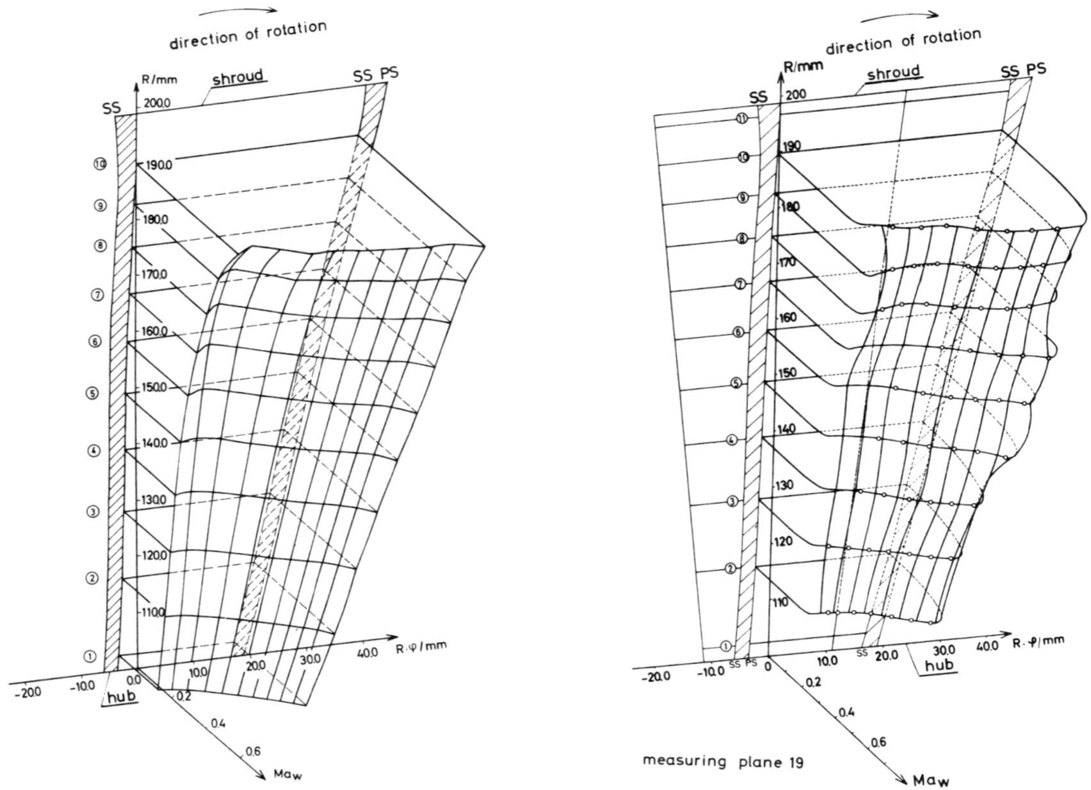

Figure 6. (Continued) Distribution at 15% Blade Chord.

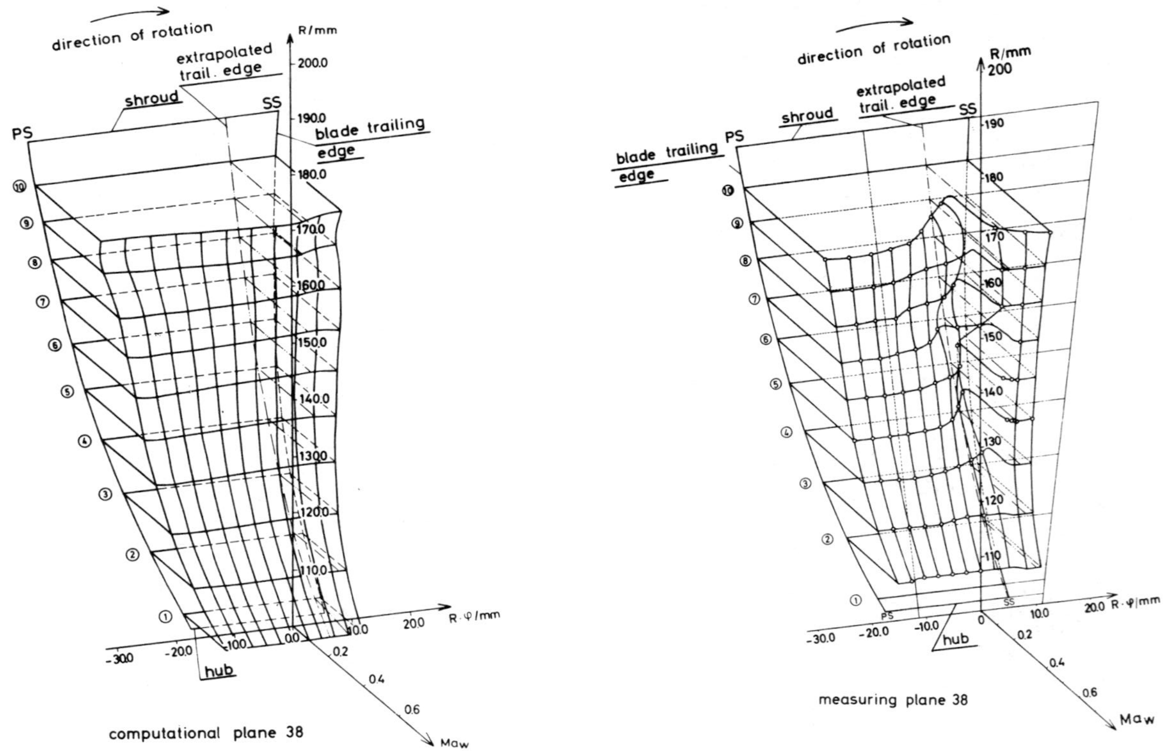

Figure 7. (Continued) Distribution Just Downstream of the Blade Row.

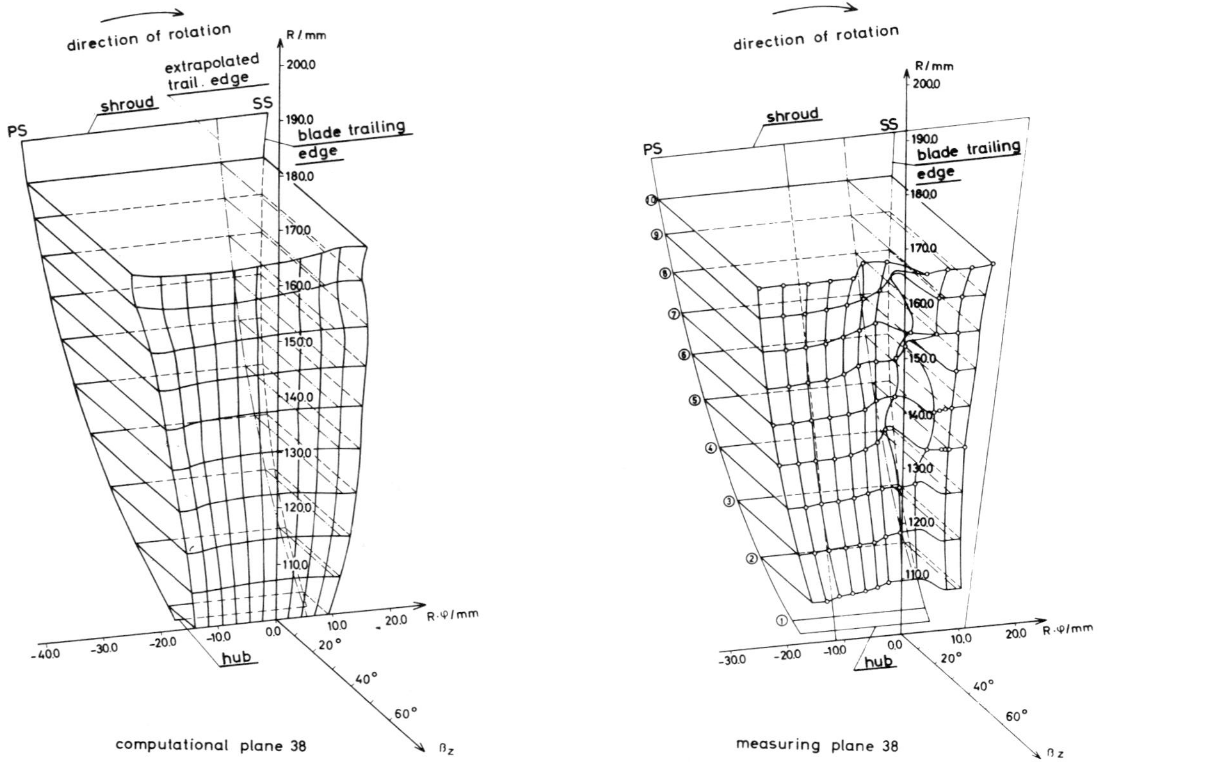

Figure 8. Perspective Plot of Computed and Measured Relative Flow Angle over One Blade Channel. (View Against Flow Direction)

VALUATION AND OUTLOOK

The optical "Laser-2-Focus" velocimeter is now in operation for investigating the internal flow in modern axial and centrifugal compressors since more than one year ago.

The obtained results up to now demonstrate the high capability of the technique for experimental studies of the complex real flow pattern in turbomachines. Three-dimensional flow processes can be analyzed point by point, effects of strong streamline curvature and of field-forces as well as the development of flow separation can be investigated, although measurements directly inside of blade - and wall boundary layers are yet difficult today. However, it seems realistic to study in detail 3-D shock waves - with respect to their position and intensity - within the blade rows.

Experimental research work like this offers new ways to accomplish step by step today's experiences of the real flow phenomena in turbomachines. It will help to prove computational techniques and to develop more realistic and reliable flow models as the basis of theoretical work.

REFERENCES

[1] Tanner, L. H. (1973): A particle timing laser velocity meter. Optics and Laser Technology, pp. 108-110.

[2] Schodl, R. (1974): On the development of a new optical method for flow measurements in turbomachines. ASME Paper 74-GT-157.

[3] Rotta, J. C. (1972): Turbulente Strömungen. Teubner Verlag Stuttgart, pp. 13-62.

[4] Schodl, R. (1975): A laser dual focus velocimeter for turbomachine applications. VKI-Lecture Series, Advanced Testing Techniques in Turbomachines, April 14-18.

[5] Eckardt, D. (1976): Detailed flow investigations within a high-speed centrifugal compressor impeller. ASME Gasturbine Conference, New Orleans.

[6] Dunker, R., Strinning, P. and Hungenberg, H.G. (1976): Comparison of experimental and theoretical results of flow studies on blade-to-blade - surfaces in an axial compressor rotor. VKI-Lecture Series on Transonic Blade-to-Blade Flows in Axial Turbomachinery.

[7] Weyer, H. and Strinning, P. (1974): Auslegung und Nachrechnung einer transsonischen Axialverdichterstufe 1. Zwischenbericht FVV, Heft 158.

[8] Strinning, P. and Hungenberg, H.G. (1974): Untersuchung der Strömung in einer transsonischen Axialverdichterstufe. 2. Zwischenbericht FVV, Heft 159.

DISCUSSION OF THE SCHODL-WEYER PAPER

KERREBROCK: What is the time resolution of the measurements?

SCHODL: With respect to what question?

KERREBROCK: How fast? What is the frequency response?

SCHODL: It's not a Laser Doppler velocimeter, it's a very special velocimeter, not measuring frequency but time differences and you can measure perhaps 10000 m/sec with 1% accuracy from the electronics part. From the optical part, and from the intensity decrease with increasing velocity, I believe, 1000 m/sec is a realistic upper limit for this technique.

KERREBROCK: What I'm asking is what's the frequency response in intensity; if the flow is fluctuating, then how high a frequency can you resolve?

SCHODL: As far as I know, there is no real method of today to measure the time response of flow fluctuations with optical methods. There are some workers I know who solved that problem but because we measure particle velocity and the particles are so particulated, so slow, that it seems to be impossible in gas flows at the moment to solve that problem.

KERREBROCK: Let me rephrase the question. You showed in this last slide, fluctuation levels. In particular, there was a 40% fluctuation level near the trailing edge of the suction side, is that a time dependence in the rotor?

SCHODL: No.

KERREBROCK: You interpreted it as having something to do with separation so that that implies an unsteadiness in the flow field.

SCHODL: Sure. Because separation leads to an increasing of fluctuation tendency of turbulence. I don't say turbulence because we measure in every blade channel and there are three things which are overlapping. First, you have the turbulent flow; second, you have recirculation regions; third, you have not the same flow in every blade channel.

KERREBROCK: It's the last that I'm asking about. Whether your technique can really tell you about blade-to-blade.

SCHODL: Yes. We have the chance to measure only one and the same channel every revolution. We can compare it and we found no significant difference to date.

RUNSTADLER: This is the same question I wanted to ask the last speaker. These are basically two component velocity measurements. I presume you will do what we hope to try to do and that is to take a continuity balance on an axial machine. Have you made a continuity balance to check the 3-D components that might be in the inter-blade row?

SCHODL: As far as I know, no, because the third component is not taken into account and there is a very great third component, especially in the tip region because the casing has a very sharp curve.

RUNSTADLER: You could do that except that you have a lot of blade lean. So you get some shadow areas which make it difficult, I presume, to make good mass balances.

SCHODL: Maybe not up to now, but a lot of results have been taken - but not everything has been calculated.

A TRANSONIC/SUPERSONIC SMOKE TUNNEL FOR THE INVESTIGATION OF CASCADE LOSS MODELS

WILLIAM B. ROBERTS, THOMAS J. MUELLER,
AND VINCENT P. GODDARD

University of Notre Dame

In the past, the High Speed Smoke Tunnels (HSST) at the University of Notre Dame have been used to visualize supersonic flow patterns by means of smoke lines [1] and to visualize shock location and intensity without probes in supersonic wakes [2].

A schematic diagram and test section photograph of the HSST set up for preliminary cascade testing is shown in Figure 1. The HSST consists of inlet screens, contraction, nozzle and test section, and diffuser. The tunnel is of the air indraft type that can be connected with up to three 1.48 m^3/sec vacuum pumps.* The cascade flow can be visualized and recorded using the smoke generator and rake in combination with regular photography, a shadowgraph system, a schlieren system and, in the near future, a Thermo Systems, Inc., 2-watt laser velocimeter. The tunnel has a contraction ratio of 100:1 between the inlet and throat and there are seven screens at the inlet to reduce the turbulence of the flow. The present cascade test section is 3.75 x 10.8 cm (40.5 cm^2) with an area ratio of 1.26 between it and the throat. The cascade is composed of six flat plate blades with a leading edge wedge angle of 15°, thickness ratio of 0.046, stagger angle of 35°, solidity of 1.25, and aspect ratio of 1.5. A streamline (i.e., steady flow smoke line)

*If three vacuum pumps are used, the test section area can be as high as 120 cm^2 for choking.

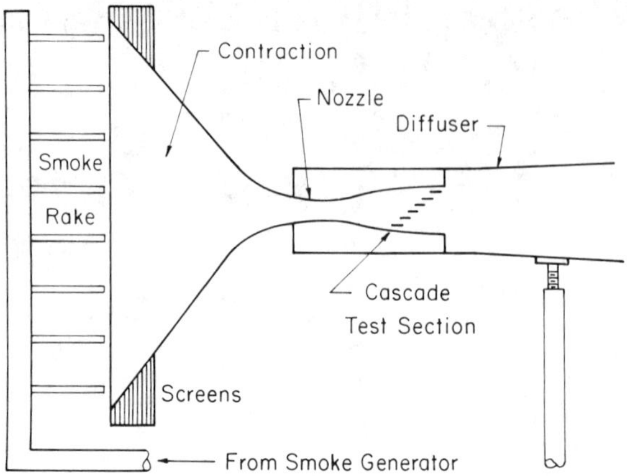

Figure 1a. Schematic of Transonic/Supersonic Cascade Smoke Tunnel

visualization photograph of this cascade is shown in Figure 2. The smoke is generated by dripping kerosene onto electric strip heaters and is forced by a blower through a vertically and laterally adjustable smoke rake placed flush with the first screen at the inlet. The

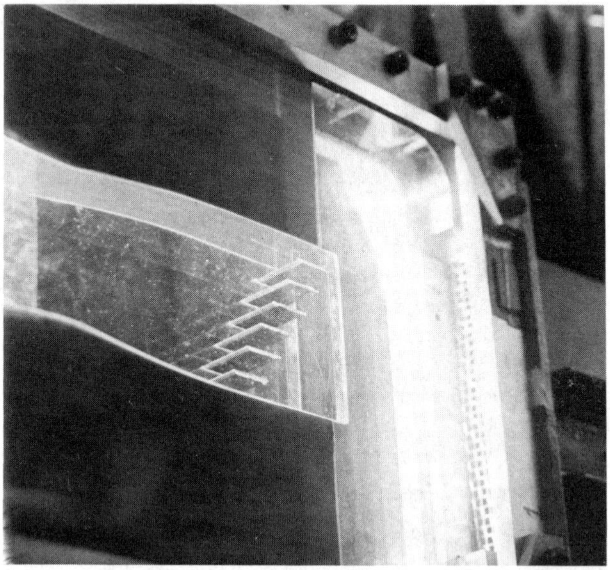

Figure 1b. Photograph of Cascade Test Section

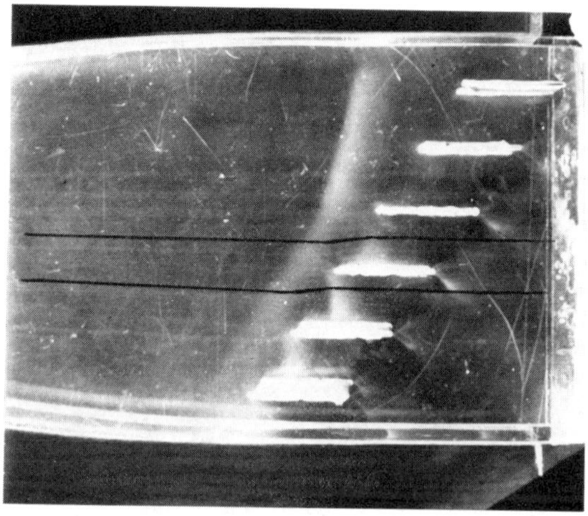

Figure 2. Smoke Line Visualization of Cascade Flow at High Transonic Mach Number

smoke is then sucked into the contraction and through the test section in continuous filaments or lines. In the photograph of the cascade section one can see the smoke filament streamlines intersecting with the passage shocks (two of the smoke lines have been darkened to enhance the visualization of the flow).

From the deflection of the smoke lines through the shocks one can have an approximate idea of the flow field around the entrance region. It appears that ~20% of the chord ahead of the leading edge there is an oblique shock nearly parallel to the cascade front, with a normal shock in the passage standing near the leading edge.* From the approximate oblique shock inclination and smoke line deflection the upstream and downstream Mach numbers are calculated to be 1.45 and 1.15, respectively. This gives a nominal blade chord Reynolds number of ~400,000.

As an illustration of the types of cascade flow visualizations that will be possible in the future, Figure 3 shows some results from a supersonic wake flow study using simultaneous direct smoke

---

*The light opaque areas shown in Fig. 2 are caused by dirt deposited where the shocks interact with the side wall boundary layers. However, because of the interaction effects, the side wall shocks are shown slightly ahead of the mid-span locations.

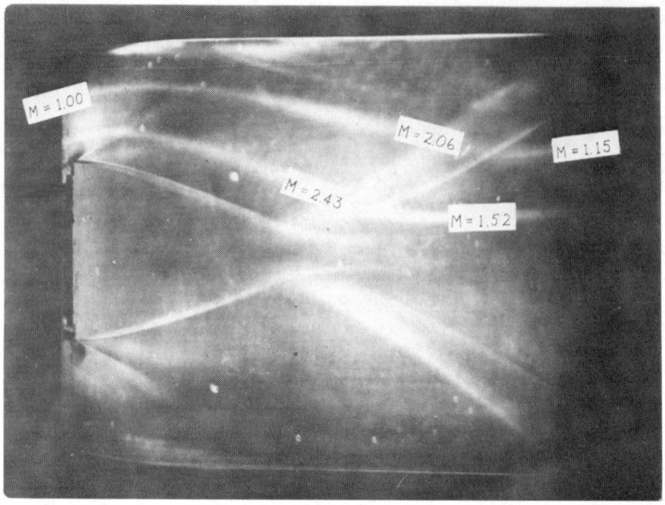

Figure 3. Simultaneous Smoke Line and Opaque-Stop Schlieren Photograph of Supersonic Wake Flow [2]

lines and opaque-stop schlieren* photography. For steady flow the smoke lines can be used in conjunction with shock wave location and model geometry to determine Mach numbers along streamlines. A correlation between this data and data measured using pitot total and static probes is shown in Figure 4 for the Mach number downstream of the last recompression shock of Fig. 3. The correlation is quite good. These visual techniques coupled with a laser velocimeter (to determine velocities along the smoke lines) will allow complete flow field investigation and determination without the use of probes. This is especially useful in the high subsonic and low transonic flow regimes.

Future research in the HSST cascade tunnel will be done on shock location and intensity, shock wave/boundary layer interaction, and wake flows for transonic axial compressor and turbine cascades.

---

*A black background is necessary for simultaneous schlieren and smoke photography. The black background is obtained using an "opaque-stop" schlieren system. Details can be found in [2].

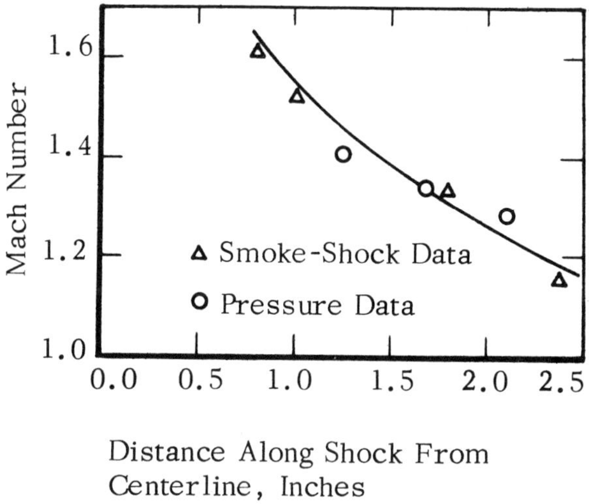

Figure 4. Correlation of Smoke Line Shock Data and Pressure Data for Mach Number Immediately Downstream of the Recompression Shock [2]

REFERENCES

[1] Goddard, V. P., McLaughlin, J. A. and Brown, F.N.M. (1959): "Visual Supersonic Flow Patterns by Means of Smoke Lines," J. Aero/Space Sci., 26:11.

[2] Mueller, T. J., Hall, C. R. and Sule, W. P. (1969): "Supersonic Wake Flow Visualization," AIAA Journal, 7:11.

DISCUSSION OF THE ROBERTS, ET AL. PAPER

MIKOLAJCZAK: I fully agreen that one can use a cascade in a limited situation to get good insight into the flow. However, I would like to comment that when you do put a cascade together, you need to take a lot of care. And from what I've seen on your slides, this cascade isn't going to be giving us any useful information.

ROBERTS: Oh, no. I agree.

MIKOLAJCZAK: If you agree with that, I have no comment.

ROBERTS: Let me comment on that. This cascade is very low aspect ratio. It was operating very near complete spill where you push a normal shock right out in front; I think the blockage was quite high.

MIKOLAJCZAK: Let me make a few specific comments. Can I have that slide on. The slide showing the cascade pack and smoke in it? First of all, you say that this is a dump diffuser and it is not really. If you want to use a dump diffuser, I suggest that you have to start dumping uniformly at the trailing edge on both sides. Otherwise, you find that the flow is not periodic at the back (yes) and the signal is propagated up front and every passage is different. That's the first point. The second is that the shock up front as you notice is not uniform and therefore it says that you have a completely nonperiodic flow in the cascade. Every passage is different. And you're going to have that even if you fix the dump diffuser problem, because you're not cancelling the shock on the topside where it's reflecting back into your cascade and again messing up the whole flow in a cascade. So I really suggest that you look through the literature on cascades; there is a lot done at Pratt and Whitney and a lot that's been done and published by Starken which will help you no end to set up, I think, possibly useful cascades.

ROBERTS: Yes, I agree. Dr. Mikolajczak has bleed – boundary layer bleed - at the top to alleviate this difficulty of shock-boundary. I'm going to incorporate a scheme to connect up bleed at the top. The purpose of this was to convince myself that I could see the smoke through the thing, and with the idea that I use that smoke for seeding. This is certainly not a cascade that I expect to get any information out of, except perhaps, to make some traces along some of the streamlines perhaps with the anemometer, as far as calibration is concerned. If we use the three vacuum pumps, we will have the capability of going up to about 120 cm^2 and if you use that area judiciously with the things that you've talked about, I think that we can have a proper cascade in the traditional sense.

WENNERSTROM: I have never seen smoke used at such high Reynolds numbers.

ROBERTS: The Reynolds number for the cascade was approximately on the order of 400,000. The length of the cascade was about 2.5 cm.

WENNERSTROM: What light source did you use?

ROBERTS: High intensity strobes that were above the cascade and below it, feeding both sides of it, keyed to the high speed polaroid system. So when you fired the strobes, you fired the shutter automatically, and obtained the maximum light intensity in it. Unfortunately, again, there was no time to completely block the clear lucite walls and there was some diffusion that you saw in the side walls as far as scratches were concerned. We hope to alleviate that also.

A LASER VELOCIMETER SYSTEM FOR SMALL RADIAL TURBOMACHINERY

PETER W. RUNSTADLER, JR., AND FRANCIS X. DOLAN

Creare Incorporated

ABSTRACT

This paper describes the design and test of a laser Doppler velocimeter to map the velocity fields in small, high speed turbomachinery. This instrument is presently being used as a diagnostic tool for the study of the basic fluid dynamics of the inducer, impeller and the diffuser regions of high pressure ratio, small, centrifugal compressors.

The LDV instrumentation has been optimized to permit the measurement of instantaneous velocities up to approximately 500 m/s, measured in absolute coordinates, within a rotating impeller.

The results of this program have shown that the laser Doppler velocimeter (LDV) can make non-disturbing gas velocity measurements and surmount the difficulties found in applying other types of aerodynamic instrumentation, e.g., hot-wire anemometers, to this measurement problem.

The work reported herein has been supported by the U.S. Army under NASA/USAAMRDL Contract NAS3-17860.

INTRODUCTION

This instrumentation system has been developed for use as a diagnostic tool in the study of the basic fluid dynamics of high pressure ratio, centifugal compressors. The use of a laser Doppler velocimeter (LDV) was seen as an opportunity to apply a novel technique to make possible velocity measurements within the very small passages of high pressure ratio, centrifugal compressors. The non-contact nature of the LDV made it ideally suited for velocity measurements in the critical flow regions of the centrifugal compressor where small aerodynamic probes introduced sufficient flow disturbances to significantly alter the compressor stage performance [1].

The details of the LDV system design were tightly constrained to optimize the system performance in the specific compressor and test facility in which it is to be used. The LDV equipment was designed specifically for use in the low speed modeling (LSM) compressor test facility at Creare Incorporated. This facility maintains similarity of geometry, Mach number, Reynolds number and ratio of specific heats, but utilizes lower rotational speed by using as the working fluid a gas with a lower speed of sound than air. High speed compressors can be tested at rotational speeds approximately 0.65 to 0.70 of the value that must be used in air. The basic advantage of the LSM facility for LDV work is that lower absolute fluid velocities in LSM testing reduce the design constraints placed upon the LDV instrumentation.

Some of the important variables considered in specifying the LDV system were the size of the sampling volume relative to the compressor scale, available signal processing methods, data sampling in a flow field with spatial and temporal fluctuations and the positioning of the probe volume within the compressor hardware. The design and development program had, therefore, to consider the following technical areas:

1) the LDV optics design needed to probe the compressor hardware and test facilities,

2) the LDV mechanical subsystems design to permit proper positioning of the probe measuring volume and to permit remote operation of the equipment during compressor test,

3) the electronic equipment design required to provide for data sampling, validation, and storage, and

4) ensure that the equipment design met the specifications for the aerodynamic research needs.

The following sections of this paper discuss these areas and

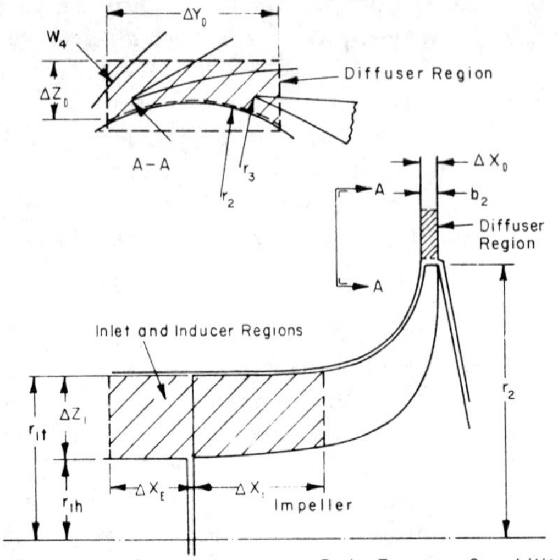

Compressor Geometry		Probe Traverse Capability	
r_{ih}	= 21.8 mm	ΔX_E	= 6.35 mm
r_{it}	= 43.7 mm	ΔX_I	= 15.3 mm
r_2	= 79.8 mm	ΔZ_I	= 21.9 mm
r_3	= 85.85 mm	ΔX_D	= 3.22 mm
b_2	= 3.22 mm	ΔY_D	= 50 mm
W_4	= 10.5 mm	ΔZ_D	= 20 mm

Figure 1. Compressor Regions of Interest in LDV Measurements

display the results thus far obtained with this LDV system.

OPTICAL AND MECHANICAL SYSTEMS

The principal objective of the LDV system is to make mean velocity measurements in the centrifugal compressor in the regions of interest shown in Figure 1. This objective requires that:

1) the laser probe volume be moved to any position within the measurement space and done so remotely while the compressor is under test on the test stand,

2) the probe volume dimension be small (compared to the compressor dimension and expected velocity gradients) to obtain good spatial resolution,

3) the highest anticipated velocities occurring within the compressor be measured, and

4) the velocity data be obtained when the impeller is at a known rotational position.

The LDV measures fluid velocity in a fixed reference frame. The compressor hardware constraints permit only the measurement of the flow velocity vector in a plane normal to the LDV optical axis. The optical axis is oriented in a radial direction in the inducer and in the axial direction in the impeller exit/diffuser entry region. Therefore, the velocity vector is measured tangent to circumferential planes in the impeller exit/diffuser entry region; however, because the actual mean flow velocity vectors are expected to lie primarily in these planes, this is not considered a major problem in understanding the flow fields in these two regions.

From Figure 1, it is seen that the minimum passage size of interest in these studies is the depth of the impeller tip/diffuser entry region b_2. Thus the passage must be probed with the axis of the input optics parallel to the compressor axis in order to map this essentially two-dimensional velocity field. In order to measure the velocity profile across the depth b_2 of the diffuser channel, the probe volume length ℓ should be kept considerably smaller than b_2.

At the outset of this work, the predominant frequency measurement systems developed and used by others consisted of automatic frequency trackers or automatic frequency counters. Primarily because of the possibility of signal drop-out in highly fluctuating flow fields, the frequency tracking method was eliminated from consideration as the possible signal processor. A review of the frequency counting methods shows a decrease in the accuracy of measurements as the signal frequency approaches the time base frequency of the processor. The error derives from the need to count time base periods against an integral number of signal periods (or vice versa). This results in a digitizing error because a fraction of a period must always be missed. A commercial frequency counting system most readily available at the beginning of this program was utilized. This system has a time base frequency of 100 MHz and requires 8 cycles of Doppler signal for period measurements. This unit also has a limit on the input Doppler frequency of about 100 MHz.

The basic frequency measurement uncertainty and the limitation on band width of the available Doppler frequency processor dictated a limit to the maximum Doppler frequency, produced by a particle traversing the probe volume, of approximately 80 MHz.

However, this maximum frequency, at the highest expected flow velocity (~ 400 m/s), implies a minimum fringe spacing of 5μm. This is larger than the desired optimum fringe spacing of 2μm. This 5μm spacing implies a probe volume longer than desired if about 10 fringes are to be available for data processing by the processor unit.

Since the probe volume length was considered to be a major design constraint, it was decided to allow an upper Doppler frequency of 160 MHz. This would result in a shorter probe volume. But, this high frequency is well above the band width of the processing electronics (~ 100 MHz), and also results in count digitizing errors as high as 20% on individual velocity measurements. This problem was solved by incorporating a frequency divide-by-two circuit ahead of the processor, thereby shifting down to 80 MHz, the maximum frequency which must be counted.

After careful consideration of all design tradeoffs, the optical arrangement for the LDV system has design parameters as displayed in Table I. In summary, this design meets the major specifications and design constraints of probe volume size minimization, yet is compatible with available electronic processing instrumentation.

Figure 2 shows an unfolded, two-dimensional view of the optical train of the LDV system. In order to position the probe volume anywhere within the compressor hardware, part of the optical system was made traversable in three mutually perpendicular directions.

The selected laser is one of the argon ion type with a maximum output of about 1400 mw at 514.5 nanometers. The output beam diameter D_i is 1.5 mm at the $1/e^2$ relative intensity points and has a divergence of 0.5 milliradians.

This entire mechanical system has been designed to provide reliable positioning and robust support of the delicately aligned optical components and simple, accurate traversing of the probe volume relative to the compressor.

ELECTRONIC DATA ACQUISITION

A remaining major specification of the LDV system is that it be capable of acquiring velocity data for known and fixed positions

TABLE I - PROBE VOLUME AND LDV SYSTEM CHARACTERISTIC PARAMETERS

Parameter	Description	Value
D_L	laser beam diameter at input focusing lens, at the $1/e^2$ relative intensity points	0.875mm
f	focal length of input lens	78.5 ± 0.2mm
f/D_L	"f/number" of input beam and lens	90
ϕ	beam intersection angle	11.2 ± 0.07°
D_A	diameter of intersecting beams at focal point of lens	0.0588mm*
ℓ	probe volume length, intersection length of two cylinders of diam-diameter D_A, at angle ϕ	0.602m*
w	probe volume width	0.0591mm*
S_f	spacing of fringes in probe volume	2.636μm*
N_f	number of fringes in probe volume width w	22*
$\Delta f_D/v_x$	velocity to frequency conversion factor	0.379 ± 0.003* MHz/(m/s)
	focal length of collecting lens	98mm
	pinhole aperture diameter	125μm
λ	wave length of radiation from laser	514.5nm
	probe volume positioning uncertainty	± 0.15mm and ± 1 degree

of the impeller. This has been accomplished by strobing the data acquisition system with a pulse synchronized from an impeller position indicator. By acquiring and storing data only for that impeller position, a velocity probability distribution (histogram) is built up which measures the velocity for repeated rotations of the impeller only when the impeller is at a known and fixed position.

The data window position and window width are set by a gate control system. The control circuitry is designed to sample once per blade passage for each third impeller revolution. During the first revolution, the impeller speed is measured with a high frequency clock. The next impeller revolution is used to calculate (based on the measured wheel speed) the desired delay time before the gate is opened. The gate is actually opened and closed during

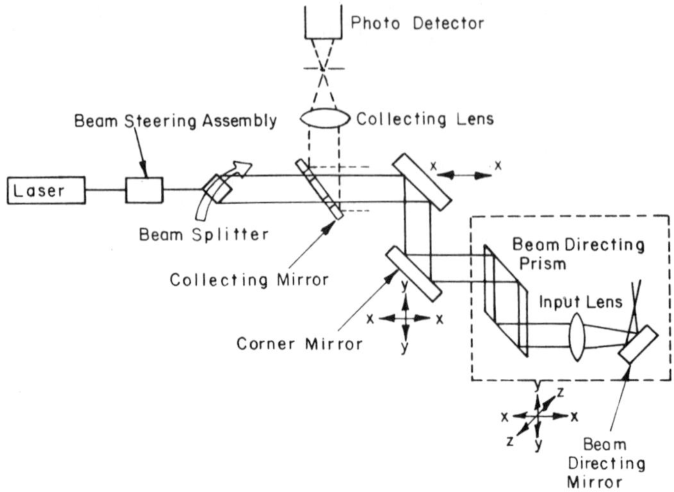

Figure 2. LDV Optical System - Unfolded View

the third revolution. With one velocity measurement per gate opening, and a gate opening for each blade channel (there are 19 blade channels in this particular impeller design), the maximum theoretical data rate is about 5700 Hz when operating in this fashion. The resolution of the gate opening position is 0.1% of blade pitch, with window width ranging from 60 ns to 6 ms (0.1% blade pitch to 100 blade pitches at 100% design speed).

The total electronic systems which detect, measure and process the Doppler signal are shown in Figure 3. The spectral response of the photomultiplier tube (PMT) is S-20 with a cathode sensitivity of 50 ma/W at 514.5 nm. Output from the PMT is taken to an instrumentation console from the test facility via an 8m coaxial cable. The signal is continuously monitored on an oscilloscope and is parallel directed to either the Doppler frequency processor (DFP) or to the divide-by-two electronics (the output from which is then fed to the DFP). The output from the frequency processor may be selected as a measure of either the time required to process 8 Doppler cycles or its inverse, which is proportional to the velocity of a particle traversing the probe volume. This data is presented at the output terminals in both analog and TTL compatible digital form along with a coincidence pulse of about 1 μs duration.

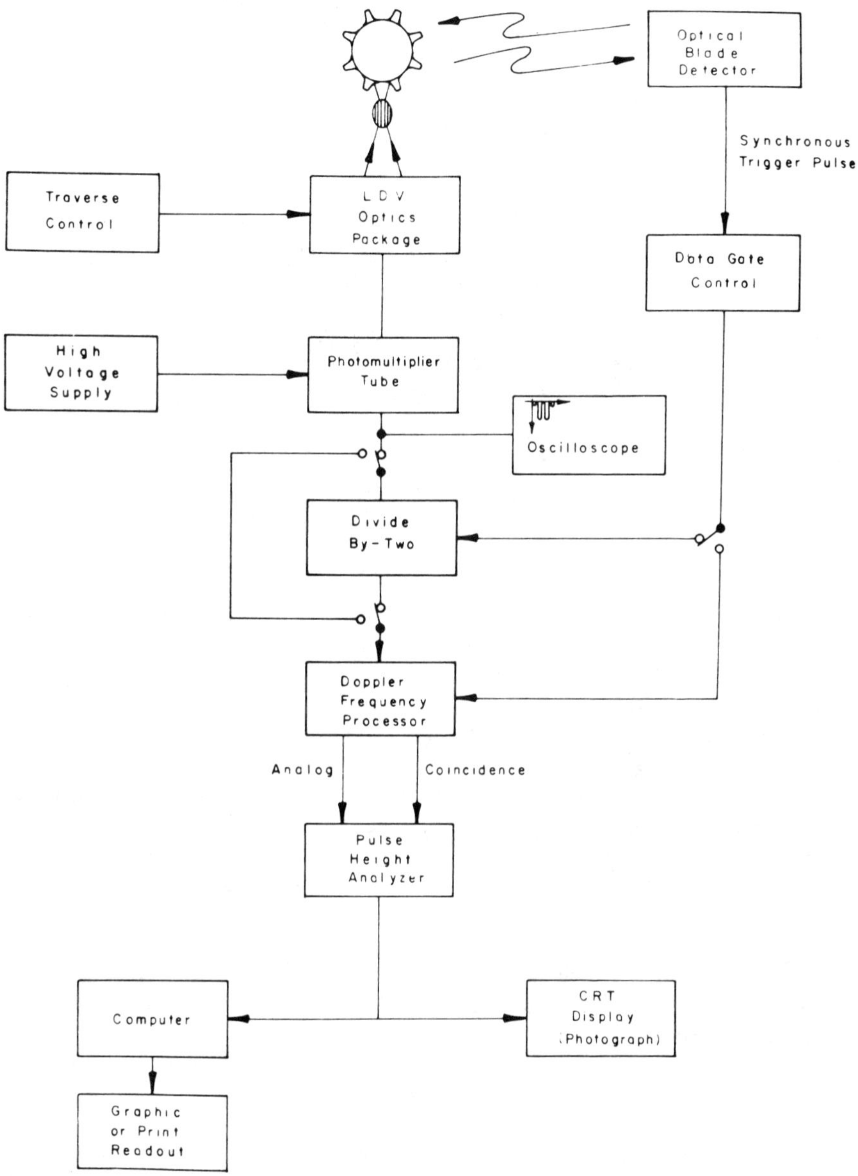

Figure 3. Electronics System

Figure 4. Assembled LDV Hardware on Seismic Bed.

In order to record and store the velocity data of a point velocity measurement, a pulse height analyzer (PHA) is used to store each output pulse from the DFP. The PHA accepts the analog voltage signal and the coincidence pulse, and logs one count into an appropriate channel corresponding to the voltage (velocity or time) level. Using the PHA, a velocity probability distribution or histogram is built up. This histogram displays the number of counts versus channel number or velocity. The information is displayed on a CRT for a real time analysis and is stored in the PHA memory for later output onto punched paper which is used for ultimate data reduction.

The complete optical/mechanical system is shown in place in Figure 4 installed on its seismic bed. The top of the bed plate is just below laboratory floor level to provide clearance under the inlet plenum, which is wheeled over the assembly for compressor operation.

The complete optical train is shielded from dirt contamination and possible damage by a cover shown in place in Figure 5. The traversing periscope, shown protruding out of the large flat box, is sealed to the box by a flexible rubber "boot." The small opening

Figure 5. Assembled LDV Hardware with Covers in Place

on front of the periscope head provides the output path for the focused beams to explore the turbomachinery region of interest.

CALIBRATION TESTS

A set of calibration tests were made to measure the transverse velocity profiles in a freely expanding jet, and to compare the results with pitot static and hot-wire anemometer measurements. Results are plotted in Figure 6 as non-dimensional velocity U/U_m where U_m is the mass flow averaged velocity at the jet exit plane, as a function of radial position, z/R. As the measurement plane is moved further from the jet exit plane (y/D increasing), the velocity profile becomes more peaked as the jet core begins to dissipate via turbulent mixing.

The unsteady velocity components corresponding to the mean velocities of Figure 6 are shown in Figure 7. For the LDV data, $(\bar{u}')^2$ was taken as the standard deviation of the velocity probability

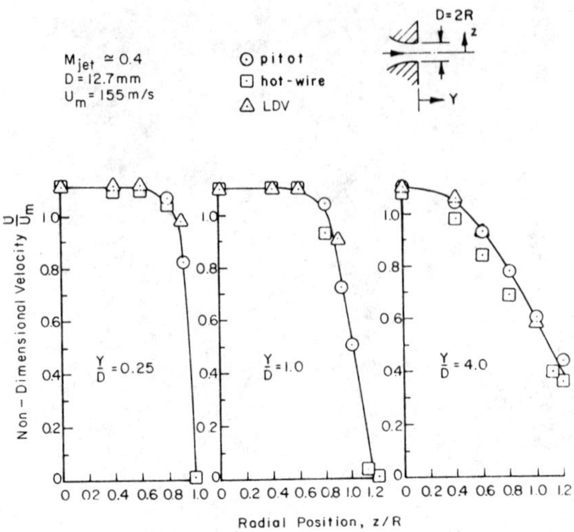

Figure 6. Non-Dimensional Velocity Profiles as Function of Axial and Radial Position in Jet

distribution. The data is compared with values of velocity fluctuation obtained from the hot-wire data.

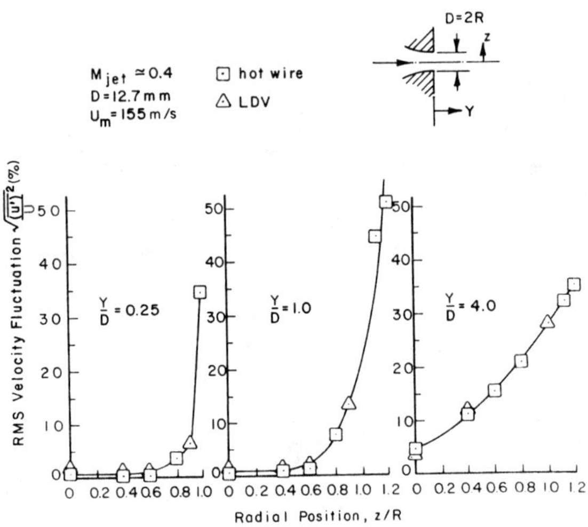

Figure 7. Turbulence Fluctuation in Jet as Function of Axial and Radial Position

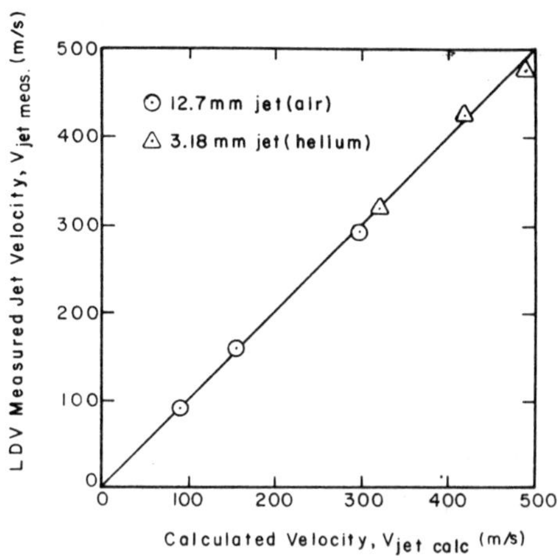

Figure 8. LDV Measured Jet Velocity as Function of Calculated Velocity

To further verify the LDV performance, a high velocity helium jet was used. This facility was used to generate velocities to nearly 500 m/s from a jet 3.2 mm in diameter. The maximum frequency measured was 179 MHz, corresponding to a jet velocity of 475 m/s. Figure 8 compares the measured and calculated (isentropic jet nozzle conditions) jet velocities over a range of 100 to 500 m/s for air and helium jets. For these tests with velocities above approximately 200 m/s, the divide-by-two system was used to shift the frequency range to a level compatible with the Doppler frequency processor.

COMPRESSOR CHECKOUT OF THE LDV SYSTEM

The LDV system was installed and aligned in the compressor test facility and a series of checkout tests were conducted to prove the feasibility of the velocity measurements in the compressor. In these tests, silica-coated, alumina particles were used for the seed. The number mean diameter of the powder is about 0.75 μm. Dispersion of this seed is accomplished using a fluidized bed, with an air-driven vibrator to prevent bed channeling. The transport fluid is LSM gas, taken from a high pressure point in the closed loop test facility.

Velocity measurements were made in the inlet/inducer region and in the diffuser entry region at compressor rotational speeds from 40% to 80% of design in air and low speed-of-sound modeling (LSM) gas (30,000 to 42,000 RPM actual).

A selection of the results of these tests are shown in Figure 9

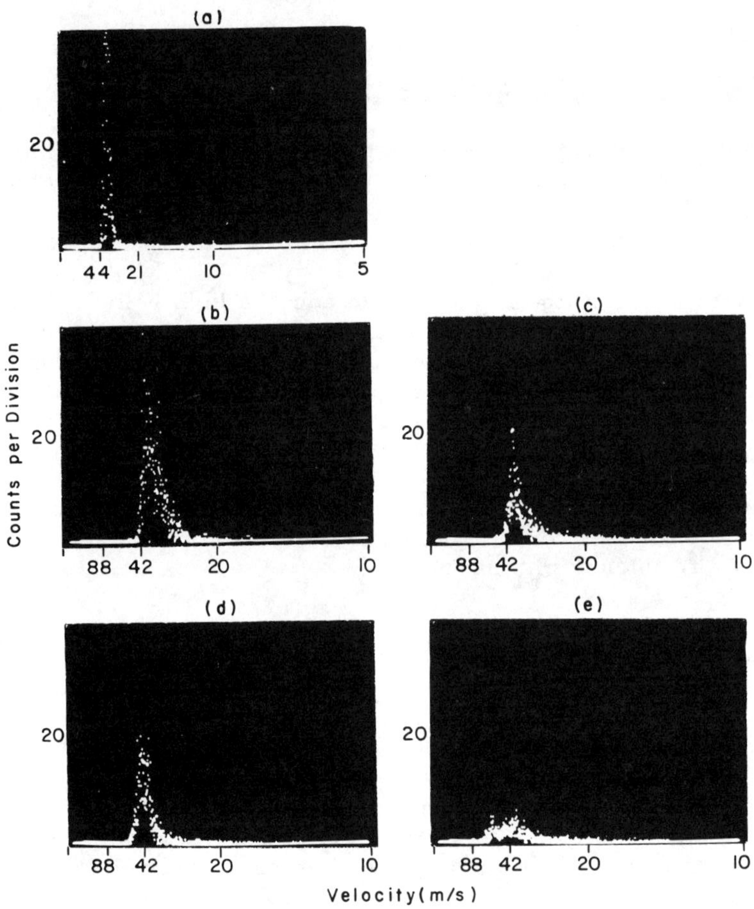

Figure 9. Velocity Histograms from Pulse Height Analyzer (Inlet/Inducer Region Data)

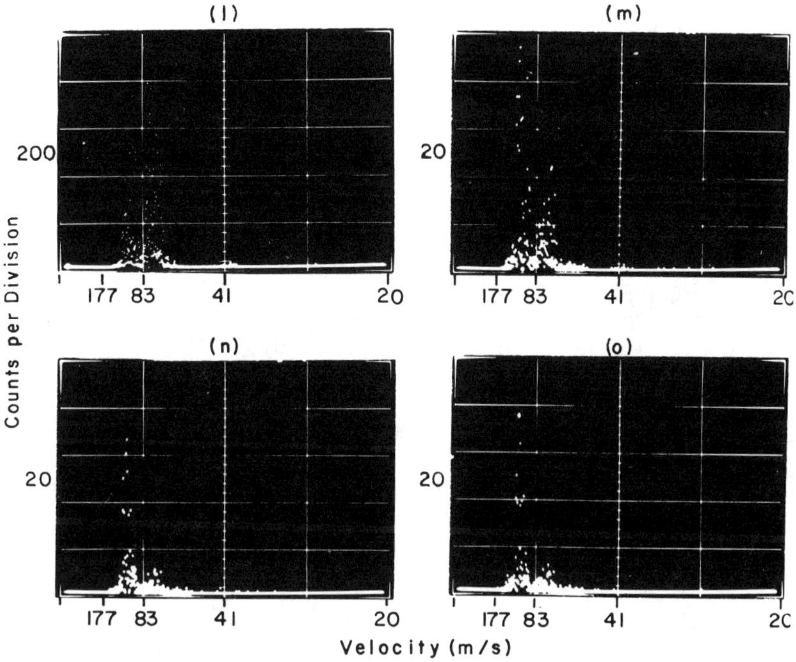

Figure 9. Continued (Diffuser Region Data)

as velocity histograms photographed directly from the pulse height analyzer CRT display. The abscissa is the channel number (1024 full scale) corresponding to the time period for a velocity measurement, while the ordinate shows the number of occurrences at each channel (velocity). Parts a through e are from the inlet/inducer tests; parts ℓ through o are from the diffuser region tests. Table II provides a key to these data photographs, including pertinent probe location and data gating parameters.

The data from this brief checkout program has verified the operation of the LDV system under realistic compressor test conditions, and has pointed out the value of the velocity histograms for interpretation of the data.

TABLE II - INLET/INDUCER AND DIFFUSER REGION LDV MEASUREMENTS
(Reference: Pulse Height Analyzer Photos -Figure 19)

Figure No.	Gas	Speed (%N_D)	Impeller Tip Velocity (m/s)	Gate Position(1)	Gate Width(2)	Measurement Plane(4)	Velocity (m/sec)	Peak Channel Counts	Total Integrated Counts	Sample Time (sec)
a	Air	40	-	(3)	(3)	1	33.2	100	2938	39
b	Air	40	-	(3)	(3)	1	36.1	100	3718	20
c	Air	40	-	(3)	(3)	2	38.1	50	1536	356
d	Air	40	-	19	22	1	42	50	2071	172
e	Air	40	-	19	22	2	36.9	15	939	500
ℓ	Air	40	251	(3)	(3)	1	73	1000	20982	92
m	Air	40	251	50	50	1	93	100	2092	177
n	Air	40	251	50	10	1	101	100	913	184
o	Air	40	251	50	5	1	101	100	974	317

Notes:

(1) Gate position is the location at which the data acquisiton gate is opened. For the inlet/inducer data (Figures a through e), it is a percentage of blade pitch, following the blade suction surface. For the diffuser region data (ℓ through o), it is the location of an impeller blade tip relative to the pitch between diffuser vanes (see Figure 20).

(2) Gate width is the duration of the data gate. It is expressed as a percentage of blade pitch rotation for the inlet/inducer data, and as a percentage of blade rotation relative to diffuser vane tips for the diffuser data.

(3) For indicated data sets, the data acquisition system was run in a continuous mode; i.e., a time-averaging mode.

(4) The measurement plane is the orientation of the probe volume with respect to the compressor hardware. The two planes are mutually perpendicular.

CONCLUSIONS

This program has carried out the design, analysis, fabrication and test of a laser Doppler velocimeter (LDV) system to be used in the study of small, high speed, turbomachinery fluid dynamics. The important characteristics of this instrumentation, which are of interest to its application in the research and development of small turbomachinery, are as follows:

The LDV instrumentation has been optimized to permit the measurement of instantaneous velocities up to approximately 500 m/s, measured in absolute coordinates, within a rotating impeller.

The LDV equipment accomplishes the velocity measurement with a non-contact process. One of the most important motivations for using LDV equipment in this application is that the non-contact process does not disrupt the flow. This is important since even the smallest probes used for conventional measurements in this type of hardware have been shown to seriously disrupt the aerodynamic performance of the machine.

The optics design has resulted in a probing volume in the form of an ellipsoid of revolution with a minor diameter (width) of approximately 0.06 mm. The small probe volume necessitates the use of small seed material on the order of 0.75 μm number mean diameter in order to accurately measure flow velocities and to obtain high data rates.

The system design has been configured so that the probe volume can be positioned accurately relative to the turbomachinery hardware. Electronics strobing equipment has been designed and fabricated to permit repeated measurements of the flow velocity at any fixed point within the flow region, and at a fixed position relative to the rotating impeller. The strobing system allows less than 0.5% of blade pitch rotation of the impeller for a single velocity realization using the 19-bladed impeller, rotating at 53,000 RPM. Measurement times required for a single velocity realization are approximately 0.1 μs at 200 m/s and approximately 0.08 μs at 500 m/s.

The output of repeated LDV measurements at a point permits the development of a histogram of the flow velocity vector located relative to the rotating impeller. Analysis of this histogram permits the calculation of the mean velocity of flow at this point and the evaluation of some properties of the unsteady and fluctuating flow.

The work accomplished has demonstrated the feasibility of making detailed velocity measurements in small, high speed turbomachinery.

REFERENCES

[1] Block, J. A. and Runstadler, P. W., Jr.(1973): Modeling of a High Pressure Ratio Centrifugal Compressor using a Low-Speed-of-Sound Gas; USAAMRDL Technical Report 73-18, U.S. Army Air Mobility Research and Development Laboratory, Fort Eustis, Virginia.

[2] Runstadler, P. W., Jr. and Dolan, F. X.(1975): "Design, Development, and Test of a Laser Velocimeter for High Speed Turbomachinery," Proceedings of the LDA Symposium, Copenhagen.

DISCUSSION OF THE RUNSTADLER-DOLAN PAPER

DODGE: In your window arrangement in the diffuser area, is that actually a window on both sides of the diffuser passage?

RUNSTADLER: That is a one-sided window. Everywhere in the impeller we use a backscatter radiation technique. In the diffuser we actually use a back reflected forward scatter, so there's a mirror on the back door.

DODGE: You're depending on reflection, then? Do you expect to have reflection troubles?

RUNSTADLER: We use the reflected signals in the diffuser primarily because, even though we are using seeding, it's much easier to seed at the right place, in particular at the entrance to the impeller. When you get up into the diffuser, because of the circulation effects of all that seed going through the wheel, your density along any given streamline reduces greatly, so that what we want to do is use that forward scatter radiation which is an order of magnitude larger in intensity than a back scatter radiation. We get the signal to noise ratio up to where we don't have to spend ten minutes taking some of the data points.

DODGE: Have you had any experience and do you expect any difficulty with oil and dirt and all the other things when you go through a centrifugal compressor?

RUNSTADLER: We've had all the experience with all the dirty things in the compressor and we don't expect it to give us any problem.

DODGE: Is that an overhung rig?

RUNSTADLER: Yes. It's in a closed wind tunnel system. I didn't describe that but it's in a low speed of sound closed loop system which we have to keep dirt and oil contamination from coming into, anyway.

DODGE: Quite often in a centrifugal compressor, particularly when you get to these kind of pressure ratios and if you have a design that's quite efficient, you'll discover that you have back flow of the diffuser region on the walls. Are you biasing your fringes so that you can see which direction the flow goes?

RUNSTADLER: No, we're not. We could do that, but we didn't go to that extravagance. So we are going to incorporate our interpretation of the measurements we made on this and other information that we have such as seed particles that collect out on the walls and after we run a sufficiently long time at a given stat you can, with a magnifying glass, see residual flow patterns on the walls. We can't detect that in the back flow.

DODGE: You don't know how far away it's going to be?

RUNSTADLER: Yes, that's a problem.

DODGE: Perhaps you could see where it crosses over, where it goes 90 degrees. Is that inducer designed to facilitate the velocimeter measurement? It seems to me to be a little longer than conventional and more axial than the standard centrifugal diffuser.

RUNSTADLER: Yes. It's a radial element design.

DODGE: The inducer is almost axial and very long, which is unusual for the centrifugal diffuser.

RUNSTADLER: It was only specifically designed to keep the tip radius axial for some distance. We would have had a slight wrap on that tip radius. The turning of the blading is not specifically tailored to do anything other than what we would normally have done.

DODGE: You wanted to keep a flat window?

RUNSTADLER: I wanted to keep a flat window in there. And we did change that aspect of the design to accommodate that.

ROBERTS: Could you comment on the seed material you used?

RUNSTADLER: Yes. Again to get a high signal to noise ratio we are seeding. The seeding material we're using is a silica-coated aluminum particle, with a number mean diameter of 3/4 of a micron. Very narrow distribution. The silica coating is done by a commercial firm that we obtain the particles from by a process that they consider proprietary. What it essentially does - it keeps the particles from agglomerating. We've checked this the best way we can with the flow system; we're even using a Coulter counting technique to measure what that would say the number of mean diameter is. It comes out o. k.

KERREBROCK: Do you think that the 12% stagnation pressure loss between the impeller tip and the outlet of the diffuser can be understood in terms of the steady flow model?

RUNSTADLER: No.

KERREBROCK: Then what's the utility of the LDV measurements?

RUNSTADLER: Because I believe that the unsteadiness you are going to pick up there probably is a function of the blade to blade passing frequency. Now we can and we are triggering off the blade rotational position. We can make measurements relative to actual design speed; we're down to 1/10 of a change in blade position, that is, a percentage of the blade to blade pitch. What we intend to do with these measurements is to actually trigger the measurements we are making relative to various positions of the blade position. In fact, one of the slides shows preliminary measurements and you do, indeed, if you're say averaging over 30% rotation of the impeller relative to its full blade to blade rotation, pick up a double velocity distribution. Very distinct. We can eliminate that bias to one or the other, depending on where we decide to close down the measuring point.

KERREBROCK: So you think there are important unsteady effects due to periodic passing?

RUNSTADLER: I do feel that they are and, again, that's based on what limited information I have and all of the data we've taken including high response Kulite data, but also some of the old work that Boeing Company did with Schlieren. Looking in detail, frame by frame, at some of the high speed movies that have been taken, I think I can see some unsteadiness relative to the blade passing frequency. That is a problem that you are properly addressing here. The LDV technique, if you're not using a trapping system, I don't think is practical today for the kind of thing we're talking of doing here. If you have an inherently unsteady flow which is at a periodicity or at a randomness which you don't know, then basically you're always going to get a heuristic average of what you are measuring. It's going to be difficult to separate out that type of phenomenon.

LASER DOPPLER VELOCIMETER MEASUREMENTS IN A TWO-DIMENSIONAL TRANSONIC FLOW*

H. DOYLE THOMPSON

Purdue University

ABSTRACT

A comparison is made of the sonic line location for a series of two-dimensional nozzles as determined by static pressure measurements, Laser Doppler Velocimeter (LDV) velocity measurements, and an analytical solution based on a series expansion in terms of the channel geometry in the throat region. For every geometry considered the analytical sonic line location is slightly upstream of the experimentally measured sonic line based on static pressure measurements. The sonic lines based on LDV measurements fall between the analytical and the static pressure sonic lines. The agreement is excellent. The work demonstrates both the extreme sensitivity of transonic flow fields to small changes in the boundary geometry and the potential of the LDV for making very accurate and meaningful velocity measurements in highly sensitive flows.

---

*The research reported herein was a team effort performed under the sponsorship of the U.S. Army Missile Command, Redstone Arsenal, Ala., under Contract DAAH01-72-C-0089. Those contributing to the research include my colleague, Prof. Warren H. Stevenson, and students, Ronald D. Flack, Michael Pedigo, Ronald Zammit and Karl Owen.

INTRODUCTION

The interest in transonic flow at Purdue dates back many years. The original interest was in the transonic flow region in thrust nozzles. Although little work has been done on transonic flow in turbomachinery per se the problems encountered in thrust nozzles, particularly multi-stream annular thrust nozzles are similar in many ways. Much of the work is reported in [1] - [8].

RESULTS

Figures 1, 2 and 3 are typical of the comparison between the LDV measurements, the static pressure measurements and the analytical solution for three slightly different two-dimensional geometries.

The three nozzles examined were two-dimensional nozzles with circular arc contours. They are designated as T6-6F, T4-8, and T2-2F. The T designates two-dimensional nozzles, the numbers 6-6, 4-8 and 2-2 refer to the non-dimensional radius of

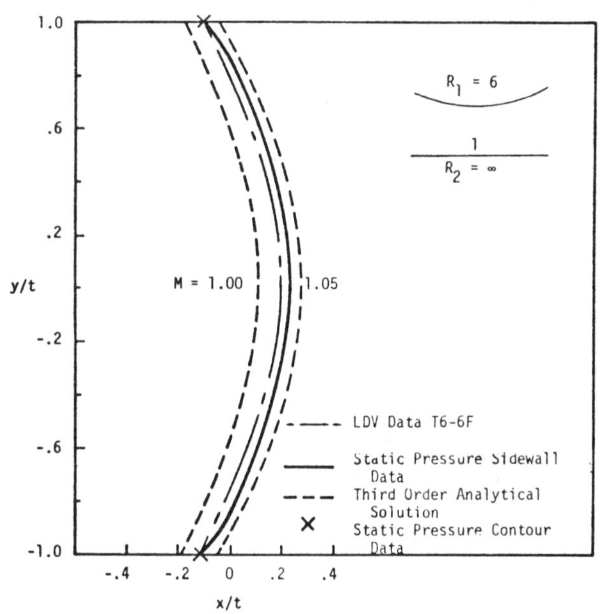

Figure 1. Comparison of Experimental and Analytical Sonic Line Location in Transonic Nozzle T6-6F

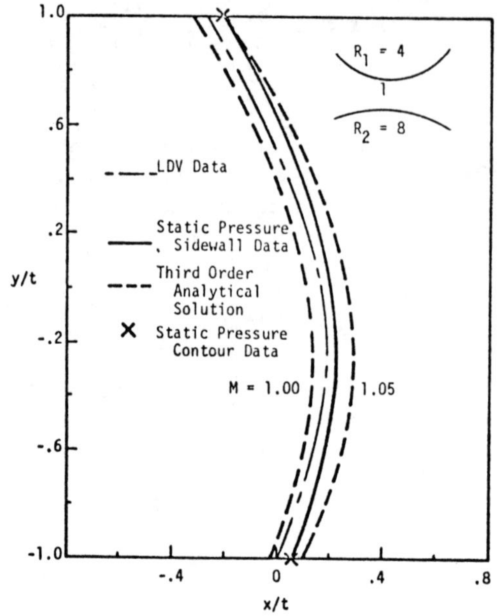

Figure 2. Comparison of Experimental and Analytical Sonic Line Location in Transonic Nozzle T4-8.

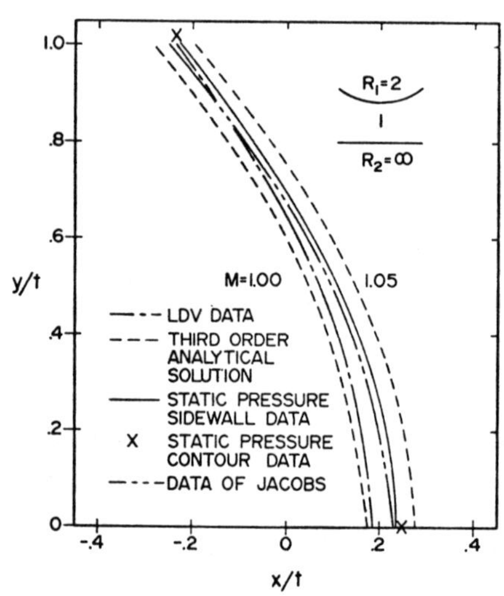

Figure 3. Comparison of Experimental and Analytical Sonic Line Location in Transonic Nozzle T2-2F

curvature of the upper and lower circular arc contours respectively, and the F denotes a half-nozzle in which the lower contour is a flat surface at the plane of symmetry.* The nozzle radii of curvature are non-dimensionalized to the half height of the throat passage. All nozzles were one inch wide.

For the LDV measurements the axial velocity component was measured in the throat region, the velocity was converted to Mach number, and isomachs were constructed by linear interpolation between measuring points. Fifty-five locations were probed in nozzle T6-6F, while 25 and 31 locations were examined in nozzles T4-8 and T2-2F, respectively. Histograms are available for all of these flow points and all are similar. A typical histogram is presented in Figure 4 where the percent of total signals processes is plotted versus Mach number. The interval in Mach number is .005. This typical histogram is fairly symmetrical and has a standard deviation of approximately .012. The data rate was 283 samples/sec. The spread in the histogram is typical of LDV results for the internal nozzle flows and is attributed to several factors, including:

a) a low level of turbulence;

b) a variation in particle lag due to a distribution in particle size;

c) electronic noise;

---

* For example, if the centerline of nozzle T-6-6 (two symmetrical boundaries with non-dimensional radii of 6) is substituted with a flat plate, the nozzle is designated T6-6F.

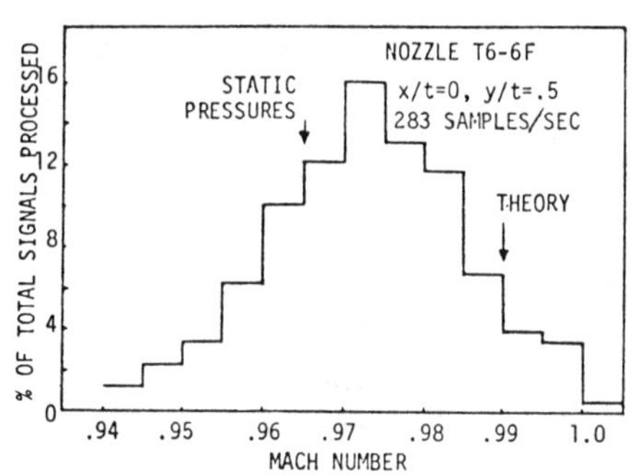

Figure 4. Histogram of Typical Transonic Flow Point

d) a finite tolerance on the 4 and 8 comparator in the signal processor;

e) some randomness in the direction of the particle path relative to the probe volume; and

f) optical distortions such as non-uniformities in the Plexiglass sidewalls which produce slight distortions in the virtual fringe pattern.

From peak values on the histograms Mach number profiles were drawn. By linear interpolation of these plots sonic lines for the three nozzles were constructed. In the figures the LDV sonic lines are compared to static pressure data and an analytical solution. For the analytical solution, isomachs of 1.00 and 1.05 are shown. The LDV sonic line for nozzle T2-2F has also been compared to static pressure data of Jacobs [9].

The analytical sonic lines and Mach 1.05 lines shown in Figures 1, 2 and 3 were calculated from a transonic flow computer program developed to calculate the transonic flow field in both annular axisymmetric and two-dimensional choked flow passages. The calculation procedure is based on a series expansion of the perturbation velocities around M=1 in terms of the curvature of passage boundaries in the vicinity of the minimum area cross section. The passage boundaries need not be symmetric. Perturbation velocities of the first three orders are calculated. The flow passage is assumed to be choked and to have uniform inlet conditions. The flow is also assumed to be inviscid, steady, and irrotational. The details of the mathematical model and the computer program are contained in [3].

Pressures from seventeen .018-inch diameter staggered pressure holes in the stainless steel nozzle blocks and twenty-four .018-inch diameter pressure holes in the plexiglass sidewalls spanning the throat region of each nozzle were measured with a 48-part Scanivalve. The plenum pressure was also measured and Mach numbers were computed assuming isentropic flow. The details of the experimental program are contained in [1], [2], and [4].

For all three nozzles the LDV data fall between the static pressure data and the analytical solution with less than two percent difference between the LDV data and the static pressure data.

The agreement between the LDV measurements, the static pressure measurements and the theoretical results are individually and collectively excellent. Certainly, the inherent uncertainties in experimental measurements could account for all the differences as could the higher order effects that are ignored in the analytical solutions.

The results presented in Figures 1, 2 and 3 confirm that the LDV is a valuable tool for making accurate mean flow measurements and that it is especially valuable in highly sensitive flows, such as transonic flows, where it is essential that the measuring device does not disturb the flow.

REFERENCES

[1] Flack, Ronald D. and Thompson, H. Doyle (1973): "An Experimental and Analytical Investigation of Internal Transonic Flow," U.S. Army Missile Command Technical Report RD-73-20.

[2] Flack, Ronald D. and Thompson, H. Doyle (1974): "Transonic Flow Measurements - A Comparison of Static Pressure Measurements, Laser Doppler Velocimeter Measurements and Analysis," AIAA Paper No. 74-15.

[3] Thompson, H. Doyle and Flack, R. D. (1974): "Transonic Flow Computation in Annular Two Dimensional Nozzles," U.S. Army Missile Command Technical Report RD-73-21.

[4] Flack, Ronald D. and Thompson, H. Doyle (1975): "Comparisons of Pressure and LDV Velocity Measurements with Predictions in Transonic Flow," <u>AIAA Journal</u>, <u>13</u>:1, pp. 53-59.

[5] Flack, Ronald D. and Thompson, H. Doyle (1974): "Preliminary Results with Purdue's LDV System," <u>Proceedings of the 2nd International Workshop on Laser Velocimetry</u>, March 27-29, Purdue University (Editors: H. Doyle Thompson and W. H. Stevenson).

[6] May, Robert J., Jr., Thompson, H. Doyle and Hoffman, Joe D. (1974): "Comparison of Transonic Flow Solutions in C-D Nozzles," AFAPL-TR-74-110.

[7] Thompson, H. Doyle, Stevenson, Warren H., Flack, R. D. and Zammit, R. E. (1974): "Laser Doppler Velocimeter Measurements in High Speed Flows," U.S. Army Missile Command Technical Report RD-CR-75-2.

[8] Flack, Ronald D., Jr. (1975): "The Application of a Laser Doppler Velocimeter (LDV) in Interpreting Turbulent Structure," Ph.D. Thesis, Purdue University.

[9] Jacobs, W. (1954): "Geschwindigkeitsverteilungen in zweidimensionalen gekrümmten Lavaldusen," Jahrbuch wiss. Gesellschaft Luftfahrt, pp. 57-62.

DISCUSSION OF THE THOMPSON PAPER

SOUTH: Aren't there any relaxation solutions for transonic nozzles? That would seem to be an easy problem for Murman's scheme.

THOMPSON: When I started this work most of the schemes that are presently being used were not available. I was aware of time-dependent schemes and things like that. I think there probably are some (relaxation solutions) for some of these (nozzles). The relaxation scheme - the boundary geometry, input geometry, can have some very sharp curvatures and, to the best of my knowledge, those schemes don't really give you much better results than anything else. A number of people have used time dependent schemes on these, too, to obtain a steady state solution. The problem becomes one of getting a fine enough grid on that type of thing and in reasonable time to give you a reasonable solution. Some of them are fairly good. We've compared those kinds of things in some annular nozzles and some regular converging-diverging nozzles, and find that those schemes don't do much better than this, which is considerably faster and easier to program. It is easier because it's already programmed, maybe.

REAL TIME MEASUREMENTS IN A TRANSONIC COMPRESSOR

R. P. SHREEVE

Naval Postgraduate School

INTRODUCTION

The purpose of this brief presentation is to report work in progress at the Turbopropulsion Laboratory at the Naval Postgraduate School to determine experimentally the behavior of the flow in a transonic compressor stage.* The intent is to develop an input into the design method which accounts, in some way, for the three dimensional and periodic nature of the flow. While this has been the goal of continuing research efforts, no significant change in the basic design model has been reported. The essential three dimensional nature of transonic flow is acknowledged, but no design method is three dimensional. Within the pseudo two dimensional analytical framework, secondary flows are known to become more significant as the blade loading is increased, and wall effects become important as the passage depth is decreased. While the effects are qualitatively known to be significant, the understanding that comes from favorable comparisons of observations with analyses is lacking. First, definitive measurements from different stage designs are needed to define the relative importance of different effects and thus provide the basis for a representative analytical design model. In order to achieve higher performance from a new design without a costly empirical development process, a design model which more nearly represents the actual flow in the blade rows is essential.

---

*Supported by Naval Air Systems Command, Code 310, under the cognizance of Dr. H. J. Mueller.

For these reasons a single stage transonic compressor was designed and built at the Turbopropulsion Laboratory at the Naval Postgraduate School [1] and is in operation in a new compressor test rig [2]. Conventional measurements of the performance of the stage were made to 65% of the design speed, at which sonic relative flow occurred in the rotor [3]. Surveys of the flow downstream of the rotor were also made using different types of calibrated pneumatic probes [4].

Instrumentation has been added to determine case wall pressure signatures using Kulite transducers. Preliminary data has been obtained at the rotor tip in the compressor using a single transducer channel. Techniques for the acquisition and analysis of real-time measurements are being developed using a periodic flow generator.

In this presentation, the goals of the program are stated and results of time-averaged flow measurements made in the initial test program, which are considered to be relevant to the program of real-time probe measurements, are given. The development of techniques to be applied in the acquisition of real-time test data is also reported. The purpose of the presentation is to communicate work in progress.

PROGRAM GOALS

The primary goal of the real-time measurements is to determine the flow at the wall and from the rotor passages from measurements made with fast response, but stationary sensors. Figure 1 is a schematic illustration.

The least ambiguous of the measurements shown in Fig. 1 is the wall static pressure pattern. Kulite pressure transducers are mounted flush with the case wall to record in real time the local pressure signature. The signature in the frame of the rotor is extracted from the time history through a change in coordinates using the rotation rate and wheel diameter. The wall pressure signatures indicate shock patterns at the rotor tip and the measurements themselves are unambiguous. The interpretation of the measurements however must allow for shock interaction with the case wall boundary layer and tip leakage flows.

The indications of sensors positioned downstream of the rotor are more difficult to interpret. Ideally, the time history of the

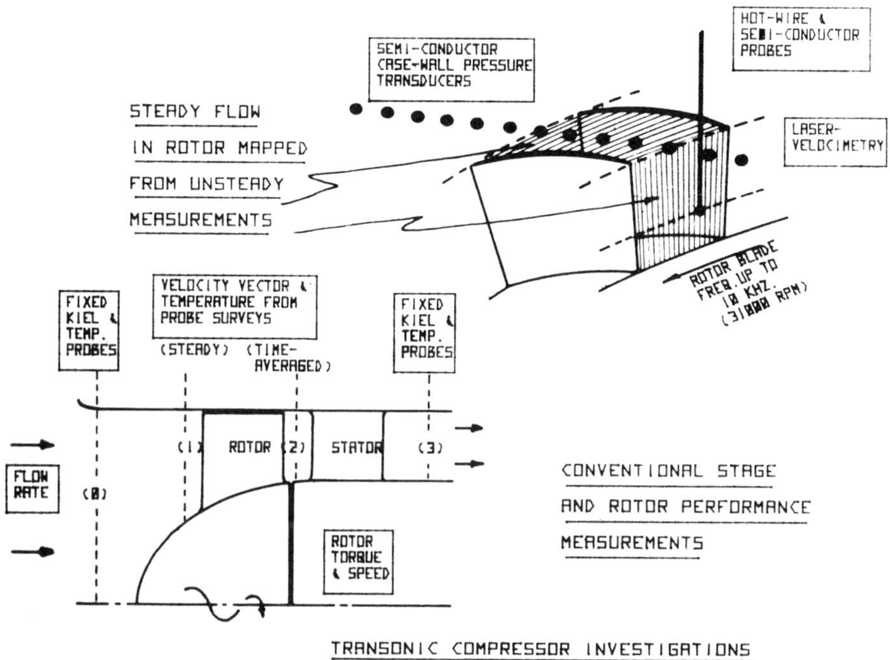

Figure 1. Instrumentation to Determine the Flow in a Transonic Compressor

velocity vector is required. If the velocity vector could be recorded in real time as a function of radial position, the flow relative to the rotor could be obtained by the change in coordinates described above. However, no single instrument is capable of measuring air velocity at a single point in space at an instant in time. Consequently, a multiplicity of measurements must be made and the interpretation of the measurements must be shown to be self-consistent. Hot wire and semi-conductor sensors have been selected for use at the rotor exit. Probes incorporating these sensors must first be calibrated in known flows. Since assumptions must be made when using probes calibrated in steady uniform flows to measure unsteady and non-uniform flow fields, redundancy is again necessary in the measurements. Laser velocimetry will be applied following calibration experiments in steady and unsteady flows.

COMPRESSOR AND TEST RIG

The single stage transonic (impulse) fan is driven by an air turbine in a compressor test rig [2] designed for 450 horsepower. The arrangement is shown in Figure 2. The compressor, (referred to as the TRANSX compressor) has a rotor diameter of 11 inches and a hub-to-tip ratio of 0.5 at the rotor face. The design pressure ratio is 1.6 at a referred flow rate of 19 lbs. per second and a speed of 30,460 r.p.m. The compressor is throttled at the intake. The flow from the stator enters an axial flow straightener and turns 90° through a diffuser to exit radially to atmosphere. A flow nozzle in the intake duct meters the flow rate and the torque is measured by flexures against which the stator assembly is free to rotate. Fixed probes have been used to measure the flow into and out of the stage to map the stage performance. Surveys of the flow into and out of the stage and downstream of the rotor have been carried out using multiple-sensor pneumatic probes and combination temperature and pressure flow probes [4].

RESULTS OF TIME-AVERAGED MEASUREMENTS AT THE ROTOR EXIT

Results of pneumatic probe measurements at the rotor exit are of interest because effects were found which could be present but remain undetected in real-time probe measurements. These effects have variously been described as "interference effects," "probe blockage effects" or "passage effects." For whatever reason, a departure occurs during an application of a probe from the behavior established while calibrating in uniform flow. Other investigators have attempted to solve this problem by measuring impact pressure, yaw angle and static pressure in separate surveys with different probes. In the present case, the axial clearance between blade rows and access through the compressor wall limited the choice of probes.

Surveys of the flow at the exit of the rotor were carried out using two different probes which could be mounted within the available access. A combination temperature and pneumatic probe developed by NASA [8] was selected because it could in principle determine velocity and loss distribution where the pitch angle was known to be small. A United Sensor Corporation 5-hole pneumatic probe was used to measure the velocity independently [4]. Both probes were calibrated in a 7-inch diameter free jet.

A comparison of the radial distributions of Mach number which

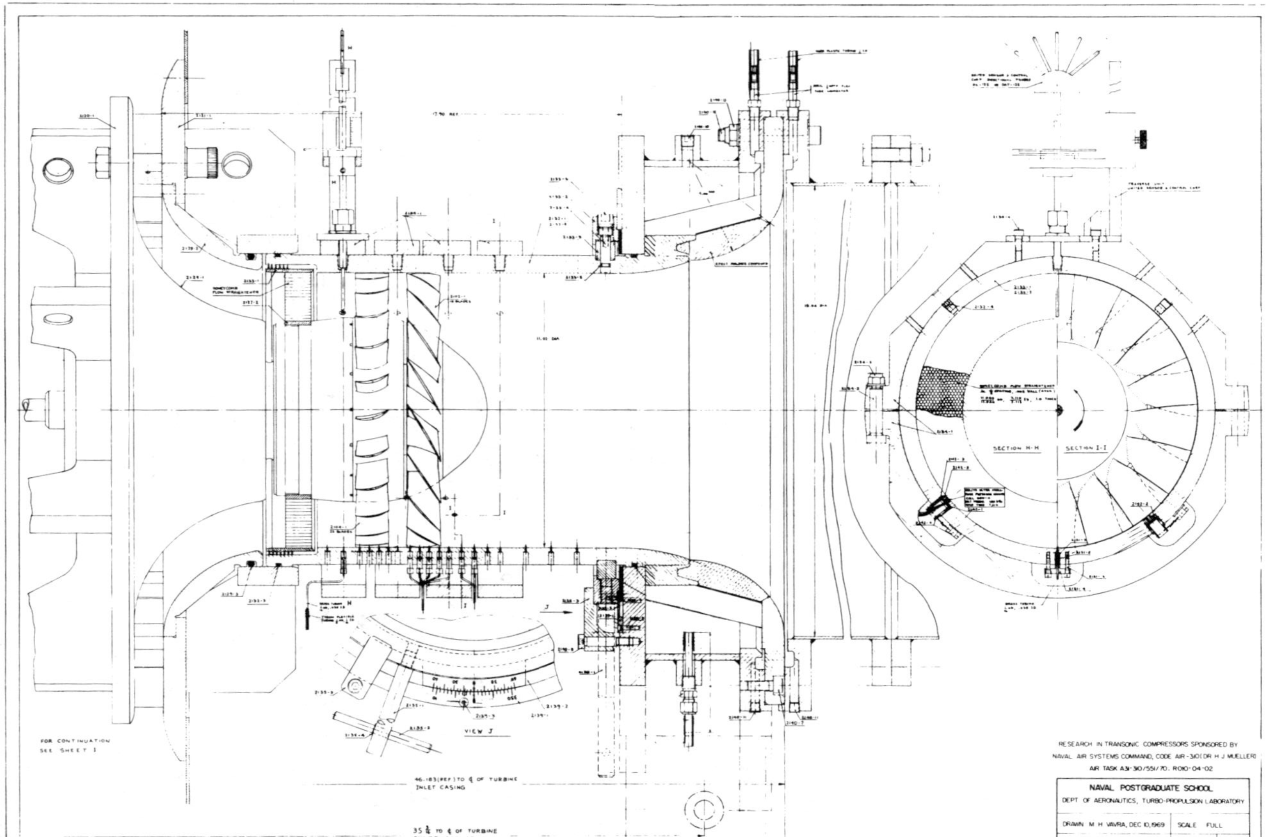

Figure 2. Layout of the TRANSX Compressor

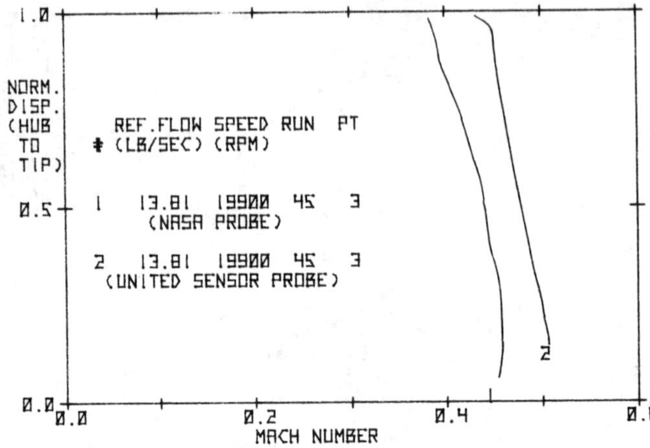

Figure 3. Radial Distribution of Mach Number at the Rotor Exit Determined Using Different Probes Calibrated in a 7" Free Jet

were measured downstream of the rotor using the two probes with the calibrations established in uniform flow, is shown in Figure 3. There was seen to be a disagreement of about 10%.

To determine the cause of the disagreement, surveys were made with both probes of simple pipe and free jet flows, and comparisons made with wedge, Prandtl and cone probe surveys. It was found that the two probes were subject to significant errors as a function of distance from any flow boundary. In particular, the combination probe [8] required an immersion depth of fully 3 inches into a uniform flow before uniform conditions were indicated. Since the corrections required for boundary effects were different, an ad hoc method of correcting the compressor measurements was attempted. The two probes were recalibrated in an annular axial flow channel having dimensions similar to those of the compressor annulus. However, the calibration was established over a range of Mach number at a number of radial displacements and the results were represented as a function of position, normalized with respect to the passage depth [4].

The result of applying the new calibration to the rotor exit surveys is shown in Figure 4. Good agreement is observed over the outer 60% of the profile, which is now more nearly uniform. Where the two distributions diverge, the error is thought to be in the combination probe since the radial static pressure distribution

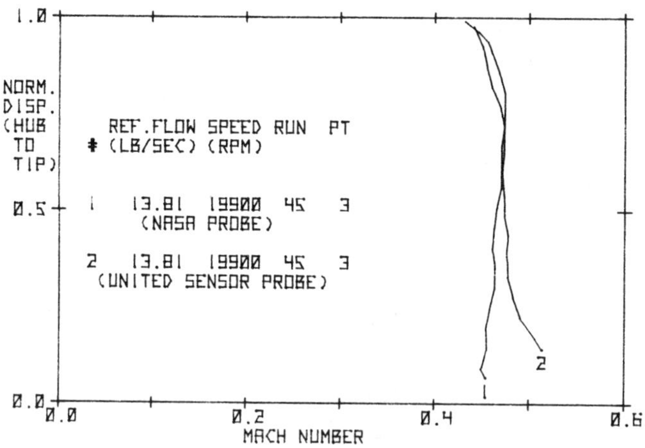

Figure 4. Radial Distributions of Mach Number (as in Figure 3) Following Probe Calibrations in an Axial Annular Channel

derived from the 5-hole probe calibration agreed very closely with the values measured at pressure taps at both the inner (hub) and outer (case) walls. The flow rate obtained by integrating the probe measurements agreed with the measured flow to within 2%.

The conclusion to be drawn from these results is that probes incorporating fast-response sensors for real time measurements must be carefully applied to measure well known flows before measurements of an unknown flow can be accepted. Calibration in uniform flow is not enough, and redundancy in measurements is clearly desirable.

REAL-TIME MEASUREMENT TECHNIQUES

Tests in which case wall static pressure distributions in the TRANSX compressor are measured using Kulite pressure transducers are underway. A brief review is given here of techniques to be used in these and other real time measurements.

Data Recording and Analysis

A 16-channel digital data system has been built and the system programmed for real-time data acquisition in single channel or sequential modes. Figure 5 shows the components of the system.

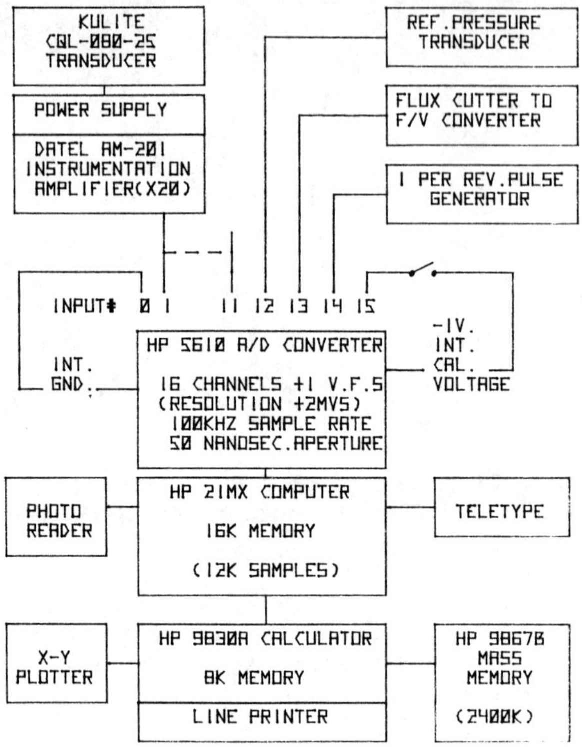

Figure 5. Real Time Data Acquisition and Recording System

An analog-to-digital converter accepts up to 16 analog signals in the range $\pm$ 1 volt. A microprogrammable computer controls the sample sequence and frequency at up to 100,000 samples per second. The digitized data is transferred from the computer memory through an HP 9830A calculator to a rigid disk mass memory device. Data is later analyzed on the HP 9830A and output on a line printer and x-y plotter. The recording and playback processes were verified using wave forms input from a sine wave signal generator. An illustration of the recording and playback accuracy is shown in Figure 6.

Transducer Response

An extensive program of real time measurements in compressors is being conducted by Gallus [5]. Gallus refers to the work

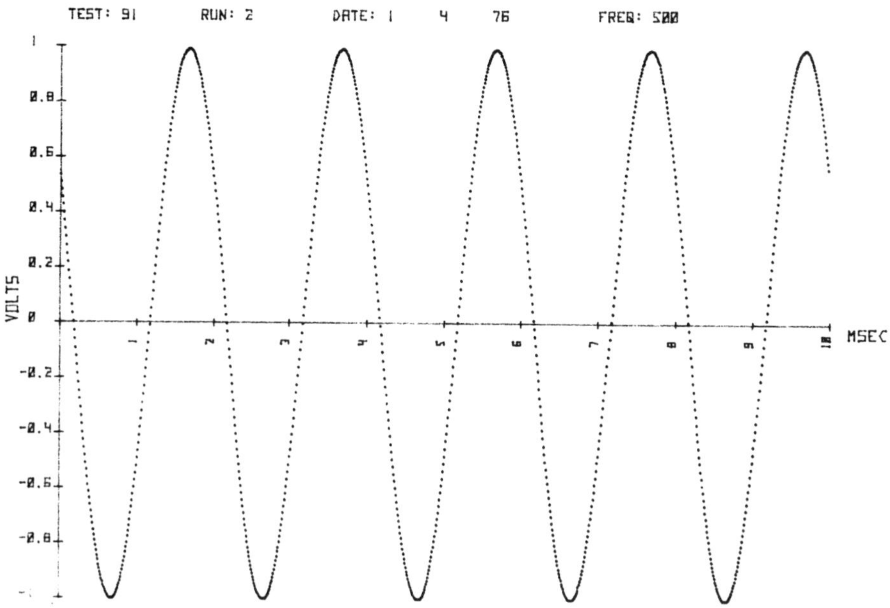

Figure 6. Comparison of the Digitized and Recorded Signal with the 1 KHZ Sine Wave Input from a Signal Generator. (• Data Recorded and Plotted using System Shown in Figure 5. — Sine Function Plotted for 1 KHZ Frequency and 1 Volt Amplitude)

of Weyer and Schodl [6], who carried out static and dynamic calibrations of Kulite transducers and attempted to simulate compressor wave forms at high frequencies.

Two observations are made here with respect to this and other published information on the response of Kulite semi-conductor transducers. Firstly, measurements of the response of the transducers to large pressure amplitudes at frequencies above 5 KHZ have apparently not been documented. The linearity of the transducer at large amplitudes must for the moment be accepted. Secondly, published conclusions concerning the variation in the slope and intercept of the transducer output with temperature have generally been confirmed in preliminary calibrations in the present work. However, a temperature-dependent hysteresis effect was measured which would make the intercept dependent also on the recent temperature history.

Measurements of the response of Kulite sensors to large

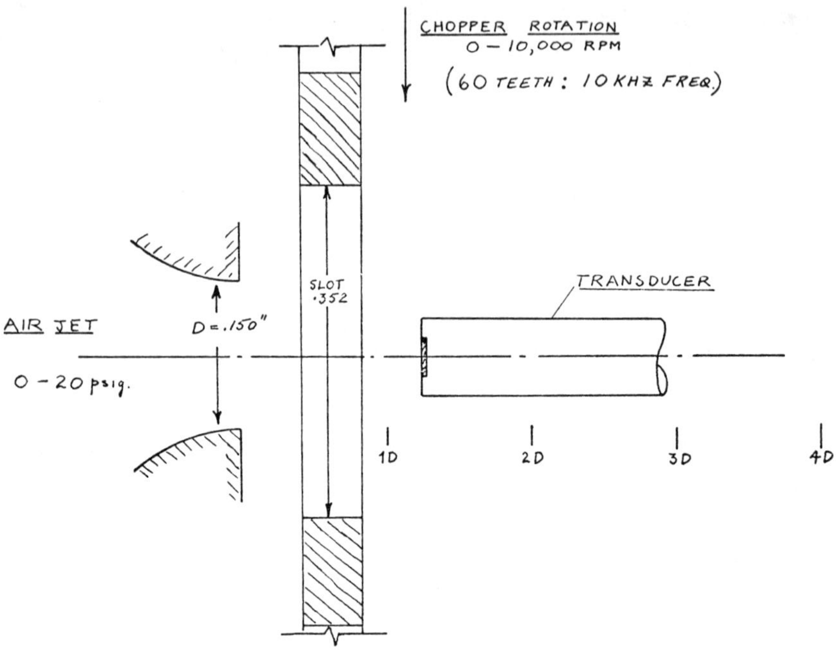

Figure 7. Arrangement of a 10 KHZ Periodic Flow Generator to Develop Measurement Accuracy and Techniques

periodic pressure amplitudes are being carried out using the periodic flow generator shown schematically in Figure 7. An example of the digitally recorded and reconstructed wave form is shown in Figure 8. The flow generator is being used to develop the following data acquisition techniques for compressor tests.

On-Line Calibration

Since the temperature of each transducer is not known and varies with the compressor operating condition, an on-line calibration technique is required. It is assumed that the transducer material does not follow the imposed temperature fluctuations, and that the time-average of the pressure being measured does not change as the calibration and test data are acquired. If the transducer is linear, the output voltage can be represented as

$$E = A_o(T) + A_1(T)(p - p_r) \tag{1}$$

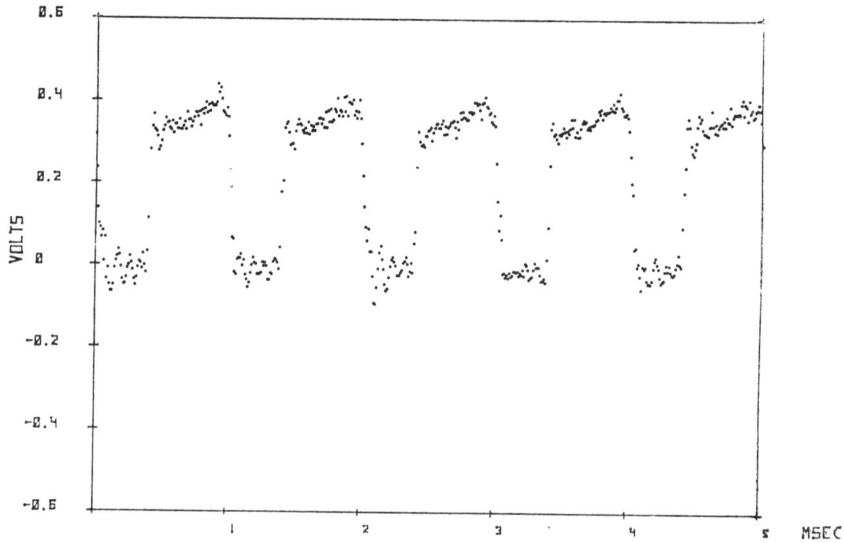

Figure 8. Sample of Real-Time Recording of Impact Pressure Generated as Shown in Figure 7. (Supply Pressure ≈ 10 PSIG. Frequency = 1 KHZ)

where

E = output voltage,

$A_o(T)$ = intercept, which is a function of temperature,

$A_1(T)$ = slope, which is a function of temperature,

p = pressure at the transducer face,

p_r = reference pressure in the transducer cavity.

Writing

$$p = \bar{p} + p' \tag{2}$$

where $\bar{p}$ is the time-average and p' the time dependent pressure fluctuation, from Eq. (1),

$$\bar{E} = \frac{1}{t_1} \int_0^{t_1} E \, dt = A_o + A_1 (\bar{p} - p_r) \tag{3}$$

If a "continuous" sample is taken of the transducer output over a time t_1, at each of three different known values of the reference pressure, the time integrations given by Eq. (2) for the three samples at first appear to be sufficient to determine A_o, A_1, and $\bar{p}$. However, Eq. (3) can be written as

$$\bar{E} = (A_o + A_1 \bar{p}) - A_1 p_r$$

or

$$\bar{E} = A_2 - A_1 p_r \tag{4}$$

wherein there are only two unknowns, A_2 and A_1.

Thus, by using two reference pressures and integrating over time the slope A_1 can be determined explicitly, and the quantity A_2 (which combines the unknown intercept, A_o and unknown time average pressure $\bar{p}$) can also be obtained. However, a separate determination of A_o or $\bar{p}$ is not possible, so that a measurement of the time average of the pressure is necessary to complete the on-line calibration procedure.

Results from different methods of determining time-average pressures have been reported by Weyer [9]. In the present work provision has been made for the simultaneous use of time-averaging detectors and Kulite sensors at similar locations through the machine.

An example of measurements made with a Kulite transducer located at the rotor leading edge (but with the sensor surface downstream of the plane of the leading edges) is shown in Figure 9. The data shown were taken as "continuous" samples at 10 μsec intervals at 50% of design speed with open throttle. The data were reduced using the on-line calibration technique described above with the assumption that the pneumatic pressure tap at the corresponding axial station at a peripheral displacement of 45°, recorded the time average of the pressure. The data from 18 periods of the wave form were plotted over a single period to show the mean and deviation from the mean pressure distribution between adjacent blades of the 18 bladed rotor.

Synchronized Sampling

"Continuous" data in real time can be recorded on magnetic

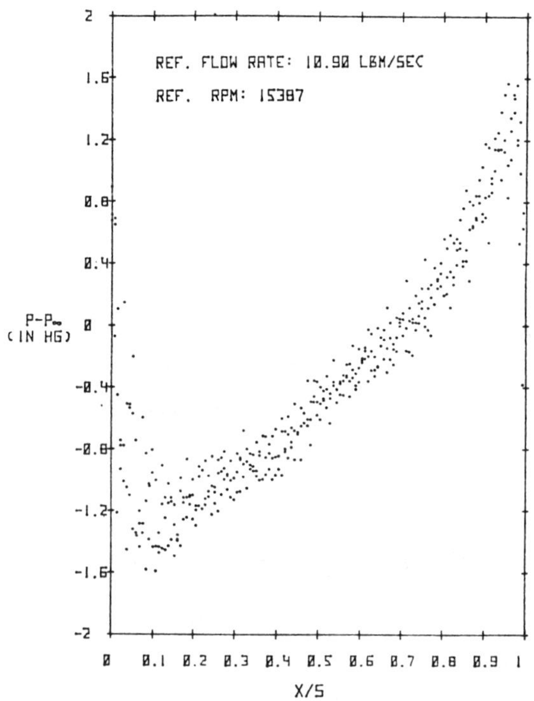

Figure 9. Case Wall Static Pressure Signature Near the Rotor Leading Edge from Continuous Digital Sampling at 100 KHZ

tape or digitized on-line and recorded at the maximum sample rate (100,000 samples per sec). Very large volumes of data have been recorded by other investigators using this approach. Unfortunately, much of the data is later found not to be in a usable form.

An alternative approach is to acquire the data in the form that it is needed while controlling the experimental conditions. The synchronized sampling technique illustrated in Figure 10 allows the properties of the flow to be determined at a point which is fixed with respect to the rotor. By triggering the sample acquisition using a once per revolution pulse and controlled delay time, a point in the rotor frame can be sampled as many times as is necessary to determine the steady and nonsteady properties.

The frequencies and time delays required for synchronized sampling in the TRANSX compressor at design speed are shown in

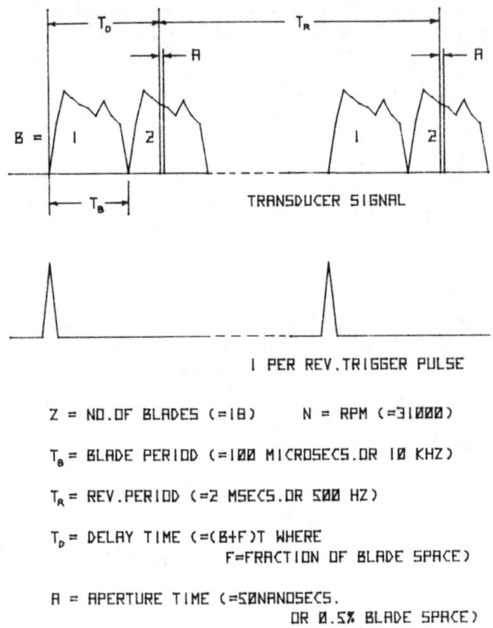

Figure 10. Schematic Showing Requirements for Synchronized Sampling

Figure 10. It should be noted that because of the aperture time of the A/D converter (Fig. 5), the "spatial" resolution of the resulting measurement in the peripheral direction is limited to about 0.040.

CONCLUDING REMARKS

In this presentation, techniques being developed for real-time measurements in a small transonic compressor have been described. In the compressor tests now underway, "continuous" sampling is used with an acquisition rate of 100,000 samples per sec. The "synchronized" sampling being developed and described above is thought to provide a promising method of analyzing a complicated periodic flow by acquiring less data, but with closely controlled parameter variations.

REFERENCES

[1] Vavra, M. H. and Shreeve, R.P. (1972): A Description of the Turbopropulsion Laboratory in the Aeronautics Department at the Naval Postgraduate School, Naval Postgraduate School Technical Report NPS-57Va72091A.

[2] Vavra, M. H. (1973): Design Report of Hybrid Compressor and Associated Test Rig, Naval Postgraduate School Technical Report, NPS-57Va73071A.

[3] Shreeve, R. P. (1974): Flow into a Transonic Rotor, Part I - Analysis, Naval Postgraduate School Technical Report NPS-57SF74081.

[4] Anderson, D. J. (1975): Velocity Measurements in a Transonic Compressor using a Calibrated Pressure Probe, MS Thesis, Naval Postgraduate School.

[5] Gallus, H. E. (1975): "Results of Measurements of the Unsteady Flow in Axial Subsonic and Supersonic Compressor Stages," AGARD Conference Preprint No. 177 on Unsteady Phenomena in Turbomachinery.

[6] Weyer, H. and Schodl, R. (1971): "Development and Testing of Techniques for Oscillating Pressure Measurements Especially Suitable for Experimental Work in Turbomachinery," ASME Paper No. 71EE28.

[7] Armentrout, E. C., On-Line Calibration of High-Response Pressure Transducer During Jet-Engine Testing. NASA TMX-71599.

[8] Glawe, G. E., Krause, L.M. and Dudzinski, T.J. (1968): A Small Combination Sensing Probe for Measurement of Temperature, Pressure and Flow Direction. NASA TN D-4816.

[9] Weyer, H. B. (1975): "Pressure Measurements in Fluctuating Turbomachinery Flows," von Karman Institute for Fluid Mechanics Lecture Series - Advanced Testing Techniques in Turbomachines, April 14-18.

DISCUSSION OF THE SHREEVE PAPER

KERREBROCK: I think that you've raised a very interesting philosophical question and somebody ought to respond to it. This synchronized sampling technique that you have is, in my view, a systematic way of throwing away information. If one says that one is going to do real time unsteady measurements in a compressor, and one uses a synchronized sampling technique like that, what you're doing is impressing upon it the assumption that the flow is periodic in blade passing. And that may or may not be true. Since you raised the question, I just wanted to reply.

SHREEVE: I'll respond to that. It's just the opposite, you see. If you sample that blade passing frequency, you measure the unsteadiness with respect to the rotor frame.

A COMPARATIVE EVALUATION OF NUMERICAL AND EXPERIMENTAL CASCADE DATA

R. A. DELANEY

Detroit Diesel Allison Division General Motors Corporation

Time-marching numerical methods have received considerable attention in recent years for solving transonic cascade flows. Many publications have been written concerning the application of these methods to both turbine and compressor cascades [1 - 5].

The author's experience with time-marching methods has shown that they yield good results for turbine cascade flows; however, in the high Mach number compressor cascade application, comparable accuracy is not achieved. The primary reason for this appears to be inaccurate modeling of the compressor airfoil leading edge region. The usual approach in solving for the leading edge point of compressor airfoils has been to treat the blade as if it were infinitely thin at the leading edge and to assign the flow direction there along the blade mean camber line.

In started, high Mach number compressor cascades, the uncovered portion of the blade suction surface has a strong influence on the entering flow field. In fact, the upstream flow angle is uniquely determined by that part of the blade suction surface lying upstream of the left-running Mach line intersecting the leading edge of the airfoil (unique incidence principle). A schematic of the entrance region flow field, showing Mach lines and shock waves, is presented in Figure 1. Shock waves generated at the blade leading edges coalesce and weaken as they propagate upstream of the cascade to yield uniform conditions at upstream infinity. Normally, mechanical integrity requires that the airfoils have finite leading edge radii and shock waves generated at the leading edge are

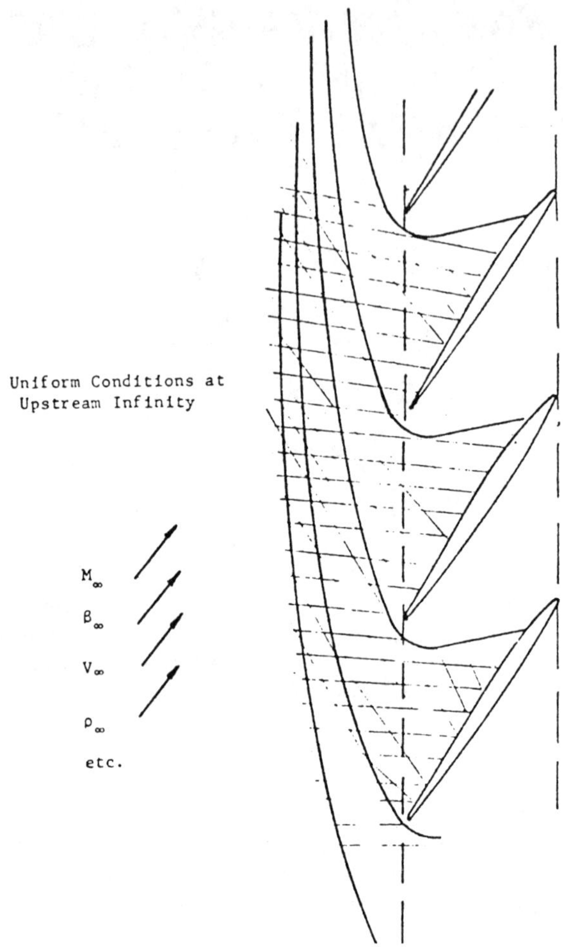

Figure 1. Cascade Entrance Region.

detached as shown in Figure 1. Entropy gradients developed in the flow field as a result of shock detachment can have a significant effect on the flow field.

A study was undertaken to quantify the effect of leading edge radius size on the flow field. A steady-flow method of characteristics (MOC) analysis for started supersonic compressor cascades was employed in this study. This analysis accounts for entropy gradients in the flow and incorporates the Moeckel [6] bow wave analysis at the airfoil leading edge.

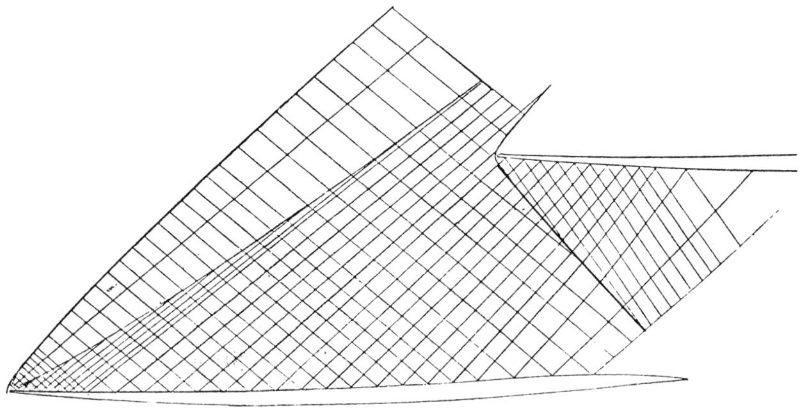

Figure 2. Characteristic Grid for MCA Airfoil Cascade. σ = 1.26, γ = 66.4, M_∞ = 1.62.

To demonstrate the accuracy of the MOC analysis, results for two multiple circular arc (MCA) cascades are compared with experimental data. The numerical results for the first cascade are shown in Figures 2 and 3 where the characteristic network and blade surface pressure distribution are shown, respectively. Also

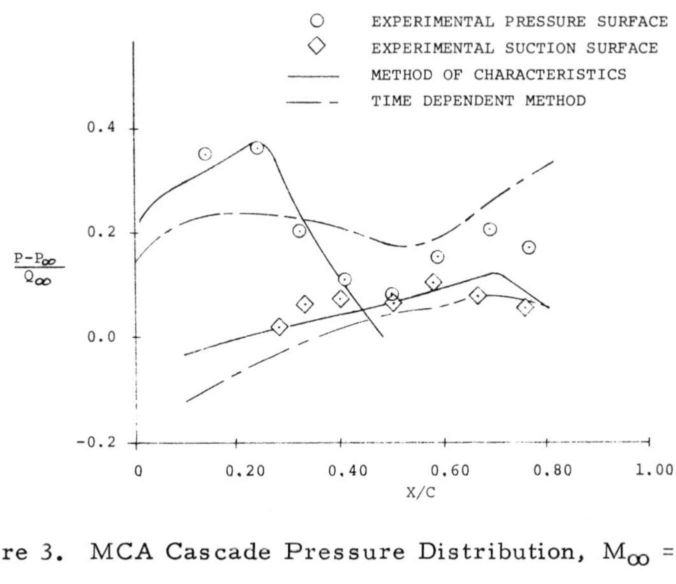

Figure 3. MCA Cascade Pressure Distribution, M_∞ = 1.62.

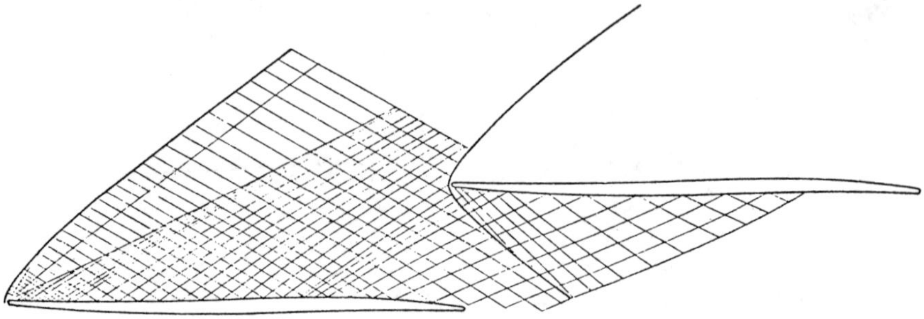

Figure 4. Characteristic Grid for MCA Airfoil Cascade. $\sigma = 1.0$, $\gamma = 74.8^{o}$, $M_{\infty} = 1.97$.

shown in Figure 3 are the solution from a time-marching method and experimental data obtained from the Detroit Diesel Allison supersonic tunnel facility. Excellent agreement is shown between the experimental data and the MOC data on the airfoil suction surface. Also, good agreement between the data is shown on the blade

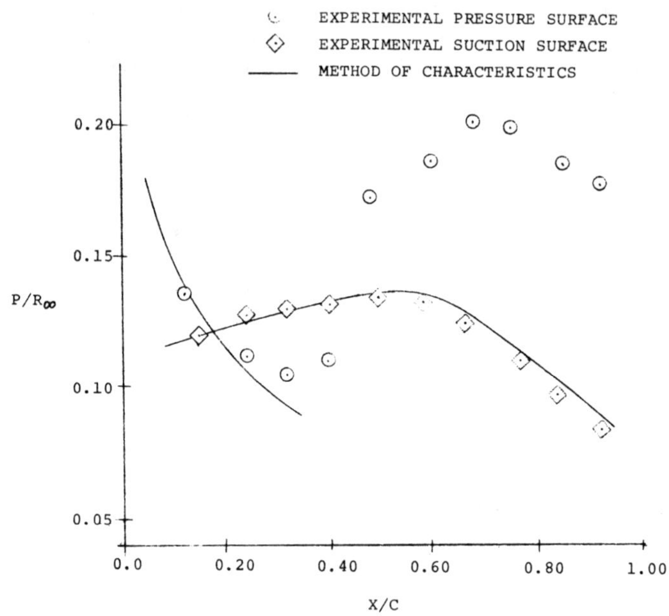

Figure 5. MCA Cascade Pressure Distribution, $M_{\infty} = 1.97$.

pressure surface up to the mid-chord point on the airfoil where the solution was terminated (a trailing edge shock wave impinges on the surface at this point and, as yet, has not been accounted for in the analysis). The results from the time-marching method, on the other hand, show poor agreement with the experimental data over most of the airfoil surface and indicate more accurate treatment of leading edge is needed.

The second MCA cascade analyzed is shown in Figure 4, overlaid with the characteristic network. The blade surface pressure distribution for this cascade is shown in Figure 5, along with experimental data. Again, there is good agreement between the MOC solution and the experimental data.

A double circular arc cascade was used to study the effect of leading edge radius size on the upstream flow field. The results of this study are presented in Figures 6 and 7 where upstream flow angle and mass averaged total pressure ratio across the upstream shock wave system are plotted versus upstream Mach number for different leading edge radii. Both figures show a significant effect of leading edge radius size on the flow. From Fig.

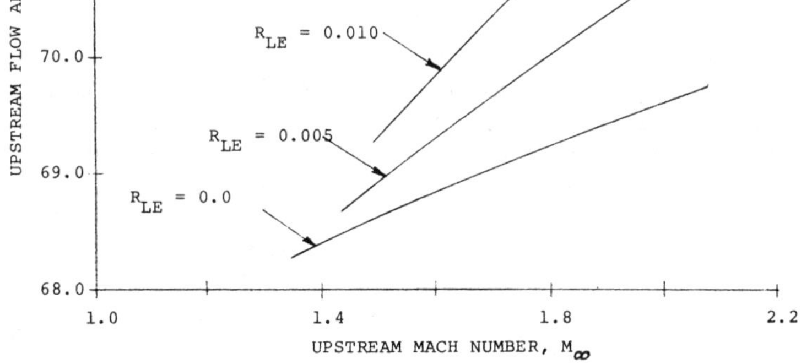

Figure 6. Effect of Leading Edge Radius on Upstream Flow Angle.

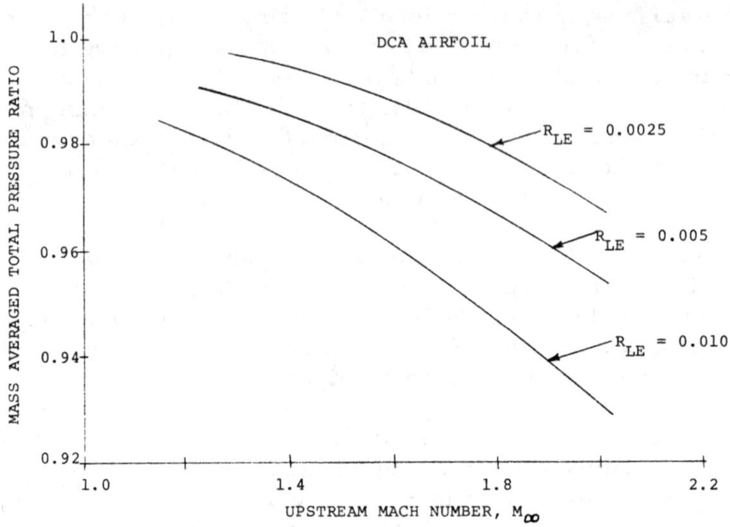

Figure 7. Cascade Entrance Region Total Pressure Loss.

6, the upstream flow angle increases with increasing leading edge radius. The explanation for this trend is that as the leading edge radius is increased, the effective flow area is reduced; thereby decreasing the axial velocity component and increasing the inlet flow angle. As shown in Figure 7, the total pressure loss across the upstream shock wave system increases with increasing leading edge radius. This result is clear, since as the leading edge radius is increased the strong shock wave losses generated at the leading edge represent a greater part of the total shock loss.

In summary, it has been shown that the size of the leading edge radius has a significant effect on the upstream flow field in high Mach number compressor cascades. It is concluded that more accurate treatment of the leading edge radius in time-marching analyses is needed to improve the overall accuracy of the method.

REFERENCES

[1] Gopalakrishnan, S. and Bozzola, R.(1971): A Numerical Technique for the Calculation of Transonic Flows in Turbomachinery Cascades. ASME Paper No. 71-GT-42.

[2] Gopalakrishnan, S. and Bozzola, R. (1973): Numerical Representation of Inlet and Exit Boundary Conditions in Transient Cascade Flow. ASME Paper No. 73-GT-55.

[3] McDonald, P. W. (1971): The Computation of Transonic Flow Through Two-Dimensional Cascades. ASME Paper No. 71-GT-89.

[4] Kurzrock, J. W. and Novick, A.S. (1975): Transonic Flow Around Rotor Blade Elements. ASME Paper No. 75-FE-24.

[5] Delaney, R. A. and Kavanagh, P. (1975): Transonic Flow Analysis in Axial-Flow Turbomachinery Cascades by a Time-Dependent Method of Characteristics. ASME Paper No. 75-GT-8.

[6] Moeckel, W. E. (1949): Approximate Method for Predicting the Form and Location of Detached Shock Waves Ahead of Plane or Axially Symmetric Bodies. NACA TN-1921.

DISCUSSION OF THE DELANEY PAPER

DODGE: Is that a Moeckel method that you're using to calculate those shocks?

DELANEY: Right around the nose, yes.

DODGE: How do you establish the stagnation streamline to calculate that?

DELANEY: The stagnation streamline is located by requiring a continuous bow shock from both sides of the airfoil. This procedure involves a modification of the Moeckel model for asymmetric bodies.

DODGE: The other question applies to your general statement: you're basically doing a rotational method of characteristics calculation where you have total pressure loss behind your summed up bow waves and if you get a large number of them it gets to be significant. Would you give an assessment of how important the two effects are? One effect you've already mentioned is the effect the leading edge radius has on the total blockage and hence, it raises the unique incidence angle. That effect would be predicted - should be predictable - by any inviscid method which has adequate resolution of leading edge region even though you pointed out that it may not be practical. However, the rotational part is the effect which you can't take into account unless you truly do have entropy changes across the shock waves. Have you a feeling for what the ratio of the two effects are?

DELANEY: The shock loss vs. the flow angle? (yes) I don't have a feeling for that.

MC NALLY: You stated that you thought it would be prohibitive to try to resolve the flow around the leading edge nose using some finer resolution of the time marching method. Why do you feel the mesh would have to get so fine in order to adequately pick up the flow there that it would be impractical?

DELANEY: As an example, if you had a leading-edge radius (normalized by chord) of 0.005 and with constant grid spacing, you need a couple hundred panels inside the passage in order to get one point on the leading edge. The other approach is to magnify that region as was discussed yesterday; but I still feel that if done properly by coupling the magnified region solution to the overall cascade solution, the run times would be too large to make this a practical approach.

MC NALLY: Why couldn't you have a refined mesh just within that one region?

DELANEY: Again, I think it would be prohibitive. Introducing a fine mesh there which is sufficient to resolve the property gradients in that region seriously limits the allowable time increment and therefore a very large number of point calculations are required.

GLIEBE: I had a question concerning your method of characteristics solution. How many cascades do you have to stack before you achieve periodicity?

DELANEY: Normally, it's on the order of four or five.

GLIEBE: My second question is really a comment. It's been my experience comparing solutions with the time dependent method vs. method of characteristics for a cascade where flow is all supersonic, that the time dependent method will locate the shocks reasonably well. The problem of leading edge radius - there are ways to handle that if you carefully place your grid points in front of the leading edge such that your unique incidence will be satisfied by the suction surface flow angle.

DELANEY: You're not intending to add numerous points in that region apparently. You say you still calculate unique incidence?

GLIEBE: Yes, that's right. It's primarily set by the suction surface angle and the effect of the leading edge radius is pretty much confined to regions close to the blade. I was wondering if you had any comparable experience.

DELANEY: It's been our experience that we don't calculate unique incidence - at least the way we handle it. Now, what we've done is to essentially assign the flow angle at the leading edge of the blade. I know a lot of codes that do that. There are probably better ways of handling that situation.

GLIEBE: In our technique, we don't specify flow angle; we specify upstream angular momentum and allow the solution to find its own flow angle. We do, indeed, generally find that the flow angle solved for does correspond to unique incidence.

DELANEY: What do you do right at the leading edge of the blade?

GLIEBE: It's a kind of modification of the blade leading edge points such that you don't spuriously put mesh points on the leading edge radius and generate unduly large surface angles for a grid that is so coarse that it just can't be handled numerically. You kind of have to allow the blade surface to transition from the flow angle far upstream onto the blade surface. I admit that this is a kind of crude approach in the design system where you're trying to analyze a design you're contemplating building that works very well. In the off design situation I'm not too confident. I think that if you want to account exactly for the leading edge radius effect, it probably could be handled in the time dependent framework by a suitable grid refinement just around the nose, similar to the Yoshihara type calculation for isolated airfoils.

DELANEY: I agree, but the figures he gave yesterday for computing time were somewhat shocking. I'm afraid that's the same situation in this case.

THE EFFECT OF LEADING-EDGE THICKNESS ON THE BOW SHOCK IN TRANSONIC ROTORS

F. A. E. BREUGELMANS

von Karman Institute for Fluid Dynamics

INTRODUCTION

The importance of thin leading edges is well recognized in all supersonic flow problems. The application of this principle in the transonic compressors led to very good performance. These bladings became very sensitive to foreign object impact and erosion, causing a serious performance penalty. The trade-off to make is between an improved mechanical resistance and a slight increase in the losses due to the thicker leading edge.

Such a study leads to the bow wave analysis as treated in [1] and [2] for the shape and loss prediction. It turned out, during a discussion at the ASME-Houston Meeting in March of 1975, that many people were using this analysis, but nobody could get good agreement with the observed loss levels.

A new modified approach is proposed which provides a closer agreement with some available experiments. This analysis has been applied for DCA, MCA and S-shape bladings. A typical DCA section is chosen for this contribution and compared to experimental data.

A. REVIEW OF THE THEORY

The Moeckel Method

An approximate method for the detached shock wave prediction has been proposed in [1]. The method is based on a simplified expression of the continuity principle and assumes a hyperbolic shape between its foremost point and the sonic point, and being asymptotic to the free stream Mach lines. The sonic line between the shock and the body is straight and its inclination depends only upon the free stream Mach number. The shock location relative to the body sonic point becomes a single-valued function of the Mach number.

A simple geometrical method is also proposed for the body sonic point and the shock location prediction. In [2], Love reexamines the methods for shape and detachment distance prediction and tries to extend the range of application through proposed modifications.

The compilation of the data on the detachment distance provides us with refinements for the nose shapes.

The study of the shock shapes indicates that the Moeckel approach is very satisfactory at the supersonic speeds and that the proposed hyperbolic shape remains valid at distances far beyond the sonic point on the shock wave.

The Proposed Approach

The cascade configuration is basically different from the models considered in [1] and [2]. The periodic nature of the configuration, the multiple shock pattern crossed by the fluid from infinity up to the leading edge and the additional compression or suction surface expansion requires some modification to the analysis of [1] and [2].

The asymptote, which is computed as a function of the Mach number and has a well defined location with respect to the leading edge, is not valid in a cascade. The value at infinity is the inlet Mach number and the relation between the inlet Mach number and flow angle can be computed by the Levine method. This computation provides the location of the Mach line at infinity which is the asymptote.

The location of this Mach line on the stagnation streamline is different from the previously computed values and will therefore produce a different shock shape.

The shock Mach number is also different from the inlet Mach number due to the suction surface curvature. The shock loss and shape have to be computed on this basis with a piecewise adaptation of the Mach number as one proceeds along the shock wave.

The analysis of the supersonic cascade flow shows that, as one proceeds along a streamline from far upstream to the leading edge plane, there occurs a crossing of an infinite number of shock waves. The losses in such a process suffer from a multiplication effect and are larger than the integrated loss along one detached shock wave.

The detachment distance is computed on the basis of the leading edge Mach number and the relations of [1] and [2].

These principles are applied in a computing procedure of which some results, obtained on a DCA cascade, will be discussed.

B. APPLICATIONS

The above mentioned analysis has been applied to a DCA blade section of [5], to MCA and S-shape blade sections of [6].

The nose thickness effect, applying a factor of 3, has been analyzed on the DCA configuration.

This blade section originally comes of rotor 2D, [9], and is within one degree of the 30% span blade section of rotor 14 [8].

The Blade Section and Cascade Geometry

The blade and cascade characteristics are given in Table I.

The nose thickness variation is obtained by cutting back the leading edge plane and assuming a circular nose shape. No blade surface modifications are assumed. The blade leading edge and the blade surfaces do not entirely follow these assumptions during an erosion process. However, a thicker leading edge in the design for better erosion resistance can be treated in this way.

TABLE I

Camber	:	$9^{o}57$
Thickness/Chord	:	4,66%
Solidity	:	1,4618
Stagger	:	$51^{o}35$
Leading edge radius	:	$6\% (t/c)_{max}$
Inlet angle	:	$56^{o}14$
Outlet angle	:	$46^{o}57$

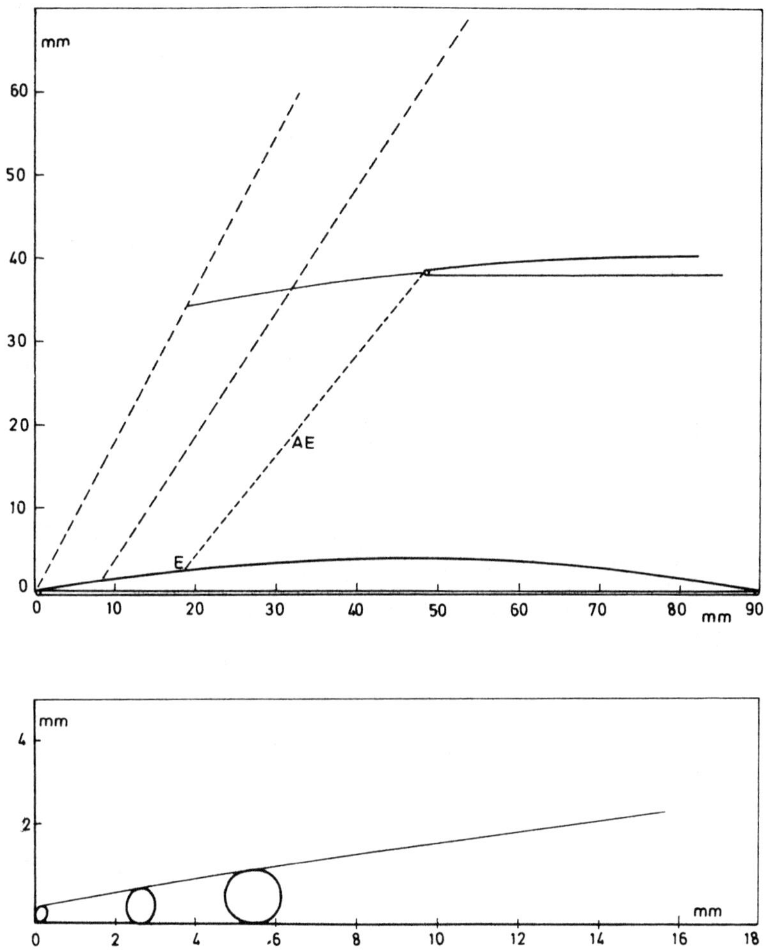

Figure 1. The Cascade Geometry & Leading Edge Thickness Models

The Figure 1 presents the cascade lay-out and the various nose thicknesses.

The Total Pressure Ratio in the Leading Edge Plane

The Moeckel analysis and the new one have been applied for the complete Mach number range and the three leading edge thicknesses.

The results for the $M_\infty = 1.58$ case are compared in Figure 2 where the blade to blade total pressure ratio is presented.

The slower decay of the shock strength, compared to the Moeckel shape, is observed for the different nose thicknesses. The total pressure ratio at the leading edge (stagnation streamline) corresponds to the normal shock value.

The variation in nose thickness produces a different suction surface expansion and explains the different values, although the cases have the same relative inlet Mach numbers.

This procedure has been applied for all Mach numbers up to the limit of axial sonic flow in the leading edge plane and the pitchwise integrated pressure ratios are presented in Figure 3 as a function of the inlet Mach number.

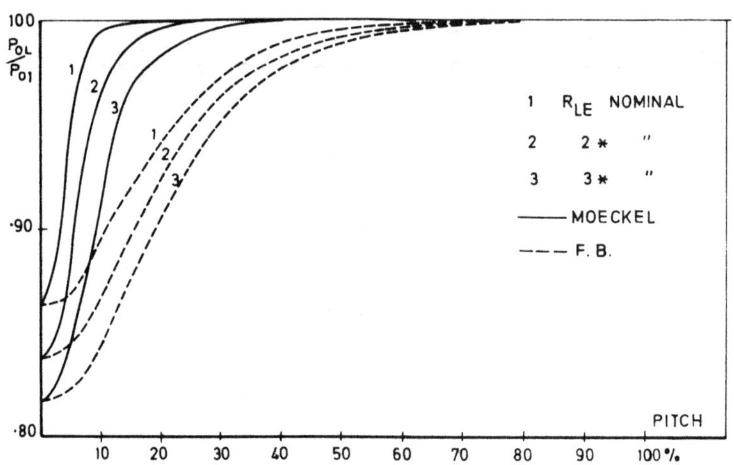

Figure 2. Pitchwise Total Pressure Variation

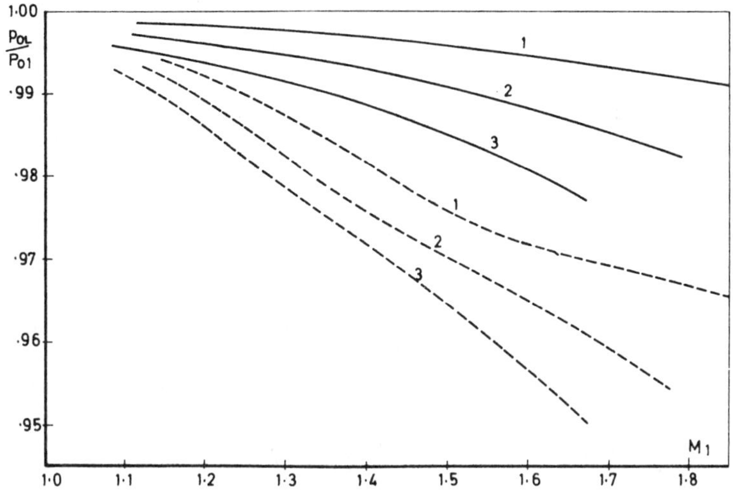

Figure 3. Pitchwise Integrated Losses in the Leading Edge Plane

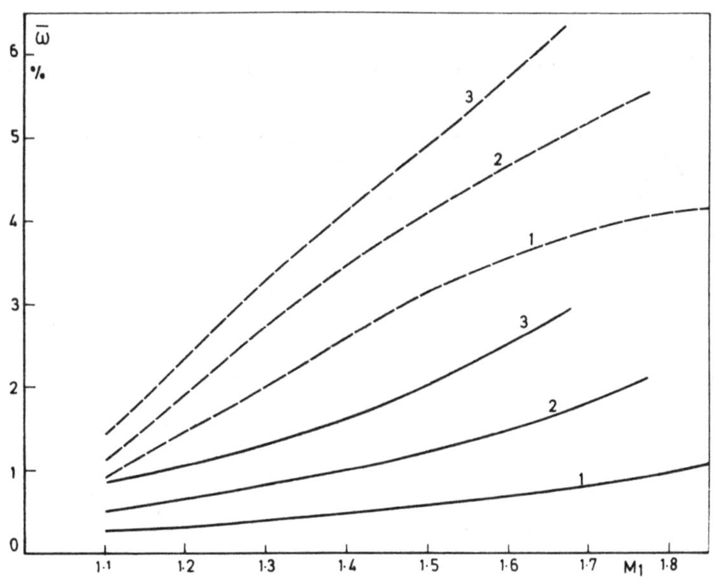

Figure 4. The Bow Shock Losses - Total Pressure Loss Coefficient

The presentation as a total pressure loss coefficient is given in Figure 4.

The Total Pressure Ratio at Different Axial Stations Upstream of the Leading Edge

The total pressure ratio at different axial stations allows a comparison with experiments, performed closely to the rotor leading edges.

The pitchwise integrated values for the DCA cascade at M_{∞} = 1.58 are presented in Figure 5. The axial distance is chosen as a percentage of the blade chord. It is shown that these losses are concentrated very near to the leading edge plane and cannot be detected during the classical inlet flow exploration. Only experiments within a few mm from the leading edge plane can detect the total pressure loss due to the bow shock structure.

The analysis of the bow shocks at a constant axial distance upstream of the blade leading edges indicates:

first, an increase of the total pressure loss when the bow shock is generated;

afterwards, a further increase, with smaller slope, due to the

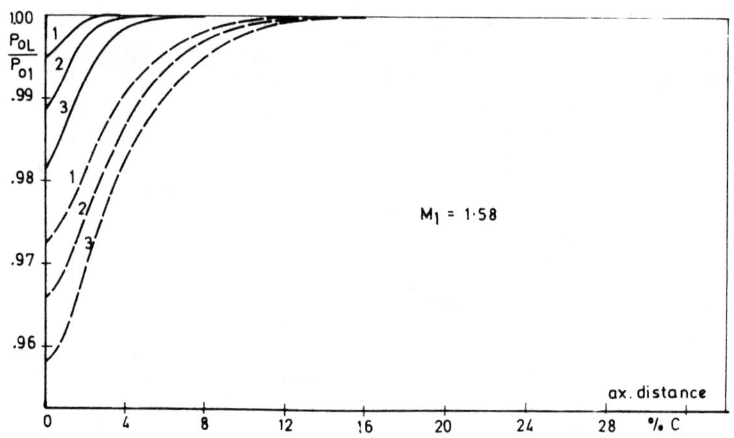

Figure 5. Variation Pitchwise Integrated Losses at Various Axial Upstream Distances

increasing Mach number;

and, finally, a decrease when the shock becomes more oblique and moves towards the leading edge plane.

This Mach number dependent trend is confirmed by the calculations at 2, 4 and 6% axial distances upstream of the leading edges (Figure 6).

Comparison with Measurements

No direct measurements on a rotor model are available for the blade section that has been analyzed, but different indirect data are available in order to cross-check the analysis.

The wall pressure measurements with fast response instrumentation do suggest a stronger shock than predicted by the Moeckel method [9]. It is recognized that wall effects mask the real shock structure.

The blade element data of [7] indicate a much stronger loss than expected from the original bow shock analysis. The DCA cascade of the analysis is very close to the 30% span section of Rotor 14 from [7]. The prediction of Figure 14 in [7], is compared

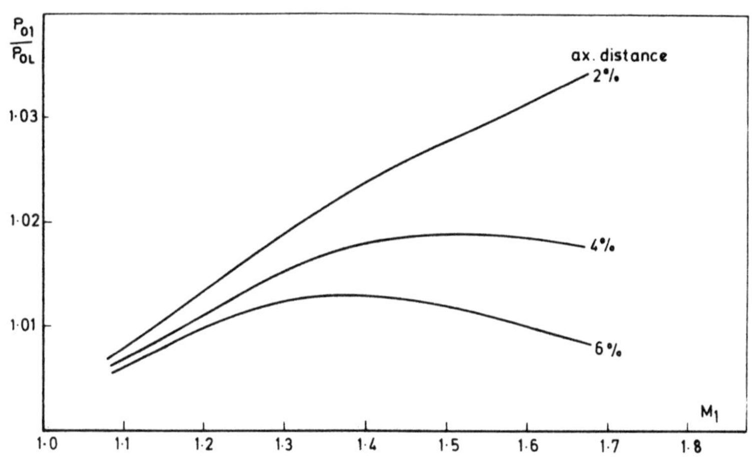

Figure 6. Integrated Losses in Diff. Axial Planes Upstream of Leading Edge

with the modified Moeckel analysis and the new one. The loss increase of the rotor experiments is confirmed. A 3.5% increase in the total pressure loss coefficient is observed for a leading edge blockage increase from 3% to 6% (Figure 7).

The measured pressure loss level and Mach number dependent trend of the supersonic rotor of [6] have also been confirmed. The total pressure loss variation of Figure 10 from [6] could be recomputed by the new analysis. An exact agreement is not yet obtained since the back pressure influence on the bow shock position is not fully incorporated.

Finally, some confirmations of these loss levels have been obtained from supersonic cascade investigation.

CONCLUSIONS

A better agreement with the experimental observation is obtained when taking into account the specific cascade features.

The analysis suggests a non-negligible loss level for the bow shocks in the transonic compressors.

The axial extension, as computed in the new approach, can provide a guideline for the blade row interference problem.

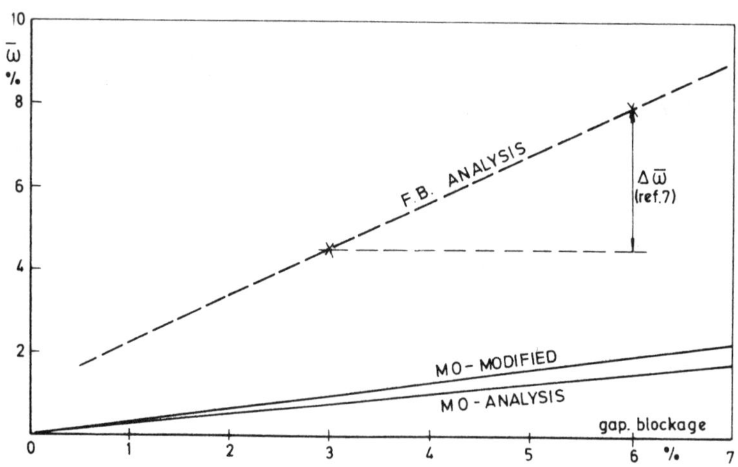

Figure 7. The Bow Shock Losses - Total Pressure Loss Coefficient

REFERENCES

[1] Moeckel, W. (1949): Approximate Method for Predicting Form and Location of Detached Shock Waves Ahead of Plane or Axially Symmetric Bodies. NACA TN-1921.

[2] Love, E. (1957): A Reexamination of the Use of Simple Concepts for Predicting the Shape and Location of Detached Shock Waves. NASA TN-4170.

[3] Klapproth, J. F. (1950): Approximate Relative-Total-Pressure Losses of an Infinite Cascade of Supersonic Blades with Finite Leading-Edge Thickness. NACA RM. E9L21.

[4] Hartmann, M., and Miller, G. (1958): Experimental Shock Configurations and Shock Losses in a Transonic Compressor Rotor at Design Speed. NACA RM. E58A14b.

[5] Starken, H. and Breugelmans, F. A. E. (1973): "The Cascade and Rotor Section Performance of a 9°57 Cambered DCA Airfoil. VKI-LS 59.

[6] Breugelmans, F. A. E. (1975): "The Mach 2 Axial Flow Compressor Stage." ASME Paper No. 75-GT-22.

[7] Urasek, C. and Reid, L. (1973): Effects of Increased Leading-Edge Thickness on Performance of a Transonic Rotor Blade. NASA TN-D-7489.

[8] Krabacher, K. W. and Gostelow, J. P., Single Stage Experimental Evaluation of High Mach Number Compressor Rotor Blading. Part 4. Performance of Rotor 2D. NASA CR-54584.

[9] Weyer, H. and Hungenberg, H. (1975): Analysis of Unsteady Flow in a Transonic Compressor by Means of High-Response Pressure Measuring Techniques. AGARD CP-177.

SESSION V

Chairman:

MAX F. PLATZER

Navel Postgraduate School

REVIEW AND DISCUSSION

REVIEW OF SESSION I ANALYSIS: BASIC FORMULATION FOR TRANSONIC FLOW PROBLEMS IN ROTORS

WILLIAM D. McNALLY

NASA Lewis Research Center

First, a few comments about the individual papers; it's difficult to comment on some of the individual presentations without seeing the entire papers and I found that particularly true with regard to the first speaker, Dr. Oswatitsch. But, in his paper, Prof. Oswatitsch did cite some early references which show that some of the equations which are now getting very active treatment because of the advances that have come along recently in computer technology, have really been around for quite a good number of years. He showed us some results in which there was not very good agreement between theory and experiment, particularly at blade leading and trailing edges which pointed up some of the difficulties which we have to be aware of in using these methods in transonic flows - difficulties which are primarily brought about because of the adjacent blade rows which we have in compressor cascade geometries. Then he pointed out the difficulty in making use of transonic similarity principles in the general transonic 2-D or 3-D blade row problem because of the large number of interconnecting variables and influences which one has in such problems. He then warned of the rapidly changing conditions that we have in flow passages when flows are near Mach 1, which again warns us that we must be extremely careful in applying proper techniques when we're trying to analyze this region. I do look forward to reading Professor Oswatitsch's complete remarks which will be covered in much greater detail than we could go into in the time that we have here.

With regard to Jim McCune's talk, the old-fashioned aerodynamics that he used seems to me to get a little bit better every year. He keeps adding more and more effects into his analysis and I get an uneasy feeling that someday he's going to come and report at one of these meetings that he has the entire problem analyzed, using his analytical techniques, and the rest of us are going to be out of work. These methods, and I include within that category the work that Bill Rae presented also - these more analytical type approaches, I think, should be applied in situations where, within the boundaries of the assumptions that have been made, they can give us useful information to use in designs. Certainly, there are limitations and if you recognize them ahead of time, I think we can make very good use of these extremely rapid techniques compared to some of the other techniques that work. Probably they're getting more heavy use from day to day. Another advantage is that they can be applied to handle both the design problem and the analysis problem. I was surprised at the results Jim showed in comparing with the runs on the MIT compressor. They really showed, I thought, quite good agreement on his axial and tangential and radial Mach number components. I was quite surprised at the results that he showed compared to the real world. I personally anxiously await the comparison of Thompkins' 3-D solution with Jack Kerrebrock's latest results and with McCune's analysis. I'd like to see all three aspects of that MIT work brought together and I think they've all reached the point now where they can be brought together so that we can weigh the various techniques against each other.

Dr. Katsanis indicated to us a viable and a simple procedure for calculating the choking mass flow that he got for a portion of the turbomachinery passage. From what I know of the technique, it seems that it requires two fundamental things: First of all, the passages do have to be guided; I believe you have to know not only the wall curvatures from the beginning to the end of the orthogonals that you're integrating across, but you have to know the streamline curvatures in between. Apparently, it does require this guided passage assumption and also the curvature and geometry which you need is sometimes difficult to obtain without doing extensive layouts and that sort of thing. But, apparently, he has all this process automated from other analysis codes that he uses in these blade passages so that it adds up to a very simple technique to estimate what the choking mass flow is through transonic blade rows. I don't know if it's quite so easy in compressors where you don't intend to have that guided passage type of geometry. There is only a very small portion of the compressor blade row where you can taken an orthogonal from one surface and go across and find another blade on the other side.

I thought Dr. Tom Adamson's paper was quite interesting. He showed us very nicely that for biconvex airfoil problems with inlet shear flow that he had this difficulty with a forbidden inlet mass flow region which we have all believed, but he demonstrated it very nicely. Obviously, there would be a similar situation in a real compressor blade row. I was sorry that Dave Oliver wasn't here yesterday to comment on a question that came up. Is he here now? (yes) You started your solution with a uniform profile upstream and then increased the shear as the calculation proceeded. Adamson asked the question whether or not the wavelets you found in Mach number distributions were due to choking. Would you comment on it?

OLIVER: I think there's no doubt that what he's talking about is real. It has to do with the choking that takes place in a nonuniform transonic flow. The conditions for which the calculation was run are close to choking. At the time, we did the calculation, we had no idea what exactly the choking limit would be. We could do it by running different cases to find what it would be. I tried to estimate that we would be on the safe side and that's how we set the operating conditions that we did use. What Tom needs to decide (the question), using his asymptotic estimate, is in fact the actual passage dimensions that we used for calculation. I don't have a copy of the report that goes with the paper that he has, which has the exact dimensions in it. I'm estimating that it was 1:2, 3 but I have to look it up when I get back to Cambridge. So there's agreement that the effect he's talking about is real. In the calculation of the transonic shear flow that was done, I tried to use some intuition about whether or not we would be choked or unchoked. Tom had a question about whether or not there was still an upstream running wave, in that calculation, that had not come to equilibrium although things were quite steady. In the vicinity of the blade there was no doubt that we were at a steady state. But, perhaps the upstream still had some reflections to go through. The upstream running wave is a slow running wave which you have to wait a good bit of time for, and it is possible that we could have been in the choking range that he calculates and we would have found, had we run a lot longer, an inlet interaction that would have altered the inlet conditions. Does that answer the question?

MC NALLY: In speaking of 3-D solutions, that brings us to Tony Ganz and I was pleased to see that somebody else was hassled with 3-D problems besides the MIT people. I know it's a big chore. I realize from your presentations that you're getting underway with it although there's still some development to be done. There is some question about the finite difference methods and particular treatments of the boundary conditions which were used, but I'm sure as the work progresses that these things will be studied and reported on. I was glad to see another effort like that getting underway.

With respect to John Erdos' work which, of course, I've been quite involved with since we're the ones who asked John to do that work, I can say that I think they have spent a good deal of time in trying to work on all three of those codes. There is the basic blade to blade stream surface solution - the transonic blade-to-blade solution using time marching methods for either a rotor or a stator or the combination; now that is the first code I know of using a blade to blade solution with the time marching approach in which the rotor and the stator are taken into account together. The treatment of the boundary conditions at that interface was quite complex, but apparently, it has all been worked through and it is working quite well. Now it is our intention that we will not only use that code ourselves but that we will publish the code and document it quite well and make it available through COSMIC so that it can get wider use; these publications are in the works right now. I expect that code to be available to the general public in a matter of a few months. This is also our intention with the other two codes he is working on. The second one is the meridional plane solution the S2 surface, and the third one is an extension of the first. The third one is a blade to blade solution for a single blade row but the gimmick here is that it is able to handle inlet flow distortions. The same sort of logic that was applied in the interblade region in the first code to handle the rotor stator interaction - that same sort of logic is going to be applied to handle the rotor interaction with the distorted upstream flow. So all three of those codes will eventually be available and we'll try to have them well documented. Right now the blade to blade codes do have the sharp leading edges assumption built into them and in the initial runs we've made with them, we feel very strongly that that needs to be improved. We need to take the leading edge into account more explicitly so that we intend to extend those codes so that they'll be able to handle the leading edge radius problem.

I'd like to make one more comment with respect to codes being available and that is with regard to the 3-D MIT code. An awful lot of work has gone into the development of that one, also. I think it would be good eventually if that code were made available to the general public. We have that code now at Lewis; there is still development being done on that code. There are later versions which exist at MIT; they would need some further development in order to be able to handle a general hub-shroud profile. Right now the way the code stands, it is limited, I believe, to the hub-shroud profile that they have at MIT, particularly downstream of the blade row. With a few more extensions to that code, it's our intention that we would eventually make it available also through the same process of documenting it and sending it to COSMIC.

Now, some general impressions: I thought there were a wide variety of techniques discussed, anywhere from purely analytic to full 3-D viscous solutions and these obviously take anywhere from a few seconds on small computers to hours on the 7600. It brings up the topic which I think we might discuss a little bit and that is - if we're going to build these codes into our design systems, somewhere we have to be awfully careful to make sure that we're not overkilling the problem. I think we have to take a strong look at what codes are available and what assumptions have been built into them, and try to use the simplest code to do the job that we want to do. As we've moved into the time marching codes, we've learned a lot about proper application of upstream and downstream boundary conditions and stability and accuracy questions and differencing schemes and shock treatments and that sort of thing, much of which, now, we can try to apply in the development of other codes. It seems that the mathematicians have - that is, the impression I got from listening to the talks at this meeting - are really having their heyday now with the relaxation solutions. This is the process that always seems to happen. The mathematicians - and thank God that they're there - grab the problem and kind of beat it to death. Finally, when they get some of the hard questions answered, then it gets down to the applications stage and people design working codes that really solve our day to day problems. I think the relaxation procedures are getting a lot of work right now mathematically and I think the time has come for them to be applied much more extensively now in real blade rows. The only one I really know doing that sort of thing is Paul Dodge. Perhaps there are others that I'm not aware of. I just personally think the time is right to really try to push for that kind of method for real blade rows.

One more comment: Besides the relaxation procedures which are a way of cutting computing time from the time marching methods, I think another procedure that has to receive some strong consideration and some discussion here is the use of implicit methods.

Dr. Yoshihara implied that but I think that warrants some discussion. I think these explicit schemes are just so bound to their coding conditions criterion which leads to large computer time, that we really need to try to look at what implicit schemes can be used to cut this computer time down. I really didn't hear anything much about that in this conference. I think we ought to discuss that a little bit.

REVIEW OF SESSION II ANALYSIS: COMPUTATIONAL METHODS

GINO MORETTI

Polytechnic Institute of New York

Let me get rid of three papers, not because I want to get rid of them, but because there are two mainstreams which probably need a little bit more discussion. The three of them do not belong to the mainstream and this is why I want to talk about them first - also because I don't have anything to say - I'm not familiar with what they are doing. One paper is concerned with the hodograph method. I believe that there is much intrinsic worth in hodograph methods and as a former mathematician I love them. But I've not been working with these things for too many years and so I lost contact with them. So I admire them from a distance. I wish people would not admire them from a distance or not admire them at all. I think these methods are valuable and interesting and should be pushed more and perhaps the combination with other methods - such as the usual computer grinding methods - would bring some truth. So I was happy to see these presentations by Dr. Holt, as much as I was in Göttingen when I saw many beautiful things made with a hodograph method. I wish we had more of this. The technical thing which I'm not familiar with is the parameter differentiation method presented by Dr. Harris and I have to confess I am at a loss. There, too, perhaps there are possibilities which escape my knowledge and are outside my sphere of competence, so if somebody else wants to discusss those things it is all right but I cannot do that. The third item, which gets closer to what I'm familiar with but does go in the opposite direction, is design techniques such as the one presented by Dr. Korn. I am more familiar with testing methods. I don't ever remember whether these are the direct problems or the inverse problems. One of them is direct and one is inverse. Direct is the one I believe we are dealing with generally; you give the geometry and try to get the flow field. Or, vice-versa! It's a matter of opinion, you know! It is beautiful to have a design method. But I think that there are very few in the world, nowadays and most of them are only for isolated airfoils. I think there are Nieuwland, Boerstol,and Garabedian and Korn, Now the fact that Dr. Korn is presenting his method as applicable to cascades, etc., seems to me to be a very interesting advance, and we have to keep that in mind, and I cannot say anything else of course. There is no competition, so we have to be glad that we have one.

A region in which there is a lot of competition and also a lot of controversy is the region to which most of the papers of yesterday and today belong. I do not intend to go over each paper in particular, but I would like to give my general impression.

Here we have two essentially different problems - one is the steady state and the other is the unsteady flow. Which one we are interested in solving is the first thing we have to ask ourselves before saying which technique we want to accept and develop. So let's talk about the steady state first.

I'm not particularly fond of steady state patterns. But many people have told me repeatedly that having steady state patterns is very useful. So if you really want to have a steady state pattern, I would say, and I think Dr. Oliver said the same thing yesterday, there is only one way of doing that thing nowadays and that is relaxation. There is absolutely no doubt that so-called time dependent marching techniques are out of competition with the relaxation method, and the reasons are very well known. One is that only one unknown has to be determined so that the number of equations to solve are ridiculously small - if you take one out you don't have any equations to solve anymore. That's very pleasant. Besides, the algorithm can be made very skinny and very fast. Besides, relaxation has the goal of reaching steady state and couldn't care less for the intermediate steps. So if you can do the calculations in one step you have solved the problem in the perfect form. If you can solve it in ten steps, it's not that perfect but acceptable. If you use 1,000 steps, something is wrong. Now we know that. Every improvement - every way of accelerating these processes is welcome and we have such improvements now, thanks to the efforts of people at NASA-AMES and Tony Jameson and others. I should mention what Jerry South has presented yesterday, because it belongs to the general comment on this session. I was not aware of that technique and I don't know how it works exactly, but it really intrigues me and I think I'm going to take a good look at it. I suggest other people who are interested do the same. It seems to be very efficient and very clever. So, we have these acceleration procedures and therefore I'm sure that in a year or so these computations will be performed in zero time! We are getting close, right? Very close!

Another thing which has practically been solved already and on which there is no controversy anymore is what is the domain of the dependence for these calculations, what kind of grids do we use, etc. Yesterday we saw a couple of contributions. For example, I was not aware of Dr. Rae's idea of taking that special grid; that seems to be somewhat different from what Jameson has been doing and has to be watched. These are facts which are accepted and which seem to be extremely important to get the good results we want to have.

Another thing which is not that much accepted yet seems to be the shock fitting in connection with relaxation procedures. I'm sure that what Dr. Hafez presented this morning is very nice. But it's just a personal feeling. I hope I'm right. The only thing which bothers me a little bit is this: why is it that many other people who have been dealing with relaxation methods for a long time and whose name I shouldn't even mention but I do - Murman and Jameson, for example - are a little bit reluctant to fit the shocks? If they have an explanation for that I would like to have it. Because it looks like what Hafez is suggesting is so simple that they should say, "ah, well," and do it immediately! If they don't do it it is probably because they have some good reason not to do it. Not being a dealer in this business, I would like to know.

Another thing which has to be mentioned as part of the improving state of the art I would say is a more extended usage of mapping and stretching. I saw with pleasure that lots of them are here. I enjoyed seeing what Jameson presented yesterday about these analytical mappings. I have some mappings for some problems of my own which I am enjoying tremendously these days and therefore seeing the log (1 - sinh z) made my heart bounce with joy! I think that these mappings are going to be very useful. Apparently, then, we are in very good shape as far as relaxation methods are concerned. It's a pleasure - only three years ago we didn't even know that relaxation could be used - three years - four years ago. That's beautiful.

Using a time dependent marching technique to do the same

thing - to achieve a steady state - seems to me is a waste of time. I try to be sincere because I started with that thing years ago. So I don't think that there is any future in that. If somebody comes out, as I heard would be possible, in the near future with some idea of making those computations much faster, it is simply because some artificial step is taken to overcome the difficulty of the Courant-Friedrichs-Lewy condition which after all is a physical condition. So either you want to follow the physics, or you do something which is not physical. Now if you are interested in getting the final result which is steady, and you are not describing the unsteady physics, it is all right with me, but let's not call that thing time dependent calculation; it's another way of relaxing.

Let's go on to unsteady flow. Unsteady flows, being unsteady, have to be described by time dependent techniques and there's nothing else we can do. It would be a contradiction of terms. So, unfortunately, there we have the problem of running times. That problem is far from being settled. Running time, as we heard yesterday, are outrageous, I would say. I'm sorry Dr. Yoshihara is not here at this moment, but he knows that I use that word. Evidently, not too much can be done to reduce the running time except by inventing a completely new machine, but not the ILIAC IV. The only thing which can be done is to reduce the number of points to be computed. In order to do this, one has to be very careful not to introduce big truncation errors and not to introduce other numerical disturbances. This brings in the necessity for shock fitting, of course, and other careful evaluations, for example, of external boundary conditions and the like.

Let us also mention one thing which was not mentioned yesterday and that is initial conditions. Problems which are mixed subsonic and supersonic have the bad habit of maintaining the initial condition for the rest of their life inside the computational region. So let's be careful with that. These are things which have not been discussed here, and generally are not discussed at all. People tend to shy away from these things, because nobody knows what to do with them. It is too bad because one of these days we are going to end up with some results which do not make any sense and we believe they make sense. That will be too bad!

I am impressed with Erdos' effort to get this blade to blade interaction and his results. I wish him well. I hope that we will see more and more of these things. I'm sorry if I'm invading your (McNally) field - that was a paper which was in your session. I want to talk about it because there was nothing like that in my session. It's a pity! It seems we stopped having something which starts making some sense in a very exacting, difficult problem. I don't know about the running time of those calculations. Perhaps since I'm happy you should not make me unhappy! Anyway, this running time business has to be straightened out, but it's not here that we can do that. It will take a few years, I believe.

One thing which I do not like to hear mentioned is nonlinear instability because that does not exist, and therefore I don't want to hear of artificial viscosity because that has been invented to eliminate the nonlinear instability! I said that in the other building seven years ago. And I keep repeating it. Let's try to do better in working in cascade flow, because in other problems things are not that good yet.

Finally, another partition which we can see in these flow problems is between inviscid flows and viscous flows. We have seen something along the inviscid lines - let us take relaxation, and Erdos' calculations as representative of something which is promising in inviscid flow. Now what about viscous flow? I'm not that happy about the viscous flow yet. But I believe, and this is probably a personal feeling and it may give you the full extent of my ignorance in aerodynamics - I believe that most of the phenomena which occur and in which we are interested are mostly inviscid, and viscous effects are confined to small regions. That is the old idea of the boundary layer, etc. What I mean is that it is not necessary to bother Doctors Navier and Stokes for these things. Some simpler approach probably will suffice, and we have seen such calculations, again not in my session - some boundary layer plus inviscid flow calculations. I think that that is the way to go, at least for the time being. If we really want to put everything together not in a patched up form, but in a well combined form - let's do it! But with this boundary layer idea in mind, one would simplify the equations, trying to march on as much as possible. If we don't want to do that, then we have to solve gigantic problems with another machine - but we don't have the time, and most of us do not have the money.

I think that this more or less summarizes my impression of the session. I must say, to conclude, that I enjoyed every minute of it.

DISCUSSION SESSIONS I AND II

JAMESON: I just want to make a quick comment about the shock fitting. I think that shock fitting looks quite practical for the simpler flows but I don't know how to do it when you have merging shocks like that triangular shock pattern that I showed you on the swept wing. I'm waiting for someone else to tell me and I'll be very happy to put it in my programs.

MORETTI: Tony, I think that you are clever enough to think of something!

SOUTH: Dr. Moretti made a comment about the fast acceleration methods for relaxation. For the benefit of most of the audience, I'd like to put in some words of caution. In the relaxation for the steady state solutions, I guess the most popular method for accelerating covergence is just the old SOR (successive over relaxation), which is a good example because in a linear program if you use any relaxation factor greater than zero and less than two, it is in fact stable; it may not work very well but it is stable: 1.999 will not blow the calculation but if you try to do a transonic calculation with 1.999, you won't last more than about two cycles.

MORETTI: I know that.

SOUTH: O.k. I want to say that the same thing applies with this new, well, this old method, to which we are now giving some attention. Multigrid. For example, I've done some rather high Mach number cases, about .95, but I've found you have to be a little careful because the supersonic region will grow so rapidly that the changes are no longer small. You can't talk about small amplitudes. The calculation can just blow off the computer. The same thing happens with ADI which Tony Jameson has done and I'm sure the same thing can happen with the other acceleration method which H. K. Chang and Jameson and Caughey have done with extrapolation. So don't be too discouraged if you can't push these methods to the ultimate in all cases because the nonlinearity will just absolutely take you off the computer. Just remember that much caution should be exercised.

MORETTI: Jerry, I didn't intend to put too much optimism in this, but perhaps I failed; I was too optimistic for a change. The fact is that a couple of years ago we would not have thought of this possibility. This thing came out very quickly and very efficiently, in a certain limited range perhaps. But good people are working on these problems and there is no reason why these methods could not be extended in order to solve more difficult cases. Where they work they work beautifully, and accurately.

I have heard or seen something about some other work at NASA-AMES, which is along those lines more or less. Allegedly, it makes in ten steps what should be done in 200 with the SOR. The Jameson method also works like that, sometimes. There are limitations, indeed, but who doesn't have limitations? At least we have cases which work and that's good.

SICHEL: Even in the first session that was theoretical, I think that about half of the papers did turn out to be numerical in the end. So I almost think in some ways this meeting was a numerical meeting with a few analytical papers thrown in. I would also like to say that I think it's worthwhile for people also to find simple problems in which they can get an analytical solution even if the problem doesn't necessarily represent a real cascade flow per se. But I think that flows in nozzles like the problem that Tom Adamson has done with the shear flow are very valuable in showing what the physics of the problem is and, in helping the people who do the numerical computations, I think those are the very things to do. I think it will be interesting to have some problems where there is an analytical solution, where there is a numerical solution, and where perhaps there is experimental measurement. One of the things that struck me was the relatively simple configuration that H. Doyle Thompson described toward the end of the session which again is the type of configuration for which I think one could apparently do experiments; one could do various numerical techniques and one can also do some analysis.

MORETTI: What do we want to do in this context? We want to grow able to make very complicated analyses of very complicated geometries and very complicated flows in those geometries, right? Because industry wants that. These are things which just <u>have</u> to be done. What you would like to see instead cannot work there because no mathematician has invented yet that kind of analysis which we can use under those circumstances. I am afraid that we'll have to wait too long to see it. We have to resort to numerical work. This is why not just this meeting but all meetings today are numerical meetings. The point I would like to try to make is this: The emphasis in my opinion should be put on analytical work in preparation of the numerical work. Analytical in the sense of putting one's brain to work to analyze what's going on - what has to be done. Nobody should plunge into a complicated problem without knowing his fluid dynamics well and also his numerical analysis well. This is the problem we are facing today: There are too many people who do not know enough fluid mechanics and who do not know numerical analysis enough and want to compute. So there are too many papers published and too many million dollars spent in things which could be better done. I agree with what you are saying but I would shift the emphasis a little bit, saying that our analytical methods should be re-examined in the light of their numerical implications and we should learn to interpret those analyses numerically. When we know that, well, then we can make a much better numerical analysis - we will waste less time, spend less money and get better results.

KERREBROCK: I think this is a very profitable discussion but I disagree slightly with Professor Moretti. From the viewpoint of someone who watches people do computations and theory and does experiments, I would pose the problem this way: One has the experiment which is the real world; one has the 3-dimensional computation which, if we take the example of Dave Oliver's and Phil Thompkins' work is, except for numerical inaccuracies, an exact inviscid solution to the boundary value problem. When one compares the two, one finds very large differences. There are two ways you can go from that point: One can try to put the viscous corrections into the computational scheme and this is surely very difficult; it will take a long time. Or you can do an analysis which relates one part of the inviscid problem to another part. This is the way I would interpret McCune's theoretical work - in the sense that he has developed an inviscid theory which is capable of predicting some of the aspects of the blade to blade variation in cascades, given some other aspects which must be taken from experiments. Now to be specific about that. If you take his theory and you put into it the experimentally measured variations of circulation in a spanwise direction, from this you can predict some aspects of the blade to blade variation - not all of them, but some of the important parts. I think there's a very, very important role for theory, Marty [Sichel], in relating parts of the problem, so that one doesn't have to solve the whole problem in one piece. By working at the thing from all directions like this, I think we can probably develop a modeling technique which is practical.

ERDOS: I want to introduce one point that I'm surprised Professor Moretti didn't bring up. There are several surfaces of discontinuity in the flow fields that we're looking at in unsteady aerodynamics. One is the shock wave; the other is the slip stream which is generated by an airfoil undergoing pitch, or blade rows subjected to periodic disturbances. The theoreticians, I believe, a long time ago recognized the importance of the motion of this vortex sheet on the unsteady forces on the blade row. We're doing these things numerically now. We've accepted, I think, that you can get something like a shock wave that's captured in the finite difference scheme in the conservation form of the equations. But the conservation form of the equations, if you look at a contact surface, simply tells you that the pressure and normal velocity have to be continuous; they don't tell you what the magnitude of the jumps are. And they have to be determined by boundary conditions. The point is: we fit the contact surfaces and we believe that their motion is important to the unsteady solution, but we accept a capturing technique for the shock waves because there at least we believe we get the jump condition correctly and the motion of the shock correctly.

MORETTI: Well, John, you are teasing me now. You're right, I forgot. I should have said that; that's another one of my pet studies. I'm surprised I didn't mention it. You know why, because a number of the others spoke about that yesterday and so - you didn't mention that yesterday to that extent. So you're actually fitting the contact discontinuities. That's nice. I don't know if you do the right thing, though, because, of course, having the intention of doing it doesn't mean that one does it. Just consider the following. I have found that if one uses my pet set of unknowns, that is pressure, entropy and velocity angles - not velocity components, then dealing with contact discontinuities is extremely simple and they come out very nicely and there are no problems. You don't even notice that you have to compute the discontinuity. The logic of that is minimal. If you use the other unknowns, and you are bound to use other unknowns if you want to use the conservation equation, then you're in trouble because you really have to make a discontinuity there to appear with its special logic and that is another burden put on the problem. The fact is that (in two dimensions) the direction of the velocity is continuous across the discontinuity and pressure is continuous as well and the only thing that is not continuous is the entropy. So if you use that set, only the entropy is the one you have to worry about. That's very easy to cope with. I'm very much aware of the necessity for that. There are cases in which there are other discontinuities which are important and those are the gradient discontinuities. Sometimes even those discontinuities should be considered and computed explicitly. In other words, our inviscid flow fields which describe the mechanics of continua are not continuous at all; they are full of discontinuities. We would probably be much better off if, instead of solving the partial differential equation, we would just solve for discontinuities and let the rest be uniform flow in between. Think of that. Perhaps it's an idea.

FARN: We are fully aware of the advantages of going to steady-state formulation in saving of computing time. The reason we went to unsteady formulation is the following. For a fully choked turbine flow, the mass flow rate is unknown. If you go to steady-state formulation, you have to use boundary conditions which you don't know. This was our first difficulty - why we didn't go in that direction. The second difficulty is: If you fully choke the flow, then the flow downstream of the sonic lines is not unique. It depends on the back pressure. We don't know how to implement [that] if we take a steady state formulation. I wish somebody here could answer this question.

MORETTI: I'm not going to answer that question! I'm trying to come out with a little paper on a very, very simple one-dimensional analysis trying to figure out this business of the boundary conditions upstream and downstream once and for all in a form which satisfies me; but I'm not finished yet. Which means I'm not satisfied. In certain cases, you see very well that if you start with a certain set of initial conditions, you are going to have them there inside your solution. You never get a final solution which makes any sense. So if you want to make a time dependent calculation, be very careful with the initial conditions. There are cases which are funny. For example, you can think of an infinite duct, cylindrical at both ends but with different diameters and with a transition in between. Think of a steady flow there, inviscid, and think of that as subsonic with the particles moving from the narrow end to the wide end. The pressure has to increase in the direction of the flow (that is a steady motion). Now, that's physically impossible. But mathematically it is possible. Physically, though, you can get it as an asymptotic solution of an unsteady flow, by properly starting it from rest. You will find an unsteady flow which, in that transition and for a length of the duct as large as you want, looks like a steady flow with an increasing pressure. The fact is that you have an expansion downstream due to the initial motion which takes care of compensating the pressure rise and produces the motion. So that you have steady flow in a region bounded by two unsteady flows. Now that's a perfectly legitimate solution and you come out with that solution numerically if you want, very nicely, and it satisfies a mathematical condition in that region as though it was a steady flow. The Mach number distribution there is correct. But the "entrance" Mach number there is not the Mach number which an observer would see at infinity. One has to define and qualify exactly what one is talking about when making this calculation. What initial conditions are there, etc., etc. All of this is a long story and very hard to reduce into discussion so now let's postpone that for another occasion.

ADAMSON: If I can twit you a little bit, Professor Moretti, it sounds like this little paper you are talking about is maybe an analysis that you are making of a simple problem that will help later with numerical work!

OLIVER: One last comment on this same theme. I often think of the time that I use the computer as the time that I do Babylonian fluid mechanics and with due respects to Gino Moretti as a Roman, I often think of Babylonians and Greeks as the way of thinking about these two things. The Babylonians calculated the number of pi, I think, to 27 significiant figures; the Greeks proved that it was an irrational number.

PLATZER: I wonder if I could get a statement on the merits of potential flow vs. nonpotential flow computation. Murman yesterday cited very nicely Mike Varra's book showing the conditions for potential flow. I think some of you are aware of Professor Kerrebrock's work which was presented at the Aerospace Sciences Meeting in 1974 which showed that in rotating flow - in flows which have rotation - you cannot use the approach which we otherwise are used to - that is, to superimpose potential flow disturbances on the mean flow. Rather, in these flows, the shear disturbances and entropy disturbances are of the same order of magnitude as the potential flow disturbances. Now we have heard a number of approaches yesterday where potential flow was used and I wonder whether I could get Professor Kerrebrock to make a statement for the record. Or anyone else who wants to comment on it.

KERREBROCK: I really hesitate to approach this subject at this late hour. It's a very confusing subject. Let me see if I can say what I think you've attributed to me. The statement is that in a strongly swirling flow, that is, in a flow with a strong rotation about some axis such as you always have in a turbomachine if it produces a substantial amount of work, then one does not have the situation where the fundamental small disturbances in the flow can be separated into three noninteracting types. Now, in the nonrotating flow they do separate into entropy, vorticity and pressure modes. These are noninteracting so that one can superimpose potential flows on turbulence and on entropy and they satisfy separate boundary value problems and everything is nice and easy. Now, in contrast to this in a strongly rotating flow, such as one has in a turbomachine, a priori these three modes are all coupled; they satisfy coupled differential equations. In particular, one finds, if one looks at the details of it, that the so-called turbulence mode is, in fact, capable of exhibiting oscillatory propagating behavior. All turbulence fields have associative pressure fields to the first order. Let me get a little bit more precise. The wakes coming off blades for example, have strong pressure fields associated with them which then have to be taken into account in calculating the pressure field of the blade.

JAMESON: Perhaps you could elaborate a bit more. Suppose there was just one fan stage, and you didn't have to worry about wakes coming from an earlier stage. Would the disturbance potential superimposed on a rotational flow be a reasonable model or not? I think Earll Murman seemed to produce some analysis to suggest that it might be, but it's one of the points I wanted to find out by coming to the meeting - what the general opinion was.

KERREBROCK: Let me respond just very quickly. I think the answer is that it depends very much on the structure of the mean swirl flow. But if the mean swirling flow has a large, let's say, solid body rotation component (in other words, if it is strongly rotational), then one finds that almost all of the disturbances which would normally be characterized as vorticity, have propagating characteristics so they have strong pressure fields associated with them which have to be taken into account, in the sense that you have to add their pressure field into that of the potential flow, if you want to back calculate the effect on the rotor.

MC CUNE: I think I can give what I regard as a splendid example of what Professor Kerrebrock is talking about, if we go back to old-fashioned aerodynamics again for a moment. Suppose we model the viscous wakes behind a set of rotor blades with momentum decrements, by two counter-rotating vortex sheets. Now associated with those vortex sheets in the presence of rotation, because of the asymmetry associated with the rotation, is a very definite pressure field. Or, alternatively if you like, an induced angle flow at the rotor blade. In order to stick to my old-fashioned aerodynamics, and to put in a model of the viscous effects that are seen in the MIT blowdown facility and, of course, in all other turbomachines, one approach which we are now taking is to compute the convected vorticity with this model in a rotating flow and compute the induced velocities. We find that we are getting effects, which are again superimposed on the mean flow, of the same order as those associated with a normal trailing vorticity due to nonuniform loading. And also the same order as the famous secondary flow effects. This is a beautiful example, I think, of what Kerrebrock showed in 1974.

OLIVER: Maybe being a bit in the middle, I'll try to answer Tony Jameson's question. I am not an accepting - sort of musing about it - practitioner of the relaxation method. To my understanding, it has not thus far been possible to implement a relaxation method for the Euler equations. That doesn't mean that there have been insurmountable difficulties; it just means that it has not been done yet. I don't think there's any reason to suspect that it's an impossible problem in any sense at all. But the fact is that the relaxation method has not been perfected for the Euler equations. Maybe somebody can tell me if that's not true. In an actual turbomachine rotor, I think, when you consider rotor-stator interactions, inlet distortions, and other phemomena that are peculiar to turbomachines, you are in a regime of effects that do, I think, distinguish that field from external aerodynamics. I think, ultimately, the need is for being able to treat the full Euler equations without making the potential flow assumption. However, in a specific case, I think, of an inlet flow that obeys certain properties that Earll Murman tried to lay out, and for that case of inlet flow in a rotating reference frame, a velocity potential will exist and the potential flow procedure, I do believe, can be carried out. For rotor-stator interaction, inlet distortion, and the consideration of other kinds of wakes and gradients, in particular, some of the disturbances that Jack Kerrebrock was alluding to earlier, I think the situation requires a more general approach and would ultimately be best served by a relaxation procedure using the full Euler equations.

JAMESON: I think that means we still can have some fun in fundamentals. I have been having a great time for the past five years. It looks like there's something still to do.

REVIEW OF SESSION III VISCOUS EFFECTS IN TRANSONIC FLOWS

ROBERT E. MELNIK

Grumman Aerospace Corporation

I'm not sure there's much left to be said. We've been talking about viscous and inviscid flows. Let me say a few words. Instead of just summarizing the papers in the order presented, I think a few comments about the general problem of boundary layer effects should be made. In external aerodynamic applications, it's been clear for some time that for real loaded airfoils, that carry a lot of real loading, the effect of boundary layers on things like the lift can be very large even at large Reynolds numbers. Comparisons with Ottawa data have shown that the effect of the boundary layer on cambering over the last few percent of the airfoil tends to reduce the lift by as much as 30% - 50%. It's a big effect. You have to put it into the calculation in order to get meaningful results in both the design and analysis problems. David Korn, Garabedian and Bauer and others have been doing that. It's clear that if you make reasonable corrections to the boundary layer near the trailing edge or near the shock wave, you can get very accurate results. I agree a little bit with Gino that perhaps Navier Stokes solutions aren't completely necessary and may not be available for these complicated 3-D problems for a long time. However, I think there is a role for them. It remains to be seen how accurate they'll be. I am now talking about Deiwert's work. There is definitely a need to improve the Reynolds stress models that are used in the calculations and I understand that Steve Deiwert is going to be spending his time on that. There's also a need to improve the overall detail that the calculations provide. I think he's got to take a finer streamwise mesh and then we can have something that perhaps takes a lot of time to compute, but we end up with an experiment in which we can turn off various effects and have a very useful device for checking out more simple models. I think that might be the role. We can do things with the Navier Stokes equations like doing an airfoil in a wind tunnel, and doing a very careful calculation a few times to get a better way of handling wind tunnel wall interference and checking boundary layer calculations. That might be the role for Navier Stokes computations. It's clear that this work should continue. The question is whether or not there should be widespread massive efforts all over the country. I think working with the Navier Stokes equations requires people who know what they're doing. I think the AMES group does and they are making progress.

As far as engineering calculations are concerned, either an aircraft company or an aircraft engine manufacturing company, I think, for the foreeseeable future, will be restricted to boundary layer corrections and three of the papers were along those lines. It seems clear at the high Reynolds numbers - you're talking two-three million or perhaps higher in turbines or compressors, 20 million in aircraft - that dropping very small terms in a differential equation makes a lot of sense. And as I've said, the experience has been that if you do the boundary layer correction wisely, you get useful and accurate results. It's not clear at this time whether we know how to do the shock interactions correctly but progress is being made. I would expect in the next year or so that they will be integrated into existing boundary layer/inviscid computations. I did not have time in my talk to say much about the trailing edge problem. I believe that not all the boundary layer corrections are accounted for in the trailing edge calculations and that we haven't been doing that right for about 30 years. What's been left out is the thing that Jim McCune just mentioned: behind a lifting airfoil is a curved wake and there is a momentum deficit in the wake. That momentum deficit leads to pressure drops in the wake that cause a downwash on the airfoil. Some of the work that we've done has shown that that effect is at least as large as the displacement effect we have very carefully been trying to put in over the last couple of decades. I would say some effort has to be directed to taking into account wake curvature effects in the boundary layer model. That work is going on and I expect again in a year or two that this effect will be completely integrated into a boundary layer code. This is all, of course, in 2-D. The thing that's impressive about all the difficulty is that the real problems that we face in this area are 3-D. It's clear that both the boundary layer and the Navier Stokes people are going to do reasonable viscous calculations that make sense in these machines. I have hopes for the local interaction approach in that the flow will be locally 2-D even though it's a complicated 3-D field; the streamwise gradients are small, and it makes sense to think in terms of at least the local interaction regions near trailing edges and through shock waves being pseudo 2-D flowfields. There's some hope that we can do complicated flowfields by patching things together.

The whole question of flow separation comes up over and over again in trying to do calculations with turbulent separated flows. It's clear that the thing that we know the least about is how to model Reynolds stresses in separation bubbles with closed streamlines. In fact, at the Washington meeting, we had thought about having an invited speaker, someone like Anatol Roshko, to give a talk based on the question: are turbulent separated flows inviscid? Can they be treated as inviscid solutions to the Euler equations where the separation bubble just comes about becuase there's a varying total pressure? I think numerical experiments have been done in the past and they look very promising. Perhaps numerical solutions of the full Euler equations taking into account all the vorticity in the flow field in the boundary layer and coming into the inlet and generated by the upstream element is all we might really need - some way of prescribing vorticity distribution in the closed bubbles. This is going back to the Batchelor model. It seems that that kind of approach has some promise of working. Apparently, it's clear from the number of papers that have been given at the meeting, there is progress being made on the inviscid flow. People are directing attention toward the important problems in that area. The lack of attention to the viscous interaction problem, the apparent lack of attention, has me puzzled; I don't understand. It seems clear that the boundary layer effects are large. If you do an inviscid calculation, we're not going to have very good agreement with experiments, particularly for highly loaded blades. It's surprising to me that there apparently hasn't been very much attention given to this problem. I would expect at a meeting at some point in the future, there will be.

DISCUSSION SESSION III

OLIVER: A good bit of viscous work is incompressible. The people who are working on it, work in very low Mach number ranges, and are doing some very fine work there - perhaps not discussed at a meeting like this because of the transonic character of it. In particular, I just want to mention that there is a calculation in progress that I'm aware of, being done by Steve Llewellen and Coleman Donaldson, that involved a separated bubble and a shoreline boundary layer, i.e., an atmospheric boundary layer. Obviously, it's a very low Mach number problem which is being treated incompressibly. The bubble is being calculated and turbulence is being calculated according to the second order closure model that Donaldson has been promulgating. The results look very good, very promising. I think some of those techniques could could be very profitably used by the people at Ames. Just a note on that: There is some very fine and fairly contemporary turbulence modeling which is being done in which the second order closure of the Reynolds stresses and a rather careful scale equation is being used (which is very important in the separation problem). Those results are just beginning to come in. But it doesn't have anything to do with transonics and so there isn't the kind of communication that would put this meeting in touch with that work.

ADAMSON: I just wanted to put my oar in again and say that I think it's clear in the local interaction calculations, that this is one place where analysis is a great help to future computational work.

PLATZER: One thing which I found rather interesting: As you know, we did eliminate unsteady flow aspects from our considerations to the extent that they affect flutter, aerodynamic stability, etc., but there is definitely the aspect of the effect on performance - turbomachine performance. As you recall, in this room about four months ago, we had an AGARD meeting on unsteady phenomena in turbomachines where we tried to review the current status of this field. What was obvious in that meeting was that as far as transonic calculations methods are concerned, in general, one could say the discussion of transonic phenomena was rather limited. Now we have seen in Dr. Deiwert's paper very nice shock boundary layer oscillations - oscillations induced by shock boundary layer-interactions, and, of course, we are all familiar with the aileron buzz and self-excited oscillations on space vehicles, etc., etc. Similar unsteady effects, it seems to me, ought to be taken into account in turbomachines. My question is: To what extent are oscillatory shock boundary layer interactions going to be worked on in the future, and what is the importance of these unsteady effects in general as far as the transonic turbomachine is concerned? I'm wondering if I can get somebody to comment on this or is it too late already to bring out unsteady effects?

MELNIK: Of course, we have unsteady effects also in external aerodynamics and that's buffeting on aircraft. It's clear it comes from the separation bubble under the shock interacting with a small separation at the trailing-edge. In aircraft companies, that's something to be avoided. There's really very little attempt to try to compute something on unsteady separated flow - instead one avoids it. I think that's probably true in rotating machinery. I don't think that computation is going to tell you very much about how to avoid it.

REVIEW OF SESSION IV EXPERIMENT

ARTHUR J. WENNERSTROM

Wright Patterson Air Force Base

In the area of experimental methods, we basically touched on about four different methods of measurement: the high response pressure measurements, optical velocity measurements, and then two unique areas, the density and temperature measurements using fluorescence techniques, and finally the use of smoke studies for transonic flow visualization. I'd like to cover some of the highlights of these in turn.

In the area of high response pressure measurements, these have now been applied over the rotor giving us shock patterns and approximate Mach number distributions in the tip region of a rotating high speed rotor. They've been used successfully for measuring blade surface pressures both on rotors and stators to get the nonsteady flow effects, and they've also given apparently reliable results for instantaneous probe measurements, as exemplified by some of the MIT experiments in the blowdown facility. A few years ago, the big problem with these transducers was simply survival. Often you didn't have enough of them lasting long enough to do anything useful. Today, the transducers themselves have improved enough and installation techniques have improved sufficiently that quite a variety of good measurements seem to be possible at a reasonable economic level.

There's obviously very heavy activity going on right now in the optical velocity measurement area. It's about the only experimental technique to come along which is potentially capable of essentially making a direct quasi-steady velocity field measurement within a high speed rotating blade row, for example. Also, even in a stationary situation where noninterference is important (which is obviously vital in the transonic area), the accuracy now appears to be on the order of one to two percent in many practical situations. This is probably better than most analytical methods that are currently available in the presence of strong shock waves and viscous effects. This should be about the best tool going, to evaluate certain things with respect to 3-D viscous computer codes - the shock definition for one thing appears to be excellent.

Some of the problems related to seeding, required where the turbulence levels are high and the histograms spreading, leads to certain difficulties in identifying just what the mean velocity is. As the Mach numbers go up, there are certain limitations with respect to testing time required in frequency response. Work in these areas seems to be moving along at a fairly useful rate. Also, we have the unique dual beam system which Mr. Schodl commented on, which seems to be somewhat less sensitive to seeding problems, for example, than the Doppler approach due to the higher signal strength applied, if I understood that correctly.

Going to the fluorescence technique, this certainly offers an interesting new method for getting direct density measurements. The accuracy has already been somewhat proven to be $\pm$ 5% now, with the potential for going to $\pm$ 2% which is in the same ball park as the LDV systems, and in principle at least, it appears this method can be used to determine temperature fields. This offers an instantaneous picture somewhat analogous to an interferogram but certainly is a much more practical thing to achieve in a real turbomachine.

Smoke studies have not generally been applied to higher Reynolds numbers or high velocity flows. We saw today that you can potentially see something useful under transonic conditions. This looks like a possibly useful extra visualization technique to be considered in conjunction with other methods.

Some final points: It was commented upon that redundant measurements or checks are very important. I couldn't agree with that more strongly. Murphy's law is still very frequently encountered in experimental work. With the variety of methods available, I think in most situations it is possible to create a redundant measurement situation. I think it's important not to overlook some of the more fundamental flow experiments such as described by Rannie and Thompson and Adamson. Some of these exotic new measurement techniques put us in a position to really learn more about what's going on inside a turbomachine than we ever have before, but these very same methods used in conjunction with these more basic models, may provide information which in some instances is more directly useful for crosschecking some of the numerical techniques.

DISCUSSION SESSION IV

JAMESON: I am not trying to be provocative - but why can't you people make a decent test in a cascade? Let me tell you what I mean. You do all these fancy laser velocity profiles, etc., but when we had our cascade tested - it had six blades - not one of them had the same force on it as any of the other five. It's supposed to be periodic with the same force on every blade. Well, it's not very encouraging for a man like Dave Korn, trying to design a blade by a really fancy piece of mathematics, if he can't get a reasonable test of it.

WENNERSTROM: The only comment I would make on that is that some of our experimental methods are more accurate than the environmental conditions present in some of the test facilities in which they're being used. Five to six blades, for example, is probably somewhat marginal in many situations, to obtain periodic flow - just to pick one point. Manufacturing tolerances play a big role in precisely the way you mentioned. In order to have a high confidence level in measurements, no matter how good the instrument is, these effects are very important factors which can't be overlooked.

JAMESON: Well, of course - why don't you go in and fix these things? You've got the fancy instrumentation; let's have some tests now - good ones.

KERREBROCK: There's a point of philosophy here, which is that a cascade is, at best, an approximation to something that you're interested in. Nobody ever built a compressor out of rectilinear cascades, least of all with five blades. If you really want to have a precise test of your calculations, then what you have to do, is do an experiment on the real machine. That is, on the real geometry. Now, as I think Mikolajczak has said over and over again, cascade experiments don't represent what happens in a blade row in a real machine, unless some constraints which the machine imposes on the cascade are adequately represented. The most important one is the mass flow density variation to the cascade. Periodicity is another one.

JAMESON: I'd like to elaborate a bit further on what I had in mind. It seems to me that we're trying to develop, let's say, theories in numerical methods in which we'd like to establish some confidence. But we can't get that unless we can have some coordinated tests in which the tests test the same thing the calculations try to do. So you can't really find out whether a theory for designing cascades is valid unless you test a cascade. It's true that when you want to built a jet engine you have to test for that one, too. But you do need some experiments that are designed to match the conditions of the calculations as closely as possible in order to try to establish some reasonable confidence in the calculations. If we got that, then it gives us some base to move on, to some more complicated type of calculation, maybe.

ERDOS: I would just like to offer a simple minded proposal. It might be easier to calculate the cascade of flat blades in a tunnel than it is to make a rotating test.

WENNERSTROM: I think your encouragement to those doing that kind of work is well founded. Actually, there probably has been some more accurate work done but which has not yet reached the public domain. I think that it would be highly advantageous if some of that work could manage to reach the open literature before too long.

OLIVER: I'm not an experimentalist, but it seems to me clearly in the case where a cascade is involved in rotating stall, you have a situation which is not periodic. I don't think it's a matter that people don't run experiments properly or that blades aren't manufactured to the right tolerances, although that's relevant - but isn't it possible that in fact in the rich variety of viscous interactions, etc., that genuine nonperiodic phenomena occur in a real cascade even if you machined it to the perfect tolerances, even if you ran it with the perfect inlet? Does anybody want to comment about that?

JAMESON: I'm quite bothered. Dave Korn designed a cascade blade but with very fancy mathematics. The tests that have been made so far haven't been very helpful. It's quite possible that there are difficulties due to, let's say, separation of boundary layers and things that make the theory not that good. But the tests we've had so far tell us nothing at all. It's clear that we didn't meet the minimum conditions for simulating what the calculation was trying to do. We really need a good coordinated test that will try to find out whether the theory is a good theory or not.

SHREEVE: I have to chip in, with so many theoreticians commenting on the experimentalists' papers. I think the job of setting up a controlled experiment of the type you're suggesting is probably as difficult as solving the analytical problem. I think that setting up the experimental situation under control is equally difficult. It might be expected to take as long.

JAMESON: I believe it is difficult for one to do it. I don't think it's easy, I think it's difficult. But we should try to cooperate on these matters. We shouldn't have two groups of people not talking to each other.

MIKOLAJCZAK: This presents quite a tough problem because the particular design we're looking at requires that we have some very accurate measurements on the blade surface. Secondly, it then requires that we simulate Reynolds numbers very accurately. But this meant that we had to go to a very long chord. This in turn put a strain on the size of the blowdown facility that we have. What it meant is that we finally finished up with an aspect ratio of about one and a half. At this point, one has to ask - is this cascade going to give us a ρC_x ratio of 1 - which means there is no convergence from the end walls? At that kind of aspect ratio that's virtually impossible. So obviously the attempt is to try and simulate this kind of ρC_x ratio in that kind of facility and eliminate the 3-D effects that are not in the analysis. This proves to be quite difficult. Added to which we had a very highly loaded blade which didn't make things any easier. What it appears, is that we may have a facility to do this job. What we may now see as a possible place is the old NASA-Langley tunnel which is operating very effectively at DFVLR in Germany. That may be the place where we have to go for such a facility. That is the kind of state we are in in this country - that about the best cascade tunnel we have is at

Pratt and there's another one at Allison. That's the limit of the available capability in this country. This is the reason why we find it rather difficult at this point to model a design very sensitive to any changes like the Korn airfoil is, right now. We're trying.

PLATZER: I would like to summarize by making just two observations. I think it is safe to say that as far as transonic flow problems in turbomachines are concerned, that the future is secure. There are plenty of problems for the next twenty years. I think that is a comforting observation in times of budget cuts - job security is very important! Secondly, I don't mean it as a mere compliment when I say that this has been the most patient and cooperating audience I have ever observed. I thank you all out of my heart for the wonderful cooperation you have given us in sticking with us to the end, from eight o'clock to six-thirty. I wish to thank in particular the speakers for their contributions, and the session chairmen for the wonderful job they have been doing.

ADAMSON: I add my thanks also.

CONCLUDING REMARKS

CONCLUDING REMARKS

It is quite apparent from the preceding papers and discussions that two major developments are greatly contributing to the advancement of the state-of-the-art in transonic turbomachinery aerodynamics, i.e.,

i) new computational capabilities and techniques

ii) new measuring probes and techniques, especially non-intrusive optical techniques.

Both techniques are beginning to yield a wealth of information which would have been considered quite unattainable only a short decade ago. Indeed, the spectacular progress which was achieved during this period in the physical understanding and the theoretical prediction of the two-dimensional transonic airfoil characteristics provides considerable hope for similar advances in the transonic turbomachinery field.

Unfortunately, the flow through a multi-stage transonic turbomachine presents a much greater challenge to its detailed understanding and prediction than the airfoil problem. The two-dimensional modelling of the flow, either experimental or theoretical, is an approximation of only limited value, as is also true for any predictive method in which complex reflected shock wave geometries cannot be considered. Thus, the results of cascade tests have to be supplemented by the more difficult and expensive tests of actual turbomachines. An accurate and efficient computation of the three-dimensional flow field, however, is still in its infancy. Furthermore, it is now generally recognized (see, e.g., the recent Project SQUID and AGARD Meetings on Unsteady Flows in Turbomachines) that the oscillatory flow characteristics will have to be accounted for if a true understanding and prediction of the machine's performance under various operating conditions is to be achieved. This adds yet another dimension to an already overly complex problem. Finally, the well known complexities introduced by shock-boundary layer interactions - especially in highly unsteady and three-dimensional flows - are further complicating aspects.

One of the more important problems made apparent by these papers and dicsussions is the lack of sufficient base line data for testing computing methods, or from another viewpoint, a lack of testing facilities, in which improvements suggested by numerical studies can be tested. It is very difficult to give rational explanations for lack of comparative agreement when for both the computation and the experiment, the test cases are at one and the same time extremely complex and yet different kinds of approximation to an actual engine flow. More test cases in which the analysis and the experiment are designed to test each other rather than to approximate actual engine conditions would appear to be of value.

Thus, the progressive reduction and eventual substitution of the present largely empirical design elements by a rational prediction method is certain to require a sustained and systematic effort over a broad spectrum of possible approaches for many more years. Therefore, government and industrial organizations concerned with turbopropulsion research will continue to be faced with agonizing cost/benefit considerations concerning the value and level of sustained research versus the expense involved in ad hoc fixes from case to case. This question of resource allocation for long-term research and its relevance in the face of ever more urgent demands for the solution of near-term problems is a very critical one indeed. It is hoped that the preceding pages are not only a source of information and inspiration for active workers in this field but also serve as a convincing demonstration of the value of sustained research.

We are very pleased to be able to conclude these proceedings with remarks by Professor Oswatitsch summarizing his vast experience on the subject of research.

SOME REMARKS ON PRESENT AND FUTURE RESEARCH TASKS IN FLUID MECHANICS

KLAUS OSWATITSCH

February 1976

There is a widespread tendency among managers responsible for research and development funding to differentiate between the academic problems to which a student will generally be exposed during his undergraduate and graduate studies and his postgraduate training at academic institutions, and the complex flow phenomena which are typically encountered in various fields of engineering, ranging from, say, aerospace and turbomachinery problems to noise and pollution control. A sufficiently detailed, or, ideally, a complete computer solution of these flow problems usually is a primary objective in modern engineering.

One is, therefore, dealing with the question of academic research and instruction, on the one hand, as opposed to industrial research and development, on the other. In this connection, the unfortunately false argument is often advanced by administrators of government research and development agencies - with which I hasten to disagree - that basic research belongs in universities and that applied research and development belong in government and industrial research laboratories.

Such questions of research planning and organization assume a great importance because of their close connection to public expenditures. Wrong concepts in research management ultimately lead to ineffective use of public funds.

In the research intensive branches of the natural sciences, of which fluid mechanics represents one example among many, there is a broad field of endeavor lying between the object of university education and the scientific tools needed to solve the problems of industrial development, which I would like to characterize as the recognition, analysis, and mastery of fundamental processes.

Examples of such fundamental processes in the area of fluid mechanics are, e.g., the following: flow about airfoil leading-edges, boundary layer separation, shock wave formation and decay, shock wave-boundary layer interactions, shock detachment, subsonic and supersonic stator-rotor interference, generalized Prandtl-Meyer expansion, sonic leading edges, multi-phase flows, fluid dynamic stability, numerical stability of computational methods, etc.

The solution of such problems is achieved in various ways, experimentally as well as theoretically, and in the latter case, in part analytically and in part numerically. Seasoned experts, i.e., senior university faculty or research laboratory staff members, are generally required to provide the necessary guidance if thesis projects are to produce meaningful contributions and advances. Therefore, it would be self-delusion to assume that substantial progress can be achieved by merely increasing the number of M.S. and Ph.D. theses without a corresponding increase in scientific supporting staff.

The need to study the "fundamental processes" often arises only in connection with certain externally imposed requirements, such as the need to increase airplane flight speeds or engine power outputs. Hence, it cannot be expected that the majority of these fundamental processes will be fully mastered in the foreseeable future. However, it is characteristic of these fundamental processes that they are encountered again and again in rather diverse projects. The mastery of fundamental processes is thus an indispensable requirement for a complete understanding and solution of development problems. At the same time, considerable savings in time, money and effort will result on a specific project if the computation of certain local details can be omitted because these are known from previous studies with which the research staff is familiar.

In summary, it can therefore be said that a not insignificant part of the work of scientific staffs should be concentrated on the solution and mastery of these fundamental flow processes. However, this work can form the subject of university education to only a minimal degree since training at universities can be devoted to in-depth studies of individual subjects to a much lesser degree than was previously the case. Rather, universities must take into account the increasing growth and specialization of technical knowledge.

Efforts to master as many fundamental flow proceses as possible will always be rewarding because the scientific tools of the expert are thereby extended, because the needed expenditures for this work are quite small compared to typical development project costs, and, finally, because this type of work will not only benefit a specific industrial development project whose fortunes or misfortunes are difficult to anticipate, but will pay welcome dividends on numerous other projects.

WORKSHOP PARTICIPANTS

Professor Douglas E. Abbott
Department of Mechanical Engineering
Purdue University
Lafayette, Indiana 47907

Professor T.C. Adamson, Jr.
Department of Aerospace Engineering
The University of Michigan
Ann Arbor, Michigan 48109

Dr. William F. Ballhaus
U.S. Army
Department of Defense
Moffett Field, California 94035

Professor Oscar Biblarz
Department of Aeronautics
Naval Postgraduate School
Monterey, California 93940

Dr. Riccardo Bozzola
AVCO Lycoming Company
Stratford, Connecticut 06497

Dr. Eugene F. Brown
Virginia Polytechnic Institute and State University
Blacksburg, Virginia 24060

Mr. Pierre Bry
Research Engineer, ONERA
92320 Chatillon, France

Dr. J. Casper
United Technologies Research Center
400 Main Street
East Hartford, Connecticut 06108

Mr. K.S. Chang
University of California
Berkeley, California 94720

Professor H.K. Cheng
Department of Aerospace Engineering
University of Southern California
University Park
Los Angeles, California 90007

Dr. Thomas Coakley
NASA Ames Research Center
Moffett Field, California 94035

Dr. D.J. Collins
Department of Aeronautics
Naval Postgraduate School
Monterey, California 93940

Dr. George S. Deiwert
Research Scientist
NASA Ames Research Center
Moffett Field, California 94035

Dr. R. Delaney
Research Laboratories
Detroit Diesel Allison Division
General Motors Corporation
P.O. Box 894
Indianapolis, Indiana 46206

Dr. L.W. Dickson
SPV Aero Methods Research
The Boeing Company
P.O. Box 2707
Renton, Washington 98055

Dr. Paul R. Dodge
Garrett-Airesearch Manufacturing Company of Arizona
402 South 36th Street
Phoenix, Arizona 85034

Dr. John Erdos
Advanced Technology Laboratories
Merrick and Stewart Avenues
Westbury, New York 11590

Mr. C.L. Farn, Senior Engineer
Research & Development Center
Westinghouse Electric Corporation
Pittsburgh, Pennsylvania 15235

Dr. T. Ganz
Pratt and Whitney Aircraft Company
East Hartford, Connecticut 06108

Dr. B. Grossman
Aerophysics
Grumman Aerospace Corporation
Bethpage, New York 11714

Dr. M.M. Hafez
Flow Research, Inc.
1819 South Central Avenue
Kent, Washington 98031

Mr. Gordon L. Hamilton
Flow Research, Inc.
1819 South Central Avenue
Kent, Washington 98031

Mr. James L. Harp, Jr.
President, Thermomechanical Systems Company
7252 Remmet Avenue
Canoga Park, California 91303

Professor W.L. Harris
Department of Aeronautics and Astronautics
Massachusetts Institute of Technology
Cambridge, Massachusetts 02139

Professor M. Holt
University of California
Berkeley, California 94720

Dr. A. Jameson
Courant Institute
251 Mercer Street
New York, New York 10012

Dr. Theodore Katsanis
NASA Lewis Research Center
Turbine Branch
Cleveland, Ohio 44135

Professor Paul Kentzer
School of Aeronautics and Astronautics
Purdue University
Lafayette, Indiana 47907

Professor J. L. Kerrebrock
Department of Aeronautics and Astronautics
Room 31-264
Massachusetts Institute of Technology
Cambridge, Massachusetts 02139

Dr. Wilhelm Kordulla
M.S. 233-1
NASA Ames Research Center
Moffett Field, California 94035

Dr. D. Korn
Bell Laboratories
Holmdel, New Jersey 07733

Dr. D.D. Liu
Structural Dynamics Department
Northrop Corporation
3901 West Broadway
Hawthorne, California 90250

Dr. E. Dale Martin
NASA Ames Research Center
Moffett Field, California 94035

Dr. W.J. McCroskey
USAAMRDL
Ames Research Center
Moffett Field, California 94035

Professor James McCune
Department of Aeronautics and Astronautics
Massachusetts Institute of Technology
Cambridge, Massachusetts 02139

Mr. H. McDonald
President, Scientific Research Associates, Inc.
P.O. Box 498
Glastonbury, Connecticut 06033

Mr. William D. McNally
Compressor Branch
NASA Lewis Research Center
Cleveland, Ohio 44135

Dr. Robert E. Melnik
Director, Aerophysics
Grumman Aerospace Corporation
Bethpage, New York 11714

Dr. A.A. Mikolajczak
Assistant Chief Engineer
Pratt and Whitney Aircraft Company
East Hartford, Connecticut 06108

Professor Gino Moretti
Polytechnic Institute of New York
Route 110
Farmingdale, New York 11735

Dr. H.J. Mueller
Research Administrator
Code AIR-310
Naval Air Systems Command
Washington, D.C. 23061

Dr. Earll M. Murman
Senior Research Scientist
Flow Research, Inc.
1819 South Central Avenue
Kent, Washington 98031

Professor S.N.B. Murthy
Director, Project SQUID
Purdue University
West Lafayette, Indiana 47907

Professor David A. Oliver
Department of Aeronautics and Astronautics
Massachusetts Institute of Technology
Cambridge, Massachusetts 02139

Mr. Robert E. Olson, Manager
Fluid Dynamics Laboratory
United Technologies Research Center
400 Main Street
East Hartford, Connecticut 06108

Professor Dr. Klaus Oswatitsch
Institut für Strömungslehre
Technische Universität Wien
Wiedner Hauptstraße 7, A-1040 Wien

Mr. James R. Patton, Jr.
Office of Naval Research
Power Program Code 473
Department of the Navy
Arlington, Virginia 22217

Professor M.F. Platzer
Department of Aeronautics
Naval Postgraduate School
Monterey, California 93940

Dr. William J. Rae
Aerodynamic Research Department
Calspan Corporation
Post Office Box 235
Buffalo, New York 14221

Professor Duncan Rannie
California Institute of Technology, 205-45
Pasadena, California 91109

Dr. A. Ricci
NASA Ames Research Center
Moffett Field, California 94035

Dr. William B. Roberts
Aerospace and Mechanical Engineering Department
Notre Dame University
Notre Dame, Indiana 46556

Dr. P. Runstadler
President
Creare Incorporated
Hanover, New Hampshire 03755

Mr. R. Schodl
Institut für Luftstrahlantriebe
Deutsche Forschungs-und Versuchsanstalt für Luft- und Raumfahrt
Porz-Wahn, 5000 Köln
West Germany

Dr. M. Scully
USAAMRDL, HQ
M/S 207-5
Ames Research Center
Moffett Field, California 94035

Profesor R. P. Shreeve
Naval Postgraduate School
Monterey, California 93940

Professor Martin Sichel
Department of Aerospace Engineering
The University of Michigan
Ann Arbor, Michigan 48109

Lt. Col. Robert C. Smith
AFOSR/NA, Building 410
Bolling Air Force Base
Washington, D.C. 20332

Dr. J.C. South, Jr.
MS 360
NASA Langley Research Center
Hampton, Virginia 23665

Professor John R. Spreiter
Applied Mechanics
Stanford University
Stanford, California 94305

Dr. S. Stahara
Nielsen Engineering and Research, Inc.
510 Clyde Avenue
Mountain View, California 94043

Mr. J.M. Thiaville, Head
Compressor Design Department
SNECMA
Centre de Villaroche
77550 Moissy Cramayel, France

Professor H.D. Thompson
Department of Mechanical Engineering
Purdue University
Lafayette, Indiana 47907

Dr. Joseph M. Verdon
United Technologies Research Center
400 Main Street
East Hartford, Connecticut 06108

Dr. Arthur J. Wennerstrom
Group Leader
U.S. Air Force
Department of Defense
Wright Patterson AFB
Ohio 45433

Dr. A.D. Wood
Office of Naval Research Branch Office
495 Summer Street
Boston, Massachusetts 02210

Dr. Hideo Yoshihara
Engineering Staff Specialist
Convair Aerospace Division
General Dynamics Corporation
Post Office Box 80847
San Diego, California 92138

INDEX